Storey's Horse-Lover's Encyclopedia

Storey's

Horse-Lover's Encyclopedia

An English & Western A-to-Z Guide

Edited by *Deborah Burns*

The mission of Storey Communications is to serve our customers by publishing practical information that encourages personal independence in harmony with the environment.

Written by Deborah Burns, Lisa Hiley, Greg Linder, and Laura McBride
Edited by Deborah Burns, Marie Salter, Annique R. Cohen-Wichner, and Robin Catalano
Cover design by Meredith Maker
Front cover photograph of Lady Sunshine, a chestnut Saddlebred mare, taken by Gemma Giannini for Grant Heilman
Text design by Carol Jessop, Black Trout Design
Text production by Susan Bernier
Photographs by Corbis: 74, 232, 446; Digital Stock: ii, iii, 254, 322, 446; Lisa Helfert: 416, 422; Photodisc: 318 342; Giles Prett: viii, 40, 98, 164, 180, 222, 228, 276; William Shepley: 12, 120, 208, 270, 388, 430; Tom Sterns: 284
Illustrations edited by Ilona T. Sherratt
Original illustrations by Johanna Rissanen-Welch and Alison Kolesar; a complete list of illustrators and illustration credits can be found on page 449.
Indexed by Susan Olason/Indexes & Knowledge Maps

Printed in Canada by Transcontinental Printing
10 9 8 7 6 5 4 3 2

Library of Congress Cataloging-in-Publication Data

Storey's horse-lover's encyclopedia: an English and Western A-to-Z guide / Deborah Burns, editor.
p. cm.
ISBN 1-58017-336-5 (alk. paper) — ISBN 1-58017-317-9 (pbk. : alk. paper)
1. Horses—Encyclopedias. 2. Horsemanship—Encyclopedias. 3. Horse breeds—Encyclopedias. I. Title: Horse-lover's encyclopedia. II. Burns, Deborah.
SF278 .S76 2001
636.1'003—dc21 00-046329

contents

preface

Perhaps there's a horse who recognizes your step and welcomes you with a soft nicker, his breath warm and steamy on a frosty morning. Or perhaps you have just started riding, and your pleasure in the sport has quickly turned into a passion. Or perhaps you are a seasoned competitor and know the thrill of partnering with a horse to pursue a goal. Whatever the reason, you are a horse lover: You exult in the grace and spirit of horses, and you are not alone.

In my case the passion for horses began before I can remember. Every week my father took my sister and me riding at a nearby stable. We started in the ring on the most amiable of horses, learned quickly, and soon went out on the trail. Over time we graduated to more challenging mounts. We refined our skills at camp and as teenagers took care of friends' horses during summer vacations. Trail riding was our favorite activity. We eventually knew every hiking trail and logging road in our part of the Berkshire mountains in northwestern Massachusetts.

For a couple of summers I took care of a plucky buckskin mare named Missy. She and I often climbed the wooded, long-abandoned, 18th-century Boston–Albany Post Road to the top of the Taconic Range that divides Massachusetts and New York. Emerging from the woods we would gallop north along the Taconic Crest Trail through blueberry meadows, high above the world. Below on our right were the thickly forested mountains of Massachusetts. On our left, the hills of New York rolled west to the Hudson River and the Catskill Mountains. Missy was willing, responsive, and eager to explore. My memory of her is forever linked with a sense of sunlit freedom and adventure. We were companions, seeking our fortune, and we were on top of the world.

The air of heaven is that which blows between a horse's ears.

—Arabian proverb

The air of heaven, says an Arabian proverb, is that which blows between a horse's ears. Now an editor of horse books, I have met many people who would agree that paradise can be found on a horse's back. *Storey's Horse-Lover's Encyclopedia* is meant to celebrate this magnificent animal and help readers become more knowledgeable. I hope you will keep this book on your tack-room shelf and refer to it often.

After all, the horse world is vast and growing, with dozens of disciplines and hundreds of breeds. It contains numerous sub-worlds, or galaxies, such as jumping, barrel racing, polo, and endurance riding, which have developed completely different riding styles, tack, attire, and language. People in one discipline can be completely isolated from, and ignorant of, what is going on in other disciplines.

Yet at the center of all these worlds stands one creature: the horse, with its eloquent eyes, burnished coat, arched neck, delicate legs, soft muzzle, and sweeping mane and tail. The horse has the same beauty and nobility whether it is carrying jockeys, rodeo team-ropers, Russian Cossack trick-riders, or Olympic three-day-eventers. The instincts of a horse are the same whether the horse is an Arabian, a mustang, a Morgan, or a Percheron. The potential health problems of a horse are the same. The ways one speaks to and cares for a horse are the same. Most of all, the profound bond between horse and human is the same, no matter what the breed or riding style.

Around the globe, horse people of diverse backgrounds have recently begun to come together to share their knowledge of horses: insights, training secrets, solutions. This wonderful development will enrich our relationship with horses. A champion Western reiner, for example, may have an insight that will solve a problem for someone training in dressage. One purpose of this encyclopedia is to facilitate communication between the various equestrian galaxies, to make it easier for us to understand and learn from one another by defining words and phrases.

Over the centuries of partnership between horse and human, an enormous amount of wisdom and general lore has accumulated. Modern "horse sense" has evolved from the experience of horse people of many cultures, including Arabian and Native American trainers; British Thoroughbred breeders; the cowboys and herders of Spain, Mexico, and the American Southwest; the military horsemen who laid down the fundamentals of precise horsemanship that later became dressage; and countless other breeders, trainers, and instructors. Every horse lover can benefit from studying equine history, physiology, and psychology.

In the end, however, the very best way to learn about horses is to spend time with them, in and out of the saddle. Find an hour just to watch your horse as he grazes, interacts with other horses, works in the ring, or simply dozes in the sun. Take long relaxed rides. Savor the moments when you are grooming your horse, talking to him, braiding his mane. This is when the true and precious bond between you and the horse is built.

Horses bring out the best in us. No matter what the rest of our lives are like, with horses we learn to be sensitive, courageous, responsible, and honest. Horses are sensitive, honest beings with whom we can share deep experiences, as I was fortunate enough to share joyous explorations with Missy on a mountaintop.

We at Storey Books wish you a fair road and a willing steed!

Deborah Burns, Equestrian Editor

Show me your horse and I will show you what you are.

—Traditional British saying

RAMBO®

AAEP

➢ *See American Association of Equine Practitioners*

Abrasion

An irritated, rough, or worn away area of the skin.

Abscess

An inflammation surrounding a concentration of pus. An abscess can occur in various parts of the horse's body, but is most common in the hoof.
➢ *See Hoof*

Acetylpromazine maleate (Acepromazine)

A tranquilizer that is injected into a horse's muscle and is sometimes used before stressful situations such as transportation by trailer; often called ace. Sometimes an unethical dealer might give a horse ace to make him appear well-mannered and calm. A veterinarian should always administer this medication, because different horses respond very differently. Generally, the product takes effect in less than half an hour and lasts a few hours.

Never use ace on an excited horse, since it may make him more upset.

Action

How a horse moves. Conformation classes evaluate action at the walk and trot, as well as overall body structure. In performance classes a horse's action is judged as a key ingredient of quality, performance, and brilliance. Judges look for straightness, easy landing of hooves, length of stride, and flexion and raising of knees and hocks.

Different breeds emphasize different types of action. For cxample, American Saddlebred horses should exhibit sparkling, animated action. Dressage horses, on the other hand, are valued for relaxed, supple, balanced, light, and precise movement. Western Pleasure horses should move in a natural, relaxed, easy, comfortable style.
➢ *See also individual breeds*

Acupuncture

An ancient Chinese healing art, an alternative therapy used to treat horse health problems from arthritis to heaves to colic. It is considered beneficial when treating painful or chronic conditions that have not responded to standard medications. The treatment, administered by a licensed practitioner, consists of applying a thin needle or pressure to certain external points on the body in order to stimulate the flow of energy or release blocked energy. In addition to needles, weak electric currents, cold lasers, and ultrasound may be used. A veterinarian may combine acupuncture with traditional medicine to maximize benefits to the horse.

Acupressure uses acupuncture points to achieve these metabolic "connections," but instead of inserting needles, pressure is applied to the skin above the relevant acupuncture points.

ACCIDENTS, PREVENTION OF

The following rules will help you avoid accidents:

- ❑ Learn about horse nature.
- ❑ Get to know your own horse's habits, moods, and reactions.
- ❑ Provide a safe, sturdy, well-ventilated home for your horse, and keep it well maintained.
- ❑ Pay attention around your horse.
- ❑ Use sturdy, well-fitting tack.
- ❑ Always wear a safety helmet when riding.
- ❑ Make sure you receive good instruction from a qualified teacher.

See also Safety

Acute

A description of a condition that comes on suddenly and worsens rapidly. Essentially, any sudden onset of illness is in the acute stage when it first appears. It does not become chronic until it stabilizes and proves that it is a condition that will either remain an issue for the horse on a daily basis, or will recur with varying frequency.

Acute laminitis

➢ *See Laminitis*

"Adopt a Pony" programs

Programs offered at some stables, in which a sponsor pays for purchase and upkeep of ponies so that youngsters who otherwise could not afford to can learn to ride.

African Horse Sickness

An equine disease endemic to much of Africa and parts of the Middle East that is transmitted by biting flies and mosquitoes. Symptoms include a high fever lasting several days and clearly apparent discomfort, with swollen eyelids and jugular veins. The disease can be fatal.

Age

Thanks to improvements in health care, most horses now live into their twenties and, increasingly, into their thirties. The breed, the quality and consistency of care, and the attitude of horse and owner can still directly affect the animal's longevity.

Older, experienced horses in their twenties can thus be valuable teachers, helping young riders develop their skills.

Horses are slow learners but have excellent memories. They reach their "prime" at varying ages, depending on individual genetics, training, and a variety of other factors. Many show horses, including dressage horses, hunters, and polo ponies, are at their best when they are at least ten, and some can continue winning for many more years.

How old should your first horse be? The golden rule is "The younger or less experienced the human, the older or more experienced the horse." Bottom line for safety and satisfaction: *Never* pair an inexperienced or uneducated person with an inexperienced or untrained horse. Such a match could result in life threatening injuries. A mature, well-trained horse can make the difference between building a solid foundation of positive experiences as a novice, and a negative and injury-filled ending to a brief trial period with the sport.

A fifteen-year-old family horse can help children learn to ride safely for many more years. For an adult, a middle-aged horse can be the perfect partner, if he is calm, sensible, and well cared for. In most cases, you should avoid buying a horse in his twenties, because you will have only a few years of riding before you have to retire him. It is difficult to sell an old horse, so you will have to decide — before you purchase him — whether you can take care of him once he is retired.

What a Horse's Teeth Tell

You often can tell a horse's age by his teeth. Look for wear, angle of teeth, and the appearance of Galvayne's groove. Remember, however, that horses are individuals and their environment and genetics may affect the appearance of their teeth.

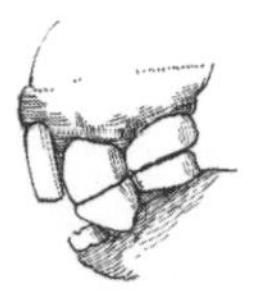

At 2½ years **the permanent central incisors appear, but no wear is visible.**

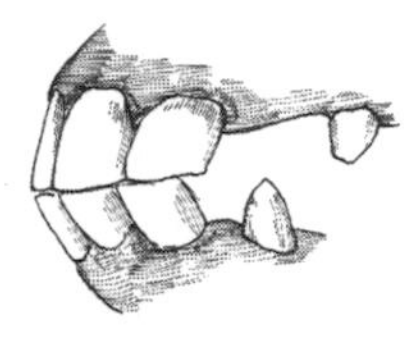

At 5 years **all permanent teeth are in, and the central incisors show wear. In male horses the canine teeth have appeared.**

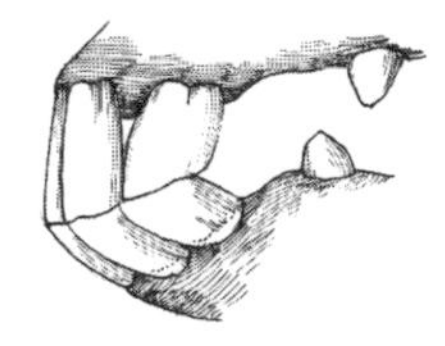

At 10 years **Galvayne's groove appears on the top corner incisors, and the teeth are increasingly slanted.**

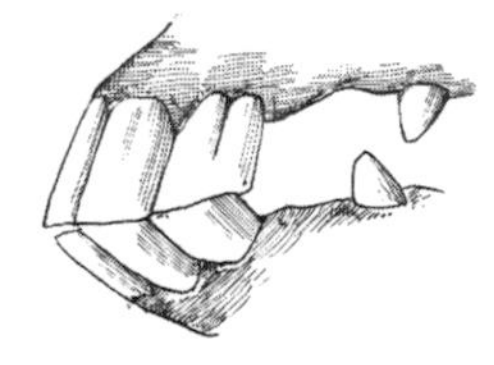

By 15 years **Galvayne's groove has extended halfway down the tooth. From the front, the lower incisors appear shorter than the uppers.**

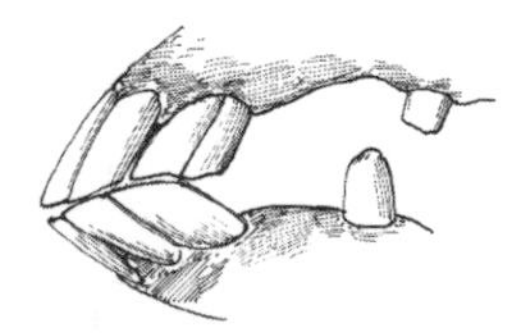

At 20 years **Galvayne's groove has extended the length of the corner incisor. The teeth have space between them, and, in some cases, the lowers may be worn almost to the gum line.**

Although different breeds can vary in maturity, five is about the youngest a first or second horse should be. By six years old he should have settled into his true personality, attitude, and manners. If he has behavioral problems at that age, it is too late for a novice rider to correct them.

If you own an old horse, keep him moderately active to maintain optimal health. Keep the activity level constant; once he is out of shape it will be hard to bring him back to peak fitness.

Horse Birthdays

Horses' first birthdays are officially celebrated on January 1 following their birth year. Between birth and weaning a foal is called a **suckling.** Once he has been weaned he becomes a **weanling.** After January 1 of the year following his birth he is called a **yearling.** The following year he is a two-year-old. A male foal is a **colt** and a female is a **filly** until they are four years old. At that point, the filly becomes a **mare** and the colt becomes a **stallion,** if he is uncastrated, or a **gelding,** if he has been castrated.

AI

➢ *See Artificial insemination (AI)*

Aids

Cues or signals from the rider that tell a horse what to do. There are two types of aids. Natural aids are the signals sent by the hands, legs, voice, and weight. Artificial aids are tools such as crops, whips, and spurs.

Aiken

A jump used in hunter classes, constructed of vertical wooden posts topped by evergreen branches.

Alfalfa hay

A high-protein, legume hay. Most horses love it, but it should not be fed alone because it is very rich and can cause problems, such as weight gain, loose stools, mineral imbalance, or even colic. Alfalfa actually contains more protein than an average horse requires. A horse that has foundered should not be offered alfalfa, because it could make him founder again. Alfalfa is best fed in a blend, mixed with clover and grass. Third-cut alfalfa is usually the most nutritious but may contain blister beetles.

➢ *See also Blister beetle poisoning; Colic; Feeding and nutrition; Hay; Laminitis*

All-in-one

A multipurpose fencing tool combining pliers, hammer, staple puller, and wire cutter.

Alternative therapies

Treatments used instead of or to supplement traditional medical remedies. Examples include acupressure, acupuncture, aromatherapy, chiropractic therapy, herbal treatments, and massage.

➢ *See also Acupuncture; Chiropractic treatment; Massage therapy*

AGGRESSIVE HORSES

Aggressiveness can range from biting to full-blown attacks with front or hind hooves. The key to defining an act of aggression is the horse's intent. Horses may hurt humans unintentionally as a defensive act or they may intentionally and aggressively do harm or try to do harm. Many, and some say most, horses that display aggressive behaviors have been inadvertently taught to do these things by untrained or inattentive humans. The majority of aggressive behaviors can be corrected using the proper techniques and equipment. It is best to work with a humane trainer to determine the most efficient approach for your horse's individual needs.

ALFALFA HAY

Alfalfa hay should not be fed alone to most horses and should be mixed with clover and grass.

Amble

A gait in which all four feet hit the ground separately. It is considered smoother to ride than the trot, beautiful to watch, and is often very fast.

This extra gait shows up naturally in certain breeds instead of or in addition to the trot or canter, and the gait has different names, depending on the breed. The Icelandic horse has a natural amble called the **tolt.** The five-gaited American Saddlebred ambles at two speeds — the slow amble is called the **slow gait,** and the fast version is known as the **rack.** Paso horses have three gaits — a walk, a slow amble, and a fast amble; they do not learn to canter. Tennessee Walking Horses have a special extended amble, in which the hind legs take particularly large steps. This "Tennessee walk" (also called a running walk) can be as fast as a canter, and is extremely comfortable for the rider.

Like the American Saddlebred, Tennessee Walkers and Rocky Mountain Horses were developed in the 18th century in eastern Kentucky, where Spanish and English horses were crossed. The Spanish blood is thought to have contributed the extra gait.

The amble is not the same as a **pace,** in which the legs on the same side strike the ground together.

➢ *See also American Saddlebred; Gaited horse; Pace; Paso Fino; Peruvian Paso; Rack; Slow amble or slow gait; Tennessee Walker; Tolt*

BREED CLOSE-UP

The American Cream Draft Horse, the only draft horse to originate in the United States, can be traced to an unusually colored mare in Central Iowa in the early part of the 20th century. Her offspring consistently showed the pink skin, amber or hazel eyes, and light, medium, or dark cream color that became the breed standard. Standing 15 to 16.3 hands high, Creams have refined heads, powerfully muscled bodies, and sturdy legs. Like most draft horses, they are kind, willing, and trustworthy.

American Association of Equine Practitioners (AAEP)

A nationwide organization of veterinarians specializing in the care of horses, ponies, mules, and donkeys.

➢ *See also appendix*

American Horse Shows Association (AHSA)

The national board, established in 1918, that oversees all equine competition in the United States. Its mission is "to inspire, encourage interest in, and regulate equestrian competition." The AHSA regulates shows, works with the United States Equestrian Team (USET) to select team members, and trains officials for competitions.

➢ *See also United States Equestrian Team (USET); appendix*

American Saddlebred

American Saddlebred, American Saddle Horse

Originally bred for a comfortable stride and the endurance to go all day, Saddlebreds and the closely related Tennessee Walking Horses combine the qualities of the colonial ambling horse with the blood of the Thoroughbred, Morgan, and Arabian. Tennessee Walkers remain closer to the original "model" and make splendid pleasure horses, while Saddlebreds have become extremely refined in appearance and are primarily show horses, though many people also use them on the trail. Saddlebreds come in two types: the three-gaited horse, shown at the walk, trot, and canter; and the five-gaited horse, which also performs the slow gait and the rack.

Sometimes referred to as "the peacocks of the showring," Saddlebreds are flashy lookers, usually standing around 16 hands tall with a strong arch to the neck and tail and a finely sculpted head. They tend to be dark in color (black, bay, brown, or chestnut), and splashy white facial and leg markings are common. Though generally good-natured, Saddlebreds are so highly trained and finely tuned that they have a reputation for being "fiery."

The Saddlebred's appearance does not come completely naturally. The extremely high set of the tail is achieved by a procedure that has been banned in some states: the muscles at the top of the tail are cut in such a way that the tail remains permanently deformed when it heals. The tail is placed in a tail-set when the horse is not in the show ring, to keep the arch pronounced. An additional aid to the arched tail is the practice of "gingering" horses before they enter the ring; that is, placing a piece of chewed ginger in the rectum as an irritant.

➢ *See also Amble; Five-gaited horse; Pace; Rack; Racking horse; Slow amble or slow gait; Tennessee Walker; Three-gaited horse*

American Standardbred

A breed of horses that trot or pace (some do both, but most have a preference) at great speed without breaking stride. The breed is named after the practice of requiring registered horses to meet a timed standard (originally a mile in 2½ minutes, though today's trotters are faster — some as fast as 1:50). Most Standardbreds trace back to a foundation stallion named Hambletonian, whose foals were natural trotters.

➢ *See also Pace; Standardbred; Trot*

BREED CLOSE-UP

American Standardbreds are the backbone of the harness racing industry, but they also make fine endurance or trail mounts, although they are not suited to jumping or galloping sports, such as barrel racing. Because retired racing Standardbreds are used to the commotion of the track, they are very safe mounts and are often retrained by police departments.

Measuring an average 15 to 16 hands, these horses are generally dark in color and less finely built than Thoroughbreds, with plainer heads, heavier bodies, and stockier legs. They are valued for their toughness and stamina as well as their pleasant, calm personalities.

See also Harness racing; Pace; Speed, of horse

American Warmblood

➢ *See Warmblood*

Ammonia

A pungent, alkaline gas formed as urine and manure decompose. Ammonia is the major cause of an unpleasant odor in a horse barn. If urine is left in stalls, ammonia can cause lung and skin damage.

American Standardbred

Amniotic fluid

The fluid surrounding an embryo within the amniotic sac inside the mare's uterus.

Anaphylactic shock

The severest form of allergy, which is a medical emergency. Symptoms include a sudden drop in blood pressure and difficulty breathing, which can be fatal. If treatment is given immediately with steroids and epinephrine, the condition is reversible. For this reason, it is advisable to keep these drugs in the barn medicine cabinet and discuss with your veterinarian ahead of time the proper ways to administer them.

ANATOMY OF THE HORSE

The structure of a horse; essentially the basic "map" of a horse's body parts. The term anatomy is objective, with no judgment of the correctness or quality of how those parts fit together. In contrast, a horse's conformation can be considered good or bad, depending on breed standards and the absence or presence of conformation defects.

See also Conformation; Eye; Foot care; Hoof; Judging; Mane, care of; Tail

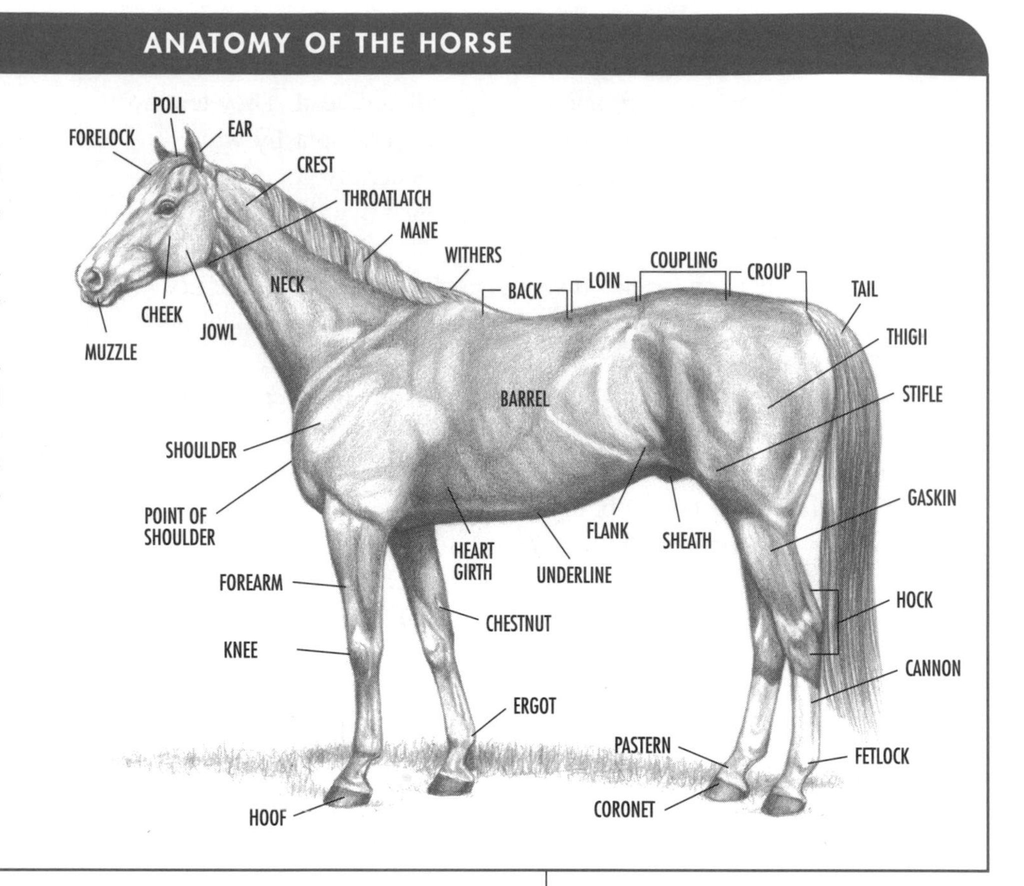

Andalusian

A lightly built riding horse with elegant lines, the Andalusian is a versatile show and pleasure horse often seen in the dressage ring. The breed is also popular as a circus performer. Primarily grey in color, these horses have very attractive heads with expressive eyes and a rounded muzzle. They are of medium height (15–15.2 hands), with low, rounded shoulders and backs. The Andalusian, a Spanish breed, is known for its lively personality, intelligence, and willingness to work.

➢ *See also Barb*

Anemia

➢ *See Equine infectious anemia (EIA)*

Anestrus

The period when a mare does not come into heat.

Anglo-Arab

A horse that is half Thoroughbred, half Arabian.

➢ *See also Arabian; Thoroughbred*

Anglo-Morgan

A horse that is half Thoroughbred, half Morgan.

➢ *See also Morgan; Thoroughbred*

Anhidrosis

Inability to perspire. This condition is most likely to occur in hot, humid climates. If a horse is confined in an airless stall during the hottest time of the year his sweat glands can overwork and ultimately shut down. The horse will be very hot but dry to the touch. If not discovered, the condition can be fatal. The remedy is to bring the horse's temperature down to normal (99°–100.5°F) by bathing him in cool water and placing him in a shady, cool spot. A fan may be necessary to cool him off. Call a veterinarian if this condition is new to you.

Leopard Appaloosas have spots all over their bodies.

Animated horse

A type of horse exhibiting brilliance, charisma, and flashy, elevated gaits. The animated horse lifts and holds back his head and neck and raises and folds his front legs in a highly collected manner. Because the gaits are exaggerated, animated horses are often suited only for competition in performance classes.

Appaloosa

A breed of spotted, dramatically colored, athletic horses developed by the Nez Percé Indians. Every Appaloosa is uniquely marked, but all share certain characteristics, such as mottled skin, mottling around the eye, and striped hooves.

The breed has two distinct genetic color patterns. Leopard Appaloosas are spotted all over like their namesake. The background color of this horse's coat is usually white, and the spots may be brown, black, bay, chestnut, or golden. The horse's eyes may be brown or blue (called a glass eye). On Blanket Appaloosas, the spots appear primarily on the horse's rump, in a blanket-shaped area. The background color of the horse may be black, gray, red, dun, or Palomino.

➢ *See also Pony of the Americas (POA)*

With the Blanket Appaloosa, the spots are confined to the hindquarters.

Appendix Horse

A horse with one Quarter Horse parent and one Thoroughbred parent, as listed in the appendix of the Quarter Horse registry. An Appendix Horse may compete in all Quarter Horse shows. In order for its offspring to be registered as an Appendix Horse, it must be bred to a permanent registered Quarter Horse. Alternatively, if the horse has earned a certain number of points in Quarter Horse shows or races before being bred to a Thoroughbred, the foal can be registered as an Appendix Horse.

➢ *See also Quarter Horse; Thoroughbred*

Approaching a horse

A horse or foal should be approached calmly and confidently from the front or the side, never snuck up on. The halter and lead rope should be ready at hand. It may be helpful to avoid eye contact while approaching the horse, because the contact may make the animal move away.

Arabian

A breed of horses valued for its beauty, spirit, and stamina. The key characteristics of the Arabian are the dished face, enormous eyes, "floating" gaits, arched neck, and short back. The breed is quite versatile, having international champions in a wide range of disciplines.

Half-Arabs can be registered by the International Arabian Horse Association, and they can compete in the association's shows.

➢ *See also Anglo-Arab; Caspian pony; National Show Horse*

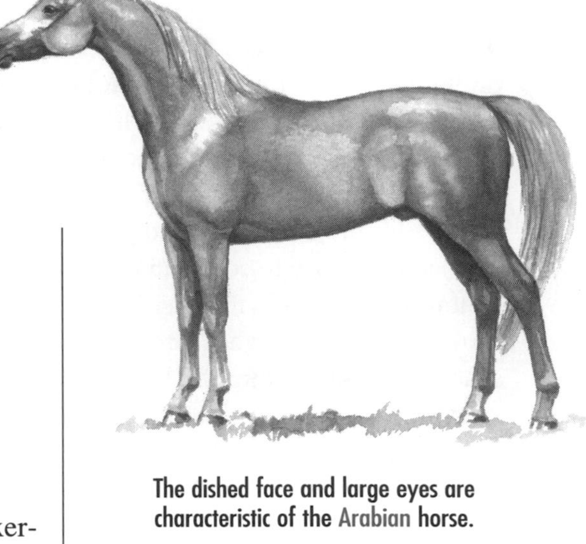

The dished face and large eyes are characteristic of the Arabian horse.

Arena events

Competitive events held in an arena; usually Western events.

➢ *See also Breakaway roping; Calf roping; Cattle penning; Cattle roping; Team penning; Team roping*

Arena work

Training in an enclosed pen or arena. Many mounted and ground exercises can be performed in an arena to improve skills and communication between horse and rider. Such exercises can provide a solid foundation for safe and effective horse-human interaction outside the arena.

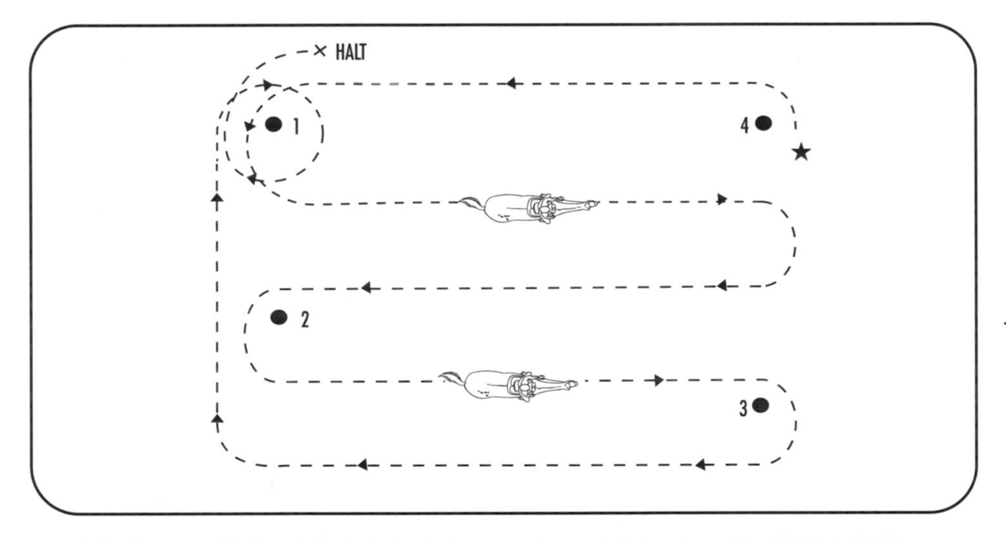

This map shows a serpentine exercise that can be performed in an arena. Beginning at the star, the rider guides her horse around four cones, at all three gaits, to improve suppleness, focus, and relaxation.

ARENA DIMENSIONS

USE	SIZE REQUIRED
Dressage (small size)	20 meters x 40 meters (66 feet x 132 feet)
Dressage (large size)	20 meters x 60 meters (66 feet x 198 feet)
Calf roping	100 feet x 300 feet
Team roping	150 feet x 300 feet
Pleasure riding	100 feet x 200 feet
Barrel racing	150 feet x 260 feet
Jumping	150 feet x 300 feet (depending on number and type of jumps and type of course)

From *Horsekeeping on a Small Acreage* by Cherry Hill

Artificial gait

An exaggerated style taught to some horses, notably American Saddlebreds. The two common artificial gaits are the **slow gait** and the **rack.** These are natural gaits, however, in horses that are genetically predisposed to them.

➢ *See also American Saddlebred; Five-gaited horse; Gaited horse; Pace; Rack; Slow amble or slow gait; Three-gaited horse*

Artificial insemination (AI)

Impregnation of a mare by other than natural means. Some breeders choose this method when a mare is difficult to breed by natural service, or to extend the semen of a valuable or older stallion. It allows a stallion to service far more mares each season than he could naturally, since one collection of semen can often impregnate more than a dozen mares.

The easiest way to collect semen is to let the stallion mount a mare in heat or a "decoy" mare (a breeding "dummy" designed for this purpose), and divert his penis into a specially designed artificial vagina.

AI is frequently used with Standardbreds and occasionally with Arabians and others, but is prohibited on registered Thoroughbreds. The reason for this is so that Thoroughbred bloodlines will remain natural and pure. Other breed associations also ban its use to preserve the integrity of their breed. The greatest fear is that the technology would allow a popular stallion, owned by breeders with strong financial resources for promotion, to flood the market, thus diluting the gene pool for the breed.

Ascarid *(Parascaris equorum)*

A large parasitic intestinal worm also known as a **roundworm.** Unless numerous enough to cause intestinal blockage, the adult worms are rarely harmful. The migrating larvae cause the most damage. Adult females lay up to a quarter of a million eggs per day, which pass out of the body in manure and can withstand freezing and drying.

Treatment usually consists of oral anthelmintics, but this must be continuous and prolonged to be effective. Good sanitation will go a long way to reduce problems, with daily cleaning of stalls and holding areas, weekly pasture rotation, periodic raking or plowing and reseeding of pastures, and protecting the water and feed supply from fecal contamination.

➢ *See also Deworming; Parasites*

ASCARID LIFE CYCLE

When the eggs are ingested by the horse, the larvae hatch in the intestine, penetrate the intestinal wall, and are carried through the bloodstream to the liver and, eventually, to the bronchi of the lungs. Then they are coughed up, swallowed, and passed into the small intestine, where they develop into adults and the cycle continues. The larvae can pass through the placenta and infect a fetus. Foals can also develop scarring of the liver from ascarids. Symptoms in the foal include a potbelly, a rough coat, failure to gain weight, and frequent coughing.

Aspiration

Pulling back on the plunger of a syringe to draw fluid into the chamber. This step must always be done before giving an intramuscular or subcutaneous injection. If the needle has penetrated a blood vessel, blood will enter the chamber of the syringe. In this case, withdraw the needle slightly and repeat the procedure, or choose another site.

Aspiration pneumonia

A type of pneumonia that develops when moisture or foreign particles collect in the lungs and cause an infection. A horse must never be tied so

that he cannot lower his head to cough and relieve a blockage in the windpipe. Horses confined this way, such as in a trailer, are at risk of developing aspiration pneumonia.

ASTM/SEI

An acronym meaning American Standard for Testing Materials and Safety Equipment Institute. Riding helmets will often carry a label stating that they comply with ASTM/SEI guidelines, indicating that they meet or exceed all ASTM safety standards.

➢ *See also Helmet, safety*

Astrohippus

A one-toed ancestor of the modern horse.

➢ *See Eohippus; Evolution of the horse*

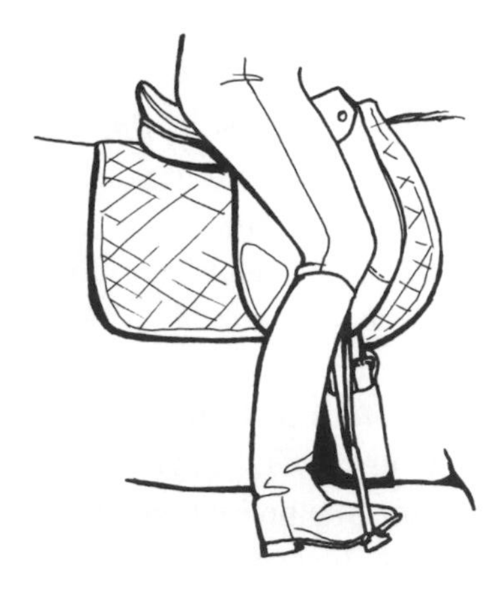

The "at the girth" position

At the girth

A phrase used to describe where a rider's legs should fall against the horse's sides when the upper and lower body are properly balanced and aligned. A straight line should connect heel, hip, shoulder, and ear.

➢ *See also Behind the girth*

Attire

Clothing; specifically, appropriate clothing for competitive events. Different classes have different guidelines (see box). Events vary in formality, but in general, English classes are more conservative than Western, and evening events are more formal than daytime events. The clothing of the rider should suit the class, complementing the horse and not distracting or detracting from his performance. Riders with long hair should wear it pulled back in a neat bun; hair nets are recommended.

For all riders, safe riding attire includes an ASTM/SEI-approved helmet and sturdy shoes or boots with a heel to keep the foot from sliding through the stirrup.

➢ *See also Bola or bolo tie; Boots, riding; Breeches; Chaps; Choker; Hat; Helmet, safety; Jodhpurs; Stock (as item of clothing)*

A rider's attire should include a safety helmet and shoes or boots with a heel. (English attire is shown.)

Australian cheeker

A device designed to keep the bit high up in a horse's mouth and prevent his putting his tongue over the bit. Also called the Australian noseband, it is made of rubber circles that fit around both ends of the bit, with strap extensions that join on top of the horse's nose and continue between the eyes and ears to the crownpiece.

Azoturia

A disease characterized by cramping of muscles, more accurately known as Paralytic Myoglobinuria; commonly called "Monday morning sickness or disease." It was named such because it occurred consistently shortly after horses started back to work after having a weekend off; despite this "vacation," the owner feeds the same high-level ration given

GUIDELINES FOR ATTIRE

Refer to specific show rule books for exact details, as these may differ.

Hunter
Conservatively colored coat; light-colored breeches or jodhpurs; white or pastel shirt with white stock tie or choker; tall black hunt boots; helmet

Jumper
Black or (in FEI classes) scarlet coat; light or white breeches; otherwise same as Hunter

Dressage
Lower levels of competition: Black hunt coat; white breeches; white shirt with white stock tie or choker; hunt cap, bowler, or derby hat; black gloves; black hunt boots with dressage top; spurs optional

FEI level: Formal coat with tails; top hat; white gloves; spurs mandatory; otherwise same as at lower levels of competition

Saddle Seat (different breeds may have different requirements)
Three-piece suit in conservative color; dress shirt with tie; top hat in formal or evening classes; derby or homburg in less formal classes; jodhpur boots; gloves; whip and boutonniere optional

Western Pleasure or Horsemanship
Vest or jacket optional; plain Wrangler jeans or Western pants; fitted chaps may be optional; neat, long-sleeved Western shirt; Western hat, preferably felt; cowboy or roper-style boots; necktie, scarf tied in square knot, or bolo; belt, gloves, and spurs optional

Western Showmanship
Coat or vest preferred; starched jeans or dress slacks; felt Western hat; no chaps or spurs; otherwise same as Western

Cutting
Batwing chaps; spurs; otherwise same as Western

Reining or Cow Horse
Shotgun chaps; spurs; otherwise same as Western

Barrel Racing, Pole Bending, and Other Timed Events
Jeans; boots; Western shirt; Western hat or safety helmet; whip, bat, or quirt; shin guards worn under jeans; gloves

Roping Events
Jeans; boots; Western shirt; Western hat; cotton glove recommended for roping hand

during times of work. It is characterized by sudden onset of semiparalysis of the hind legs; usually occurring within 15 minutes of resuming work. The muscles of the hindquarters usually stiffen and become hard to the touch.

The cause is not fully understood, but is thought to be related to the buildup of glycogen in the muscles of an idle horse that is fed a high-energy ration. Upon exertion, this glycogen is used up rapidly and an increase of lactic acid in the muscles follows. This causes swelling and degeneration of the muscle fibers. Another theory supports the idea that the disease is related to a deficiency in selenium and/or vitamin C.

Treatment requires veterinary care and absolute rest.

➢ *See also Rhabdomyolosis*

The seat on a horse makes gentlemen of some and grooms of others.

—Miguel de Cervantes, *Don Quixote*

b

Back at the knee

A foreleg condition in which the front of the knee is insufficiently extended, while the portion of the leg behind the knee appears overextended. Horses with this conformation fault are prone to knee fractures. Also known as calf knees, this is a serious fault in horses being considered for competition or breeding.

➢ *See also Conformation*

Back combing

A technique of combing shorter mane hairs back so that longer hairs can be trimmed evenly.

➢ *See also Grooming; Mane, care of*

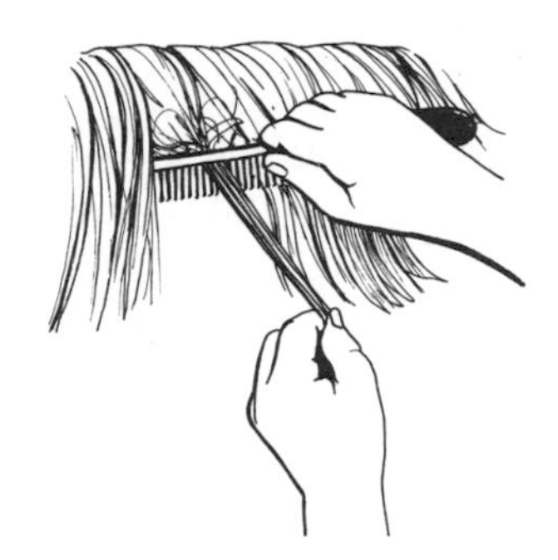

Back combing helps handlers trim the mane evenly.

Backing up

Inducing a horse to back up to a desired area or position.

Backing a Standing Horse

The handler generally stops the horse and stands in front of it, facing the shoulder. While holding the lead rope and whip in the left hand, put the tips of your right-hand fingers on the point of the horse's shoulder. Give a "BAAAACK" command while applying light pressure with the fingertips.

After a number of repetitions, the horse will respond automatically to the command. Many handlers teach a horse to back straight up by creating a lane out of poles, hay bales, or other materials and backing the horse down that lane.

A properly trained horse will back up in response to fingertip pressure and a voice command.

Backing while Riding

Your horse should be able to back up several steps without fighting the bit. Before asking him to back up, make sure he is moving forward well and is relaxed and responding to the bit. Then from a halt, pick up light rein contact and squeeze with your thighs to ask him to back up.

Without pulling on the reins, maintain even contact. Reinforce your leg aid by bumping him with your heel or tapping him with your whip on the inside shoulder. He should relax his jaw and give in to your request. Correct any drifting to either side with your leg or whip.

Don't let him move forward again until his head and jaw are soft and in the correct position. Practice moving forward at the walk or trot, then stopping and backing up.

Back rail

In riding competitions, the section of the show pen that is behind a judge's back.

Back splice

A piece of rope folded over itself and fastened together to form a small loop at one end. Back-spliced ropes with bull snaps are used for in-hand work with horses.

A back-spliced rope with bull snap

Balance

Balance can refer to a horse's anatomical structure or to the posture and position of a rider.

Balanced Horse

A balanced horse performs efficiently, moving in a smooth, even fashion. To determine the degree of balance displayed by a horse, note the animal's skeletal structure. The horse has both vertical and horizontal balance. Vertical balance is determined by dividing the horse into three parts with two vertical lines, perpendicular to the ground. The first is drawn through a point at the base of the horse's neck; the second is drawn through the point of the hip. A horse is said to have vertical balance if the three parts are relatively equal, with the forequarters being the widest and the midsection being the shortest. (The neck/head are the balancing arm of the horse's balance and thus should be long. The midsection includes the horse's back; a short back is stronger and can support more weight than a long back.) There is very little discrepancy between breeds with regards to breed ideals for vertical balance.

The "ideals" for horizontal balance tend to be more individualized according to requirements of different breeds and/or disciplines. However, there is a uniform method for evaluating such balance. Again, it involves drawing lines. This time, it's only one line, drawn through the elbow and stifle joints. This line should result in approximately equal upper and lower halves. (Ideals vary; for example, race horses are more successful if they are "leggy," and cutting horses are more agile if they're "low to the ground.") The line should be roughly parallel to the ground. If it runs downhill in front, the horse will tend to be heavy on its forehand, might be difficult to train for competition, and could be prone to tripping and falling while carrying a rider.

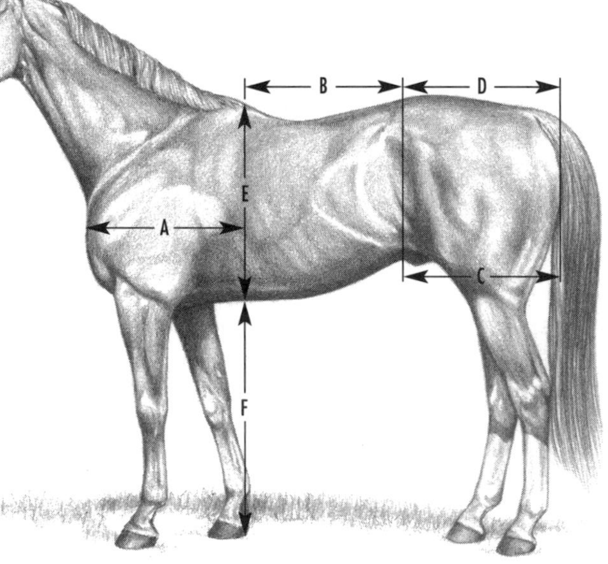

A balanced horse exhibits vertical balance (A = B = C = D) and horizontal balance (E = F).

Balanced Rider

The rider's position and bearing affects a horse's performance and comfort. A balanced rider sits straight in the saddle without being rigid. A vertical line is formed from ear to hip through knee and ankle. The rider's legs hang at the girth in a relaxed manner, and the body is positioned directly atop the horse's center of balance. The rider generally looks straight ahead and pushes down on the stirrups with his or her feet. This balanced position communicates the rider's skill and ensures an efficient, comfortable ride for both participants.

Balanced seat

A fundamental approach to riding, wherein the rider's center of gravity corresponds to the horse's center of gravity. The balanced rider also aligns his or her body in such a manner that a vertical line could be drawn from the ear down through the shoulder, point of the hip, and back of the heel. The rider is able to stand in the stirrups while his or her calves remain motionless. An out-of-balance rider may fall off the horse or cause the horse to stumble or tire prematurely.

Bald-faced

A description of a horse whose face is predominantly white.

➢ *See Face markings*

Baling hay

Hay growers bale their crop when it reaches the proper maturity. In most situations, hay should be baled before late-bloom stage, as this helps the hay retain its leaves. Bale size is generally determined by the size of the bale wagon and the strength of the person handling the hay on a daily basis. Tight bales are easier to stack and handle, but overly wet hay that is baled tightly may generate mold and heat. Heat reduces the nutritional value of the hay and can also trigger fires; moldy hay can lead to respiratory problems in the horse.

➢ *See also Feeding and nutrition; Hay; Heaves*

Ball mount

A towing device that connects a horse trailer to a truck chassis.

➢ *See also Trailering a horse*

Ballotade

A spectacular leap performed by Lippizaners. From a rearing position, the horse leaps straight up, tucking his hind legs up to his belly.

➢ *See also Lippizaner*

Bandage

Gauze or cotton that protects a horse's wound from infection. It also holds the wounded area in the proper position for healing, prevents inflammation, keeps the area warm, prevents it from drying out, and protects tissues from bumps and bruises. Equine bandages are multilayered and are changed every one to three days on a fresh wound. They are also used to protect a horse from cuts and bruises when trailering the animal over long distances.

Bandy-legged

A bandy-legged horse appears bowlegged when observed from the rear. The condition can adversely affect the horse's hocks, which are the joints above the fetlocks on the rear legs, and stifles, which are the joints above the hocks.

➢ *See also Conformation*

BANDAGING KIT

The typical wound bandaging kit includes:

- ❑ 14" x 30" Cotton batting (disposable diapers do well)
- ❑ 12" x 30" Cotton leg quilt (or cotton diaper)
- ❑ 8" x 3" Sterile dressings
- ❑ 4" x 4" Gauze squares
- ❑ Bandage cutting knife
- ❑ Small brush
- ❑ Bandage scissors
- ❑ 4" x 5 yds. Stretchable crepe bandage
- ❑ Povidone iodine ointment
- ❑ 4" x 5 yds. Elastic adhesive tape
- ❑ 4" x 3 yds. Clinging gauze rolls

Adapted from *Horse Health Care* by Cherry Hill

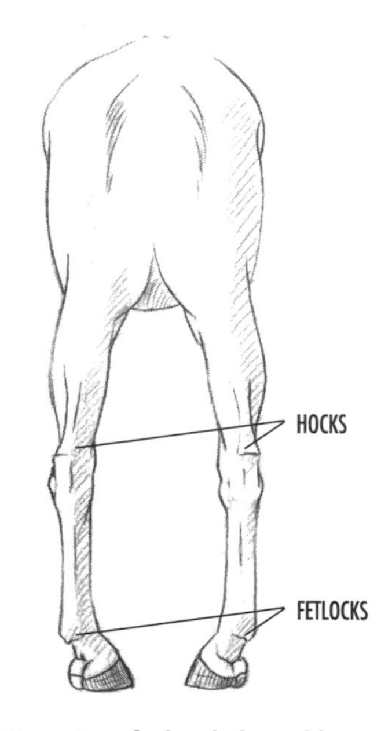

Rear view of a bandy-legged horse

Barb

An early ancestor of the horse that evolved to survive in the deserts of central and southern Asia. This animal later migrated to North Africa and, ultimately, to Spain. The Barb is the ancestral father of the Andalusian horse, which is associated with Spain.

➢ *See also Evolution of the horse; Spanish Barb*

Barbed wire

Any of half a dozen forms of wire consisting of two strands twisted together and armed with sharp points or barbs. Barbed wire should *not* be used to confine horses, as it is likely to inflict serious injury. In many cases, an entrapped horse will panic and attempt to twist loose, thus aggravating the injury.

➢ *See also Fencing materials*

Bareback riding

Riding without a saddle. Many instructors believe that bareback riding helps beginning students gain a sense of balance and stability. It also offers riders a unique opportunity to move in concert with the horse and feel the action of the animal's muscles.

Rodeo Event

Bareback riding has been a judged rodeo event for many years. Typically, the rider sits atop a bucking horse for an eight-second ride. The rider must hold the bareback rigging with one hand and control the horse through the use of spurs. A rider who touches the horse, the equipment, or his or her own body with the free hand is disqualified.

➢ *See also Rodeo; Western riding*

Barley

A high-energy grain that can be a good choice of feed for a horse that needs extra nourishment. Because it has a hard, prickly hull, barley is usually fed in crushed or rolled form. This enhances digestibility and nutritional value. Barley is also commonly mixed with bran or oats.

➢ *See also Feeding and nutrition*

Barley is a high-energy grain often found in feed mixtures.

Barn

A well-built horse barn is a pleasure for both owner and occupants. Typically, a horse barn consists of an appropriate number of stalls, a tack room, a feed storage room, and a work area or center aisle. The barn should offer sufficient ventilation and lighting; secure, nonslippery flooring; and good drainage.

Drainage

A barn should be situated on relatively high ground, so it won't become inundated with water after every storm or snowmelt. When constructing a new barn, the site can be built up so it provides adequate drainage.

The shelter provided by a well-designed barn is an important part of good horsekeeping.

Flooring

Dirt is usually considered the best barn floor. Packed dirt provides sure footing and, in most cases, adequate drainage. Floors made of concrete become slippery when wet and are poor for drainage. They are also hard on the horse's legs. Wooden floors are softer, but they are high maintenance and can also become slippery. Rubber stall mats are recommended where you want better traction or an easier-to-clean surface, or if your horse may be lying down frequently (such as a mare about to foal).

Lighting

Well-lit barns are safe and comfortable. Ideally, a light should be placed in each stall. Lights are also useful when placed at intervals above the center aisle and above tack and feed areas. They should be at least 8 feet off the ground so the horses cannot reach and break them. Housing lights in a wire cage provides additional protection. Skylights, windows, large sliding doors, and roof panels and vents can be excellent sources of natural light. Because they reflect light rather than absorbing it, light-colored walls and ceilings also help barn owners take advantage of available light.

Stalls

Ideally, each stall should accommodate the horse for which it is intended. Ponies are comfortable in a 10-foot-by-10-foot space, but full-sized horses generally require a stall that is 12 feet square. For foaling or nursing mares, a double stall is recommended. Stallions also benefit from a large stall.

Stall walls must be durable to withstand the kicking, rubbing, and general abuse that horses inflict upon them. Walls are commonly made of cement block, wood, or metal (but a horse can injure himself by kicking cement or by putting a hoof through metal). They should be smooth to prevent possible injury to the animal. Protruding splinters, nails, bolts, or sharp edges of any kind should be removed or covered. Experts recommend stalls at least eight feet high, with doors about four feet wide. Stall floors should be durable but relatively soft and comfortable. Among the preferred flooring materials are well-tamped dirt, clay, limestone, or a bluestone base.

If bars are used at the front or sides of the stall, the spaces between the bars should be no wider than four inches, so the horse cannot get caught in the bars.

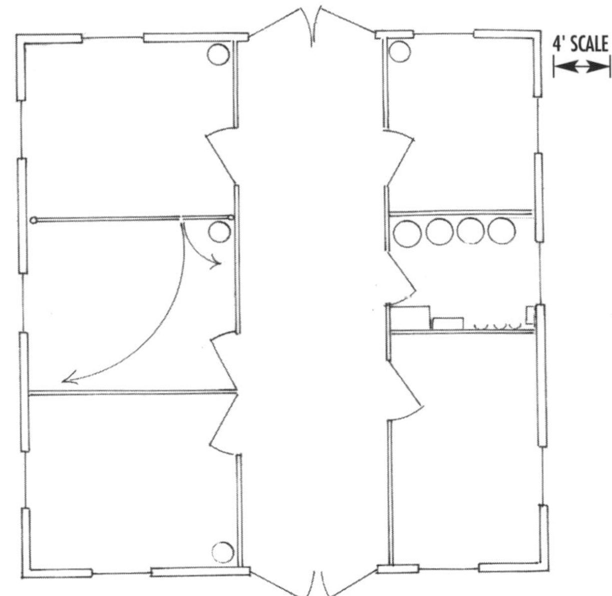

This barn floor plan includes a double stall with a movable partition.

Ventilation

Ventilation is the movement of air through a barn. An average horse may produce 50 pounds of manure and 10 gallons of urine in a single day. Without ventilation, the barn air will become saturated with ammonia and other products that are released when waste materials decompose. These odors, as well as dust and humidity, pose respiratory threats to confined horses, but so can cold drafts. The trick is to build or adapt a barn that is well ventilated but does not expose the horses to excessive drafts.

A window in each stall can simultaneously provide ventilation and light. Large, sliding doors at each end of a barn can do much the same. Roof vents and exhaust fans are additional ways of enhancing the exchange of air.

➢ *See also Bedding; Facilities; Stall; Tack room*

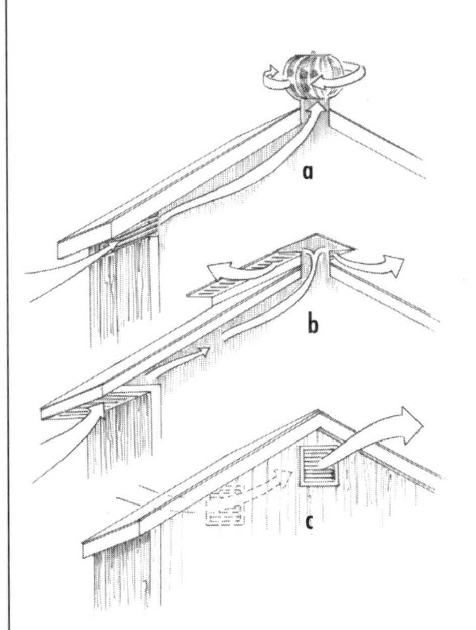

Barn ventilation can be improved through the use of (a) a turbine vent; (b) soffit louvers and vents; or (c) gable louvers.

BARN-BUILDING TIPS

- ❑ Talk with the owners of existing barns to learn about the strengths and weaknesses of different layouts.
- ❑ When constructing a horse barn, use long-lasting materials that require little or no upkeep. Painted vinyl, aluminum siding, or treated plywood can be advantageous choices.
- ❑ Shop around for materials and services. Prices vary dramatically.
- ❑ Consider building the barn yourself or hiring low-wage students.
- ❑ Build on firm ground that will not shift or collapse.
- ❑ To facilitate drainage, choose a site that is higher than the immediate surroundings.
- ❑ Invest time and thought in designing the barn, so it will help you work efficiently and ensure the horses' comfort.
- ❑ In temperate climates, place doors and openings so they face east and south. This will help keep the barn warm in winter.
- ❑ Cover all windows with bars to eliminate glass breakage.

Barn sour

A description of a horse that is reluctant to leave the security of the barn or the herd.

Barrel racing

A Western riding event in which competitors must ride in a cloverleaf pattern around three standing barrels. The winner is the rider who completes the course in the fastest time.

➢ *See also Rodeo; Western riding*

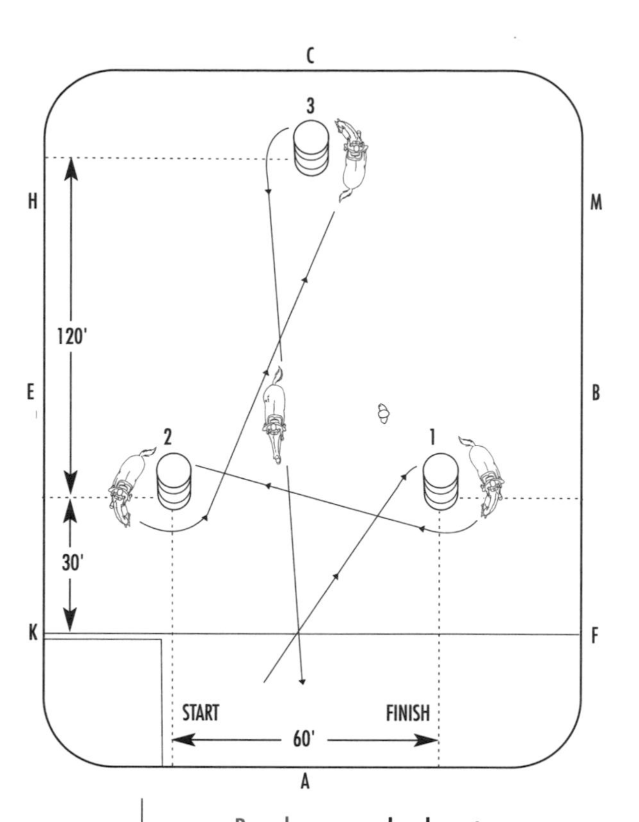

Barrel racers make sharp turns around three barrels, then gallop to the finish line.

Bars

The toothless, gummed section of the horse's jawbone on which the bit rests.

Base-narrow and base-wide

Seen from the front, a base-narrow horse has feet that are too close together. A base-wide horse has the opposite characteristic — its feet are too far apart. The forelegs are straight on a horse with desirable conformation.

➢ *See also Conformation*

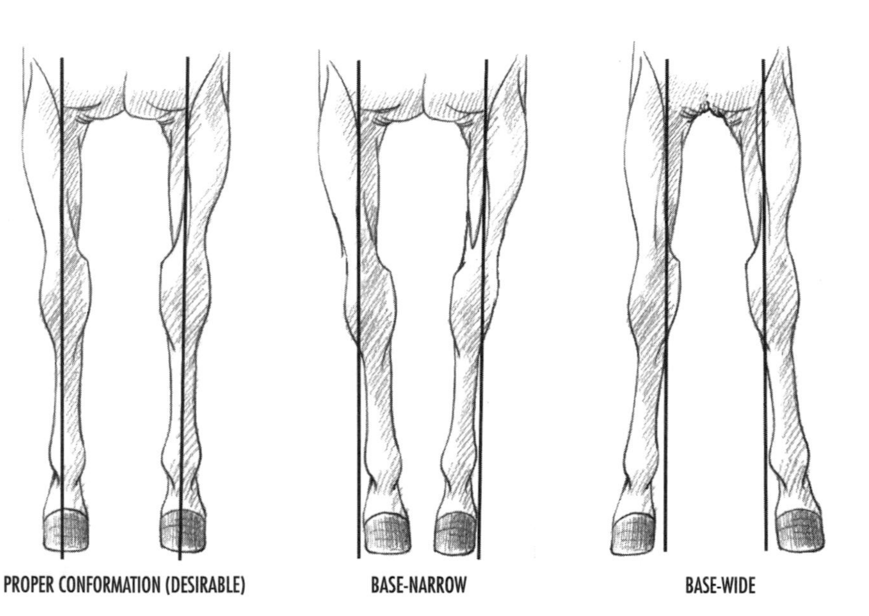

Bathing horses

For a horse, as for humans, a bath is a mixed blessing. On the one hand, the bath keeps the animal clean and can help prevent infection. On the other hand, the bath dries out the horse's hair and skin by removing helpful oils. The general rule is to bathe a horse often enough to keep it clean without exposing the animal to excessive dryness or bacterial infestation.

For pleasure horses, two to three baths per year is a sufficient number. Handlers suggest a late spring bath soon after the horse has finished shedding; a midsummer bath; and a bath in fall just before the horse's winter coat begins growing. The temperature in all cases should be greater than 50°F.

Many horse owners use an indoor wash rack whose texture helps the horse maintain his footing. Washing outdoors is fine also, preferably on a large rubber mat. The washing site should offer a convenient place to tie the horse.

Supplies needed to bathe a horse include warm water, shampoo designed for horses, and (if desired) a conditioner. The shampoo should be diluted as instructed on the bottle. Conditioner is added after shampoo and a rinse, in order to restore shine to the coat and moisture to the skin.

THE SEVEN-STEP EQUINE BATH

1. Wet the horse thoroughly. Start from the hoof and work your way up slowly so the horse gets used to the water temperature.

2. Shampoo the horse one section at a time.

3. Rinse each section as you finish shampooing it. Use a sweat scraper or your hands to remove excess water.

4. Condition the coat by pouring diluted conditioning solution all over the horse.

5. Rinse the conditioner out if necessary. (Check the product label to determine whether this step can be skipped.)

6. Dry the horse. Use a sweat scraper or your hands to remove water from heavily muscled areas. Use a cooler so he won't get chilled while drying.

7. Finish the horse by making sure all of the mane is lying on one side of the neck. Apply a finishing spray (optional) to the mane and tail when they are slightly damp.

Adapted from *Your Pony, Your Horse* by Cherry Hill

Bay coloring

A body coloring that varies from tan to reddish brown. A bay has a black tail and mane and sports black coloring on its lower legs.

➢ *See also Coloring*

Beat

A single step in a gait, taken by one leg or by two legs simultaneously.

Bedding

Bedding for a stall or shed provides a horse with a comfortable surface on which to rest. Bedding can also preserve warmth in the stall, protect the horse from abrasions, and absorb much of the horse's urine.

Materials used for bedding include straw, wood shavings and sawdust, shredded newspaper, and sand. Peanut hulls are used in some southern areas, where they are available at a reasonable cost.

Each bedding material offers advantages and disadvantages, but straw is the perennial favorite. If spread deeply enough, it is soft, warm, and highly absorbent. However, it can be expensive. Wheat straw is preferable to oat or barley straw, as a horse is less likely to eat large quantities of wheat straw bedding. (Doing so can cause colic, which can be fatal.)

Horse owners who use wood chips, shavings, or sawdust as bedding should avoid hardwoods. Some hardwoods — particularly black walnut, but also oak and hickory — can trigger toxic reactions in horses.

Whatever material is used, it's important to keep bedding clean and free of dust, mold, and sharp edges.

For foaling purposes, clean straw provides safe and comfortable bedding.

Behavior in horses
➢ *See Body language, of horses; Pecking order, in herds*

Behind the girth
One of the two fundamental leg positions for a rider. Positioning one's legs at the girth means placing them directly over the horse's center of balance. Positioning legs behind the girth means moving them back two to four inches from the center. The behind-the-girth position is considered a leg aid used by a rider to control the horse.
➢ *See also At the girth*

In the "behind the girth" position, the rider has moved her legs back to allow for different pressure points.

Belgian
A popular, heavy-bodied breed of draft horse known for its strength and patience. Most Belgians are roan- or chestnut-colored.
➢ *See also Coloring*

Bell boot
A boot, often made of rubber, that fits over the top of a horse's hoof. The boot protects the area from injury when the horse is traveling or working, and is especially vital for horses that overreach.

Bell boots can be used over shipping wraps when a horse is being trailered.

Bench-kneed
A defect in which a horse's cannon bones are not centered directly under the knees. The defect, also known as offset cannons, can hamper the horse's knee movements and cause enlargement of the area where the splint bone attaches to the cannon bone.
➢ *See also Conformation; Leg, of horse*

Bending
The curving of a horse's spine to the left or right, also observable in the horse's neck. Bending customarily occurs while riding in a circle or making a turn.
➢ *See also Flexion*

Bending the Horse
A series of actions taken by a rider that causes a horse to bend, most often during turns, but also in higher-level movements. Bending the horse is often necessary to maintain the balance of horse and rider and to prevent falls (see box).

HOW TO BEND A HORSE

1. Position your inside leg at the girth. This is your active leg.
2. Position your outside leg behind the girth.
3. Sit on your inside seat bone.
4. Turn your shoulders toward the bend and to the degree of bend.
5. Create the bend with your inside hand and give equally with your outside hand.

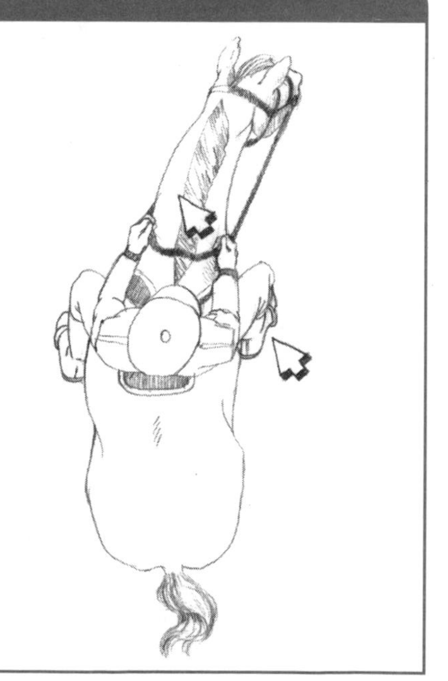

From *Getting the Most from Riding Lessons* by Mike Smith

Betadine

➢ *See Tamed iodine (Betadine)*

Birth defect

Foals are subject to just as many birth defects as any other mammal. Below is a listing of birth defects that are fairly common in foals. Some faults are deadly, some can be surgically corrected, and some are self-correcting.

- cleft palate
- crooked head from lying against the uterine wall in an odd position
- crooked legs from being cramped in the womb
- incomplete digestive systems, especially with the anus closed before reaching the rectum
- leaky navel; the canal connecting the bladder with the umbilical cord fails to seal at birth
- heart defects
- turned-in eyelids
- retained sheath and penis

Of these, the last two are the most easily corrected surgically. Crooked head and legs often right themselves in several days. The other defects may or may not respond well to repair.

Bit

A metal mouthpiece that aids a rider in controlling the horse. It rests on a toothless, gummed portion of the horse's jawbone known as the bars. Manipulating the horse's reins moves the bit, causing a trained horse to respond as desired.

Selecting a Bit

It's best to use the mildest bit that works effectively with an individual horse. Although hundreds of models are available, there are three basic types of bits. **Snaffles** are considered the mildest of bits, and are most often recommended for beginning and pleasure riders. They are designed to apply pressure directly to the corners of a horse's mouth. Many snaffles are jointed at the center, but others feature multiple joints. Still others, known as straight snaffles or straight bits, are jointless. **Curbs** are bits that work by applying leverage directly to the bars of the horse's mouth, used with a shank of varying lengths. The longer the shank, the greater the leverage and the more severe the action. Curb bits are used in conjunction with a chain or strap designed to fit beneath the horse's jaw. **Combination,** or Pelham, bits offer the rider the option of exerting pressure on either the corners of the mouth or the bars.

Bits come in a wide range of sizes and thicknesses. The width of the horse's mouth at the corners should be used to determine the width of the bit. Five inches is the average width, but young horses and smaller breeds may need narrower bits, while large horses often need wider bits.

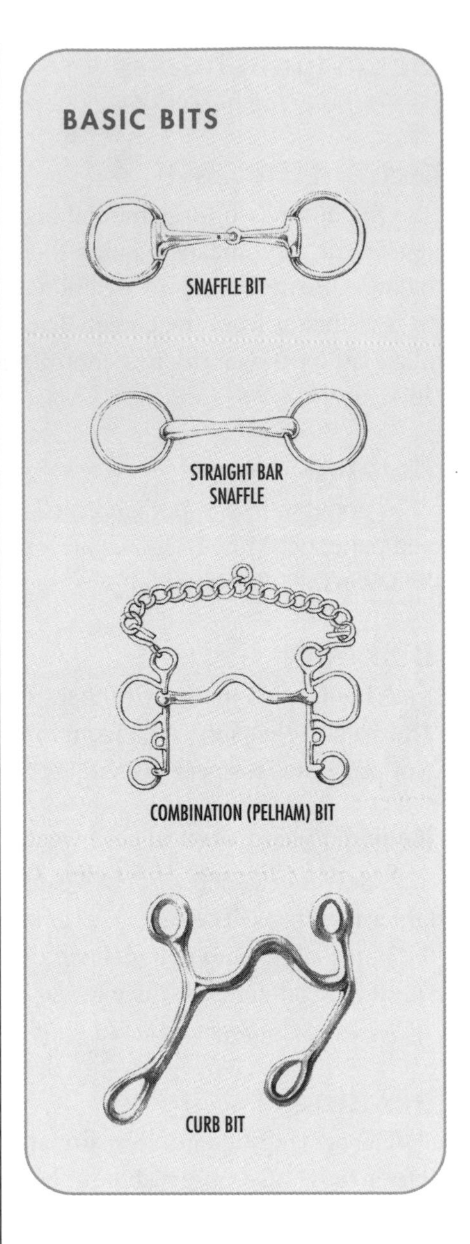

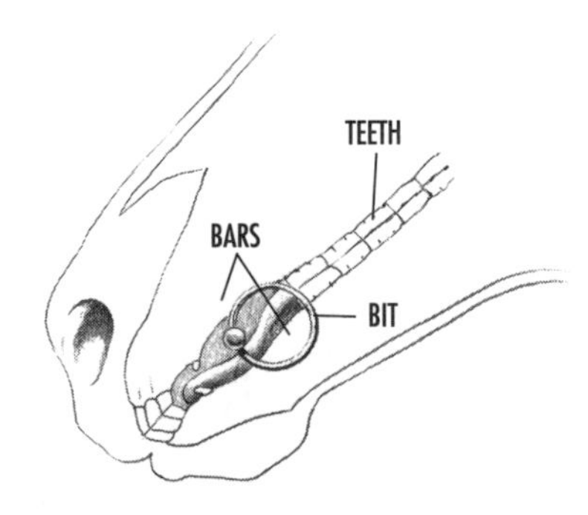

The bit rests on a horse's bars, a toothless portion of the jawbone.

As a general rule, a thicker bit is milder and more comfortable for the horse. However, excessively thick bits may not fit comfortably into a given horse's mouth. The texture of a bit — its smoothness or sharpness — can also greatly influence the bit's comfort. Many horse owners have bits professionally fitted to ensure proper function and to prevent possible injury.

BLANKETING A WET HORSE

Blanket a wet horse only if the blanket is wool. Wool wicks moisture away from the skin. Synthetic fabrics cause moisture to penetrate the natural protective oils in the horse's skin.

Bit guard

A temporary device to keep a bit from pinching a horse's cheeks or lips. Two flat rubber circles fit around the ends of a bit between the skin and the bridle rings.

Black coloring

To be considered black, a horse's entire body (except for white markings on the leg and face) must be colored true black. This coloring is uncommon.

➢ *See also Coloring*

Blanket clip

A body clip pattern that leaves hair on the legs and the area covered by a blanket. This clip is suitable for horses doing medium to hard work in cold weather.

➢ *See also Clipping; Hunt clip; Trace clip*

Blankets and blanketing

Horses kept at pasture year-round may never need blankets. In colder climates, they grow a thick winter coat that conserves heat and protects them against wind, moisture, and cold air. However, an outdoor horse that has been clipped or is in poor health will need blanketing. Horses that become excessively cold or wet may also need blankets. **Turnout blankets** are recommended in such cases. These blankets are entirely waterproof, fit snugly, and are cut in a way that allows freedom of movement.

Stalled horses that are ridden in cold weather also benefit from winter blankets that help them cool down gradually after a ride. Any horse whose coat has been clipped must be blanketed if he will be exposed to cold or wet weather, and horses should be blanketed after they have been given a bath in cold weather.

Blankets are often used when trailering a horse to protect against drafts. However, overheating can occur in a closed trailer, so the horse's body temperature should be monitored. Blanketing the horse in two or more lightweight layers allows the handler to regulate the horse's temperature by removing or adding blankets one layer at a time.

It's a good idea to monitor a horse's body heat any time a blanket is used. A blanketed horse can become dangerously hot even in winter, if the day is warm or sunny or if the blanket does not wick moisture away from the body effectively.

A smaller blanket and saddle pad are used beneath Western saddles. These saddle blankets are often decorative, but they also protect the horse's back from chafing and saddle sores.

Blanket Care

Horse blankets should be washed with soap and disinfectant. If the blankets are used frequently, they should be cleaned at least twice during the winter. Blankets that are regularly muddied should be washed more often. Blankets should also be checked regularly to ensure that the fabric is intact and the straps are working properly. When not in use, winter blankets should be stored in trunks or closets that provide protection from rodents, insects, and moisture.

Choosing Blankets

The traditional horse blanket is made of wool, which is relatively warm. However, wool blankets are not ideal if a horse must be protected from water and wind. Among the alternatives are blankets that feature an outer shell made of waterproof, windproof synthetic fabric. The middle layer of these blankets consists of insulating material that keeps the horse warm, and the layer next to the horse's skin is designed to be smooth and non-irritating. These blankets are lightweight, durable, and comfortable. Outer shells may also be made of canvas, which is notable for its durability but provides less-than-total protection from rain and snow.

Sheets and **coolers** are special lightweight blankets that are used to help horses cool down gradually or to keep them clean.

Blaze

A white marking or stripe down the center of a horse's face that extends beyond the sides of the bridge of the nose. A blaze is wider than a stripe. If it extends around the sides of the face, the horse is said to be **bald-faced.**

➢ *See Face markings*

Blind spots

Areas not visible to horses unless they turn their bodies or move their heads. Handlers should be aware of these blind spots to avoid startling or spooking their horses.

A horse's blind spots include the area behind his tail; the portion of his spine from immediately behind the head to his mid-back; the area directly beneath his head and in front of his forelegs; and a small area immediately in front of his forehead. Horses have excellent vision otherwise. They can spot small movements at a considerable distance, and they see well at night.

➢ *See also Vision, of horse*

Blister beetle poisoning

Several blister beetle species carry a substance in their bodies called cantharadin. If a horse ingests these beetles, cantharadin toxicity will result. The ingestion of one or two beetles (dead or alive) can cause severe

Measuring for a Blanket

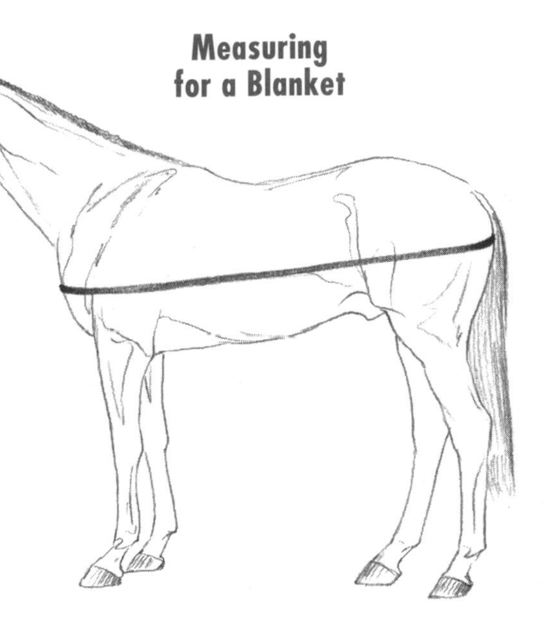

To determine a horse's blanket size, measure from the center of the horse's chest to the center of the tail. An ill-fitting blanket can cause sores and abrasions. It could also fall off or become tangled around the horse's legs.

BLINDERS AND BLINKERS

Blinders are leather flaps used on a racing or driving bridle to shield the horse's vision from side distractions. *Blinker*, an older, synonymous term, also describes the cloth hood worn by some racehorses.

blistering of the horse's digestive tract. Ingestion of as few as three beetles could kill the horse.

The beetles feed on the pollen in alfalfa hay blooms, and ingestion often occurs when a horse eats hay harboring these small flying insects. The symptoms of blister beetle poisoning are similar to those of colic, which can make the condition difficult to diagnose in a timely manner. Drugs and intravenous fluids can be used to minimize absorption of the toxin if treatment is begun soon enough. The best preventive measure is to feed first-cutting alfalfa (because the beetles generally do not mature until midsummer) or hay that has been cut prior to the bloom stage.

Blister beetles may contain a toxin called cantharadin.

Bloating

An action in which a horse fills his stomach with air, then holds his breath. The horse may do this when a handler is attempting to tighten the girth, only to release the air later. This can result in a dangerously loose saddle. Signs of bloating include a pinched-shut mouth and nostrils.

Blocks, salt and protein

Salt and protein blocks provide nutrients that supplement a horse's primary diet. Most horses benefit from unrestricted access to a trace mineral salt block. Molasses protein blocks are primarily useful for horses whose diets are unusually low in protein; access should be regulated. A horse that overindulges in protein blocks can be subject to colic, dehydration, or diarrhea. Horses can also benefit from calcium and phosphorous blocks, depending on their age and diet.

Bloodworms

As the name implies, bloodworms are blood suckers. They are technically known as strongyles, but they are also commonly labeled red worms or palisade worms. Whatever the terminology, these parasites are dangerous to a horse's health.

Three species of bloodworms infest horses. The smallest species, *Strongylus vulgaris*, is the most destructive. It can inhabit the large arteries that lead away from the heart and vessels that supply blood to the intestines. A relatively light infestation might cause pain and colic. A heavy infestation can rupture an artery, diminish intestinal blood flow, and inflame, scar, and weaken blood vessels that supply the intestine. These developments can, in turn, trigger clots, aneurysms, or death of an intestinal section. Fatalities sometimes result.

Fortunately, horses can be protected from bloodworm infestation through the use of a safe, effective deworming drug called ivermectin. This drug paralyzes the parasites, rendering them unable to feed or reproduce. Ivermectin should be used at least twice a year. A newer drug that is also effective against strongyles is called moxidectin.

Symptoms of a bloodworm infestation may include diarrhea, loss of appetite, lethargy, weakness, weight loss, dull hair coat, and episodes of colic or lameness.

➢ *See also Deworming; Parasites*

Bluegrass

A grass hay that can provide vitamins, calcium, and protein in amounts sufficient for most horses. Bluegrass can be fed in combination with legume hays, such as alfalfa, if horses need supplemental nutrition.

Blue roan coloring

A body coloring that is an evenly distributed mixture of black hairs and white hairs. However, the head and legs of a blue roan horse are often darker than the remainder of the body.

➢ *See also Coloring*

Boarding stables

Some boarding facilities simply offer a space in which to keep a horse. Others offer a full range of services including feeding, grooming, routine health care, training, and exercise. In general, the more services provided, the higher the boarding fee. Under the most common arrangement, the owner is charged with responsibility for the horse's grooming and exercise needs.

A **self-care facility** simply rents the owner a stall or pasture space. The owner must care for the horse in all respects — including, in some cases, providing water.

A **field-board facility** provides pasture area for a boarded horse. Basic services may be provided, and a shelter is generally available for the horse in times of émergency or bad weather.

A **full-board facility** offers a range of services, including water and food, but in many cases the horse owner remains responsible for grooming and providing the horse with exercise. There may be an extra charge for special services such as medicating, bathing, and blanketing the horse.

A **complete-care facility** attempts to meet all of the horse's needs, leaving the owner free to visit when he or she chooses. In addition, the stable may offer a range of amenities to the horse owners. Such facilities are generally the most expensive option.

Choosing a Stable

In addition to boarding costs, horse owners who are shopping for a stable should consider facility safety and cleanliness, as well as feeding, pasturing, and watering practices. Other factors to be considered include whether adequate riding space is available, the credentials and qualifications of the manager, the distance of the facility from the owner's home, nighttime security provisions, and the willingness of the stable owner and employees to help boarders learn to care for their own horses. It is a good idea to ask for references and speak with other boarders about their experience with the facility.

Body brush

A small brush with soft bristles that is applied in long strokes to remove the finer dirt particles from a horse's coat.

➢ *See also Grooming*

BLUE HOOVES

A hoof is described as a blue hoof when the horn has a dense, blue–black color. Other horn colors are dark, pale, or mixed. When the horn is mixed, the contrasting colors always run vertically.

Body language, of horses

Horses reveal much about themselves through nonverbal cues and signals. Riders and handlers who know their horses intimately also know that being alert to these signals is the best way of preventing (or staying out of the way of) bites, kicks, shoves, and other indignities. Understanding a horse's body language is also an effective way to strengthen the cooperative bond between horse and rider.

Like humans, horses experience and express moods. Squinted eyes and a pinched mouth can indicate equine ill humor. A horse that tips his ears back and shows the whites of his eyes is probably feeling hostile or angry. Conversely, a horse that extends his ears and head forward is probably demonstrating friendliness and good humor. A horse that shakes his head while being ridden may be feeling playful or rebellious — unless he's simply trying to rid himself of pesky flies or communicate other discomfort.

Signs a Horse Is About to Kick

The warning signals that many horses display prior to kicking are particularly important. These signals commonly include pinning back the ears, rolling the eyes, backing toward a handler, stomping a hind leg, presenting the rear end, and tucking or swishing the tail. It is especially unwise to enter a stall in which the horse greets you by presenting his rump. Refrain from entering the stall until the horse turns around.

Signs of Ill Health

A horse's body language conveys information about his health as well as his temperament. A normally alert horse with his head down, eyes dulled, and drooping appearance is likely to be experiencing a fever, abdominal pain, or other discomfort. A horse that is tensing his stomach muscles and humping up his back is either in severe pain or has suffered an injury. A horse standing on three legs or pointing with one foreleg for a prolonged amount of time may be suffering from a sore leg or foot. Significant changes in behavior, posture, or physical movements should be taken seriously as indicators of possible health problems.

Learn Your Horse's Language

Many equine expressions and postures are subtle and ambiguous. For example, a facial expression that signals anger in one situation might indicate fear in different circumstances. Because horses are individuals, body language also varies from one horse to the next. The best way to gain understanding of a horse's body language is to spend a great deal of time with that particular horse.

Bola or bolo tie

A narrow, braided string tie held in place with a decorative clasp or ornament, worn as part of Western attire.

HORSE TALK

Here are some common examples of horse body language and what they mean:

Ears pinned back, head reaching toward you. "Stay back or I'll bite you."

Ears forward, head high. "I wonder what that is over there?"

Pawing with front feet. "I wonder if there is anything here to eat" or "I want to get out of here."

Swishing tail (not at flies). "I'm irritated" or "My stomach hurts."

Swinging hindquarters toward you. "I'm afraid" or "I'm getting ready to kick you."

Lifting or stomping one hind leg (not at flies). "I might kick you" or "I have a stomach ache."

Ears forward, head reaching toward you. "Hi, pal."

Ears back toward you when you are riding. "I'm really concentrating and listening to you."

Ears pinned back flat against the head. "I'm getting ready to buck."

From *Your Pony, Your Horse* by Cherry Hill

Bolting

Behavior in which a horse unexpectedly bolts forward or ignores the handler's commands. Try to remain calm if your horse bolts; panic will make things worse.

Boots, horse

Horse boots can be useful protection for a horse's knees, fetlocks, lower legs, coronary band, and other body parts that may be vulnerable to wounds or sores. The boots come in many types, and each type is generally available in at least two sizes, ensuring an adequate fit for most horses.

Fetlock boots and hock boots are named after the part of the horse's body they protect. Splint boots utilize a reinforced plate to protect the splint bones on the inside of the legs. Rundown boots, also called skid boots, protect the rear fetlock area of a working horse from injury due to sudden stops. Bell boots cover the upper portion of the hoof, encircling and guarding the coronary band. Sport boots provide additional support to tendons and protect the inside portions of the legs. Rubber hoof boots are sometimes used to protect bare hooves.

For horses traveling in a trailer, traveling boots and shipping boots protect the coronary band, the heel bulbs, the knees, and the hocks. The exteriors of these tall, durable, and thickly padded boots are often made of nylon.

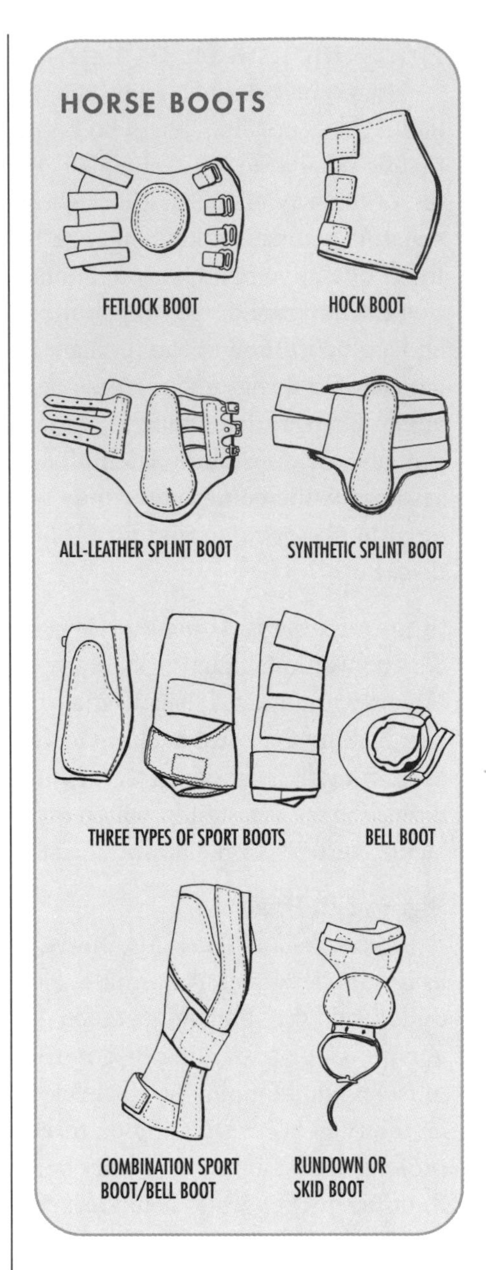

Boots, riding

Riding boots — both English and Western style — are, aside from the safety helmet, perhaps the most important items of riding attire. Boots that fit properly help secure the rider's feet in the stirrup, partly by limiting the range of motion of the ankles, and can prevent feet from becoming caught in a stirrup during a fall. They also provide additional support for tired or weak ankles, helping the rider maintain the proper posture and position. Traditional riding boots are pull-ons with thick heels, but lace-up paddock boots can also be comfortable and supportive. Looser, insulated boots are recommended for winter riding.

➢ *See also Attire; Sneakers, riding*

Bosal

A noseband made of plaited rawhide that is part of the traditional hackamore bridle. The bosal is used to control a horse in Western riding. It should be carefully fitted on a horse's head so that it will not chafe the jaw. The front part should be positioned about 4 inches (10 cm) above the nostrils, and the rear part should not touch the jaw at all. The term "bosal" is also used to indicate a bitless bridle.

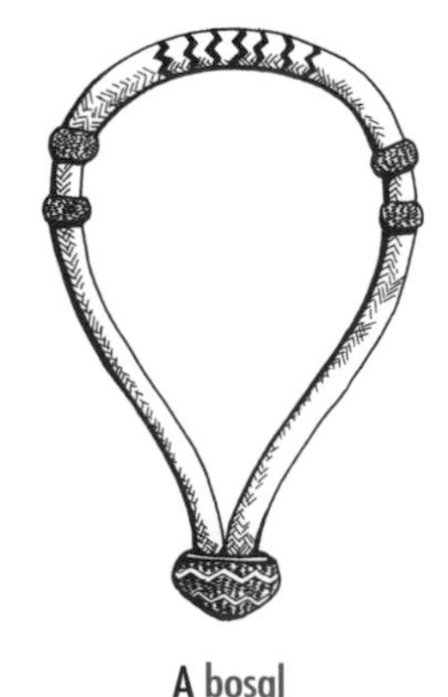
A bosal

Bot block

A coarse stone used to remove the eggs of bot flies from a horse's hair, particularly its legs.

Bot flies

The *Gasterophilus,* or bot fly, is a brown, hairy fly that resembles a bee. Bot flies commonly lay their eggs on a horse's hair; the horse subsequently ingests the eggs by licking or nuzzling itself. The larvae hatch in the warm, moist conditions of the horse's mouth. They lodge at first in the gums, and later emerge and are swallowed by the horse. The larvae mature in the horse's stomach, eventually passing out in the animal's manure to become adult flies and begin their life cycle anew.

Infestations of the larvae can cause stomach ulcers, and a severe infestation may lead to unthriftiness, colic, poor appetite, and even death. Swarms of adult bot flies can also cause a horse to bolt or strike at the flies with his forelegs, a potentially dangerous situation for handlers.

Life cycle of the bot fly: Flies lay eggs (a). Larvae migrate to mouth (b), pass to stomach (c), pass out with droppings (d), migrate into the ground to pupate (e), and become flies (f).

Controlling Bot Flies

Successful control of these parasites requires a two-pronged approach. Horses should be given an effective boticide after the first eggs appear as clusters of small orange or yellow dots on the horse's coat. The treatment should be repeated every two months until bot fly season (which varies depending on local climate) has ended. In addition to the medication, horsekeepers should regularly scrape the eggs off a horse's coat with a bot block and a bot knife. Alternatively, the flies can be controlled externally by washing the horse with warm water or applying a warm cloth directly to the eggs. This forces the eggs to hatch prematurely, and the larvae quickly die.

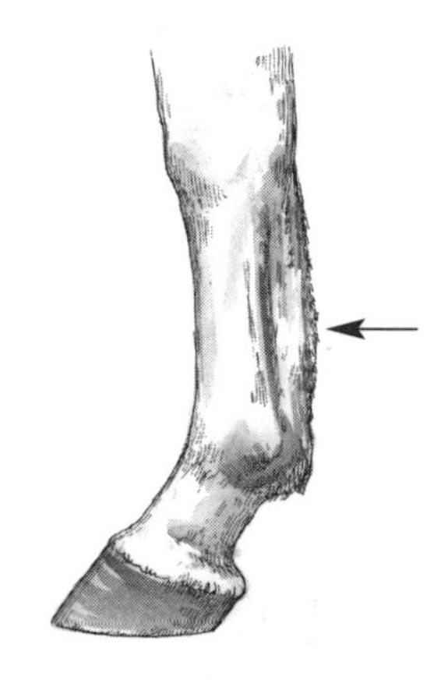
Bowed tendon is an injury to the superficial flexor tendon and/or the deep flexor tendon; note the uncharacteristic bulge.

Botulism

A serious illness caused by a spore-producing bacterium, *Clostridium botulinum,* which is found throughout the environment, particularly in soil. Horses become ill after ingesting spores while grazing or from eating contaminated food. Symptoms appear 3 to 7 days later and include paralysis and difficulty swallowing, breathing, and standing. Progressive paralysis causes death from suffocation. Unvaccinated horses almost always die, even with prompt treatment.

Mares vaccinated during pregnancy can pass immunity along to their nursing foals. Foals can receive the series of vaccinations when they are as young as 2 to 4 weeks old. Other methods of prevention are to keep food storage areas scrupulously clean and sealed. A dead bird, snake, or rodent in a feed bin or a bale of hay can spread the toxins — get rid of any contaminated feed or hay.

Bowed tendon

An injury to one or both of the flexor tendons on the back of a horse's leg. This serious condition generally requires medical treatment, followed by an extended period of rest. The injury can happen during activity, but it can also be caused by incorrectly applied bandages.

Box stall

Any stall in which a horse can move about freely. A 12-foot-by-12-foot box stall is roomy enough for standard-sized horses, but foaling mares and stallions often need additional space. Ponies and small horses are generally comfortable in a smaller stall. Stall walls must be free of sharp edges and protrusions to avoid injury to the horse; they must also be durable enough to withstand kicking, rubbing, chewing, and the effects of decomposing waste.

➢ *See also Barn; Stall; Tie stall*

Bradoon, Bridoon

➢ *See Double bridle*

Bran

A by-product of milled grains that can provide additional bulk and fiber in a horse's diet. Bran is high in protein and can serve as a mild laxative. However, it is also high in phosphorus, a nutrient that can prevent calcium absorption and trigger bone-related problems in horses that eat excessive amounts. Therefore, bran should not be fed to young horses whose bones are developing; nor should bran comprise more than 10 to 15 percent of the diet of a healthy adult horse. A wheat-bran mash, consisting of bran and hot water, has traditionally been recommended for horses that are old, sick, or dentally disadvantaged. It is also fed in subzero temperatures to ensure healthy bowel function.

Breakaway halter

A safety halter designed to literally "break away" if a horse gets caught on something while in the stall or at pasture.

➢ *See also Halter*

Breakaway rope

A rope that is tied to a saddle horn with a length of string.

➢ *See also Breakaway roping*

Breakaway roping

A timed calf roping event in which the rider ties one end of the rope to the saddle with a length of string. The rider ropes the calf as quickly as possible and stops the horse. A flagman signals time when the rope is pulled away from the saddle horn. A skilled breakaway roper can complete the event in as little as three seconds.

➢ *See also Arena events; Calf roping; Competitions, guidelines for; Roping; Timed events*

Breeches

Comfortable pants designed for riding. They are available in a wide assortment of styles and cuts, ranging from skintight to flared bottoms. Riding breeches are stretchy for easy mounting and freedom of movement when riding. Most feature reinforced patches on the inside of the calves.

BOXY FEET

A narrow hoof with a small frog and a closed heel is described as boxy.

BREASTPLATE

Occasionally a horse needs a breastplate or breastgirth to keep the saddle from slipping backward. A breastplate is a loop that goes around the horse's neck. It attaches to the saddle by a strap on each side, and to the girth by a strap between the forelegs. A breastgirth passes around the breast, horizontal to the ground, and attaches to the girth on both sides. A strap crosses over the neck to keep the breastgirth from slipping downward.

Usually, however, if the saddle fits properly, you will not need to add a breastplate or breastgirth. In either case, the device should be fitted to allow four fingers between it and the horse's shoulder.

These knee patches help the rider maintain a safe position, protect the knees from bruising, and enhance the durability of the breeches.
➢ *See also Chaps; Jodhpurs*

Breed

A group of horses with common ancestry and common inherited characteristics related to appearance, body shape, and/or coloration (see box). Organizations known as **breed registries** maintain records identifying all ancestral and living members of each breed; registries also accord registration papers to qualifying horses. Not every registered horse is a pure breed member. Many are half-bred horses, sired by a registered father but carried by an unregistered mare, or vice versa. Many others are crossbreds, the product of two selected parental breeds.

Each breed of horse has a standard to which members of that breed are expected to conform. In some cases, the establishment of a breed originally depended on physical ability in a particular area, regardless of conformation or looks. For example, the term "Standardbred" comes from having a standard time in which a horse had to trot a mile. Other breeds were carefully developed by mating high-quality horses, with the goal of combining the best qualities of each. When farmers in England began to cross the stout utility horse called the Old Norfolk Trotter with Arab blood, the result was the Hackney, which has a refined look and naturally high-stepping trot. Many breeds can be traced to several foundation sires, whose ability to pass along desirable traits created a new type of horse. The best known of these is probably the Morgan, named after Justin Morgan, the man whose tough little horse established one of the most popular breeds in North America.
➢ *See also Breed standard; Types of horses; individual breeds*

Breed association

An organization dedicated to promoting and showcasing the virtues of a specific breed of horse. A breed association generally sponsors shows open only to horses of the breed it promotes. In the United States, these associations include, but are not limited to, the American Morgan Horse Association, the American Paint Horse Association, the American Quarter Horse Association, the American Saddlebred Horse Association, the Appaloosa Horse Club, the Pinto Horse Association of America, the Palomino Horse Breeders of America, the American Buckskin Registry Association, and the International Arabian Horse Association. Many of these associations also offer programs and clubs designed specifically for youths. Most have excellent Web sites.
➢ *See appendix*

Breeding

The managed or controlled reproduction of horses. Mares are seasonal breeders; fully three-quarters of all mares do not ovulate during the winter. In most climates, the prime breeding months are May, June, and July.

COMMON BREEDS

Horse Breeds
- Appaloosa
- Arabian
- Morgan
- Mustang
- Paint
- Pinto
- Quarter Horse
- Saddlebred
- Standardbred
- Tennessee Walking Horse
- Thoroughbred

Pony Breeds
- Connemara
- Hackney
- Haflinger
- Pony of the Americas
- Quarter Pony
- Shetland
- Welsh

Draft Horse Breeds
- Belgian
- Clydesdale
- Percheron
- Shire
- Suffolk Punch

FINDING A BREED ASSOCIATION

See the appendix for a partial list of breed associations in the United States. A great deal of breed information is also available on the Internet.

Breeding Stock

Breeding to obtain the best results begins with the selection of appropriate stock — that is, choosing a mare and a stallion whose conformation, breed, genetics, physical condition, and other qualities make them a good match.

Breeding stock should strongly exhibit the characteristics the breeder would like to see in the foal. In many cases, the breed of the candidate horses is a primary consideration; however, a wide range of characteristics exists within each breed. Individual horses, including favorite animals, should be assessed impartially to determine their strong points and weaknesses.

Successful breeders choose sires and dams whose abilities and temperament are compatible with the lifestyle envisioned for the foal. In other words, a breeder looking for a highly athletic foal will choose a highly athletic stallion and broodmare.

Breeders should avoid horses with genetic imperfections, significant conformation faults, or unsoundness, which is defined as any physical defect that diminishes the usefulness of a horse.

Broodmares should be in good physical condition to withstand the rigors of carrying a foal to term. They should be fully mature (at least three years old) with healthy reproductive systems. A prospective broodmare should be given a breeding soundness examination by a veterinarian to ensure that she is fully able to conceive and carry a foal. Any mare that will be bred should be dewormed and entirely up-to-date on her vaccinations. Many stallion owners also require an intrauterine culture to test for sexually transmitted diseases.

Breeding at Pasture

Three methods are commonly used in breeding horses. Breeding at pasture is the most natural method, requiring minimal intervention. It is particularly useful when one stallion is pasturing with a small group of mares. In such a situation, the stallion will approach, tease, and breed the mares at appropriate times, which enhances the likelihood of conception. He will also understand the need to be patient and cautious around the mares, minimizing the chance that he will be injured by an unreceptive mare. Injury to an unfamiliar or inexperienced stallion is the number one hazard of breeding at pasture. The method also makes it difficult to control which mare is bred and when conception occurs.

Hand Breeding

The second method, hand breeding, involves supervision of the mare, the stallion, and all aspects of the breeding process. The method requires knowledge of the mare's heat cycles, as ovulation typically occurs one to two days prior to the end of each cycle.

A mare in estrus may show signs including an open or relaxed vulva, "winking" of the clitoris, mucous discharge, a tail that is slightly raised, frequent urination, and an atypical degree of sociability. However, some mares are prone to "silent heats." These mares may be in estrus without giving any external signals. To determine whether such mares are in heat

BREEDING SUCCESS

Of all the mares bred in the United States every year, only 60 percent produce live foals. Reasons for this can include an unsuitable stallion; a stressful breeding situation; a uterine infection; or an abnormal estrus cycle.

A mare should be in good physical condition, neither excessively thin or fat. She should be given a rest period of at least 60 days after racing or showing before being bred.

Keep records for each mare that include the following:

- ❑ Estrus cycle (dates, behavior, etc.)
- ❑ Injuries and infections
- ❑ Past breeding or foaling difficulties
- ❑ Success at lactation and bonding
- ❑ Each foal's health

— and in some cases, to stimulate stronger heat — **teasing** is often necessary. Teasing involves exposing a mare to a stallion under controlled circumstances. A breeder can often determine whether the mare is in estrus by watching her behavior during the encounter. A veterinarian can also be called in to determine whether a mare is in heat. The veterinarian can check the cervix and uterus by conducting a rectal palpation, which involves reaching into the mare's rectum while wearing a special lubricated glove. Potential reproductive problems can also be detected during this procedure.

Before the breeding act, a thorough cleaning of the mare's genital areas and a washing (with water only) of the stallion's penis and sheath are recommended. This serves to minimize the chances of disease and infection. The mare is brought to the breeding area, generally an open area with a strategically placed wall or fence that limits the mare's movements. Breeding hobbles and a tail rope may be used on mares that are particularly difficult to control or aversive to breeding. A handler leads the stallion up to the mare, approaching the female from the side to avoid startling her. Handlers control both mare and stallion during copulation, which may last for 30 seconds to two minutes. A handler may have to hold the mare's tail to one side to allow the stallion's penis to enter properly. Ejaculation is often indicated by a rhythmic flagging of the stallion's tail. Following copulation, the stallion is again washed, and the mare is returned to her stall or paddock. The mare must then be monitored closely to determine whether she has "settled" (become pregnant).

Traditionally, mares are hand-bred on the third day of heat and every 24 to 36 hours thereafter throughout the duration of the heat cycle. However, veterinarians can palpate mares to determine precisely when ovulation is occurring. This may eliminate the need to breed the mare more than once.

Artificial Insemination

The third breeding method is artificial insemination (AI), which involves collecting semen from a stallion and subsequently depositing the semen in the mare. AI often requires daily palpation of the mare by a veterinarian to determine when the mare begins ovulation. Some broodmare owners shop long-distance for the best possible stallion and have his semen shipped to them at the appropriate time. This approach requires diligence and planning, but it also allows the broodmare owner to choose from a virtually unlimited field of sires.

➢ *See also Artificial insemination (AI); Conformation; Estrus; Foaling; Genetics; Pedigree; Stud farm; Stud fee (breeding fee); Teasing*

Breed standard

A set of criteria or standards that establishes the ideal physical characteristics for each breed. The standard is determined by owners, handlers, and other experts who have a great deal of experience working with the specific breed.

Bridges

Some horses are reluctant to cross bridges. A rider may be able to deal with this fear by dismounting and walking the horse back and forth across the span several times, then riding it back and forth. Handlers sometimes construct a small "training bridge" and practice riding across it to desensitize a reluctant horse.

Bridle

A piece of tack consisting of leather or nylon straps or cords that are stitched together. Bridles, which are most frequently used in conjunction with a bit, are secured around the head of the horse, and they are designed to aid the rider in controlling the horse. English and Western-style bridles are the most commonly used types. Bridles without bits are known as hackamores or bosals.

Although bridle parts may be sold separately, the typical bridle includes cheek pieces, brow piece, crownpiece, throat latch, bit, and reins. English bridles also include a cavesson or noseband.

A bridle should fit the horse comfortably and be free of cracks, dryness, loosened or rotted stitching, and other damage that could make it fall apart during use. Although a certain degree of adjustability is built into bridles, it may be useful before buying one to measure the distance from one corner of the horse's mouth over the top of the head to the opposite corner of the mouth. This measurement can be compared to the head strap of the bridle to ensure proper fit.

Bridles

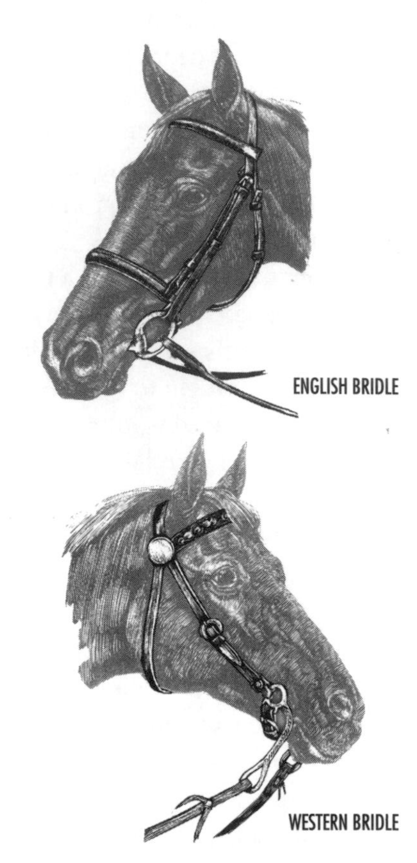

BRIDLING A HORSE

Bridling a horse seems intimidating at first but quickly becomes a routine procedure. Protective headgear should be worn until the horse becomes accustomed to the handler. Here are the basic steps involved in bridling:

1. Check all bridle components for signs of damage or weakness. Replace if necessary.
2. Stand with your right shoulder near the horse's throat. Place the bit by the horse's mouth and position the headstall over its ears. Most handlers hold the top of the bridle with the right hand, bringing the right arm over the neck and between the ears. For tall horses, it may be easier to pass the right arm beneath the horse's jaw, holding the cheek pieces of the bridle while placing the bit in the horse's mouth.
3. Insert the thumb of the left hand into the horse's mouth and press down on the gums of the bars. The horse will open his mouth.
4. Position the bit properly in the mouth by gently pulling up on the bridle with the right hand. Place the crown piece of the bridle over the right ear, then the left.
5. Fasten the throatlatch and cavesson (nose piece). If used, also adjust the brow band and earpiece.
6. Check to ensure that the bit is comfortably in place and that the bridle is distributed evenly.

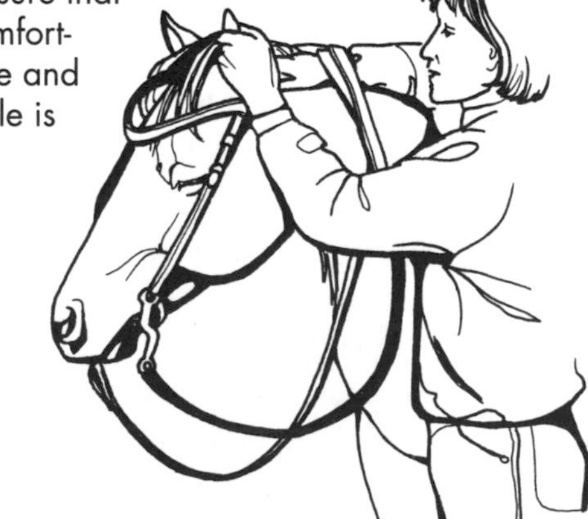

Bridoon

➢ *See Double bridle*

Brindle dun coloring

Dun (yellow or gold) body coloration with darker streaks on forearms and gaskins.

➢ *See also Coloring*

Broke

A broke horse is a trained animal that accepts being handled and ridden. The horse should accept the handler's dominance, pay attention to his human companion, and respond in predictable ways to applied pressure from legs, hands, straps, or ropes.

➢ *See also Green horse*

Broke to tie

Trained to stand quietly when tied.

Bronc, Bronco

An unbroken range horse of western North America. Specifically, a bronc is a horse used for saddle bronc and bareback riding in a rodeo. The name comes from the Spanish word *bronco,* meaning "rough" or "wild."

Broncs are bred specifically for the rodeo circuit, and some have become famous. The ideal bronc is one that bucks rhythmically, instead of twisting. The rider must stay on for eight seconds, holding on with just one hand and raking the horse with blunt spurs from front to back, and is scored on style. In saddle bronc events, the horse wears a saddle with a bucking or flank strap and a halter with a single braided rope. In bareback events, the horse wears only a girth strap with a hand loop for the rider to grasp.

Broodmare

A mare selected for breeding. Broodmares should be in good physical condition with healthy reproductive systems. They should show no unsoundness that might cause them to abort during the eleven-month gestation period, and they should be dewormed and up-to-date on all vaccinations. In addition, a broodmare should be chosen for the physical and dispositional characteristics she exhibits. A breeding soundness exam, conducted by a qualified veterinarian, can help determine whether a mare is a suitable candidate for breeding.

Broom polo

A game that is often used to introduce riders to the fundamental concepts and techniques of polo. The game is played on horseback with a broom or bamboo fishing pole and a ball that is generally 12 inches in diameter (basketball-sized). Riders form teams and attempt to hit the ball into their opponent's goal.

If a feller's been a-straddle
Since he's big enough
to ride,
And has had to sling
his saddle
On most any colored
hide, —
Though it's nothin'
they take pride in,
Still most fellers
I have knowed,
If they ever done
much ridin',
Has at different times
got throwed.

—Anonymous, from
When You're Throwed

Brown coloring

A horse considered brown may actually appear black, because his coat is a mixture of black and brown hair. He typically has a brown muzzle, mane, tail, and legs. (Black "points" make the horse **bay,** not brown.) He may also have lighter coloring around the eyes.

➢ *See also Coloring*

A soft-bristled brush removes hair and dirt from a horse's head. Special care must be taken around the eyes.

Brumby

The feral horse of the Australian outback. Brumbies are the descendants of horses that escaped from mining settlements during the Gold Rush of 1851. Hardy survivors, they are difficult to domesticate and now roam in large herds. In the 1960s they were the subject of a controversial "culling" program to reduce their numbers.

Brushing

An important aspect of grooming a horse that helps remove dirt, dry skin, grass and manure stains, and shedding hair. Handlers typically use a stiff-bristled dandy brush on the larger areas of a horse's coat. An assortment of medium-bristled body brushes can provide further cleaning, and a soft-bristled brush will remove dirt from the horse's head. After the application of a conditioner or detangling product, brushes can be used to smooth the tail and mane.

Buck fence

A rustic, durable fence consisting of a series of crossed poles (bucks) that hold the fence rails in place. The fence originated in the Rocky Mountains and is ideal for large, rocky pastures where digging postholes is difficult.

The buck fence, also known as the "jackleg" fence, is durable even where high winds and heavy snows are common. The best wood to use in constructing a buck fence is lodgepole pine.

Bucking

Springing into the air with an arched back, landing with stiff forelegs and lowered head. Some horses are natural buckers; these do not make safe riding animals but may be used successfully in saddle bronc riding events in rodeos. Others will buck if they have been left in the stall too long, so it is essential to exercise horses regularly. Sometimes a horse that has not been adequately warmed up will buck after a jump.

Buck knee

A foreleg conformation fault also known as **over at the knee.** The knees of a buck-kneed horse are bent too far forward, meaning they appear likely to buckle in that direction. Such buckling can endanger riders.

➢ *See also Conformation*

Buckskin coloring

A buckskin horse is characterized by a tan, yellow, or gold coat with black coloring on its mane, tail, and lower legs; a dorsal stripe down the spine is common.

➢ *See also Coloring*

Buggy

A light one-horse carriage. American buggies have four wheels, while English buggies have two wheels.

Bull pen

A corral used to train horses.

Bull riding

One of the classic rodeo events, in which a competitor tries to ride a bucking horned bull for 8 seconds and is scored by judges.

Bull snap

A metal snap used with lead ropes and trailer ties to assist in securing horses.

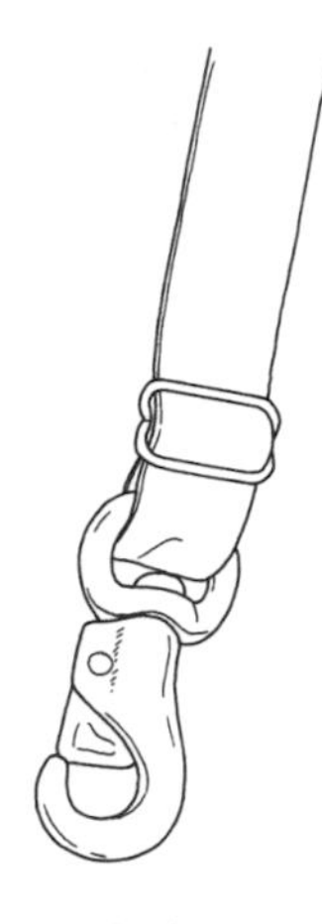

A bull snap attached to a trailer tie

Bump

A brief tug on the reins in order to make contact with a horse's mouth.

Burchell's zebra *(Equus burchelli)*

The small zebra most often seen in captivity.

➢ *See also Zebra*

In the choice of a horse and a wife, a man must please himself, ignoring the opinion and advice of friends.

—George John Whyte-Melville, *Riding Recollections*

Burner

A rawhide covering that protects the eye of a looped rope from excessive wear.

Bute

➢ *See Phenylbutazone (bute)*

Butt bar

A bar placed behind a horse when the animal is being trailered or is in stocks. Rear-end contact with the bar encourages the horse to stay in place. A horse may also lean on the bar in order to maintain his balance while trailering.

➢ *See also Restraints; Stocks (as restraints)*

Buying a horse

Because the process of evaluating a horse is a difficult one, first-time buyers should work closely with a consultant who is an experienced horse buyer. This might be a professional trainer, handler, riding instructor, breeder, stable owner, or simply a friend who can offer experience assessing and purchasing horses. The search for a suitable horse may require spending considerable time with several horses; making a wise decision could take weeks or even months. However, taking the time to reach a well-reasoned decision is far preferable to ending up with a problematic or unridable horse.

Regardless of where you buy a horse, it's not at all unreasonable to require that a clause for a "suitability guarantee" be written into the

contract. It allows the buyer a mutually agreed-upon amount of time to "test drive" the horse. Most sellers will require the horse to stay at their farm, so you can learn how to communicate with the horse, via lessons with them. Some will allow you to take the horse to another professional's barn, if you purchase mortality and loss of use insurance on the horse.

Where to Look

Novice buyers often locate suitable horses by inquiring with boarding stable owners, riding instructors, and other handlers in the local area. A school horse can be ideal for a new rider or as a family horse, because it has a track record of tolerating riders of all sizes and abilities.

Young riders who belong to pony clubs and 4-H horsemanship groups often "outgrow" perfectly good horses. These groups can be an excellent source of rider-tested animals.

Horse dealers and breeders offer horses for sale. Experienced buyers can often refer first-timers to reputable dealers, but it's a good idea to contact a dealer's previous customers for references.

Many buyers purchase horses from horse shows, auctions, ads in horse publications, and even over the Internet. However, these are not recommended venues for first-time buyers, as they usually offer little opportunity to examine or get acquainted with a horse prior to purchase.

Health

Buyers can attain a general sense of a horse's health through careful observation. The horse should appear well-fed and alert with clear, bright eyes, pink gums, and properly aligned teeth. The eyes and nose should be free of discharge. The animal's coat should be even and sufficiently thick, with no patchy, sore, or bald spots. The hooves should be well-shaped and without deformities.

The buyer should inquire about the horse's diet and determine whether he has any special needs or is taking medications. Asking for the horse's medical history, including vaccinations and deworming treatments, is appropriate. The animal's veterinarian should be consulted if any questions emerge. In fact, buyers should make any purchase contingent upon the results of a veterinary examination. This will provide assurance that the horse is fundamentally healthy and is suitable for the purposes envisioned by the buyer.

Size

An additional and sometimes overlooked factor to consider when buying a horse is the animal's size. Generally, larger horses are best for larger riders and smaller horses are best for smaller riders. As a rule, a rider should weigh no more than 25 percent of the horse's weight, although some sturdy ponies are exceptions to this rule. When a rider is mounted and the stirrups are adjusted properly, the feet should fall naturally at a level even with the bottom of the girth.

Soundness

A sound horse is one free of defects that would diminish its usefulness — particularly, problems that could lead to lameness, loss of wind, or loss of vision. The horse's **conformation** is the best indicator of its soundness.

BREED AND APPEARANCE

Breed, markings, coloration, and overall appearance can be of paramount importance to certain buyers — those who are buying a horse for show, breeding, or resale, for example. However, because the long-term suitability of a horse has more to do with practicality than aesthetics, these should be secondary considerations for most pleasure riders.

Certain breeds tend to exhibit specific qualities of disposition, are suitable for certain types of work or riding, and may be susceptible to specific injuries or health problems. However, the characteristics of the individual horse are ultimately more important than breed tendencies.

The term refers to the way in which the horse's physical features are put together. Buyers can assess general conformation by looking for a well-proportioned, balanced body, including the head, neck, withers, back, legs, and feet. The horse's legs should be straight, not crooked or mis-shapen. When standing squarely, the horse should distribute his weight evenly on all four legs. When in motion, he should move smoothly at all gaits. Trotting a horse in a circle can be a good way to detect any signs of lameness.

Temperament

A horse of beauty is a joy to behold, but a horse of even temperament is the one most pleasure riders will want to own. Evaluating a horse's disposition involves observing his behavior closely in a variety of situations. The horse should be ridden under varying circumstances, and the buyer should personally groom, feed, and tack the horse.

If the horse will be ridden primarily by novices or children, the buyer should look for a horse that is gentle, calm, and obedient while being ridden and handled. Geldings are generally regarded as having the most stable temperaments, and a settled, older horse is preferable to a young and feisty one. Horses are usually mature and settled by the age of 6 or 7, and beginning riders may fare well with horses as old as 15.

BY THE RULES

When describing a foal's lineage, it is proper to say a horse is "by" the sire, "out of" the dam.

Buzkashi

An ancient game believed to have originated with the Turkic-Mongol people, buzkashi is the national sport of Afghanistan. Literally translated as "goat grabbing," there are two different versions, but the general idea is the same. Mounted teams of riders (from six to dozens on a side), called "chapandaz," form a circle around the carcass of a goat or calf. At the signal, each rider tries to lean over, hoist the carcass from the ground, hitch it under his leg, and ride at a dead gallop away from the other players, who do their best to retrieve the goat. In the original game, the rider who kept the carcass away from the rest of the horde was the winner. A more complex version requires that the rider gallop around a distant flag and bring the carcass back to his team's scoring circle.

Not a game for the faint of heart, buzkashi is a fast and furious sport that is played by tough, highly trained horses and expert riders. Riders wear thickly padded clothing and carry short leather whips, and although kicking and hitting opponents is not allowed, pretty much anything else goes, and the game can get quite rough. The carcass, which is sometimes filled with sand, usually winds up resembling a limp leather rag by the end. The winners receive great honor and acclaim, as well as prizes of money and fine clothes donated by sponsors.

Byerly Turk

A 17th-century stallion and one of the foundation sires of the Thoroughbred breed.

➢ *See Thoroughbred*

Cadence

The rhythmic crispness of a horse's gait (see chart).

Calcium

A mineral that is necessary, in combination with phosphorus, for maintaining strong bones and teeth. Calcium deficiency can impair growth or lead to unsoundness in horses; however, too much calcium may affect the normal rate at which cartilage is converted to bone. A balance between calcium and phosphorus is recommended, and a growing horse should, ideally, be fed the same amount of each mineral. Pasture and alfalfa hay are generally high in calcium, while grains are high in phosphorus. A horse's primary diet can be supplemented by allowing the animal free-choice access to a calcium and phosphorous block. Most commonly, these blocks contain 12 percent calcium and 12 percent phosphorus.

Calf knee

➢ *See Back at the knee*

Calf roping

A competitive event in which contestants must rope a running calf, dismount, flip the calf on the ground (unless it has already fallen from being roped), and tie three of its legs together, usually in about the time it takes to read this sentence. The event originated with cowboys' desire to test and improve the skills they used on the range. A good roping horse will get right behind the calf and stay there without directions from his rider. As soon as the rope is tossed, the horse should begin to stop. When the rider dismounts, the horse keeps the rope taut so that the calf can't run away or get up.

Being smaller, calves are harder to catch than steers, so the rider must stand up in the stirrups and aim slightly downward to get the loop around the calf's neck. Once tied, the calf must remain down for 6 seconds.

➢ *See also Breakaway roping; Cattle roping; Piggin string; Team roping; "Two wraps and a hooey"*

California-style bit

A loose-shanked bit that allows riders to direct their horses with a light touch on the reins.

➢ *See also Bit*

California yellow star thistle *(Centaurea solstitialis)*

A plant found in some pasture areas that is toxic to horses.

➢ *See Poisonous plants; Toxic substances*

Calk

A tapered piece of a horseshoe that points downward and is designed to prevent slipping.

CADENCE OF GAITS

GAIT	CADENCE
Walk	4-beat
Jog/Trot	2-beat
Pace	2-beat
Lope/Canter	3-beat
Rack	4-beat
Gallop	4-beat

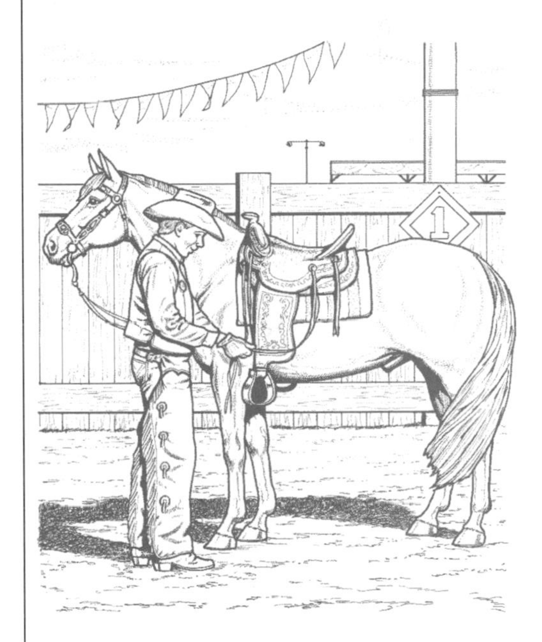

Calluses

The larger the horse, the harder it is to keep the bony parts of his hind legs (the point of the hocks) from developing calluses from lying down in the stall. In the wild, horses lie on grass, snow, and sand that has not been packed down by years of heavy animals standing on one small area, unlike a stall. Natural earth is almost always softer than what we can provide in barns. It is essential, especially with heavy horses (drafts, warmbloods, and anything over 16.2 hands with bulk), to keep the flooring well covered with bedding to prevent calluses from forming.

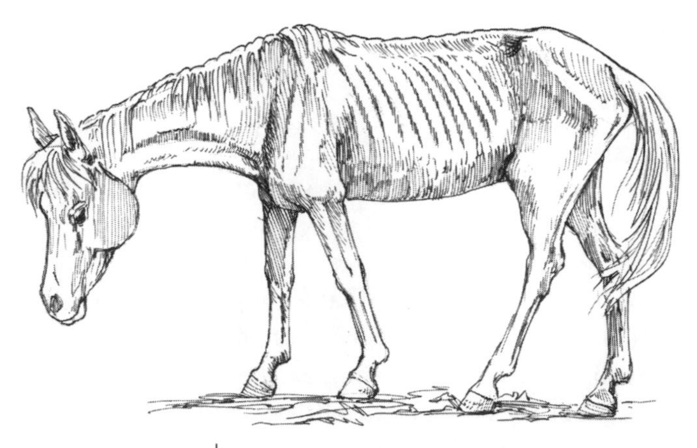

A horse with abdominal cancer may lose an excessive amount of weight.

Camped out

Term used to describe hind legs that extend beyond the point of the buttock. This misalignment can increase stress to the animal's hock, stifle, and hip joint.

Cancer

Although uncommon, cancer can occur in horses. Horses are subject to skin cancers, abdominal cancer, and cancer of the blood. Mares can also develop granulosa cell tumors, which, while not usually fatal, can prevent pregnancy and affect the mare's disposition. Many forms of equine cancer are treatable if discovered in early stages.

Symptoms of cancer can include chronic digestive problems or colic, hard lumps on the skin, unusual bleeding or discharge, a raised sore on the eyelid, dramatic weight loss, and (in the case of granulosa cell tumors) radical personality change.

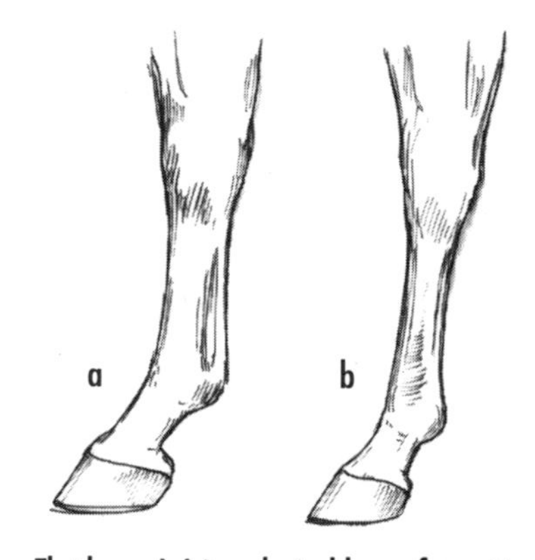

Flat bone (a) is a desirable conformation of the cannon bone that contributes to the strength and sturdiness of a horse's legs. Round bone (b) has the opposite effect.

Can chaser

An informal term for a rider who participates in barrel racing.

➢ *See also Barrel racing*

Canine bud

A small canine tooth developed by a small percentage of mares.

Canine teeth

A set of teeth located behind the incisors, primarily seen in male horses. The canines typically erupt when a horse is four to five years old. Because they get extremely sharp and can be dangerous, the canines should periodically be filed or clipped.

Cannon bone

A long bone extending from the knee to the fetlock of a horse's foreleg, and from the hock to the fetlock of the hind leg. The cannon bone should be sizable enough to support the horse; it should also be centered under the knee or hock to provide sufficient support to his body. An offset cannon will increase the amount of strain on the splint bones and knees or hocks; it can trigger the condition known as splints, in which hard, bony growths develop on the cannon itself.

An appropriately positioned cannon should make the leg appear wide from front to back — a desirable conformation known as **flat bone.** By contrast, **round bone** is a conformation fault. It occurs when a horse's tendon and cannon bone are positioned too closely together.

A canter is the cure for all evils.

—Benjamin Disraeli

Canter

A smooth, natural, three-beat gait that is faster than the trot. The typical speed of a canter is about 13 miles per hour. The gait begins with one hind leg, continues with the other hind leg and the diagonal front leg, then finishes with the leading foreleg. When a horse is cantering, he places his leading front foot farther forward than his other front foot. Western riders refer to this gait as the **lope**.

➢ *See also Gait; Lope; Speed, of horse*

The canter (left lead)

Cantharadin toxicity

➢ *See Blister beetle poisoning*

Cantle

The rear portion of the seat of a saddle, which projects upward.

Cap

A hollow, deteriorated baby tooth that should detach from the gums when a horse's permanent teeth emerge. If the caps do not properly release from the gums, a condition called **retained caps** results. Retained caps can cause problems ranging from swollen gums to sore mouth, sinus infection, and crooked or impacted permanent teeth.

Capillary refill time

A measure of blood pressure and hydration. Handlers often check a horse's capillary refill time by pressing a thumb against the horse's gum above the upper incisors. The gum tissue will initially turn white from the pressure. By counting the number of seconds before the normal pink tissue color returns, the handler can evaluate the horse's status. A refill time of one to two seconds is considered ideal. A refill time that exceeds 3.5 seconds may indicate low blood pressure, circulatory impairment, dehydration, or shock.

➢ *See also Shock*

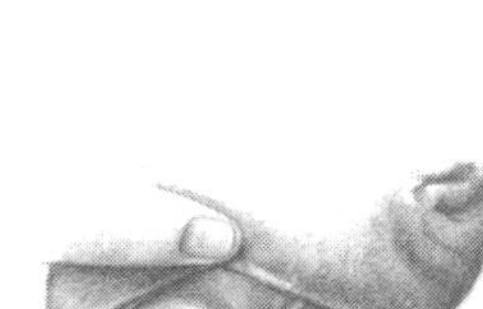

If you suspect that your horse is experiencing shock, check his capillary refill time. Press your thumb against his upper gum for about 2 seconds, squeezing the blood away from that spot. When you remove your thumb, the resulting white spot should disappear in about a second. If it takes 5 to 10 seconds to go away, your horse is having circulatory problems, a possible sign of shock.

Capped hocks

A condition in which the hocks (the joints above the fetlocks on the hind legs) have become thickened due to injury or a conformation fault.

Capriole

A spectacular leap performed in high levels of dressage. The horse leaps forward and, while airborne, flings his hind legs out behind him before landing.

➢ *See also Lipizzaner*

Captan

A garden fungicide available under the brand name Orthocide that can also be used to treat girth itch in horses.

Carbohydrates

The sugars and starches that furnish most of the energy in a horse's diet. Oats, barley, and corn may contain as much as 60 percent sugar and starch. Molasses is another concentrated source of carbohydrates. The amount of "energy food" required by a horse varies according to his age, condition, reproductive status, and activity level.

CARBOHYDRATES AND HORSES

Carbohydrates are stored in the muscles and provide instant energy whenever needed. Grains, such as oats, are the best source of carbohydrates. Grass and hay provide some carbohydrates, but not enough for work.

Carrier

In genetic terms, a horse or other animal carrying a recessive gene that may affect its offspring.

Caslick repair

A surgical procedure in which the lips of a mare's vulva are cut and sutured. The lips subsequently grow together, reducing the size of the vaginal opening. This can prevent feces from falling into and infecting the mare's vagina. A veterinarian determines whether this procedure is needed.

Caspian pony

Caspian pony

A primitive type of horse that lived in Mesopotamia and surrounding arid regions during the Stone Age. It is the ancestor of today's Arabian horses.

Cast horse

A horse that is on his back in a stall and unable to rise, often with its legs caught against a wall. If a horse remains in this position for a significant amount of time, he may sustain intestinal, circulatory, or respiratory problems.

Castor bean *(Ricinus communis)*

Castor beans are raised as a crop and as an ornamental plant in parts of the United States. The beans are quite toxic to horses; an animal that ingests a few ounces of the beans may die. Consumption of lesser amounts can cause diarrhea, sweats, and moderate colic.

Castration

The process of removing a male horse's testicles. Early castration minimizes the trauma to the animal and the amount of care he will require following the procedure. A castrated horse is called a **gelding.**
➢ *See also Gelding*

Dear to me is my bonny white steed;
Oft has he helped me at pinch of need.

—Sir Walter Scott, *The Lay of the Last Minstrel*

Cataract

A medical problem in which the lens of the eye becomes cloudy or less transparent. Some equine cataracts can be surgically removed, but the condition often leads to blindness.

Catch rope

A rope that ends in a loop and is used for catching and immobilizing cattle.

Cattle penning

A rodeo event in which riders work together as a team to herd three cows into a pen. The object is to accomplish the action in the shortest possible amount of time. In most cases, teams are allowed no more than 90 seconds to finish penning. A variation in which one rider herds one cow into a pen is known as **one-on-one.**
➢ *See also Rodeo*

Cattle roping

A rodeo event in which a rider or team of riders uses a rope to capture a fleeing cow. Most cattle roping events are timed, and the winners are those who accomplish the roping task in the shortest amount of time.

Cavalletti, Cavalettie

A series of parallel poles or rails that are placed on the ground or on supporting logs, spaced apart at a distance roughly equivalent to a horse's trotting stride. Riding instructors use the construction to prepare their students for higher-level riding. Students gain control skills and awareness by riding in a straight line down this horizontal "ladder." The cavalletti is also an exercise in agility for the horse.

A cavalletti prepares developing riders for higher-level riding.

Cavalry

Horses enhanced the fighting capabilities of warriors for many thousands of years until the advent of modern warfare made cavalry units obsolete. From the fierce hordes of the central Asian plains, horses and riding techniques spread to the Greeks and Egyptians and into Europe. By the Middle Ages, horses that had been bred for size and strength could carry a knight in heavy armor, though at the cost of maneuverability and speed.

As military technology advanced, cavalry techniques became more sophisticated. When armor was discarded, saddles became less bulky and horses were bred for agility and quickness. In the 18th century, French and Spanish riders developed methods of horse training based on reward

and encouragement rather than pain and fear. These techniques formed the basis of dressage riding as it is practiced today.

By the end of the 19th century, repeating rifles and machine guns wreaked havoc on mounted soldiers, and though many thousands of horses were used in World War I, the day of the cavalry was over. By the 1950s, both the United States and the British armies abolished their mounted units.

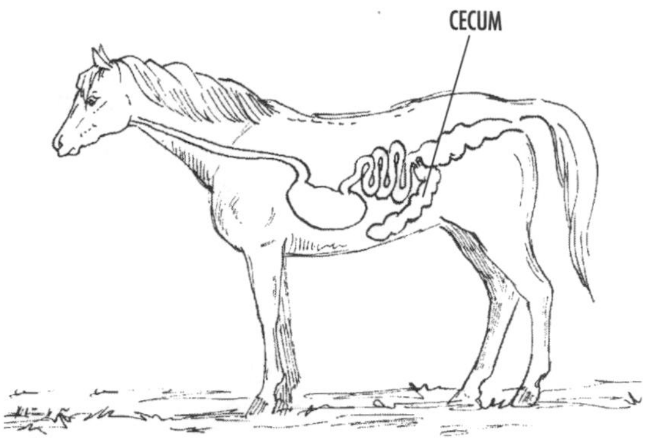

The cecum is a pouch that aids in digestion.

Cavesson

The noseband or nosepiece of a bridle, extending over the nose and under the horse's jaw. A longeing cavesson is a reinforced noseband with rings for attaching the longeing reins. When selecting a longeing cavesson, make sure it is designed so that the longe rope will not accidentally pass over the horse's eye.

Cecum

Part of a horse's digestive system, the cecum is a pouch located between the small and large intestines. The cecum's function is to "ferment" or break down roughage; it may be up to four feet long and hold five to ten gallons of ingested material.

Cellulose

The fibrous portion of roughage, which is converted into carbohydrates during the digestive process. The cellulose is converted by microorganisms that inhabit the digestive tract.

The mythical centaur was a musician, a healer, and a seer.

Celtic pony

A primitive horse that lived in the damp regions of what is now western Europe. In response to its wet, chilly environment, the Celtic pony evolved a water-resistant coat. This animal is the forerunner of many pony breeds.

Centaur

A mythical half-man, half-horse, the centaur comes from Greek mythology, where he is the only one of the fabled monsters to demonstrate traits of nobility and wisdom. The centaur Chiron was said to have taught humankind the skills of the hunt, the lore of medicine, and the love of music, as well as the ability to tell the future.

Center of gravity

A person's center of gravity is an anatomical point located within the abdomen, generally one to two inches below the navel. A rider who seeks to attain balance while on horseback should position his or her center of gravity as close as possible to the horse's center of gravity. (The horse's center of gravity depends on his level of collection.) Instructors sometimes refer to this ability as "developing a deep seat."

➢ *See also Collection*

Certificate of Veterinary Inspection (CVI)

A document that certifies the general health of an animal. It is prepared by a veterinarian after a physical examination has been conducted. Horse owners who transport their horses across state borders often need to present a recent CVI to state authorities or horse show organizers.

Certified instructor

Most people wouldn't dream of sending a child to a school that has unlicensed teachers; yet, few people consider the credentials of horsemanship and riding instructors before enrolling in formal classes. Certification by a reputable organization that holds instructors to strict standards of professionalism, integrity, and knowledge bears witness to an instructor's qualifications. Numerous regional organizations certify instructors; however, only two national-certifying organizations are widely accepted at this time: American Riding Instructor Certification Program (ARICP) and American Association for Horsemanship Safety (AAHS).

ARICP was established in 1984 to recognize and certify outstanding teachers of horseback riding who instruct their students in a safe, knowledgeable, and professional manner; the program is administered by the American Riding Instructors Association (ARIA). Eligible instructors are evaluated via written and oral testing, and videos of their teaching methods are critiqued. Instructors must attend an ARICP seminar, a four-day educational program taught by master instructors and other accomplished professionals. ARICP offers certification at three levels in eleven specialties. Recertification is required every five years.

AAHS conducts group and individual certification clinics for riding instructors, camp personnel, and others interested in horsemanship safety. Certified instructors have attended a 40-hour intensive training program taught by AAHS clinicians from a standardized curriculum. To qualify for the program, applicants must pass a mounted test. Prior to full certification, instructors must demonstrate proof of current CPR and first aid certification. AAHS certificates are granted based on levels of achievement. Not all levels require experienced horsemanship, such as Equestrian Safety Supervisor.
➢ *See also appendix*

Certified Journeyman Farrier (CJF)

The highest level of certification awarded by the American Farriers Association.
➢ *See also Farrier*

Cervix

The neck of the uterus. A mare's cervix is tightly closed unless she is in estrus or on the verge of foaling. When the mare is in heat, her cervix opens as wide as three-fourths of an inch, allowing the sperm of a male horse to enter the uterus.
➢ *See also Breeding; Foaling; Gestation; Ovulation; Pregnancy; Uterus*

"There is no secret so close as that between a rider and his horse."

—R.S. Surtees, *Mr. Sponge's Sporting Tour* (1853)

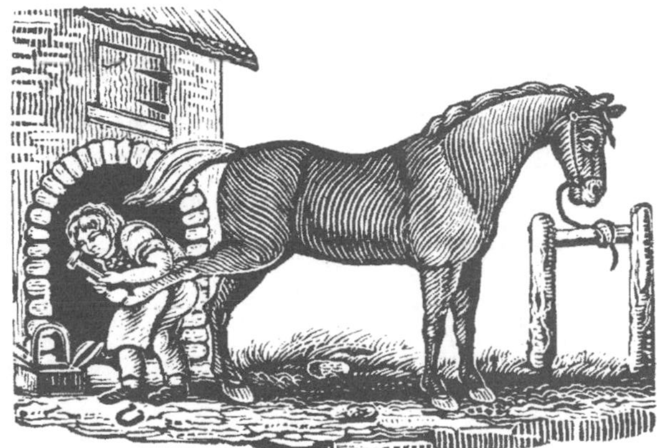

Chain-link fencing

Mesh or woven-wire fencing that is highly effective at confining domestic animals while keeping other animals out. A chain-link fence can be used effectively with horses; however, the sharp edges at the top and bottom should be covered to protect the horses from injury.

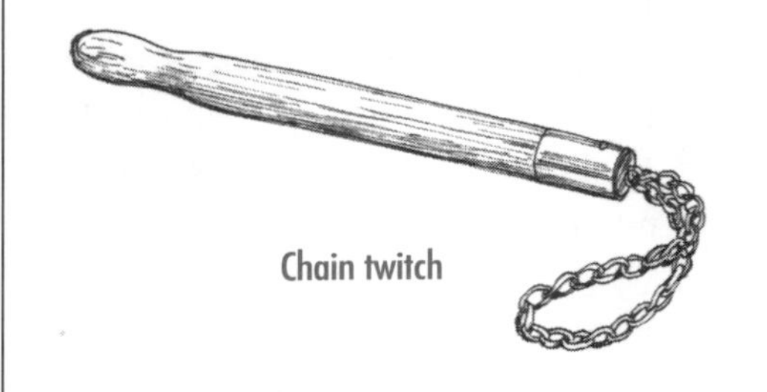

Chain twitch

Chain twitch

A device, consisting of a chain loop attached to a wooden handle, used to subdue or restrain a horse. The loop is placed around the horse's upper lip and is then tightened by twisting the handle. It typically causes the horse no discomfort. In fact, application of the twitch triggers the release of endorphins, brain chemicals that calm the horse and relieve pain.

➢ *See also Restraints*

Chaps

Along with a cowboy hat and boots, chaps are standard Western riding apparel. Made of tough but supple leather, suede, or cowhide, chaps protect the rider's legs and jeans from thorns and brush, and to a certain extent, dirt, rain, and rope burns. Chaps also help riders maintain a secure seat in the saddle. They come in three basic styles — batwings, chinks, and shotgun — and can be as plain or as fancy as the wearer likes (or can afford).

Batwings are wide, full-length chaps that drag on the ground and offer the most protection. Chinks fit loosely like batwings but come just below the knee and are usually fringed at the ends. Shotgun chaps fit snugly around the length of the leg and are zipped into place; they can be fringed or not.

➢ *See also Attire*

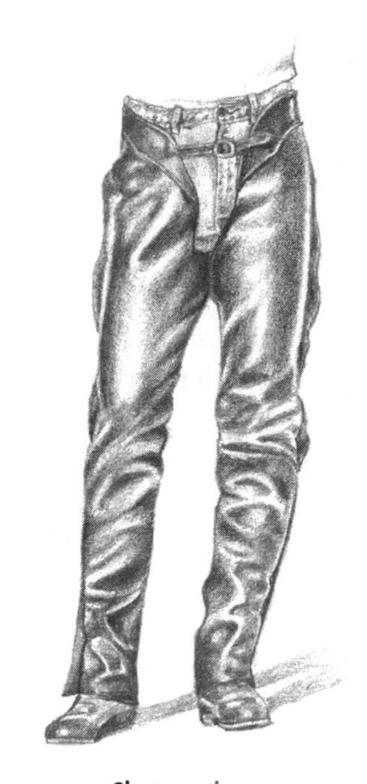

Shotgun chaps

Check

An upward snapping of the reins used to get a horse's attention.

Check rein

A short strap that runs from the bit to the top of the bridle and back to the harness or saddle, used to prevent a horse from lowering his head.

Cheek piece

Straps on a bridle that connect the bit to the crownpiece. On a halter, the cheek piece connects the noseband and crownpiece.

Cheek ring

A part of the halter to which straps are attached. Halters generally include a cheek ring for each side of the horse's head, positioned between the mouth and the eyes.

> **CHARGER**
>
> A charger is the horse ridden by a military officer.

Cherry tree (*Prunus* spp.)

At the wilting stage, the leaves of a cherry tree contain a chemical that is quite poisonous to horses. A cherry tree located near pasture or paddock areas should be removed to prevent horses from ingesting leaves.

➢ *See also Poisonous plants*

Chest bar

A chest-level bar placed in front of a horse when the animal is being trailered or is in stocks. Contact with the bar encourages the horse to stay in place within the trailer while allowing him to lower his head.

Chestnut

A tough, horny patch on the inside of all four legs.

Chestnut coloring

A coloration that includes various shades of red, ranging from bright red gold to reddish brown or auburn. The manes and tails of chestnuts often match their coats but can be flaxen with white hairs. Also called a sorrel.

➢ *See also Coloring*

Chewing

Horses rely on their molars and premolars to perform most chewing and grinding tasks. These flat-bottomed teeth appear in rows on both sides of a horse's mouth, on the top and on the bottom. With heavy use, they may become worn into sharp points that are not optimal for chewing. In these cases, the teeth can be filed using a tool known as a "float."

Chewing also refers to the bad habit some horses develop of chewing on wood — including fences, stall doors, stall walls, and scrap wood. Lack of roughage in the diet can lead to chewing, but the habit is more often attributed to boredom. It is difficult to correct a horse that has begun chewing, but the owner of a chewer should take care to ensure that the horse is not ingesting toxins, such as paint, along with the wood.

➢ *See also Cribbing*

Chi

In Eastern philosophies, *chi* (also written as *ki* or *qi*) means energy and attention. It can be either negative or positive, but for riding, you would want to develop as much positive chi as you could. Developing the ability to keep a strong central focus while performing physically in a relaxed yet controlled manner will help you improve your riding and your horse.

Chin spot

A white spot or hairs below the lower lip, visible from the front.

Chip off

A Western riding and cattle-cutting strategy in which the rider works to isolate one cow from the herd.

Chipping in

The response of a horse when he finds himself too close to a vertical jump or other obstacle to make the leap in his natural stride. The horse will break stride and jump in a steep vertical arc, in some cases failing to clear the obstacle with his hind feet. This is also known as popping.

I wish your horses swift
and sure of foot;
And so I do command
you to their backs.

—William Shakespeare, *Macbeth*

A horse that begins chewing wood may need roughage in his diet or an alternative to boredom.

Chiropractic treatment

A horse's spine is relatively inflexible, even though it contains approximately 170 joints. The horse can experience pain and show symptoms if any of his vertebra become misaligned or inflamed. Equine chiropractors use spinal adjustments to treat problems ranging from chronic back and neck soreness to lameness, stiffness, gait irregularities, and reluctance to be saddled.

Chiroptic mange

➢*See Mange*

Choke

A condition in which a piece of unchewed food, a wood splinter, or some other object becomes lodged in a horse's throat, blocking the esophagus. Symptoms of choke include straining to breathe, coughing, an extended head and neck, slobbering, and squealing. Although the condition is often self-correcting, it can lead to fatal complications such as aspiration pneumonia, so a veterinarian should be consulted when symptoms first appear. Keeping the horse calm and gently massaging the throat area may help resolve the blockage.

Chokecherry

➢*See Wild cherry* (Prunus avium *and* Prunus serotina)

Choker

Collar of the shirt known as a ratcatcher, worn during hunter/jumper and lower-level dressage competitions.

Cholla

A team roping game similar to polo, in which riders attempt to rope a six-pointed jack and drag it up and down the playing field through the goals of both teams. The jack itself is also called a *cholla* (pronounced *CHOY-uh*) after the desert cactus of the same name. It is usually made of leather or canvas stuffed with a filling material.

Chop

Hay that has been chopped, dried, and placed in bags for convenient storage. Chop provides roughage and can help prevent food bolting if added to the grain in a horse's diet.

Chouse

Cowherd's term for pursuing cattle, often to the point of exhausting the cattle.

Chrome

White markings on a horse, such as blaze, stockings, and so on; the term is most commonly used when a horse has a lot of white markings. Some breed standards consider chrome ideal; others view it as undesirable.

CHINCOTEAGUE PONIES

The Chincoteague ponies are small (12 hands high), tough, hardy inhabitants of Assateague, an island and national park off the coast of Virginia. They probably descended from horses that belonged to 17th-century colonists. Many, but not all, Chincoteague ponies are Pintos. They are the only wild pony in the United States.

The 1947 publication of the book *Misty of Chincoteague* by Marguerite Henry, followed in 1961 by the motion picture *Misty*, brought the ponies to the attention of a wider public.

On the last Thursday and Friday of July the Chincoteague Volunteer Fire Department holds its annual Pony Penning Days, a roundup of the ponies and an auction of the foals and yearlings. The proceeds go toward herd management.

The cholla is a six-pointed jack made of leather or canvas.

Chromosomes

Strands of genetic material. Each of a horse's cells contains chromosome pairs, and these chromosomes carry the genes that determine which traits are present in the animals.

➢ *See also Genetics*

Chronic obstructive pulmonary disease (COPD)

This common respiratory disorder, often called **heaves,** interferes with a horse's ability to breathe by constricting air passages, thus also limiting the horse's stamina and activity level. Like asthma in human beings, COPD is often triggered by dust or allergies. Horses kept indoors or fed dusty hay are susceptible, and relapses are likely once the first episode occurs; breeding problems occur in mares with COPD.

Symptoms of COPD included coughing, fluid discharge from the nostrils, wheezing associated with exhalation, and diminished stamina. More serious cases can result in weight loss and difficulty breathing.

The condition is generally controlled through the use of antihistamines and medications that clear the air passages. Horses suffering from COPD should also be kept in environments that are free of dust and mold.

Chute

In cattle events, a fenced and gated holding area for cattle. Crew members open the chute's headgate to allow a cow into the roping arena.

CID

➢ *See Combined immune deficiency (CID)*

Cinch

A girth strap or band used to secure a saddle to a horse, primarily in Western riding. Traditional cinches are made of mohair, but they may also be made of nylon, cotton, rayon, neoprene, or leather. Back cinches are sometimes also used to hold the back of the saddle in place. These are commonly made of leather.

Circle drill

A riding exercise that involves riding in two overlapping circles. The rider completes the first circle, changes leads while riding in a straight line through the middle of the overlap area, then completes the second circle.

Circle reining

Controlling a horse and changing leads while riding in circular patterns.

Circular pen

➢ *See Round pen*

CHRONIC

A description of a condition that lasts a long time or frequently recurs.

If you hear a wheezing sound when listening to your horse's windpipe, the horse may be suffering from chronic obstructive pulmonary disease.

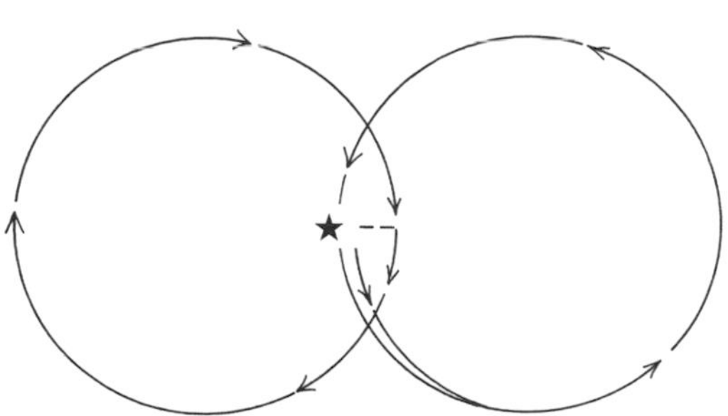

The circle drill is a riding exercise that helps riders learn to change leads smoothly. (The drill begins at the star.)

Circulatory system

Collective term for the heart and the system of vessels that circulates blood through a horse's body. The general health of a horse's circulatory system can be gauged by determining his capillary refill time and the time required for an elevated pulse to return to normal following exertion.
➢ *See also Capillary refill time*

Class

In horsemanship circles, the various kinds of riding and show events united by a set of criteria. Some classes judge the quality or conformation of the horse, others are concerned with the horse's performance, and still others evaluate the performance of the rider. Class may also be determined by the age and breed of the horse, or by the age and gender of the rider.

CLASSIC FINO

The classic fino is a collected slow gait demonstrated by the Paso Fino horse in which the hooves move rapidly but take extremely short steps.

See also Paso Fino

Classical High School

The classical training of Grand Prix level dressage horses; primarily used in Europe. The Lipizzaners are the most commonly known horses involved in Classical High School training; they perform "airs above the ground."
➢ *See also Dressage; Lipizzaner*

Stall-cleaning equipment: pitchfork, manure fork, scoop shovel, broom, muck bucket

Cleaning stalls

Ideally, an occupied stall should be cleaned once a day to ensure freshness of bedding, minimal odor, and a healthy environment for the horse. To enhance safety, turn the horse out before cleaning the stall. If this is not possible, be sure to tie the horse.

Begin the cleaning routine by removing manure and soiled bedding with a manure fork or pitchfork, leaving as much of the clean bedding in the stall as possible. Next, use a shovel to remove bedding that is wet or packed down. Now, sweep dry bedding over wet areas to soak up any remaining liquid from the floor. Then push all bedding against the stall walls and let the floor dry. If necessary, apply an odor-control product to wet spots. When the floor is dry, finish the cleaning task by adding new bedding as needed and redistributing all bedding. Place the cleanest bedding in the horse's rest area.

"Stripping the stall" by removing all bedding and replacing it with fresh material is recommended on a weekly basis.

Cleaning tack

Dirt and sweat accumulate readily on tack, making it unpleasant and in some cases unsafe to use. To clean tack, use a bucket of water and a sponge, adding a small amount of ammonia to the water if desired. Wipe all tack clean with the sponge, rinsing it often to keep it clean. When finished, apply saddle soap or a leather-conditioning product to the tack with a dry sponge.

Old tack can also be reconditioned by soaking it in neat's-foot oil, mink oil, olive oil, or baby oil. The oil should be warmed (at least to room temperature), and the tack should soak for about an hour. Be forewarned that soaking may weaken and destroy stitching. Afterward, wipe off excess oil and allow the tack to dry for 24 hours. Repeat the reconditioning process if the tack remains somewhat dry or brittle.

Cleft of frog

The V-shaped indentation in a horse's hoof, located in front of and adjacent to the heel bulb.

➢ *See also Hoof*

Click

A cross between two equine bloodlines that yields highly desirable offspring.

Clinch

The exterior end of a nail used to fasten a horseshoe to a hoof. Loose clinches can cause a shoe to drift on the hoof or even to fall off.

Clinch cutters

A tool used to cut or straighten clinches and remove horseshoes.

Clippers

Haircutting tools of various styles and sizes used to trim horse hair.

Clipping

The process of trimming a horse's hair. Areas commonly clipped include the legs, exterior ear hairs, jaw, and the bridle path. Clipper blades should be washed, brushed out, and lubricated regularly. They may also need spraying with a coolant if they become too warm after extended use.

Horsemen new to clipping should obtain assistance from someone who is confident and able to teach safe clipping methods. Horses can become quite nervous and potentially dangerous if handled inappropriately during clipping.

➢ *See also Blanket clip; Grooming; Hunt clip; Trace clip*

Clipping a pasture

Also known as mowing a pasture. It is done to control weed growth and stimulate density of grass growth. The timing of this procedure depends on the life cycle of local weeds and grasses, so consult your Extension agent or agriscience instructor for advice.

Clitoris

A mound of tissue that is part of a mare's reproductive tract. The clitoris is generally about the size of a walnut and is enclosed by the lips of the vulva. A mare may "wink" her clitoris during estrus, providing a visible signal that she is in heat.

Close

The act of tugging subtly on the reins by closing the fingers.

COMMON CLIPPING TERMS

Bridle path: Clipping hair from the mane at the top of the neck, just behind the poll. The length of bridle path varies with breed and discipline. It prevents chafing by the bridle's crownpiece and development of painful ingrown hairs, which can cause head shyness.

Stable clip: Trimming excess hair from muzzle and fetlocks and providing a bridle path. Most owners, at the very least, do this clipping to make their horse appear well kept.

Body clipping: Clipping all of the hair off the entire horse. Used most often in preparation for horse shows in spring; however, some people rely on body clipping to maintain a smooth coat throughout show season. Body clipping removes the natural oils from a horse's coat and requires fairly rigorous daily upkeep of the coat to prevent dry hair and dermatitis.

Show/sale clipping: Clipping is a major aspect of preparing a horse for competition and sale. In both cases, the goal is to make the horse look his absolute best. Clipping requirements and trends vary by breed and discipline, so it is best to learn these techniques from a knowledgeable professional.

Closed reins

A pair of reins attached to a flexible leather riding whip known as a **romal.**

Clostridium botulinum

A spore-producing bacterium that can induce botulism in horses if ingested. It is most commonly ingested through contaminated feed or spore-laden soil.

➢ *See also Botulism*

Clostridium tetani

A spore-producing bacterium that can induce equine tetanus if it is allowed to contaminate a wound. Punctured feet are particularly vulnerable to invasion by *C. tetani.*

➢ *See also Tetanus (lockjaw)*

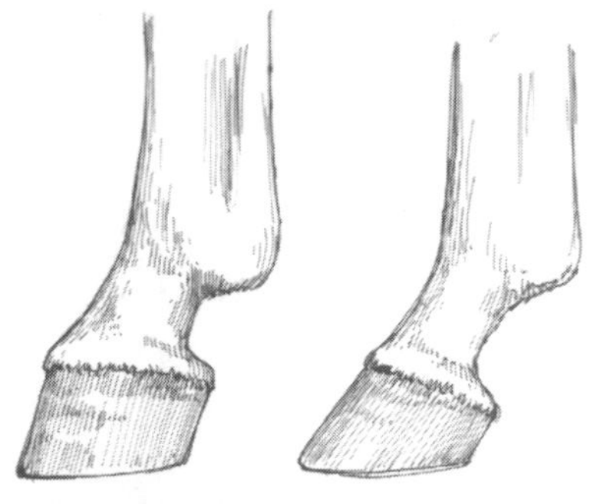

Clubfoot is a conformation fault that makes a horse less agile, with an irregular gait.

Clothing (for rider)

➢ *See Attire*

Clover

A legume hay high in vitamins, calcium, and protein. Small quantities of clover are often mixed with grass hay when feeding a highly active horse that requires supplemental protein. Horses find clover tasty; it's a great hay for picky eaters.

➢ *See also Hay*

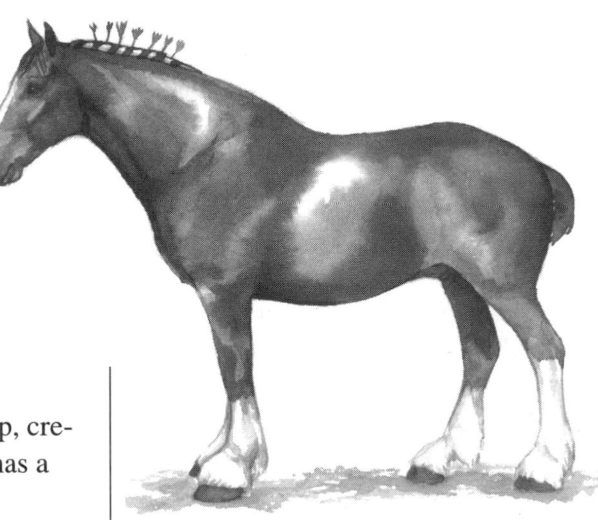

The Clydesdale is often identified by its massive, feathered feet.

Clubfoot

A condition in which the hoof-to-ground angle is atypically steep, creating a foot that is upright rather than angled. A clubfoot typically has a shorter than normal toe and a longer than normal heel.

Clubfoot can be triggered by an injury or unskilled hoof trimming, or it may be an inherited predisposition. The effects of the condition can be minimized through frequent corrective foot trimming, especially if the horse is not yet fully grown.

Clydesdale

A breed of draft horse that originated in Clydesdale, Scotland. The Clydesdale is a heavy, big-boned horse known for its strength and its elaborately feathered feet. The breed gained fame when it became associated with televised beer commercials, and teams of Clydesdales are popular attractions at horse shows.

➢ *See also Breed*

Coach

A coach is *not* the same as a trainer or instructor. Coaching involves a complex combination of sports psychology, skill development specific to the sport of riding, and competition savvy. Consider the difference between a physical education teacher and athletic coach at any educational level. The two roles require related, yet vastly different, skills and knowledge. Not all teachers make competent coaches for school athletics, and not all riding instructors or horse trainers make good riding coaches.

Select a coach carefully. Consider, first and foremost, whether the coach's definition of success is compatible with your own. If skill improvement and performing to a "personal best" standard of achievement means more to you than material reward, then seeing a list of the coach's winning students is less important than seeing evidence of positive self-esteem and teamwork when the coach's students compete. Because self-esteem relates to how you fit in among your peers, be sure the coach's students show a record of consistent improvement in relation to their experience, education, and commitment. If nothing less than championship trophies will do and you're prepared to do whatever it takes to win, then be sure to choose a coach who has had champions at the level at which you plan to compete.
➢ *See Trainer*

His neck is high and erect, his head replete with intelligence, his belly short, his back full, and his proud chest swells with hard muscle.

—Virgil, *Georgics*

Coat

The thick layer of hair that covers a horse's body, serving to help the horse stay sufficiently warm and dry under most weather conditions.

A horse's coat grows considerably longer and fluffier in winter, adding insulation. By contrast, the horse begins shedding hair in spring. Horses require an adequate amount of protein and fat in their diet to sustain healthy, shiny coats. Regular grooming also contributes to a healthy coat. Conversely, a rough, dull coat indicates potential neglect and poor health or nutrition.

Cob

A small, short-legged horse that is somewhat larger than a pony. Cobs can be good horses for riders who are too large for ponies but are a bit small for standard-sized horses. The **Norman Cob** is a light draft horse used on farms in Normandy, prized for its sturdiness and energetic action.

Cocklebur *(Xanthium strumarium)*

An annual plant with prickly fruits whose spring seedlings are toxic to horses. The cocklebur thrives in disturbed soils, lakebeds, and lowlands adjacent to creeks, streams, and rivers. Symptoms of equine cocklebur poisoning include difficulty breathing, rapid pulse, and muscle spasm in the leg and neck areas.

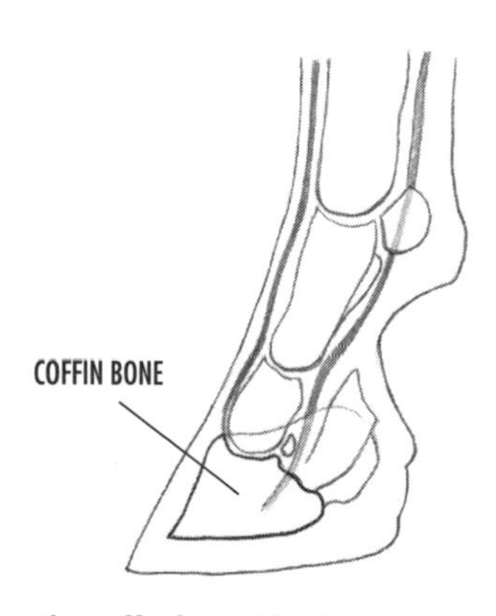

The coffin bone. This bone can rotate downward as a result of founder.

Coffin bone

A bone, the distal phalange, located inside a horse's hoof. It is comparable to the bone in the tip of human fingers and toes. If tissue surrounding the coffin bone becomes inflamed, the condition is known as **laminitis.** If the inflammation progresses, the coffin bone may rotate downward; this rotation is a symptom of the serious, disabling condition known as **founder.** In extreme cases, the coffin bone may protrude through the bottom of the foot.

An overly rich diet and excessive consumption of water after exercise can contribute to the development of laminitis and founder. Contact a veterinarian immediately if you suspect the presence of either condition.
➢ *See also Laminitis*

Coffin joint

The joint between the coffin bone and the bone above it, called the **short pastern,** or medial phalange. The joint also involves a third bone — the tiny, pyramid-shaped structure known as the **navicular bone.**
➢ *See also Navicular disease (navicular syndrome); Pastern*

Coggins certificate, Coggins test

A Coggins test involves drawing blood from a horse's jugular vein.

The Coggins test is a blood test administered by a veterinarian. It is designed to screen for **equine infectious anemia,** a serious viral disease also known as swamp fever. The test is generally repeated on a yearly basis; check your state's laws for local requirements. A negative test indicates that EIA is not present and results in the issuance of a Coggins certificate, a document required when traveling or participating in horse shows. A positive test indicates, at least, exposure to the virus; after a positive test, the horse is usually quarantined and retested. A horse infected with EIA will probably have to be destroyed; however, the disease is uncommon.

Coldblood

A term for draft or half-bred horses reflecting their tendency to move more slowly and to demonstrate stable dispositions.
➢ *See also Warmblood*

Cold-weather care

Horses are generally hardy animals: They endured quite well prior to humankind's intervention. Still, horses were not native to the most severe of the cold-climate regions. Horses that reside in these regions require some extra-special care.

Horses living in other regions require that their care be relative to the sudden shifts in conditions as seasons change. This is when they are most prone to illness and injury related to cold weather.
➢ *See also Blanketing; Feeding and nutrition, First aid; Watering*

COLD-WEATHER WATERING

Lack of water, or excessively cold water, can cause colic. A variety of systems exist to keep water at an acceptable temperature. An insulated bucket holder is useful because it requires no electricity and can keep water from freezing for several hours or more, depending on the air temperature. Other options are electrically heated buckets and immersible heaters. Any electric device should be properly installed and carefully monitored to be sure it is working safely.

Colic

Although many people believe that colic is a single disease, the word "is actually a generic term for abdominal pain. Colic can be a symptom of a number of digestive, circulatory, or organ-related disorders — some of them life-threatening. The possible underlying conditions include intestinal spasms, a twisted or blocked intestine, excessive gas in the digestive tract, fecal impaction (constipation), blood vessel blockage, urinary obstruction, pain after foaling, or the malfunction of the bladder or kidneys.

Symptoms

Horses with abdominal pain may exhibit a variety of symptoms. Among these are poor appetite, listlessness, rapid pulse, repeated episodes of getting up and lying down, foot stamping, sweating, squatting, issuing unusual groans or squeals, straining as if trying to urinate, biting at the belly or sides, and violent attempts to roll over on the ground. It is imperative to contact a veterinarian as soon as a horse shows any sign of colic.

Prevention and Treatment

Because parasites can foster abdominal pain, twisted intestines, and blood vessel blockage, one of the best preventive measures you can take is having your horse dewormed regularly. Providing your horse with plenty of clean drinking water and appropriate amounts of high-quality feed that is free of mold, parasites, and toxins is another fundamental precaution.

Treatment of colic is entirely dependent upon the condition that is provoking the abdominal pain. If excessive gas is the problem, a horse can, in many cases, be treated with an antigas medication, such as Maalox. Bowel blockage can be remedied with a lubricant, such as mineral oil, administered through a stomach tube.

In more serious cases, such as those involving twisted or blocked intestines, surgery is often necessary. The conditions collectively associated with colic (together with self-inflicted injuries that result when the animals try to relieve their pain) make colic the leading cause of death among horses.

There are several types of colic, including:

- **Spasmodic colic.** This may come from overexcitement and nervousness, but usually from eating spoiled or moldy feed, after a sudden change in the diet, or after a big drink of cold water when the horse is too hot. Spasmodic colic usually resolves itself in a few hours, although most owners prefer to call the vet, who may administer Banamine to ease the pain.
- **Twisted gut.** This can result from attempts to relieve abdominal pain. The intestine may become twisted as the horse rolls on the ground; it also can be triggered by muscle spasms or excessive gas buildup. If untreated for too long, endotoxic shock can result, making it difficult for the horse to recover even if surgery is performed or a relatively small length of bowel is involved.
- **Sand colic.** This condition results when the horse has ingested sand or small gravel over a long period. This can happen if the horse eats hay on sandy ground, or if his pasture is sandy and sand clings to the roots of plants he eats. Because sand is heavier than the horse's feed, the sand settles in the cecum, the lower part of the intestine, and may obstruct other material trying to pass through.
- **Flatulent colic.** This problem can be caused by an intestinal obstruction (for which surgery is the only cure) or from eating too much lush green grass or too much clover and alfalfa.

COLIC: WHAT TO TELL THE VETERINARIAN

When you suspect that your horse has colic, call your veterinarian. Be prepared to relay as much of the following information as possible:

- ❑ Time and place you first noticed the problem
- ❑ Behavioral and physical symptoms
- ❑ When the horse last ate and what was eaten
- ❑ Whether the gums and eye membranes are of normal color
- ❑ The horse's pulse and respiratory rates
- ❑ Rectal temperature
- ❑ Capillary refill time
- ❑ Description of any digestive noises (you can hear these by resting your ear against the abdomen in various locations)
- ❑ Appearance and frequency of recent bowel movements
- ❑ Recent changes in the horse's diet or exercise level
- ❑ Medical history, including deworming treatments

• **Impactions.** Eating dry food without drinking sufficient water may cause manure to become hard and dry, creating constipation. With a little movement, gas builds up, causing even more pain. Surgery may be necessary, but repeated treatments with lubricants and medications often resolve the condition and soften the material enough to get it to pass. Thereafter, it is essential that the horse has water sufficient to keep his stool soft.

➢ *See also Digestive problems; Torsion, as cause of colic*

A horse suffering from colic (abdominal pain) may stagger about on buckling legs.

Collection

A state of organized movement and balance that characterizes a well-trained horse. The desired configuration or posture includes active hindquarters, flexed abdominal muscles, a rounded back, a flexed poll, and head and neck held in correlation to proper elevation and degree of collection. A collected horse is agile, light on his feet, and responsive.

➢ *See also On the bit*

Coloring

Color in horses is a genetically controlled trait. The color genes come from a horse's parents, and dominant and recessive genes play a role. Certain colors exist in all breeds; other colors are unique to one breed or a handful of breeds. Breeders who wish to obtain a foal of a specific color may need considerable understanding of genetic factors, including whether the desired color is dominant, recessive, or mixed; what the parents' genotypes are; and the colors of the parents' ancestors.

➢ *See also Face markings*

BASIC HORSE COLORS

COLOR	DESCRIPTION
Bay	Tan to reddish brown coat with black points
Black	A completely black animal except, in some cases, for white leg and face markings
Brown	Dark brown to black coat with brown muzzle
Buckskin	Light brown to gold coat, with dark points, often with a darker dorsal stripe down the back
Chestnut	Entirely red or auburn animal
Dun	Yellow or gold body with black or brown points, often with stripe on back, legs, and withers
Gray	Gray or white coat, may have dapples; dark skin
Palomino	Gold to yellow coat with lighter mane and tail
Roan	Some individual hairs that are two colors — the primary body color and white. A strawberry roan has red and white hairs; a blue roan has black and white hairs.
Sorrel	Western term for red horses of certain breeds, including Quarter Horses, Appaloosas, and Paints
White	White body with pink skin and muzzle

Colostrum

The initial milk a mare produces after foaling. Colostrum contains high levels of critical nutrients, providing twice the energy and five times the protein of regular mare's milk, plus a huge helping of vitamins and minerals, as well as antibodies that confer disease resistance to the foal. It also contains a creamy fat and is easily digested. Colostrum is generally thicker and stickier than typical milk. Its color ranges from yellow to orange.

If the dam dies during foaling, her colostrum can be milked out immediately to give to the foal. If the dam simply refuses to nurse the foal, she can be twitched and milked, and the colostrum can then be given to the foal. If there is no way to obtain the dam's colostrum, use frozen colostrum, or, as a last resort, contact the veterinarian to obtain a serum to give the foal intravenously.

Colt

An uncastrated young male horse. Age varies with region; average age is four years old or younger. The term may also refer to a young racehorse (either gender) that is in its first year of training.

Combined driving

A competitive sport that tests both horses and drivers in a kind of "triathlon." Participants compete in the areas of dressage, obstacle negotiation, and cross-country driving, generally performing in one category each day for a period of three days.

Combined grazing

A technique for the efficient utilization of pasture, in which horses and other animals with complementary eating habits (generally cattle) are allowed to graze together in the same area. The horses eat certain portions of the grass, while the other species consumes different portions, ensuring that maximum benefit is derived from the pasture.

Combined immune deficiency (CID)

A genetic disorder in which a foal's immune system fails to develop fully, leading to subsequent infections and death. CID is found in Arabian and part-Arabian horses. There is no effective treatment, but the disorder can be prevented by testing for the presence of the genetic defect through a DNA analysis. Carrier horses can still be used for breeding as long as the mate is tested and found to be free of CID. However, this practice is discouraged in an effort to maintain the future integrity of the breed; a carrier can pass the defective gene to offspring.

THE CONCEPT OF COLLECTION

A horse is described as "collected" when he is driving himself forward with his hindquarters, whether at a walk, trot, or lope or canter. Hindquarter engagement frees up the front end of the horse to take care of balance and lateral movement.

Combined training/eventing

A competitive sport designed to test training and horsemanship skills in the areas of dressage, cross-country riding, and stadium jumping. Although the sport has been featured in the Olympics since 1924, there are competitions for horses and riders of all skill levels, including novices.

Coming into hand

Refers to a horse that is entirely under the control of the rider.

Command of preparation/execution

Verbal commands used by instructors to ensure that riders perform movements or drills in a coordinated manner. The command of preparation alerts riders that a movement is about to begin; the command of execution signals riders to begin the movement.

Commands

Riders employ a number of verbal commands in order to guide their trained horses through various gates and movements. Horses respond most readily to concise commands issued in a firm, authoritative tone. Commands vary as widely as horsemen and trainers. Commonly used commands include words such as "Walk," "Trot," "Canter," and sounds like the "kiss" or "cluck." Perhaps the most critical command is the one that stops all motion, most frequently, "Ho" or "Whoa."

Voice commands taught on the lead or longe-line lay the foundation for leg cues that are used while mounted. Horses learn by conditioning. Applying leg pressure consistently while using a previously learned verbal command leads to a conditioned response, which ultimately makes verbal commands obsolete. (Research indicates that most horses require 25 to 100 repetitions of such verbal/leg command combinations before they learn to respond *consistently* to leg cues alone.)

Common horse

A horse that is plain and unremarkable in appearance.

Competitions, guidelines for

Most competitions have a governing association that determines the rules and regulations of exhibitor, administration, and officiating judge conduct. These rules are as diverse as the horsemen and associations involved. It is best to seek out specific information from the organization sponsoring the competition you plan to attend as an observer or competitor.

➢ *See also American Horse Shows Association (AHSA); appendix*

Competitive trail riding

An event that involves riding a predetermined trail within a specified period of time. Both horse and rider are scored — the former on soundness, condition, and behavior (or stress levels), the latter on trail equitation and horse care. A competitive event may cover 15 to 90 miles and last 1 to 3 days. The primary goal of this long-distance event is to maintain each horse in good physical condition throughout the ride. There are events for riders of all skill levels, many conducted under guidelines established by the North American Trail Riding Conference (NATRC).

COMPETITIVE TRAIL RIDING DIVISIONS

There are three divisions of trail rides sanctioned by the North American Trail Ride Conference, as follows:

- **Novice.** For new competitors (junior, lightweight, and heavyweight classes)
- **Competitive Pleasure (CP).** For more experienced riders (no weight restrictions)
- **Open.** For advanced riders, featuring longer distances and a faster pace (junior, lightweight, and heavyweight classes)

Novice and CP riders cover about 40 miles in two days, traveling at 3½ to 5 miles per hour. Open riders travel about 60 miles in two days, at 4 to 6 miles per hour.

Complete feeds

Also known as complete rations, processed feeds that offer horses complete nutrition in a convenient package. They come in wafer, pellet, or cake form and comprise alfalfa, hay, grains, and minerals. The quality of these feeds can vary. Some crumble easily, creating a great deal of powder waste; others simply don't offer the horse sufficient "chew time," which can result in cribbing, or wood chewing.

Concentrates

Feeds that are low in fiber but high in digestible nutrients, including grains, grain by-products, and oilmeals. A horse should not be fed a diet that consists only of concentrates and water, as the animal's digestive system also requires significant quantities of fiber or roughage. The customary source of roughage is hay. Some common concentrates are barley, bran (made from wheat), commercially prepared rations (content varies), corn, cottonseed meal, linseed meal, oats, roasted soybeans, sorghum, soybean meal, sweet feed (a mixture of oats, corn, bran, pellets, and molasses), and wheat.

➢ *See also Grains*

Condition of horse

This term often refers to the fitness or physical abilities of a horse; however, it can also concern the amount of fat a horse's body contains. You can estimate a horse's condition (fat levels) by feeling his back, ribs, neck, shoulder, withers, and the area directly above the tail (see chart below).

JUDGING A HORSE'S BODY CONDITION

This chart provides rough guidelines for determining body condition. Some fat is necessary, and the ideal score is 5 or 6.

SCORE	DESCRIPTION
1 Poor	Extremely emaciated; no fatty tissue can be felt; ribs, vertebrae, and withers project prominently; bone structure easily observed
2 Very thin	Emaciated; ribs and vertebrae prominent; faintly noticeable bone structure; very thin neck, shoulders, and withers
3 Thin	Ribs easily seen; slight fat cover over ribs, neck, shoulders, and withers but they nonetheless appear thin
4 Moderately thin	Very faint outline of ribs; neck, shoulders, and withers not obviously thin; vertebrae prominent along back
5 Moderate	Ribs cannot be seen but are easily felt; area above tail beginning to feel spongy; back is level over loin; shoulder blends smoothly into body; withers rounded
6 Fleshy	Can barely feel ribs; area above tail feels spongy; may see slight crease down back over loin
7 Fat	Individual ribs can be felt, but area between ribs contains noticeable filling of fat; crease down back over loin; fat deposited along neck, withers, area above tail, and area behind shoulders
8 Very fat	Fat deposited along inner buttocks; difficult to feel ribs; thickening of neck; fat on withers, area above tail, and area behind shoulders; positive crease down back over loin
9 Obese	Can't feel ribs; bulging fat on neck, withers, area above tail, area behind shoulders, and along inner buttocks, which may rub together; very obvious crease down back over loin; flank filled in flush

From *Stablekeeping* by Cherry Hill

Conditioner (for coat)

Coat conditioners can improve the shininess, manageability, and moisture content of your horse's coat. Some products also contain sunscreen to protect the horse's skin. When using a conditioner, be sure to dilute and apply it according to label instructions. Some products must be rinsed out following application; others do not require rinsing.

➢ *See also Bathing horses*

Conditioning program

Sedentary horses, like sedentary people, are generally out of shape. Creating and following a conditioning program for your horse can be the best way of assuring that he achieves and maintains fitness. Work with an experienced trainer or handler to develop a program that is appropriate for both you and your horse.

In general, conditioning an out-of-shape horse involves gradually increasing his activity level (both duration and intensity) over a period of time. Achieving a moderate level of muscle and respiratory fitness typically requires 3 to 6 months of regular exercise. Strengthening bone is a longer process that relies on prolonged, low-intensity activities.

➢ *See also Exercise (for horses)*

Conformation

The shape and form of a horse's body, compared to an ideal or standard of perfection. A horse whose physical characteristics approach the ideal has good conformation, or is considered a "well-made" horse. A horse whose characteristics deviate considerably from the ideal is described as having conformation defects or faults. Most faults are features that could limit the horse's performance or lead to problems, such as lameness. Others, such as a "Roman" nose, are simply considered unattractive.

Few, if any, horses offer perfect conformation, so it's wise to have realistic expectations when selecting a horse. However, conformation should not be dismissed as merely a matter of aesthetics. A horse's body structure plays a critical role in how efficiently he performs, how comfortable he is to ride, and whether he is likely to remain sound over a lifetime of riding or work. Horses with poor conformation generally have higher maintenance costs due to special veterinary, farrier, and other care. Evaluating conformation is particularly important when selecting breeding stock.

Many conformation standards can be applied to virtually all horses (see chart). However, the standard for a specific breed may vary with respect to certain features. The term **breed standard** refers to the written standards that comprise ideal conformation within a particular breed.

➢ *See also Anatomy of the horse; Balance, balanced horse; Base-narrow and base-wide; Breed; individual body parts and conformation faults*

Conjunctivitis

Also known as **pink eye,** conjunctivitis is an inflammation of the **conjunctiva,** the mucous membrane that covers the front of the eyeball and lines the eyelids. It is generally caused by contact between the horse's eyes and flies or other environmental irritants.

A horse with conjunctivitis will commonly squint or develop swollen eyelids. Tears or yellow pus may be discharged from the eyes. The condition is readily treatable with eyedrops or ointment, but early intervention is important; untreated conjunctivitis can lead to impaired vision. During times of heavy insect infestation, equipping your horse with a fly mask can help prevent conjunctivitis.

Round hoof'd, short-jointed, fetlocks shag and long,
Broad breast, full eye, small head and nostril wide,
High crest, short ears, straight legs and passing strong,
Thin mane, thick tail, broad buttock, tender hide:
Look, what a horse should have, he did not lack,
Save a proud rider on so proud a back.

—William Shakespeare, *Venus and Adonis*

CONFORMATION: WHAT TO LOOK FOR IN A HORSE

Conformation standards vary somewhat by breed. They also vary according to the type of work a horse will perform. Features desirable in a pleasure horse, such as well laid back shoulders, may actually be drawbacks in a working draft horse. However, most horse aficionados will find the characteristics listed below useful in evaluating conformation. *See also Anatomy of the Horse.*

Overall Structure

Well-balanced body with weight distributed appropriately (see *Balance* for essential information)

Muscling long and lean, not bulky

Attractive general appearance

Head and Neck

Head size proportional with the rest of horse

Wide forehead to provide adequate cranial space

Eyes large enough to provide good peripheral vision

Head tapers to a small muzzle with wide nostrils

Upper and lower jaws match, upper and lower incisors meet evenly

Long, flexible neck with a slight arch or crest to serve as balancing arm for horse

Neck meets well-defined withers

Body or Barrel

Deep heart girth

Ribcage wider than shoulders (wide barrel)

Long, well-sloped shoulder; ideal shoulder slope is 45 degrees

Withers prominent enough to help anchor the saddle in place

Well-muscled, strong back and loin that is not excessively long

Back runs smoothly into well-muscled, deep hindquarters with sufficient distance from point of hip to point of buttock

Well-muscled hindquarters

Legs and Feet

Strong, straight, well-placed legs

Weight distributed evenly on all four legs when horse is standing square

Distance between feet the same as distance between forearms at chest

An imaginary straight line drawn from point of shoulder should go down through center of front legs

Knees large, flat in front, and well-proportioned

Cannon bone centered under knee, appears wide when viewed from the side

Hind legs straight when viewed from rear; an imaginary line drawn from point of buttock should go down through the center of the hind legs

Large, sturdy hock joint that appears flat on the outside edge

Pasterns of adequate length

Proper angle between pastern and foot; ideal angle is 50 degrees

Well-shaped feet large enough to support the horse

Feet pairs in exactly the same size and shape

Center of foot directly under center of fetlock

Toes point directly forward

Hooves wide at the heel, not contracted

When horse is in motion, legs and feet move forward in straight lines; hind feet travel same path as front

Conformation reflects the way a horse's body parts are put together.

Connection

The cooperative relationship that exists when a horse, the rider's driving aids, and the rider's restraining aids are working together to create a smooth, unfettered flow of energy.

Connemara ponies are known to be excellent jumpers.

Connemara

A popular riding pony known for its effortless gait and jumping abilities. The Connemara, which originated in Ireland but also has Arabian influence, was first imported to the United States in the 1950s. This relatively tall pony can grow to a height of 14.2 hands. Many Connemaras are gray or black, but others are dun, bay, or brown.

Consignor

A person who puts a horse up for sale at an auction or on a consignment basis.

Constipation

Constipation in horses, sometimes called **fecal impaction,** often results in a lethargic horse that eats little or nothing. The horse may lie down for unusually long periods of time and fail to pass normal amounts of manure and urine. Factors that can contribute to constipation include ingestion of bedding or dry feed, dehydration, and lack of exercise. Simple constipation can be cured by administering mineral oil through a stomach tube, followed up with plenty of drinking water.

Constipation is a leading cause of death among newborn foals. You may need to remove hardened fecal pellets from the rectum of an afflicted foal if he does not pass them within a few hours of birth. Administering an enema can also clear up the blockage, and severe cases are treated with a laxative.

Contact

The pressure a rider applies to a horse's mouth by using the reins.

Contest events

Competitive Western riding events — such as barrel racing, pole bending, stake racing, and flag racing — that are judged by time only.

Contest horse

A horse that has been trained to participate in contest events. Speed, agility, and responsiveness are the key attributes of a contest horse.

Contracted foot, contracted heel

A condition in which a horse's foot is atypically narrow, particularly at the heels, which grow too close together. The frog, the foot's shock absorber, also shrinks or shrivels, and the sole of the foot may become dished or concave. If left untreated, the condition can lead to permanent or long-term lameness.

Contraction of the feet is generally caused by injury, excessive dryness, or improper shoeing, including cutting out of the frog whenever the horse is shod. The condition usually affects the front feet, but the back feet may also suffer. Lengthening the toe, or front of the foot, also causes contracted heels because the extra length eliminates the slight expansion the entire hoof experiences when the foot strikes the ground. Lameness from other causes that alters the way the horse's feet strike the ground can also cause contracted heels. A certain amount of pressure on the frog is necessary for a horse's foot to grow the proper amount of heel.

The condition can often be treated through corrective trimming and the use of special shoes, but recovery may take a year or more.

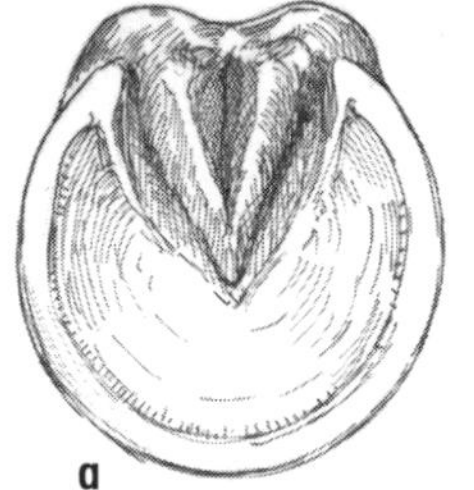

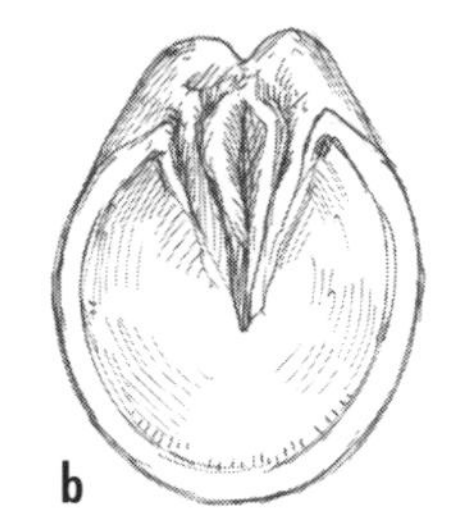

Normal foot (a) and contracted foot (b) with narrowed heels and atrophied frog.

Contracts

In this age when lawsuits are common, it is vital that all horsemen become familiar with the basics of contract law. No matter what your role (buyer/seller, stable owner/boarder, client/trainer), putting your understanding of an agreement in writing may save you thousands of dollars and a load of heartache down the road.

Essential to all contracts are the following:

- Identifying information of agreeing parties
- Description of item or service being agreed upon
- Description of what each party will give and what each will receive
- Signatures of all parties
- Date of agreement, and length of time contract covers, if service agreement
- Warranties and/or limitations of contract or agreement

Common contracts in the horse industry are: Sales, Breeding, Boarding, Lease, Training, Coaching, and Liability Release.

Control skills

➢ *See Aids; Bending; Hands, rider's; Seat; Stop*

Cooler

A blanket draped over a horse to help him cool off gradually after a bath or vigorous activity. Although relatively light, coolers are often made of wool or wool blends and can cause overheating on hot days.

Coolers help a hot or wet horse cool off gradually.

Cooling down (after exercise)

Like human athletes, horses should both warm up before and cool down after heavy exercise. Special care should be taken in hot weather to prevent overheating, and special care should be taken in cold weather to prevent chilling.

Begin the cool-down by slowing your horse to a walk during the last segment of a workout. In warm weather, your horse will benefit from a hosing or sponging-down of the neck, legs, and stomach. Use cool (*not* cold) or lukewarm water. Let the horse's back cool off naturally; once his body temperature is normal, you can sponge off the saddle marks. Use a plastic or metal scraper or squeegee to flick excess water from his coat. On a sunny day, he can dry off on his own, either in the pasture or in his stall, but in cold weather you must make sure he is dry before you leave him.

On a cool day, remove his tack and hand walk him until his body temperature is normal and he is dry enough to groom. The length of time you'll need to walk your horse will depend on several things: the amount of exercise he's had, the weather, and how fit he is. If he's extremely wet, rub him down with a towel or hay wisp to hasten the drying process or drape a wool **cooler** over his back to help wick away moisture. He can have a limited amount of cool (*not* cold) drinking water immediately after exercise if he is being walked, but wait until his body reaches normal temperature before providing him with free access to water and before feeding him any grain.

COPD

➢ *See Chronic obstructive pulmonary disease (COPD)*

Corn, as feed

A grain that is considered a **concentrate** because of its high nutrient content. Corn is particularly high in carbohydrates, which provide energy, but its protein value is low. A horse's diet should not consist exclusively of concentrates, although they can be useful as supplements to a diet of hay or grasses. Too much corn in the diet may cause a horse to become high-strung or difficult to manage, due to the abundance of energy provided by the feed. It can also lead to excessive weight gain.

Corn is commonly fed in shelled form, but some horse owners feed it directly from the cob. Younger and older horses may have trouble chewing whole corn. In such cases, the corn can be served rolled or cracked. All corn should be dry and a year or more old before it is fed, to minimize the chance that a horse will consume mold- or fungus-infested kernels. Poisoning from spoiled corn can be fatal.

➢ *See also Concentrates; Grains; Mold (fungi) in feed*

Corn, as medical condition

A corn is a bruise on the sole of a horse's foot, most commonly a front foot. It is often visible as a reddened area, and it may feel hot and tender. Corns can develop by stepping on a sharp object, but they are more often the product of poorly fitting shoes or shoes that are left on for too long a period of time. A severe corn can cause infection and lameness.

Treatment often simply involves removal of the shoe followed by a period of rest and healing. If the injury becomes infected, a veterinarian may have to pare down the area with a hoof knife to allow drainage. The hoof will then need daily soaking and continued protection until the healing process is complete.

LOCATING CORNS

Another reason to clean a horse's feet frequently is to detect corns before they become problematic. Still, corns are not always visible, except during trimming, and they're not always visible even at this time. But they are detectable with a *hoof tester,* an instrument that looks like a large pincers and is used to put pressure on the sole. If the horse flinches, you've found a corn, visible or otherwise, or something more serious, such as an abscess.

Cornea

The transparent front portion of the eyeball. A horse's cornea is susceptible to injury from tree branches or any other object that makes contact with it. The result is significant pain for the horse and a possible infection. Although many corneal injuries heal without intervention, an infected cornea should be treated with antibiotic ointment.

➢ *See also Eyes; Vision, of horse*

Corner feeder

Hay and grains can be fed with a corner feeder, which is mounted in the corner of a stall. Corner feeders should be mounted low enough to allow the horse to eat in a comfortable, natural position.

➢ *See also Feeder; Feeding and nutrition*

Coronary band

Also known as the **coronet,** this is an area of soft tissue directly above a horse's hoof, where the hoof meets the leg. An injury or bruise to the coronary band will affect hoof development and can cause a horse to become lame. It is essential for a horse's foot health to keep the coronary band in good condition. If it is injured, veterinary attention is advised to prevent the formation of proud flesh, which could hamper hoof production and cause deformity and/or lameness. Horse owners often use bell boots to cover and protect this sensitive area, particularly when trailering the animal.

➢ *See also Lameness; Proud flesh*

Coronet

➢ *See Coronary band*

Corpus hemorrhagicum, corpus luteum

The corpus hemorrhagicum is a blood clot that occurs in a mare's ovary immediately following ovulation. It can be detected by a veterinarian during a rectal exam.

As the blood clot congeals, it becomes a yellow orange mass of glandular tissue called the corpus luteum, or "yellow body." This mass enlarges for a week or so, stabilizes, then begins to shrink. It is significant because it emits the hormone progesterone throughout its active period, preventing the mare from going into estrus.

Corral

Corrals are turnout areas designed to confine horses while providing them with an opportunity for moderate exercise. They are generally square or round, and are often attached to barns for convenience.

A corral should be more than an effective containment area; it should also be a safe, open place in which to romp. A durable perimeter fence 5 to 6 feet high is the standard, with no sharp edges or protruding nails. Be aware that clutter on the ground, or rough surfaces, may invite injury. Size is also a factor; the possibility of injury increases as the size of the enclosure decreases. A typical corral is 1,600 square feet.

Hast thou given the horse strength? hast thou clothed his neck with thunder?

—Job 39:19

Corrected feet

Feet that have been trimmed and shaped to mimic ideal foot conformation. Owners of show horses sometimes engage in correction in order to make their horses appear more attractive. Some believe that corrective shoeing will fix certain leg and movement faults. However, this sort of alteration can impair the ability of the feet to absorb shock, and it may lead to chronically sore feet or even lameness.

Corrective trimming

Trimming performed on a horse's foot to change the shape, to alter the relationship between the toe and the heel, or to correct a significant conformational problem. Trimming can be quite helpful when it is done to correct a defect or help an injured foot heal. However, trimming done for purely cosmetic reasons can lead to foot problems.

Riders that compete in Appaloosa Costume Class events often wear garb that authentically recreates the dress of the Nez Percé Indians.

Corticosteroids

Natural or synthetic hormonal compounds used by veterinarians to treat equine allergies, inflammation, and pain.

Costume classes

Riding events in which the rider's dress plays a significant role. The Arabian Costume Class features riders in desert costumes that might include colorful capes, coats, pantaloons, scarves, sashes, and head-dresses. For Appaloosa Costume Class events, riders don historical costumes from American frontier and settlement days, with special emphasis on costumes modeled after the historic garb of the Nez Percé Indians. Even the gala costume events are judged primarily on the skill of the rider and the performance of the horse.

COUNTER-CANTER

Intentionally cantering on the wrong lead.

Courbette

An advanced maneuver performed by high-level dressage horses. Balancing on his hind legs, the horse leaps forward several times.

➢ *See also Lipizzaner*

Course designer

A specialist in the development of safe, performance-friendly courses for horse shows, particularly in the hunter, jumper, and trail classes. Designers must be licensed for rated or accredited competitions.

Covers

Various kinds of horse coverings, including sheets, blankets, coolers, and rugs. This term can also mean the act of breeding a mare.

Cowboy dressage

➢ *See Reining class*

Cowboy tradition

Western riding is a direct descendant of the riding, roping, reining, bronco busting, and cattle driving techniques practiced by the working cowhands of the early American West. However, that tradition was itself an evolution of the seat and saddlery techniques of the early Spanish cowboys, or *vaqueros* (which inspired the slang term "buckaroos").

The cowboy tradition came to be much admired throughout the world, embodying not only riding practices but also personal qualities, such as toughness, independence, fairness, unflinching honesty, and the ability to survive in the natural world. Many of today's Western riders emulate the techniques, dress, and personal qualities associated with the American cowboy, in a conscious effort to preserve and perpetuate the cowboy tradition.

➢ *See also Rodeo*

The cowboy (and cowgirl) tradition is a primary source of inspiration for many Western riders.

Cow-hocked legs

A conformation fault in which the hind legs appear knock-kneed because the hocks are closer together than the fetlocks or hooves. The defect can excessively stress the hock stifle and hip joints. However, a mild case of cow hocks can actually be a virtue in working horses since such conformation allows the stifles to clear the flanks.

➢ *See also Base-narrow and base-wide*

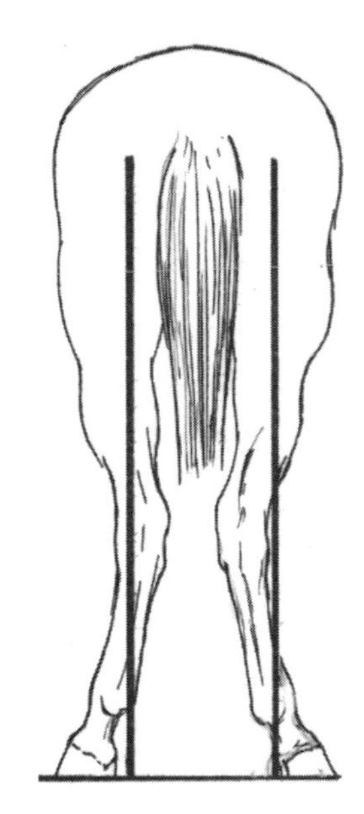

A cow-hocked horse appears knock-kneed when viewed from the rear.

Cow sense

Cows and horses interact powerfully as adversaries in many arena events. Both the horse and its rider benefit from knowing how to assess and deal with cattle. For a rider, cow sense means developing the ability to closely observe and anticipate the behavior of his or her quarry. Cow sense in a well-suited horse is best described as an enthusiasm for working cows coupled with the assertive nature and know-how necessary to control their movements. A horse that has acquired cow sense may be described as "cowy."

Cracked heel

A skin infection that can develop in the fetlock area of a horse's feet. The condition is sometimes called **scratches.** It results primarily from exposure to moist pasture areas. Initial symptoms include reddened, tender skin patches over the heels and or back of fetlocks or both. Deepening cracks with calloused edges may later appear, and the area may discharge pus or blood. In the worst cases, the condition can cause swelling of the leg and excessive scar tissue growth that affects movement of the fetlock joint.

The best preventive measure involves covering the skin with a protective coating of petroleum jelly when a horse must spend time in wet areas. If cracking occurs, an antibacterial cream or antibiotic should be applied on and around the scab.

Cradle

A collar-like device placed on a horse's neck. The purpose of the cradle is to limit movement of the head, preventing the horse from nipping or licking his body and, more commonly, from chewing on bandages or blankets.

Crazyweed (locoweed, milkvetch, poisonvetch) (*Astragalus* and *Ozytropis* spp.)

➢ *See Locoweed; Poisonous plants*

Cremello coloring

Creamy white with light mane and tail and blue eyes.

Crest

The convex line formed by the top of a horse's neck, extending from the poll to the withers. The horse's mane grows along this part of the neck, and the upward angle of the crest is known as the **arch.** A horse with a short, thick neck and a steep arch is said to be "cresty" or "cresty-necked." Stallions often have thicker crests than do mares.

➢ *See also Anatomy of the horse*

Crest release

A maneuver in which a rider moves her hands forward along a horse's crest while the horse is jumping to avoid using the reins for balance.

Cribbing

An equine bad habit, arising principally from boredom or an imbalanced diet. Cribbing involves biting on the edge of a board or clamping onto any solid object with the front teeth while simultaneously arching the neck and gulping air, producing a grunting or burping sound. Some people refer to this practice as **wind sucking,** and horses prone to the behavior are known as **cribbers** or **wind suckers.**

Although it sounds rather innocuous, cribbing may lead to problems including indigestion, colic, excessive weight loss, and prematurely worn teeth. Once established, the habit is difficult to break. It becomes addictive, in part because the horse gets a bit of a "high" from doing it; research indicates that cribbing triggers the release of endorphins, which stimulates the pleasure center in the horse's brain.

To discourage cribbing, keep the affected horse away from objects suitable for the activity and/or employ a **cribbing strap** that prevents the horse from expanding his esophagus in order to suck air.

Crop

A short riding stick or whip used to reinforce leg aids. Under most circumstances, riders do not need to carry a crop. If one is necessary, it should be held in the palm of one hand and applied only if the horse does not respond to leg aids. The crop should be dropped at the first indication of friskiness or odd behavior from the horse, as the horse may have become spooked by a glimpse of the crop.

CREASE NAIL PULLER

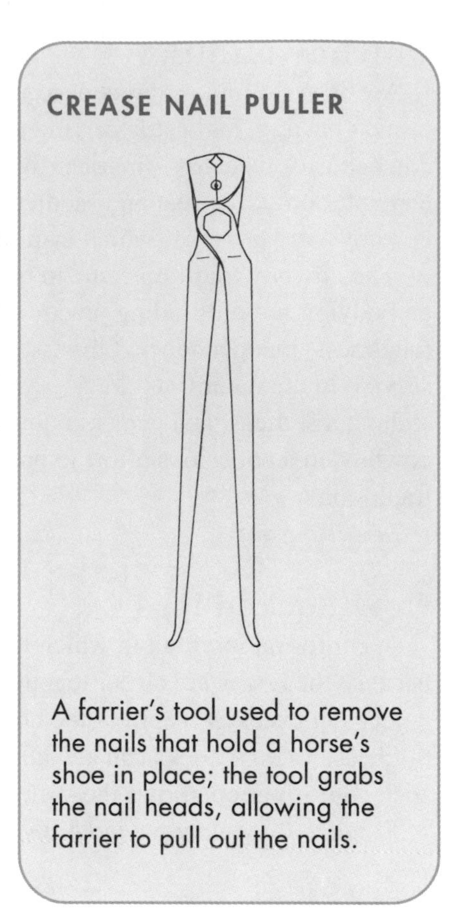

A farrier's tool used to remove the nails that hold a horse's shoe in place; the tool grabs the nail heads, allowing the farrier to pull out the nails.

Cribbing is a behavioral quirk that can cause indigestion, colic, weight loss, and tooth wear.

Crossbred

A horse whose parents are of two different breeds. In some cases, the parental breeds were carefully selected by a breeder; in others, the pairing was orchestrated entirely by the horses. Also known as **crosses.**

Crossbreeding

Breeding done across breeds, usually in order to produce a foal of a certain appearance or disposition, or one of excellent athletic ability. As might be expected, the production of an outstanding foal requires two parents who are themselves outstanding; however, the results are not entirely predictable. A thorough understanding of the traits of both parents is essential to achieving the desired results.

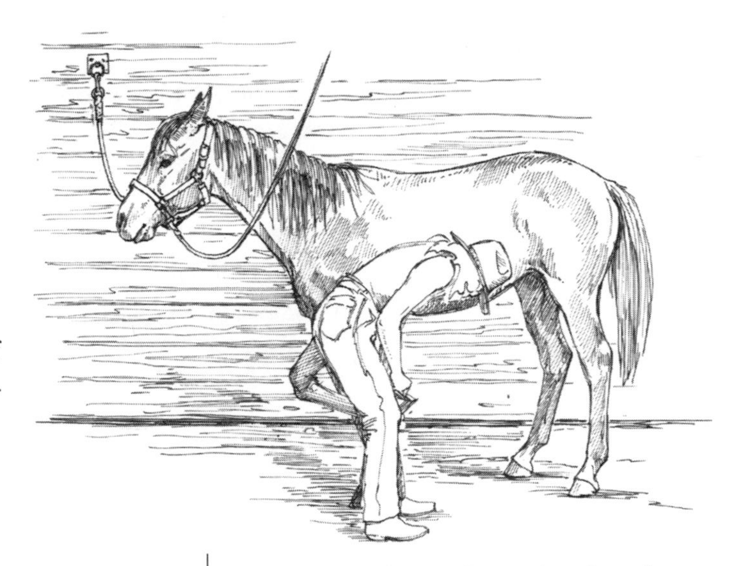

Cross-tying a horse enhances the safety of both the horse and the handler. A horse should not be left alone while in cross ties, as injuries can result if the horse panics.

Cross-canter

A three-beat gait in which the horse leads with one foot in the front and the opposite foot in the back. An undesirable habit created by stiffness in the horse's body; very uncomfortable to ride.

Cross-country riding

A riding event performed on a prescribed course that includes natural or natural-looking obstacles that must be jumped. The course can be quite challenging, incorporating uphill jumps, downhill jumps, and various kinds of terrain. The obstacles used may include logs, fences, ditches, and streams.

Cross-tying/cross ties

Tying a horse on both sides of its head with ropes or light chains. The tie on either side should extend from the halter to a sturdy wall, rail, or post, and the snaps used to attach the ropes should allow for quick release in case the horse should panic.

Cross ties are often used while working on a horse in its stall, but the horse can also be trained to stand calmly while cross-tied in alleys or barn aisles. By limiting side-to-side motion, cross ties diminish the likelihood of injury to both horse and handler, allowing the handler to groom the horse, work on its feet, provide medical treatment, or perform other procedures.

Crotalaria (rattlebox) (*Crotalaria* spp.)

A plant containing alkaloids that can devastate a horse's liver and ultimately him. Crotalaria belongs to the pea family. It is also known as **rattleweed** and **showy crotalaria.**

➢ *See Poisonous plants*

Croup

The top of the rump, corresponding to the sacral vertebrae.

Crowding

An undesirable equine behavior in which a horse that is being led pushes too close to its handler. If you are being crowded by your horse, nudge his neck with your elbow to make him back off.

Crupper/crupper strap

A strap used to hold the saddle or harness in its proper place. The crupper strap attaches to the saddle or harness, loops backward under the tail, then extends back. Crupper straps are frequently used for trail riding.

Cryptorchid

A male horse with either one or two testicles that did not fully descend into the scrotum. Castration of a cryptorchid is a major surgical procedure. The medical term for the condition is **cryptorchidism.**

Cues

Term for the aids or signals that tell a horse what he is expected to do.

Culture

A growth of living cells cultivated in a solution hospitable to the cells. In horses, cultures are tests that are often used by veterinarians to determine whether a mare's genital tract is infected. Also commonly used on cells of peritoneal fluid drawn during colic.

Curb (of hock)

Name given to swelling of and, later, a permanent enlargement or thickening of the calcaneo-cuboid ligament, which is found about four inches below the point of the hock on the inner side. The condition is commonly caused by a sprain that can result when a horse slips on wet ground in such a way that the hind legs slide forward under his belly. Horses with too much angle to their hocks, or sickle-hocked horses, are prone to this type of sprain. Treatment options vary, so contact your veterinarian.

Curb bit

A basic Western riding bit that includes a mouthpiece and shanks and is used in conjunction with a curb strap or chain. Curb bits deliver signals to a horse via leverage pressure applied to the jawbone.

Curry/curry comb

To groom a horse's coat using a curry comb. The comb itself is a round or rectangular tool usually made of rubber or plastic. It has stubby, semi-pointed "teeth" that serve to loosen dirt, hair, and dander for later removal with brushes. It also is used to remove hard-packed dirt and debris from the coat, stimulating oil glands; thus, it is a vital part of grooming for a healthy coat. A curry comb is applied with a vigorous, circular rubbing motion.

➢ *See also Grooming*

Noblest of the train
That wait on man, the
flight-performing horse.

—Cowper, *The Task*

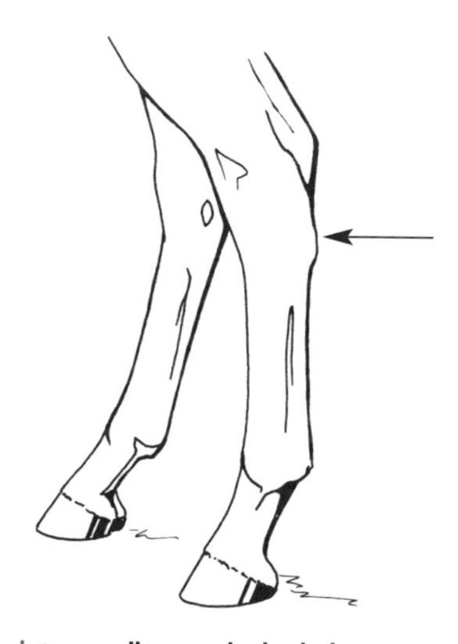

A curb is a swelling on the hock that can become permanent unless treated.

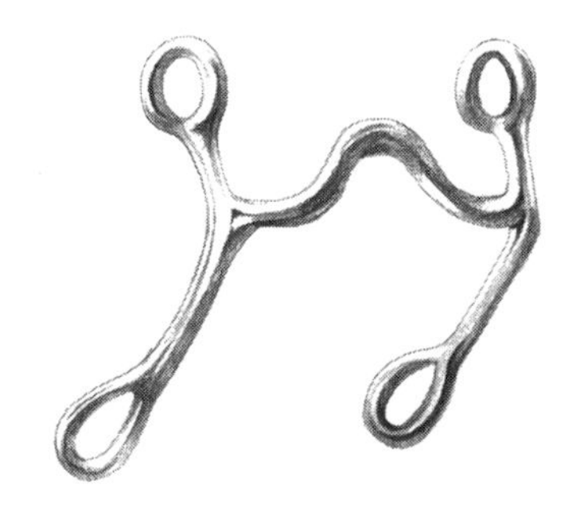

A curb bit works through leverage action.

Cushing's disease/Cushing's syndrome

A metabolic disorder that primarily affects older horses. It is characterized by overactivity of the adrenal glands, frequently in response to a tumor of the pituitary gland. The adrenal glands are responsible for secreting several hormones called corticosteroids; excessive secretion causes symptoms such as frequent urination, weight loss, persistent infections, laminitis, and a coat that fails to shed. Cushing's disease is not curable, although some symptoms can be addressed, and it is ultimately fatal.

Cutter

A rider who participates in the Western show event known as cutting; also, the first rider to enter the herd in the Western event called team penning. This rider "cuts" individual steers from a herd. The term may also refer to a horse that participates in the show event.
➢ *See also Cutting; Team penning*

Cutting

A Western show event in which a horse and rider demonstrate their skill at separating one steer from a herd and successfully blocking the steer's attempts to rejoin the herd. During this event, mounted helpers known as **herd holders** and **turnback riders** control the remainder of the herd and attempt to keep the cut steer facing the horse and rider. Each contestant performs for 2½ minutes.

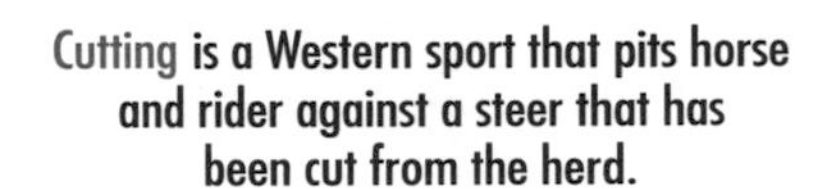

Cutting is a Western sport that pits horse and rider against a steer that has been cut from the herd.

Cutting horse

A horse with extraordinary agility and cow sense that is adept at controlling the movements of a single cow, isolating it, and then preventing it from rejoining the herd. Cutting horses are sometimes called **cutters.**

Cutting (hay)

Each harvest or crop of hay within a given year. A farmer may harvest up to four cuttings per year, and good-quality hay from any of these cuttings can serve as the mainstay of a nutritious equine diet.

Cutting teeth

The somewhat painful emergence of first teeth in a young horse. While cutting teeth, the horse may seek to relieve gum soreness by chewing on materials including wood, metal, and rope. Many behavioral problems encountered while training young horses can be attributed to pain from cutting teeth.

CVI

➢ *See Certificate of Veterinary Inspection (CVI)*

d

Daily feeding rations

➢ *See Feeding and nutrition*

DAISY-CHAIN KNOT

The daisy-chain knot is easy to tie, and it releases quickly if the need arises.

1 2 3 4 5 6

Dally

To wrap the end of a rope around the saddle horn. Dallying is a technique often used by participants in cattle roping events.

Dam

The female parent of a horse. Foals are often described as "out of" a specific dam.

Dandy brush

A stiff-bristled brush used to remove dirt, loose hair, and scurf from a horse's coat. The brush, sometimes known as a **mud brush,** is applied with short strokes and a whisking action. The bristles should be cleaned periodically with a curry comb.

Dappled coloring

Describes a horse whose coat has rings or spots of a color that is different from the rest of the coat.

➢ *See also Coloring*

Dark bay coloring

Desirable brown horse whose coat is so dark that it almost appears black; legs (above white markings), mane, and tail are black.

➢ *See also Coloring*

My horse be swift in flight.
Even like a bird;
My horse be swift in flight.
Bear me now in safety.
Far from enemy arrows,
And you shall be rewarded
With streamers and ribbons red.

—Sioux warrior's song to his horse

d

Dark horse

A relatively unknown competitor that makes an unexpectedly good showing.

A dark horse, which had never been thought of, and which the careless St. James had never even observed in the list, rushed past the grandstand in sweeping triumph.

—Benjamin Disraeli, *The Young Duke*

Darkness, riding in

The first and foremost danger of riding in darkness is the inability of both you and the horse to see obstacles, holes, changes of terrain, protruding tree branches, and other hazards. If you're riding on a road, drivers will have difficulty spotting you, and your horse could be panicked by oncoming headlights. For these and other reasons, it's best to avoid riding in darkness. If you must do so, use reflectors and reflective tape, wear light-colored clothing, and stick to familiar territory.

Deadened tissues

Tissues on a horse's body that have become numb. The condition is most often associated with a too-small halter. Pressure from the halter straps can deaden tissues, reducing or eliminating the horse's ability to respond to cues.

Deadly nightshade (*Solanum* spp.)

A vinelike weed that produces purple flowers and red berries, generally in late summer. Deadly nightshade often thrives in barnyards. Horses will usually ignore this weed in favor of more desirable forage, but a very hungry horse will eat just about anything. If a sufficient amount is ingested, deadly nightshade can live up to its name.

➢ *See also Poisonous plants; Toxic substances*

Dead-sided

A horse that does not consistently respond to cues from its rider. In many cases, the horse is simply having difficulty distinguishing between purposeful signals and other, unnecessary movements on the part of an imbalanced rider.

Dealer

A person who buys and sells horses on a professional basis. Reputable dealers are skilled evaluators of horses, and these dealers can be of service in helping buyers find horses appropriate for their needs. However, a comprehensive veterinary exam is recommended prior to any purchase.

Death camas (*Zigadenus* spp.)

A plant in the lily family that produces clusters of small blossoms and grows primarily in the western and plains states. A horse that consumes several pounds of this plant is likely to die within 48 hours. Symptoms of poisoning include drooling, weakness, respiratory difficulty, and coma.

➢ *See also Poisonous plants; Toxic substances*

POISONOUS PLANT

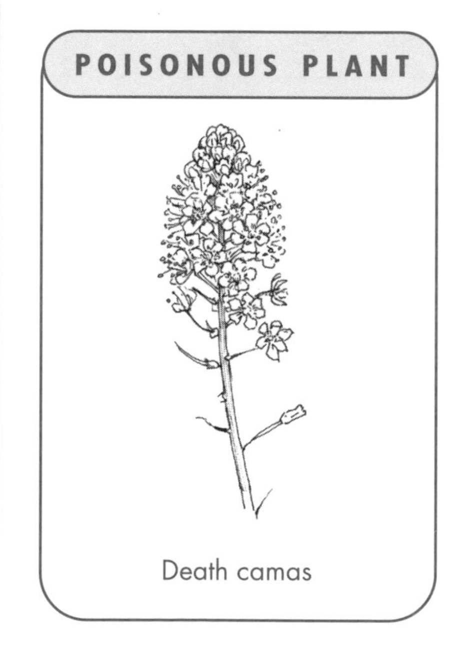

Death camas

Deciduous teeth

The "baby," or temporary, teeth that emerge in the first nine months of a horse's life. The horse's first deciduous incisor appears at birth. All of the temporary teeth have typically been shed by the time the horse is 4 to 5 years old.

➢ *See also Dental care*

Deep digital flexor tendon

A band or cord of connective tissue at the rear of a horse's leg. Like other tendons, this one serves to transmit the action of the muscles to which it is connected. It flexes the bones of the lower leg.

Deep seat

One of the fundamental abilities developed by any accomplished rider. It involves being conscious of your center of gravity, relaxing your thighs, and positioning yourself so that you can sit deeper in the saddle, affording both horse and rider more security. A rider with a deep seat will move in a natural rolling motion with the horse, decreasing the chance of being thrown from the saddle or bumped out of position.

➢ *See also Rider, levels of experience*

Deer flies

Biting parasites that can annoy or panic horses, cause skin reactions, and transmit diseases, such as equine infectious anemia and encephalomyelitis. The flies are members of the *Tabanidae* family, and the females of the species feed on blood.

Controlling deer flies is difficult. However, you can attempt to keep your horse away from the marshes and swampy areas where fly populations are heaviest, and fly repellent products can also be helpful. Use of a fly mask or fly hood can effectively keep the insects away from your horse's eyes. Because the flies prefer sunshine to shade, bringing your horse indoors on sunny days may also provide relief.

➢ *See also Horse flies (tabanids); Parasites*

Defecation

The act of eliminating waste from the bowels. By producing some 50 pounds of manure a day, a typical 1,000-pound horse excretes more than its own body weight every three weeks.

For the horsekeeper, this output poses a sanitation challenge, particularly because much of it is deposited within the confines of the barn and some of it could even end up in the horse's drinking water. Good sanitary practices serve to safeguard the health of both horse and handler, reduce odor problems, and control insects.

Sanitation practices include regular stall cleanings (twice a day is recommended, albeit unrealistically, for stable owners), use of odor control products when appropriate, and proper disposal of the waste material.

"DEEP GOING"

Soft, wet ground, called deep going or heavy footing, can be dangerous for a horse. Be careful not to take your horse at high speed through soft, difficult terrain. If a portion of your pasture tends to be wet and spongy, you can add a layer of sand or gravel to firm it up.

Steeds decked with purple and with tapestry,
With golden harness hanging from their necks
Champing their yellow bits, all clothed in gold.

—Virgil, *Aeneid*

Because horse manure is a prized organic fertilizer, composting is a common and practical means of disposal. In many instances, well-composted manure can be given away to gardeners or sold to greenhouse facilities.

To protect a horse's drinking water from fecal contamination, place large obstacles such as rocks around the water tank, trough, or barrel. This will allow the horse access to the water but prevent him from backing up to or knocking over the water container.

➢ *See also Bedding; Cleaning stalls; Manure, handling; Mucking out stalls*

Defect, conformation

A degree of deviation from ideal body structure. Few, if any, horses are entirely free of what knowledgeable observers would regard as conformation defects or faults. Some defects are matters only of appearance and are harmless in terms of performance or health. Others can lead to serious problems, such as lameness (see chart). Assessing the potential significance of a defect is an essential task for anyone who is buying, breeding, or handling horses.

➢ *See also Buying a horse; Conformation; individual defects and body parts*

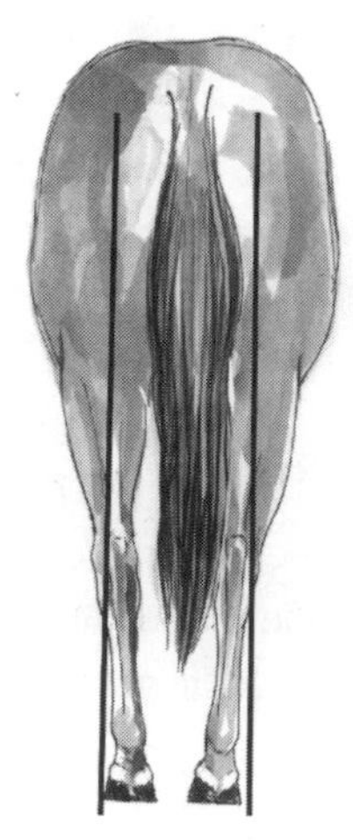

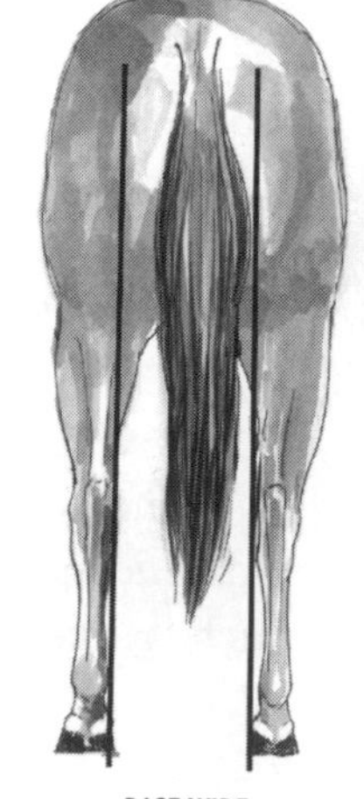

The base-narrow horse (left) has feet set too close together. The base-wide horse (right) has feet set too far apart.

CONFORMATION DEFECTS

These defects may have a significant impact on a horse's performance, condition, or health. However, the presence of a defect or fault does not necessarily predict future problems, and defects may be present in degrees ranging from minor to severe.

DEFECT/FAULT	DESCRIPTION	POTENTIAL IMPACT
Back at the knee (calf knees)	Foreleg condition in which back of leg behind the knee appears hyperextended. Cannon bone is placed too far back.	May be harmless in a pleasure horse but could lead to knee fractures if horse is used in high stress events. Dangerous in horses used for jumping. Do not breed.
Bandy legs	Hind leg defect. Horse appears bow-legged when viewed from rear.	Likely to cause significant problems in hock, stifle, or hip.
Base-narrow	Legs and feet too close together, often accompanied by pigeon toes (toes that point inward).	Horse will probably paddle when he travels, swinging his legs outward. Limits speed and agility. The defect puts additional strain on pastern and fetlock joints, which can trigger windpuffs, ringbone, or sidebone. Hooves wear excessively on outside edge.
Base-wide	Legs farther apart at feet than at forearm or gaskin. Often accompanied by splay-footedness (toes that point outward).	Horse will probably wing inward when he travels. Places greater stress on inner part of limbs, often causing windpuffs, ringbone, or sidebone. Hooves wear excessively on inside edge.

CONFORMATION DEFECTS (continued)

DEFECT/FAULT	DESCRIPTION	POTENTIAL IMPACT
Contracted heel/foot	Hooves too narrow at heel, diminishing blood supply to foot. Hoof does not spread apart normally when horse places weight on it.	If left untreated, can develop into permanent or long-term lameness.
Cow hocks	Hocks turned inward rather than pointing straight ahead, and are closer together than feet.	If moderate, may not pose a problem. If severe, could cause calcified growths on bones in the hock joint and subsequent lameness.
Large ("paddle") feet	Feet too large for horse's weight and body size.	Clumsiness in movement; tendency to trip and fall.
Mutton withers	Very low withers, with little or no "hump" to help hold the saddle in place.	Saddle may slide around, especially on uphill and downhill rides and while mounting.
Offset knees	Shoulder-to-knee bones do not line up properly with knee-to-hoof bones, generally because the cannon bones are not centered under the knees.	Knee joints may not move properly; horse may develop splints on cannon bones. Potential for lameness.
Parrot mouth	Upper and lower jaw do not line up properly (overbite) or lower jaw is malformed.	Generally harmless in stalled horses, although unattractive; in severe cases can lead to chewing problems and inability to tolerate a bit.
Pasterns long and sloping ("Coon-foot")	Pasterns slope excessively, making them low to the ground, and are overly long.	Weakens the pastern; fetlock joint may actually hit the ground when weight is placed on the leg; can lead to bowed tendons and other leg injuries.
Pasterns short and upright	Short pasterns, sometimes almost perpendicular to ground	Increases force of concussion to feet and pasterns, which can ultimately result in lameness; ride is likely to be bumpy and jarring. Navicular disease common.
Pig eyes	Eyes smaller than normal.	May diminish peripheral vision and make the horse more easily frightened by objects he is unable to see clearly.
Round bone ("Tied in below knee")	Cannon bone and tendon too close together from knee to fetlock.	Weakens the entire lower leg, can restrict free movement and lead to unsoundness.
Sickle hocks	Too much angle in the hock; back of hock and cannon bone are not perpendicular to ground.	Strains tendons and hock joint; increases likelihood of curb or bone spavin.
Small feet	Feet too small to support horse's weight. May have been trimmed for show purposes.	May lead to lameness due to navicular disease or contracted heels.

Deficiency diseases

Health problems that are triggered or exacerbated by nutritional deficiency. These include, but are not limited to, colic, respiratory ailments, rickets, hyperparathyroidism, vitamin A deficiency, dehydration, anemia, and abortion. To minimize the possibility of deficiency-related disease, feed your horse a balanced ration that is free of mold and fungi. Ensure that the horse has plenty of salt and fresh, uncontaminated drinking water. Be aware, however, that overfeeding a horse can also lead to health problems.

➢ *See also Dehydration; Diet; Feeding and nutrition; Overfeeding*

Dehydration

A condition in which a horse's body fluids are abnormally depleted. Horses need to drink 5 to 20 gallons of water per day. The amount for a given horse depends on the size, age, and condition of the horse and prevailing weather conditions. Horses tend to drink less in cold or wet weather.

Under normal circumstances, you can avoid the problem of dehydration simply by allowing your horse free access to plenty of fresh drinking water. However, a variety of health problems can adversely affect the horse's water intake or deplete his body fluids. Chief among these problems is diarrhea. Severe cases of diarrhea can dehydrate and kill a horse within a day. Consult a veterinarian if your horse begins producing watery manure.

Dehydration can be a particularly serious problem in a sick foal. The foal may stop nursing, which simultaneously exacerbates the effects of the illness it is experiencing and depletes the foal's body fluids. The early symptoms of foal dehydration include dullness, listlessness, and skin that has begun to lose its elasticity. As the condition worsens, the foal's eyes begin to appear sunken and its skin will further lose elasticity, failing to spring back into place when pinched. Shock, kidney failure, and death will result if the loss of fluids is not quickly addressed.

➢ *See also Diarrhea; Water*

THE PINCH TEST FOR DEHYDRATION

To test for dehydration, pull a fold of skin on your horse's neck or shoulder area away from his body. When you release the skin, it should return almost immediately to its normal position flat against the body. If it remains markedly peaked for 2 to 3 seconds, some dehydration is probably present. A "standing tent" of skin that lasts 5 seconds or longer indicates moderate to severe dehydration that may require the attention of a veterinarian.

Adapted from *Stablekeeping* by Cherry Hill

Delayed foaling

➢ *See Foaling*

Delphinium

A number of perennial flowering plants in the buttercup family. The varieties of delphinium commonly grown in gardens are considered poisonous to horses. Wild delphinium, or tall larkspur, can also sicken or even kill a horse if he is exercised immediately after consumption. Intake of tall larkspur impairs the horse's ability to breathe and use his muscles, and subsequent exercise may overtax the animal's system.

➢ *See also Poisonous plants; Toxic substances*

POISONOUS PLANT

Wild delphinium

Dental care

The primary means of caring for your horse's teeth includes having them examined regularly by a veterinarian or equine dentist. Horses two to four years of age should be examined two times per year, or three times if they are in training or give you cause for alarm. Other horses can generally get by with an annual exam, although twice-a-year checkups may be recommended for elderly horses, as well.

Horses are not prone to cavities, but this does not mean they are free of dental problems. A common problem in younger horses is **retained caps,** in which the horse's "baby" teeth do not properly detach from the gums and fall out. A veterinarian may need to remove the caps in order to prevent gum inflammation and create room for new adult teeth to grow in properly.

In many, but not all, young horses, premolars known as **wolf teeth** erupt along the upper jaw. These can cause the horse's lip to become painfully pinched between the wolf teeth and the bit, so they are routinely removed by veterinarians.

At about age three, a horse may be subject to **impacted teeth** — emerging adult teeth that are pinched by existing teeth and cannot grow upward. The problem is generally self-correcting, but in some cases impacted teeth must be removed surgically. A "tooth bump" on the lower jaw of a horse can indicate the presence of an impacted tooth. If the bump swells rapidly or is tender to the touch, contact your veterinarian.

Uneven wear can cause teeth to develop sharp points or ridges. Under normal circumstances, a veterinarian or equine dentist can remove these by filing the teeth with a rasp — a process known as **floating.** If left untreated, teeth with sharp ridges or points can make chewing a painful and/or inefficient activity, leaving the horse susceptible to starvation, colic, and impaction.

As a horse ages, he may begin to lose teeth or wear them down to the gums. This can also interfere with chewing and digestion. If your older horse begins to lose weight or eat slowly, consult a veterinarian.

➢ *See also Age; Dentist, equine; Floating teeth; Impaction; Teeth; Wolf teeth*

Dentist, equine

Awareness of the connection between horses' behavior problems and dental ailments has led to the development of the specialty known as Equine Dentistry. Previously, the majority of equine dental work was performed by general veterinarians. Most only knew how to float (file down) sharp edges on molars, remove wolf teeth, and pull abscessed or broken teeth. Today, there are veterinarians who specialize in dental care for horses. This means owners have far more resources when they suspect their horse has teeth problems. Be sure to request credentials and references from people who market dental services, since this area of the industry is only beginning to be regulated.

Derby

A racing or riding event held for horses of a specific age. Derbies may be sponsored by regional or national associations, and are generally reserved for four-year-old horses.

➢ *See also Futurities*

SIGNS OF POSSIBLE TOOTH PROBLEMS

- Bad breath
- Chewing difficulty or slow eating
- Drooling
- Dropping wads of food while chewing
- Hay or grain remnants in drinking water
- Head tossing or other avoidance of the bit
- Loss of appetite
- Reluctance to turn in response to bit
- Tender tooth bump on lower jaw
- Weight loss

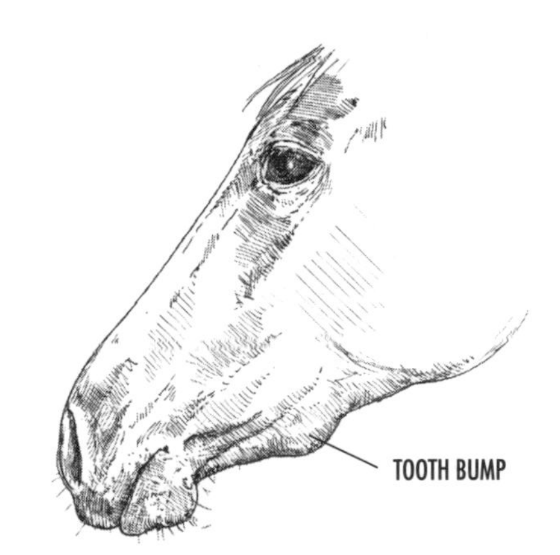

A tooth bump on the lower jaw often indicates the presence of an impacted tooth.

Dermatitis

Any swelling or irritation of the outer layer of skin. Equine dermatitis is commonly caused by infestations of parasites, such as mites; these parasites can also carry diseases. However, dermatitis may also be triggered by burns, allergies, tumors, and assorted irritants found in the environment. Symptoms typically include redness, swelling, lumps, and scratching or rubbing against fences, walls, or similar objects.

The principal way of detecting dermatitis in its early stages is to check beneath the horse's coat, looking for areas of irritation or abnormal swelling. If you locate a possible problem area, have it checked by a veterinarian before washing, clipping, or brushing the area, so you do not end up "destroying the evidence."

➢ *See also Fungal infections; Hives (wheals); Mange; Mites; Parasites; Photosensitization*

Deworming

Internal parasites pose a serious, ongoing threat to the health of every horse. Young horses are particularly vulnerable. A successful deworming program combined with effective pasture management is the best way to minimize this threat.

Dozens of parasites can affect your horse, but bloodworms (strongyles), roundworms (ascarids), pinworms, and bot flies generally cause the most trouble. There are two common strategies for dealing with these unwelcome intruders. The first involves deworming your horse every two months on a year-round basis (once a month for foals). This is supplemented at least twice a year with a purge-strength dose of ivermectin or a similar broad-spectrum drug, which provides effective protection against bot flies as well as all kinds of worms (except tapeworms).

Daily deworming

A deworming program that is carried out on a daily schedule, using a number of drugs. One strategy involves the use of pyrantel tartrate on an everyday basis to keep the horse dewormed. The drug is administered as a pelleted feed additive that also contains alfalfa and molasses.

In most cases, a daily deworming program begins by giving the horse a dose of the broad-spectrum drug ivermectin, in order to clear his system of a wide range of parasites. When followed up by daily doses of pyrantel tartrate, the program is generally quite effective. However, the daily medication does not protect against bot flies, so you will need to treat the horse with a boticide during bot fly season.

Daily removal of manure from pasture areas is another aspect of controlling parasites, as the larvae thrive and mature in the warm, moist environment provided by the manure.

➢ *See also Bot flies; Ivermectin; Parasites; Pyrantel tartrate (Strongid C); individual drugs and parasites*

DESERT HORSE

Desert horses are bred in the world's deserts to thrive in dry conditions. They can resist heat and survive with little water. They are lean (sometimes described as "dry") with little fatty tissue, and their veins stand out prominently on their skin.

See also Arab

Dexamethasone

A steroidal preparation useful in treating swelling, pain, fever, shock, and allergic reactions. Horsekeepers are often advised to keep a supply of dexamethasone on hand. Check with your veterinarian concerning how and when you may need to use this medication.

➢ *See also Steroids*

Diagonal

A diagonal pair of legs at the trot, such as the right front leg and the left hind leg. A trotting horse's legs move in diagonal pairs, creating the steady two-beat rhythm of this gait.

Riders who rise up and down in rhythm with the trot are **posting.** Since they rise and fall in cadence with the diagonal legs of the horse's trot, it is said that they are posting to the diagonal. In order to help horse and rider maintain balance while riding in a circular pattern, on the rail or in a smaller circle, riders are taught to post the "correct" diagonal. It is correct to rise with the outside front leg (leg toward the outside of the circle, or on the rail if riding close to the wall of the arena).

During arena work, **crossing the diagonal** can also be a maneuver taking the rider or driver from one corner of the arena to the opposite corner via the center.

Diagonal jump

A jump of uneven height, with one pole shorter than the other.

Diamond wire mesh

A safe, corrosion-resistant, and durable form of fencing that is often used to enclose paddock and pasture areas. It is sometimes called Kentucky wire.

➢ *See also Fencing materials*

Diarrhea

Loose, runny, or liquid stool. In a horse, diarrhea is often a minor, transient condition caused by eating overly rich forage, a dietary change, or some other factor; however, it can also be symptomatic of a more serious disorder. Watery diarrhea is always sufficient cause to contact a veterinarian, as it can rapidly lead to dehydration. Marked loss of bodily fluids can kill a horse within 24 hours.

Serious diarrhea may signify the presence of an infectious disease, so it is wise to separate a diarrhetic horse from others. Remember to clean your hands, clothing, and equipment after working with a sick horse, so you don't carry the infection to other horses or prolong your own exposure.

In foals

Diarrhea in foals is extremely hazardous, as the depletion of bodily fluids associated with diarrhea may lead to weakness that prevents the foal from nursing. This aggravates the dehydration, initiating a vicious cycle that can result in death within a matter of hours. Early treatment is essential.

Do not trust the horse, Trojans. Whatever it is, I fear the Greeks even when they bring gifts.

—Virgil, *Aeneid*

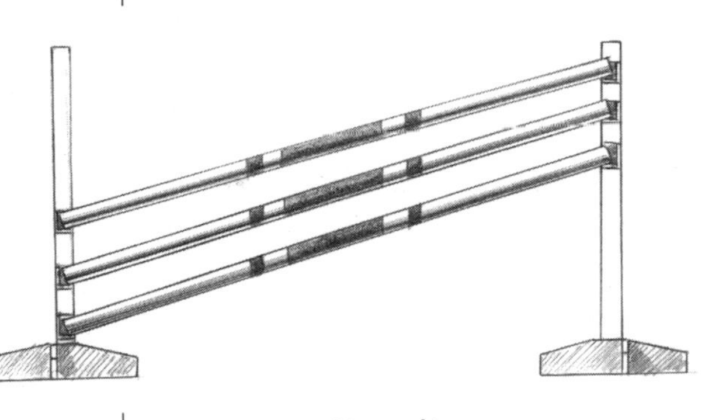

Diagonal jump

Symptoms can include listlessness, weakness, a drooping head or ears, skin that has lost its elasticity, loss of hair and skin irritation around the buttocks, and failure to nurse. In some cases you can help a foal recover from diarrhea by feeding him an electrolyte formula, but it's best to immediately consult a veterinarian. Antibiotics may also be indicated.

One common form of diarrhea in foals is not particularly dangerous. Known as **foal heat diarrhea,** this condition typically presents itself when the foal is 4 days to 2 weeks old and lasts for 1 to 3 days. The stool becomes soft or liquid, but the foal does not evidence other symptoms of illness. In most cases, the only treatment necessary is cleaning the buttocks area and protecting it from fecal matter with mineral oil or petroleum jelly.

➢ *See also Dehydration; Diseases*

Diatomaceous earth

An inert, silica-rich material that contains the skeletal remains of one-celled algae. Horsekeepers sometimes use diatomaceous earth as a feed additive because it inhibits the ability of flies to reproduce in a horse's manure. It thus serves as a fly-control product that does no harm to the horse.

➢ *See also Parasites*

Dichlorvos

An organic, phosphorus-containing compound that is used as a deworming medication. Dichlorvos is particularly effective against the larvae of the worms known as small strongyles. Several treatments given at regular intervals are generally necessary. In some cases, dichlorvos can be effective in controlling worms that have become resistant to other drugs. It is also effective against bot flies.

Diestrus

The period of time in a mare's reproductive cycle when she is not in estrus, or heat. For most mares, diestrus lasts 14 to 19 days and commences immediately following ovulation.

Diet

➢ *See Feeding and nutrition*

Digestion

The horse's digestive tract can be described as a long and winding tube that extends from the mouth to the anus. In between, it includes the esophagus, the stomach, the small intestine, a pouch called the cecum, the large colon, and the small colon.

The horse's stomach is relatively small, with a capacity of just 2 to 5 gallons. However, the bulk of digestion is accomplished in the four-foot-long cecum and the 12-foot-long large colon. These combine to provide a much larger holding capacity of about 25 gallons.

O, for a horse with wings!

—William Shakespeare, *Cymbeline*

DIASTEMA

The technical term for the bars of a horse's mouth (the area where the bit rests).

Fermentation and processing of the food is accomplished by an array of microbes that inhabit the cecum and the colon. Bacteria, protozoa, and yeasts break down cellulose and other fibrous materials, transforming them into the nutrients that maintain a horse's healthy condition. These nutrients include fatty acids, amino acids, and B vitamins. The process also generates heat, which is why feeding a horse on a cold day can help him stay warm.

➢ *See also Colic; Digestive problems; Feeding and nutrition*

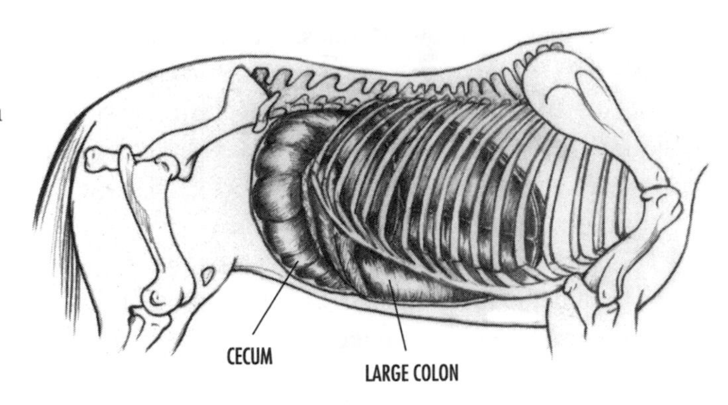

Digestion in a horse is accomplished primarily in the cecum and colon.

Digestive problems

The horse's digestive system couldn't be much simpler, but it is nevertheless subject to a variety of difficulties and breakdowns. These problems are often collectively described as **colic,** which is not a single disorder but a generic term for abdominal pain.

In some forms, colic can be fatal. These include disorders due to blockage, shock, or twisted intestines. Colic can also be relatively minor — the product, in some cases, of nervousness or overstimulation — but it's imperative to consult your veterinarian whenever your horse displays symptoms of abdominal discomfort.

Flatulent Colic

As the name implies, flatulent colic results from the buildup of gas in the digestive system. The root of the problem may be intestinal blockage or gas generated by highly fermentable feed.

Inflamed Intestines

The more technical name for inflammation of the intestines is **enteritis.** Equine enteritis is often due to ingestion — of accumulated sand or gravel, a poisonous plant, or a chemical. However, it can also be triggered by infection or a parasitic infestation in the digestive tract. An affected horse generally experiences pain and diarrhea, and he may injure himself by kicking at his belly or rolling on the ground. Enteritis should be treated by a veterinarian, particularly if it involves diarrhea.

Obstructed Intestines

Intestinal obstruction is a serious matter. Untreated, it leads to colic, dehydration, and death. The blockage is often caused by ingestion of an indigestible material such as rubber, twine, or extremely dry feed. Sometimes enteroliths (mineral stones) cause the blockage. Constipation is a common symptom, as is diminished consumption of food and water.

If treated early, the obstruction can often be flushed out of the horse's system with lubricants, such as mineral oil. However, surgery may become necessary if the lubricants are ineffective.

Spasmodic Colic

Horses with spasmodic colic are subject to intermittent cramps or pain. They may roll over, get up and down frequently, and attempt to kick their

I am that merry wanderer
of the night.
I jest to Oberon and make
him smile
When I a fat and bean-fed
horse beguile,
Neighing in likeness of a
filly foal. . . .

—William Shakespeare,
A Midsummer Night's Dream, II, 2

own stomachs for a few minutes, then resume normal behavior. You may hear rumbling or gurgling sounds emanating from the stomach of an affected horse. If the condition does not improve within a few hours, contact your veterinarian. He or she can prescribe medication to calm the digestive tract and ease the pain.

This form of colic can be triggered by nothing more than nervousness. It can also be brought by abrupt dietary change or overdrinking immediately after vigorous activity.

A horse with some form of colic (abdominal pain) is likely to roll in an attempt to relieve the pain.

Twisted Intestine

Ironically, a twisted intestine can result from your horse's attempts to relieve his own abdominal pain. The intestine may become twisted as the horse rolls on the ground. However, the problem can be triggered by any abrupt movement that involves the intestines, or even in a calm horse with muscle spasms within the abdomen. It can also be caused by excessive gas buildup in the digestive tract.

A twisted intestine puts pressure on nearby blood vessels, thus blocking the flow of food and blood. The real danger results from the absence of blood. Without blood, the affected section of intestine dies quickly, and this, in turn, triggers shock, the release of toxins, gangrene of the bowel, or intestinal rupture — all of which can kill the horse. The only corrective treatment is surgery, but in order to be effective, the procedure must be done shortly after the onset of the intestinal blockage.

The most effective preventive measures are those that diminish the likelihood of colic: careful feeding practices, humane training and handling, and an effective deworming program.

➢ *See also Colic; Digestion*

Digestive system

➢ *See Digestion*

Digital artery

An artery located on the outside and inside of a horse's leg, just above the fetlock. Because the digital artery is easy to find, you should be able to palpate it when taking your horse's pulse.

➢ *See also Pulse and respiration*

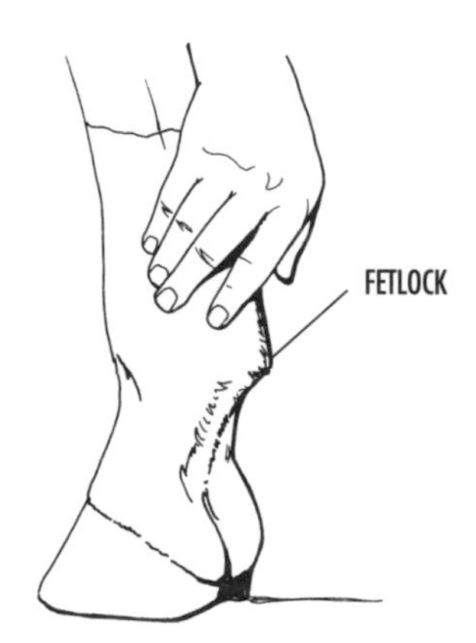

The digital artery is located just above the fetlock.

Digital cushion

A spongy area within the horse's foot, located above the frog. The digital cushion helps absorb shock when the horse is in motion. Also known as **plantar cushion.**

Digital extensor tendon

A vertical band of connective tissue found at the front of a horse's lower leg used to extend the bones in the leg.

Dimethyl sulfoxide (DMSO)

A solvent that is a by-product of manufacturing wood pulp. DMSO is also an anti-inflammatory agent useful in treating horses for a variety of conditions, ranging from the common skin problem known as scratches to the brain tissue swelling from a head injury.

Dinohippus

An early ancestor of the horse. Fossils show this to be the most horse-like of equine precursors.
➢ *See also* Eohippus; *Evolution of the horse*

Direct rein

A method of using the rein in which the rider communicates directly to the horse via his mouth. For example, to turn a horse right, the rider increases pressure on the right rein and the horse responds by turning right. In contrast, a neck rein communicates via the neck by laying the left rein on the left side of the neck, resulting in a turn to the right.
➢ *See also Reins*

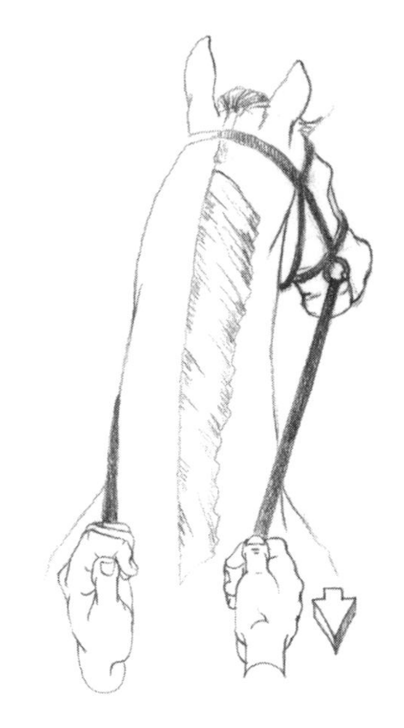

Applying a direct rein involves closing your inside hand while giving with your outside hand.

Dirt flooring

For many years, dirt was the flooring used almost universally in horse barns and stables. Dirt floors are still used frequently in many areas, and they offer some advantages: they are inexpensive; they afford good traction when dry; and they can provide a comfortable, shock-absorbent cushion for horses and handlers. However, there are also a number of drawbacks to dirt floors. They may produce dust in dry conditions and turn into a sea of mud when wet. In addition, maintaining a smooth, clean, and safe dirt surface often requires a great deal of work.

Diseases

The horse, a large and complex creature, is susceptible to a wide range of diseases (see box, next page). But there is a brighter side to the picture. The immune system of a mature, healthy horse is capable of dispatching many of these diseases on its own. In addition, a vaccination program targeted at the diseases prevalent in your area provides protection against those diseases that might otherwise cause your horse serious problems.

The two primary categories of disease in horses, as in other species, are infectious and noninfectious. **Infectious diseases** are caused by external pathogens — small (often microscopic) organisms that enter and damage a horse's body. These organisms include bacteria, viruses, fungi, protozoa, and small worms.

Noninfectious diseases are caused by a problem that exists within the horse's body, such as hormonal imbalances or genetic disorders.

Prevention

The possibility of disease cannot be entirely eliminated, but you can undertake a number of commonsense measures to maximize your horse's health and minimize his chances of contracting disease.

Fundamental to your horse's health is proper nutrition and clean, abundant drinking water. Working with your veterinarian to implement an effective vaccination program is also an essential measure. Because stress plays a role in the development of disease, keeping your horse in a safe, low-stress environment is important, too. Finally, you can help your horse avoid infectious diseases by limiting his exposure to unfamiliar horses as well as equipment and facilities (including saddles and water buckets) that have been used by other horses.
➢ *See also Innoculation; Vaccinations; individual diseases*

DISHING

When a horse does not move his legs straight forward when walking or trotting, but throws them out to the side, it is known as dishing.

EQUINE DISEASES

The diseases listed below are both serious and relatively common. See entries on the individual diseases for further information, including symptoms, treatment, and prevention.

Infectious Diseases
Anthrax
Botulism
Equine encephalomyelitis
Equine infectious anemia (EIA)
Equine protozoal myeloencephalitis (EPM)
Equine viral arteritis (EVA)
Hepatitis (infectious)
Influenza
Leptospirosis
Lyme disease
Piroplasmosis
Pneumonia
Potomac horse fever (PHF)
Rabies
Rhinopneumonitis
Salmonellosis
Strangles
Tetanus

Noninfectious Diseases
Cancer (various forms)
Chronic obstructive pulmonary disease (COPD; heaves)
Cushing's disease (Cushing's syndrome)
Hepatitis (noninfectious)
Laminitis (founder)
Navicular disease
Peritonitis
Rhabdomyolosis (muscles tying up)

Dished face

Characteristic of Arabians, a dished face is a concave profile, often combined with a dainty muzzle and large eyes.

Dishonest horse

A riding or show horse prone to misbehavior.

Disinfectant, for clippers

Clipper disinfectant products are recommended any time your horse shows signs of a skin infection, and any time the clippers you're using have been used on other horses. Disinfecting the clipper blades helps minimize the possibility of transmitting infection from one horse to another.

Disinfecting wounds

Any wound on a horse should be disinfected by treating it with povidone iodine, diluted in distilled water.

Dismounting

Dismounting properly is a fundamental aspect of safe riding. To minimize the possibility of injury, do not begin dismounting until your horse is standing squarely and at a complete halt. Grasp both reins and a portion of the horse's mane in your left hand. Next, remove both of your feet from the stirrups. While holding the pommel of the saddle with your right hand, swing your right leg over the horse's rump and bring the leg down to meet your left leg. At this point, you are supporting the weight of your body with both arms. To complete the dismount, simply slip to the ground, landing on the balls of your feet.

The **emergency dismount** is a somewhat different technique for dismounting on either side while the horse is in motion. It also involves removing your feet from both stirrups, holding onto the pommel of the saddle, and swinging one leg (either leg) over the horse's rump. However, the rider should push up and away from the horse while swinging the leg, landing away from the horse and facing forward.

Because the emergency dismount can itself result in injuries to the rider, it should be used only when absolutely necessary. It is recommended you learn how to do it properly from a certified instructor.

➢ *See also Mounting, technique*

A safe dismount is completed by sliding to the ground while holding onto the pommel of the saddle.

Display sets

Used pieces of tack that are displayed at stables, ranches, arenas, restaurants, clothing stores, and department stores. You may be able to sell your used tack to one of these venues, even if it is cracked or broken.

Disposition

The emotional characteristics, attitude, and "personality" of a horse. A horse with a high-strung disposition (or temperament) is easily upset or distracted. A horse with a bold disposition may attempt to challenge and dominate his riders. A horse with a timid disposition may frequently become afraid and skittish. By contrast, a horse with a calm, easygoing disposition is generally easy to handle and a pleasure to ride.

For most people who are interested in buying a horse or learning to ride, only the horse's physical well-being is a more important consideration than disposition. However, the rider's temperament and needs should also be considered in order to make an appropriate match. Complementary traits are often desirable. For example, an even-tempered, steady horse may be the best mount for a timid or short-tempered rider.

You can best assess a horse's temperament by observing him closely in a variety of situations. Pay particular attention to the horse's body language when you approach, groom, and ride him. Disposition is an inheritable trait, so it may become doubly important if you decide to breed the horse.

➢ *See also Body language, of horses; Buying a horse*

A good horse should be seldom spurred.

—Thomas Fuller

Distance games

Riding competitions or training exercises that emphasize the ability to cover distance. These include ride and tie races, hunter pace rides, point-to-point races, mock fox hunts, and other events.

Distemper

➢ *See Strangles*

Disunited

Describes cantering or loping on different front and hind leads.

➢ *See also Cross-canter*

Diverticulum (blind pouch)

A small pouch in the sheath of a male horse's penis, located adjacent to the urethral opening. Balls of smegma, a waxy substance consisting of dirt, skin cells, and fatty secretions, accumulate in the diverticulum. In some cases these balls (or "beans") become large enough to affect urination, so the diverticulum should be cleaned once or twice a year.

DMSO

➢ *See Dimethyl sulfoxide (DMSO)*

Dock

The flesh and bone part of the tail, located over a horse's tailbone. All tail hairs grow out of this area, so the dock and its underside should be carefully cleaned and rinsed when you shampoo a horse's tail. If itching develops, the horse may rub against fences or walls in attempting to quell the itch. This can cost him many tail hairs — and it may be the start of a tail-rubbing habit that is tough to break.

No one can teach riding so well as a horse.

—C. S. Lewis

Docking

The practice of eliminating a horse's tail by cutting it off between the third and fourth coccygeal vertebrae. The practice is now uncommon and is regarded by some as inhumane because tail-swishing is one way a horse defends itself against flies.

Dogs and horses

The prevailing wisdom can be summed up in five words: Dogs and horses don't mix. Even well-trained dogs tend to make many horses nervous, perhaps because of the longstanding predator-prey relationship between canines and horses.

Interactions between dogs and horses can present a variety of problems. Your herding dog may decide to coerce your horses into undesirable migrations — escorting them down the road to the neighbor's place, for example. A neighbor's dog might chase, frighten, or even bite your horse, turning a pleasant ride into an upsetting or dangerous one. On the other side of the coin, your horse could deliver an injurious kick to an annoying dog.

At home or at the riding stable, separating horses and dogs is the wisest course of action. If you ride in the country, you may be unable to avoid encounters with local canines eager to protect their territory. In many cases, firm voice commands will convince a dog to back off; however, you may want to carry a crop or long whip along on your rides, primarily as a warning device. Turning your horse to face an approaching dog can also convince the dog to turn around, and can help prevent the horse from kicking.

> **EQUINE EVOLUTION**
>
> Donkeys (or asses) are thought to have evolved in northern Africa; zebras, in southern Africa; and horses, in Europe and western Asia.

Dollar bill class

A show-class contest, also known as "Ride a Buck," in which each rider attempts to keep a dollar bill under his or her knee through the walk, trot, and canter. The winner is the rider who holds it there the longest. This class tests the effectiveness of the rider's leg position.

Dominance and handling

To be a successful handler, you must establish a relationship with your horse in which you are the dominant partner. However, this dominance does not typically involve harsh treatment or physical punishment, which are often harmful and counterproductive.

The first step is learning to be thoroughly at ease with your horse. This is important because horses have an uncanny ability to "read" human emotions and will quickly sense the presence of fear — even if you attempt to conceal it. Once you learn to relax around your horse, use your voice and your body language to convey quiet confidence. Unless the horse has a particularly difficult disposition, he will quickly learn to trust you and accept your leadership.

➢ *See also Handling horses; Physical force, use of*

Dominant gene

Each gene is a chemical code that has the potential to transmit a trait or characteristic. A dominant gene is one that will actually produce the trait in an offspring when more than one gene for the trait is present.

➢ *See also Genetics*

Dominant trait

Any trait represented by a dominant gene. Some traits are dominant over all others in the same category.

➢ *See also Genetics*

Donkey

The domestic ass *(Equus asinus)*. This shaggy cousin of the horse often gets neglected by horse-lovers, but he has a devoted following of fans. Hardy and gentle, donkeys are affectionate companions for people and other animals, and can serve as alert guardians of a flock of sheep, goats, or llamas. With their friendly nature, they are ideal for children and make very good mounts for therapy programs. They are strong and sure-footed, making them ideal pack animals.

Like horses, donkeys come in a variety of sizes (from miniatures standing under 36 inches to mammoths that can be 14 hands and up) and colors (gray, dun, roan, black, brown, even spotted). However, like mules, they will often "discuss" your request with you, wanting to know exactly why you want them to do it. Donkeys are fairly easy to keep, but need regular inoculations and foot trimming.

DORSAL STRIPE

A dark stripe along the back of a horse or donkey. Primitive horses exhibited these, including Przewalski's horse. In addition, some modern horses with dark skin have them. Buckskins, duns, and other horses with dark points sometimes have dorsal stripes.

Doors, stall

Stall doors can swing in or out, or they can be mounted on rollers. However, doors that swing in make it difficult for you to enter the stall safely. Most importantly, a stall door should be wide enough to permit your horse to enter and exit without danger of injury. A width of four feet is considered standard.

➢ *See also Barn; Stall*

Doors, trailer

➢ *See Trailering a horse*

Double

To bend or turn a horse sharply. Doubling can be an effective means of correcting a runaway or disobedient horse. The maneuver involves establishing a secure seat and using one rein to bring the horse's head around toward your knee. Because of the discomfort involved, the horse will align his body with his neck, eventually heading where you want him to go. Doubling should only be attempted in an area that offers secure footing and sufficient space.

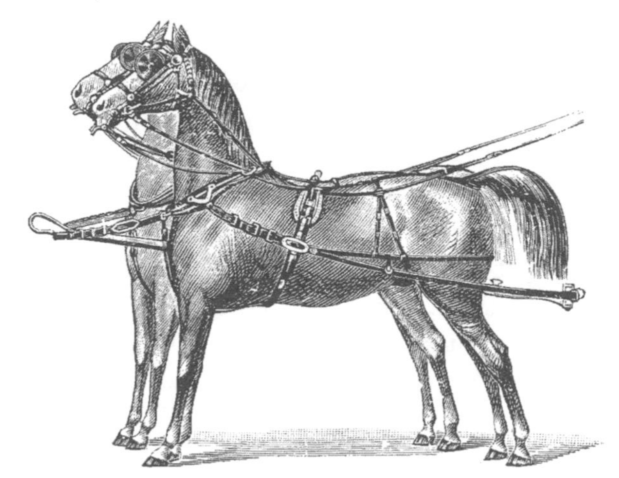

Double bridle

A bridle that enables use of two bits: a curb bit, and a small snaffle bit called a bradoon. The bradoon lifts, and the curb aids in flexion. These bridles are most commonly used on highly trained horses in the dressage and saddle seat classes.

Double-judged

Horse show events that are evaluated or scored by two judges.

Double tree

A swivel-type device used to connect two swingletrees in harness. This is used to connect side-by-side horses when hitched in multiple-horse driving teams.

➢ *See also Harness*

Down command

A voice command used by handlers in conjunction with fingertips pressed down on the top of a horse's head. It signals the horse to lower his head.

Downhill riding

For many novices — and for experienced riders in some circumstances — riding downhill can seem a bit harrowing. Visions of sliding over the horse's head onto the ground and then being trampled upon are not uncommon. However, such falls are quite rare.

To increase your safety and security when riding downhill, start by sitting tall and upright in the saddle. Your upper body should remain perpendicular to the horizon. Remember that the saddle horn is there in part to hold you in place. Because horses have a natural tendency to speed up when going downhill, use checks and releases to maintain control. Unless the slope is treacherous or has an established diagonal trail, head straight down instead of attempting a diagonal descent. If you encounter a steep hill, don't be reluctant to dismount and lead your horse. Remain at the horse's side to avoid being hurt if the horse falls.

DRAFT HORSE BREEDS

For heavy farm work, the heavy horse breeds of 1,600 pounds or more, standing at least 16 hands high, are most suitable. These include Belgians, Clydesdales, Percherons, Shires, and Suffolks. Of these five, only the rare Suffolk has been developed as a plow horse.

Sturdy saddle horses, particularly Quarter Horses, are often used for lighter work. Ponies, such as the Haflinger or the Fjord, are ideal for hilly farm work.

Draft cross

A horse that is a cross between a draft horse and a smaller breed.

Draft harness

➢ *See Harness*

Draft horse

Draft horses include various breeds of heavy, muscular horses that have traditionally been used for farm work, logging, mining, and hauling freight. "Heavy horses" date back to early Medieval times, when a horse known as the "Black Horse of Flanders" inhabited present-day Belgium and northern France.

Draft horses are considered **cold blooded,** due to their typically even temperaments and their tendency to move slowly. Many weigh considerably more than a ton. Among the modern draft horse breeds are Belgians, Clydesdales, Percherons, Shires, and Suffolks. Horses known as warmbloods are actually crosses between a draft horse and an Arabian or Thoroughbred.

There are also a number of short but sturdy breeds that are considered draft ponies. These include Fell, Fjord, Haflinger, and Welsh ponies.

➢ *See also individual breeds*

Drag

In the context of a cattle roping event, a cow that drags is one that plants all four feet and refuses to move after being roped. A person who "rides drag" is one who positions himself or herself at the rear of a column of riders.

Dress

➢ *See Attire*

Dressage

A remarkable combination of art, equitation, competitive sport, and training. It can be defined as the practice of "classical horsemanship," and further described as the art of training a horse to perform precise exercises in a balanced, supple, athletic, and responsive manner. At the competitive level, the sport involves riders in formal attire who control their horses with subtle movement of the hands and legs and slight shifts of weight.

Dressage is an art, a sport, and a method of training both rider and horse.

Dressage (which means "training" in French) has its roots in ancient Greece with the writings of Xenophon, whose enlightened insights on horsemanship still influence the riding styles we know today. Dressage initially reflected the training of horses for military purposes. The first equestrian academy was established in 1532 by Italian Federico Grisone, but dressage as a competitive discipline dates only from the end of the 19th century. It was introduced as an Olympic event at the Stockholm games in 1912.

Horses and riders of nearly all backgrounds could benefit immensely from learning the exercises and movements that constitute dressage. Horses become well-conditioned and are able to fulfill their athletic potential. Riders experience a horse that is a joy to ride and benefit from their own greatly improved riding skills. Over the course of time — and this can take a number of years — horse and rider become a highly skilled team, riding in harmony with seemingly effortless coordination.

A traditional driving bridle, equipped with blinders.

Dressage students begin with the execution of relatively simple exercises. Horse and rider learn to ride in figures of exact dimension, such as figure eights and circles. They practice lateral (sideways) movements and learn to make transitions from one gait to another. As these are mastered, the exercises become progressively more difficult.

For many students and their horses, competition begins with dressage shows sponsored by local riding schools. Many riders then progress to area and regional competitions recognized by the American Horse Shows Association or other equestrian organizations. These competitions follow national rules and are conducted in arenas of specific sizes. Each horse-and-rider team performs solo in the arena, completing the exercises prescribed for their skill level. Judges observe and score each performance.

Horses and riders begin participating in dressage at the Introductory or Training Level, progressing to First Level, Second Level, Third Level (medium difficulty), and Fourth Level; only then do they graduate to intermediate levels. The highest level of accomplishment is the Grand Prix, which is the standard applied at the Olympics and at world championships.

Dressing room

Many horse trailers come equipped with walk-in dressing rooms, which may double as tack rooms. The typical dressing room adds about four feet to the length of a trailer, but it provides a handy, secure place to store essential supplies — and, of course, a private dressing space.

Drills, mounted

Riding performances that allow students and others to demonstrate team horsemanship skills such as precision, timing, cooperation, and coordination. Drills may be directed by voice or hand signals or with whistle signals, and they are often performed to music. The riders proceed on a predetermined course or pattern. Common drills include the Figure Eight, the Crossover, the Snail, and the Tricycle.

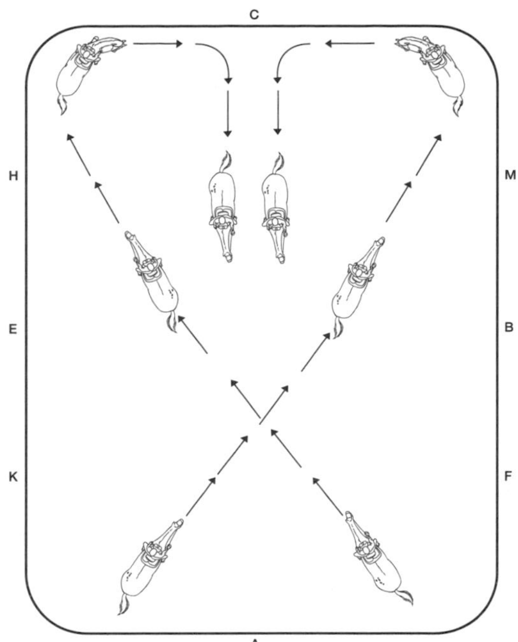

Drills such as the figure eight allow riders to improve their horsemanship skills and practice teamwork.

Drinking water

➢ *See Dehydration; Water*

Drive

The force of a horse's forward movement.

Driving

A pleasure activity and sport you might consider taking up if you love horses and horse shows but don't consider yourself a rider. It involves horse-powered vehicles ranging from small, two-wheeled carts to buggies and coaches. Driving can consist of companionable rides down country roads, appearances at horse shows, or participation in rigorous competitive events. It can involve teams of large horses or a single-passenger cart and one miniature horse. The choice is yours, and the options are many.

Exhibitors in the pleasure driving classes are evaluated for the manners and performance of their horse or horses. In turnout class, participants are judged on the appearance and appropriateness of their harness and attire. In the reinmanship class, judges assess the skill and talent of the driver. Those who participate in the competitive sport of **combined driving,** which combines driving and riding, are judged in the areas of dressage, cross-country driving, and obstacle course.

➢ *See also Combined driving; Pleasure driving class*

It's easy to participate in the sport of driving. All it takes is one small horse, one small cart, and one enthusiastic driver.

Driving bridle

A bridle designed for use when driving (rather than riding) horses. These bridles include cheek pieces, brow band, crownpiece, throat latch, and reins, and they are available in recognizably English and Western configurations. Traditionally, blinders and shadow rolls were added to driving bridles. These special pieces of tack limit a horse's field of vision, minimizing the possibility that the horse will become frightened by something he sees.

Driving pony

A pony used to pull a cart or wagon. Most breeds are well-suited for this purpose.

Driving tack

Tack used when driving horses, all of which generally fits into either the light harness or draft harness category. Driving tack can include a halter, bridle, bits, reins, and the harness, which consists of the collar, straps, and a number of smaller pieces.

➢ *See also Driving bridle; Dutch collar; Harness; Tack*

Droncit (praziquantel)

A canine deworming product that can also be used to rid horses of tapeworms. The recommended dosage should be increased to reflect the horse's weight, which is typically far greater than a dog's.

Dropped noseband

A corrective device to be used only with a snaffle bridle and by experienced horsemen. If used improperly, this noseband can cause permanent damage to the sensitive nasal bone and create breathing problems that hinder performance. It aids in keeping a horse's mouth quiet and responsive, especially when the horse tends to hang his mouth open, chew excessively, or flip his tongue over the bit.

The front should be four fingers' width above the nostrils and the back should rest in the chin groove. It is best to hold the front in place by attaching it to the cheek pieces. Beware that the noseband not be allowed to rub or pinch the corners of the horse's mouth. It should be loose enough to allow the horse to relax his jaw, but snug enough to aid in solving the problem it's being used for.

Dropping

Releasing a horse too early in attempting a jump. Dropping can cause the horse to fail to jump and the rider to fall backward.

Drop the shoulder

A riding error that involves leaning excessively to the inside while making a turn.

Drying a horse

➢ *See Bathing horses; Cooling down (after exercise)*

Duckett's dot

On a horse's hoof, the point at the center of the sole.

DRUGS AND HORSES

Appearance- or behavior-enhancing drugs are sometimes given to horses by unethical dealers or horse show competitors. The drugs may have the temporary effect of calming an unruly horse or making a sick or lame horse appear healthier. However, both the unwary buyer and the horse himself stand to suffer after the drug has worn off.

Many horse show associations attempt to deal with this difficult problem by conducting random drug testing and penalizing dishonest exhibitors. A horse buyer can take a similar approach by insisting on a veterinary exam that includes the analysis of a blood or urine sample prior to purchase.

Dun coloring

A yellow or gold body and leg coloring, often accompanied by a black or brown mane and tail. A dun often has a stripe down its back (dorsal stripe) as well as stripes on the legs and withers.
➢ *See also Coloring*

Dun coloring

Dutch collar

An alternative to the neck collar, the Dutch collar is designed to fit over a horse's chest. Variations include the breast collar or heart collar. The collar is a primary part of the driving harness.

Dutch Warmblood

A high-performance breed of sport horse produced and perfected by breeders in the Netherlands after World War II. They can be high spirited but are generally good natured and willing. Standing 16 to 17 hands high, these long-bodied horses have smooth gaits and excel at jumping and dressage. They are bay, brown, and chestnut, with some white markings allowed.
➢ *See also Warmblood*

Dwarf horse

As distinguished from a miniature horse, a dwarf horse is a genetically deformed animal that is abnormally proportioned. For example, its head may be too big for its body. Such deformity is heritable.

Dwell time

Time spent atop a horse with no objective other than teaching the horse to stand calmly and quietly.

Dystocia

A difficult or medically problematic birth.
➢ *See Foaling*

He is pure air and fire; and the dull elements of earth and water never appear in him, but only in patient stillness while his rider mounts him. He is indeed a horse, and all other jades you may call beasts.

—William Shakespeare, *Henry V*

e

Ears

A horse's ears are both receptors and communicators. In addition to channeling sound, the ears communicate much about how the horse is feeling and what he may be thinking.

An upset horse will often lay his ears back flat against his neck and show the whites of his eyes. Aggressive action is likely to follow. However, holding the ears back can also convey other states of mind: that the horse is tired, for example, or that he is earnestly listening to your voice. Context and circumstance are always important in understanding a horse's "ear talk."

Ear Anatomy

The inside of a horse's ear consists of an ear canal at the base that descends in a straight line for about two inches. The canal then turns sharply and extends to the eardrum. This bend serves to keep debris away from the eardrum. The inner ear also contains fine hairs that filter out debris and insects. Because these hairs play an important protective role, they should not be removed or trimmed.

Ear Problems

The primary problems that affect equine ears are injury to the eardrum, internal infections, and minor external problems. Damage to the eardrum will result in hearing impairment or deafness. The eardrum is well protected under normal circumstances, but penetration of the ear by a sharp object (such as a tree branch), blows to the head, and middle-ear infections can cause damage.

Middle-ear infections are caused by bacteria or viruses that enter the ear via the bloodstream or the Eustachian tube, the canal that extends from the throat to the ears. Symptoms of infection include fever, head tossing, ear rubbing, an accumulation of discharge in the ear, pain upon chewing, drooping ears, and reluctance to have the ears handled.

Horses are more commonly subject to external ear problems, but these conditions tend to be relatively minor and do not result in hearing loss. Insect infestations, frostbite, warts, skin infections, and small tumors are among the problems. Consult your veterinarian any time your horse evidences ear discomfort or you discover unusual lumps, inflammation, or fluid discharge.

➢ *See also Body language, of horses*

Earthquake

If you live in an area that could be threatened by earthquakes, don't forget your horse facility when you implement emergency preparedness measures. In the barn, secure anything that could fall and injure the horses or could burst into flames. These items include water heaters, shelving, tools, and flammable substances. Safety latches on cupboards can help keep the contents secure, and you may want to store your hay in a separate facility so it doesn't contribute to a barn fire.

EAR TALK

Here are some of the common equine ear positions and what they often reveal about a horse's state of mind. Bear in mind, though, that the meaning of these positions may vary from horse to horse and circumstance to circumstance.

I'M AFRAID

I'M ATTENTIVE

I'M ALERT AND FRIENDLY

I'M ANGRY!

I'M UNHAPPY OR IN PAIN

If an earthquake occurs, bring your horses outdoors to a safe, open area away from structures that could collapse and bodies of water that could flood. Keep the horses in this safe area until the tremors pass, and remember that aftershocks could follow. Fill containers with clean drinking water, in case your regular drinking water supply becomes unavailable or contaminated.

Eastern equine encephalomyelitis (EEE)

➢ *See Equine encephalomyelitis*

Easy keeper

A horse that requires little or no supplemental grain to maintain his health and body weight. Horsekeepers also use the term "low-octane horse," which means the same thing.

Edema

An abnormal accumulation of fluid within, or surrounding, connective tissues; also known as fluid retention. The most common location for edema in horses is in the lower half of the hind legs of idle horses, or around the udder of pregnant mares. The best prevention and/or treatment in either situation is mild exercise. Exercise increases circulation, which helps the fluids diffuse and allows the tissues to regain normal tone.

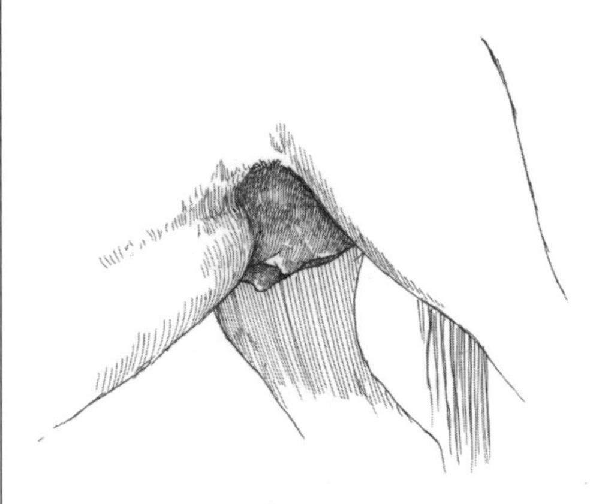

Edema causes swelling around the udder and abdomen of some pregnant mares.

Egg-bar shoe

A special oval-shaped horseshoe designed to help problem hooves. Unlike standard shoes, the egg-bar shoe provides support beneath the heel, which is especially beneficial when a horse has navicular disease, collapsed heels, underrun hooves, and flat soles. Thoroughbreds and warmbloods are prone to these problems.

Egg-bar shoes are also used to give more support on soft ground. However, if your horse is wearing these shoes, do not turn him out into a muddy area. Mud, manure, straw, and other debris easily collect inside the egg-bar shoe. For this reason, it is especially important to clean the horse's hooves daily.

Many competing horses wear egg-bar shoes in the ring. These include hunters, jumpers, dressage horses, cutting horses, and reining horses. Increasingly, pleasure and trail horses wear them as well, as owners learn of their benefits.

➢ *See also Lameness; Shoes, horse*

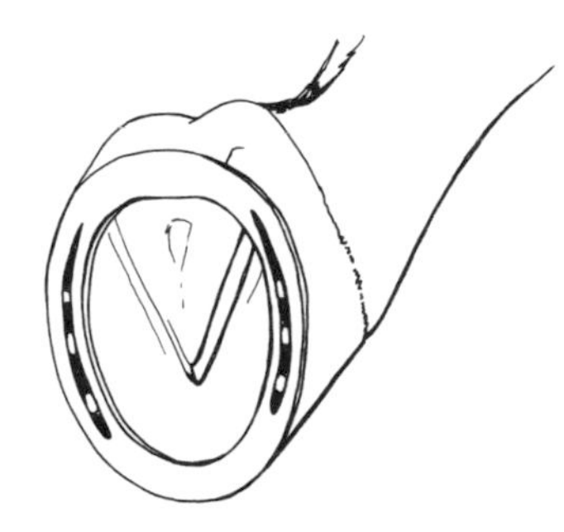

Unlike standard shoes, the egg-bar shoe provides support underneath the horse's heel.

Ehrlichia risticii

An organism that causes Potomac horse fever, a serious equine disease. The organism is spread by the larvae of parasitic flatworms known as flukes, which reside in freshwater snails.

➢ *See also Flukes; Potomac horse fever (PHF)*

EHV

➢ *See Equine herpes virus (EHV)*

EIA

➢ *See Equine infectious anemia (EIA)*

Electric fencing

As a rule, electric fencing is neither visible enough nor sturdy enough to use as a primary fence for horses. Electric "tape" solves many of these concerns, however, making electric fence a more viable option. There also are a number of secondary applications that can be useful for horsekeepers.

An electric wire strung atop a wooden fence can prolong the life of the fence.

You may want to use electric fencing in conjunction with a more substantial fence, such as a wood fence. A strand or two of electrified wire strung above a wood fence can prevent horses from chewing on, leaning on, or rubbing against the primary fence, thereby prolonging its life and reducing the need for maintenance. A strand of electric fencing along the inside of a barbed-wire fence can keep horses away from the potentially injurious barbed wire. And electric fencing can be an effective means of temporarily dividing pasture areas, keeping your horses grazing in one area while another area is allowed to grow.

For maximum effectiveness (in most situations), string electric fencing 30 to 36 inches above the ground. Immediately after turning on the electrical charge, encourage your horses to investigate the fencing and learn about the shock that accompanies contact.

It's important to check and maintain electric fencing. It is subject to shorting from contact with vegetation, metal, or just about anything else that touches it. You can use an electric fence tester to ensure that an electrical charge is present. Trim grass, bushes, and trees regularly to keep them away from the fencing and to prevent malfunction.

➢ *See also Fencing materials*

Electric horse walker

➢ *See Hot walker*

Electrolytes

The "salt" molecules or minerals normally found in the bloodstream, including sodium, potassium, calcium, and magnesium. These elements are essential to maintaining a horse's normal body functions, and they can be critical when the horse is extremely active, under stress, or dehydrated. You can address an electrolyte imbalance by feeding your horse powdered electrolyte mixes or pastes, but it's wise to first consult your veterinarian. You can also make your own electrolyte solution (see box) rather than relying on commercial preparations.

HOMEMADE ELECTROLYTE SOLUTION

To make an effective electrolyte solution for your horse, start with one gallon of warm water. Add the following ingredients and mix well:

- 4 tablespoons corn syrup or dextrose
- 2 teaspoons table salt
- 2 teaspoons baking soda

From *Small-Scale Livestock Farming* by Carol Ekarius

Embedded seeds

Sharp seeds from cheatgrass, foxtail, and other plants can become embedded in a horse's mouth, where they may trigger sores or abscesses. Symptoms typically include swollen gums or cheeks or difficulty eating. Consult a veterinarian if you encounter these symptoms.

Seeds may also become embedded in a horse's eyes. If they remain under an eyelid, the normal motion of the eyelid can damage or irritate the cornea. Have your veterinarian periodically check your horse's eyes and show you what to look for yourself. Irritants can generally be removed or flushed out.

Emergency dismount

➢ *See Dismounting*

Emergency first aid

➢ *See First aid*

Encephalomyelitis

➢ *See Equine encephalomyelitis*

Endorphins

Natural morphinelike chemicals that are produced in a horse's brain. When endorphins are released, they serve to relieve pain, decrease heartbeat, and calm the horse.

Endotoxemia

A condition in which endotoxins released by pathogenic bacteria in the intestines seep through the intestinal wall and enter a horse's bloodstream; triggered in many cases by a uterine infection or overconsumption of grain. The poisons damage blood vessels, in turn depriving body tissues of needed oxygen. Because the ends of the arteries in a horse's hooves are particularly vulnerable, endotoxemia can quickly lead to laminitis, or founder. Consult your veterinarian any time your horse gains access to the grain bin or shows signs of founder, or if you notice abnormal discharge from your mare's vulva.

➢ *See also Endotoxic shock; Laminitis*

Endotoxic shock

If a horse's intestine becomes twisted, which can have various causes (such as when a horse rolls on the ground in an attempt to relieve the pain from colic), the blood supply to that portion of the gut will quickly be cut off. The affected part of the intestinal wall will die, allowing endotoxins from the intestine to enter the bloodstream in large quantities. As the horse's body systems are shut down by these poisons, the horse will go into endotoxic shock. The condition is difficult to reverse and can be lethal.

Symptoms of endotoxic shock include an accelerated heart rate, weak pulse, dehydration, cold extremities (feet and ears), and mucous membranes that have turned pale or have reddened. Contact your veterinarian if these symptoms appear. Good feeding practices and an effective deworming program, both primary defenses against colic, also reduce the chances that your horse may suffer endotoxic shock.

What do we, as a nation,
care about books?
How much do you think
we spend altogether
on our libraries, public
or private, as compared
with what we spend
on our horses?

—John Ruskin (1865)

Endurance riding

A trail riding competition, now an Olympic event, originally inspired by horseback explorations of the North American frontier. During this demanding event, horse and rider cover a predetermined course within a maximum amount of time. The winner is the fastest rider whose horse also meets the rigorous soundness guidelines specified by the competition, making endurance riding both a race and a test of the horse's conditioning. Riders generally cover about 50 miles per day, with a maximum three-day ride of 150 miles. Veterinarians examine the horses at periodic intervals. This sort of competition is open to all breeds, and novice endurance rides of 25 to 30 miles are offered for less experienced riders and/or horses.

BEST BREEDS FOR ENDURANCE RIDING

In North America and Great Britain, Arabian horses have dominated the sport of endurance riding. Although they often stand no more than 15 hands high, their stamina, speed, and heart have helped them meet the sometimes grueling conditions of an endurance course.

Enemas (for foals)

A newborn foal may have trouble completing his first bowel movement. It's an essential accomplishment for the foal, because constipation in newborns is frequently fatal.

If your foal does not have a bowel movement within the first few hours of his birth, it's time to take action. Begin by removing the pellets that have accumulated in his rectum. This can be done by wearing a lubricated latex glove, gently inserting one finger into the rectum, and taking out the pellets.

It is likely that hard pellets remain in the foal's rectum. To soften them, administer a standard adult human enema, inserting it with an enema tube. Alternately, you can make your own enema solution by mixing one cup of mineral oil with one quart of warm water.

The enema should stimulate bowel movements, which will soon become soft and yellow. If the constipation continues, the foal may need a small amount of mineral oil (½ cup to 1 cup) administered orally, or an oral dose of magnesium hydroxide, which functions as a laxative. If this, too, proves ineffective, contact your veterinarian.

➢ *See also Constipation*

Engagement

A condition in which a moving horse has shifted his weight to the hindquarters and is striding forward with the hind legs. Engagement serves to increase the horse's energy and propel forward movement. An engaged horse has a rounded top line, lowered croup, and flexed abdominal muscles.

➢ *See also Collection; Movement, conformation and*

English bridle

There are several different types of English bridle. The most common are snaffle, pelham, and full bridles. All have similar construction that includes:

- a crownpiece (top piece of leather that runs behind the horse's ears, includes the throatlatch and upon which the cheek pieces are attached)

- cheek pieces (which attach the bit to the crownpiece)
- a noseband (styles vary with discipline)
- a throatlatch (which runs from ear to ear under the throat and prevents the bridle from being pulled off the front of the face)
- reins (plain, braided, laced, plaited — single or double, depends on horse's needs)

A snaffle bridle holds one snaffle bit, which has a single rein attached. A pelham bridle holds one pelham bit, which usually has two reins attached for more precise control; however, a converter can be used for novice riders who can handle only a single rein with confidence. A full bridle holds two bits, a small snaffle (also known as a bridoon), and a curb, with a single rein attached to each; thus, it is a bridle that requires more education and experience to master.
➢ *See also Bit; Bridle; Cavesson; Curb bit; Pelham bit; Reins; Snaffle bit*

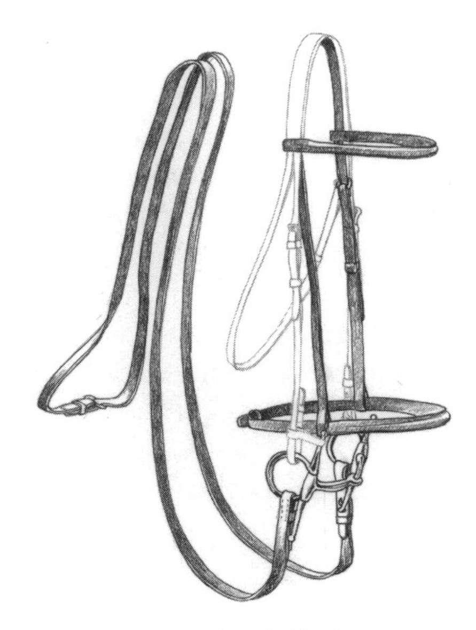
A typical English bridle

English competition

Competitive events, or horse shows, that cater primarily to riders in the English disciplines. Such shows may include classes for Hunter Pleasure or Under Saddle, Hunter Over Fences, Jumper, Saddle Seat Pleasure, Gaited Classes, Equitation for Hunt and/or Saddle Seat riders, and more.

English competition includes several "classes," and each class may be divided into a number of competitive events (see box below). In showmanship classes, judges evaluate the in-hand skill of a horse's handler as well as the "turnout" or overall appearance of the horse. In conformation classes, judges evaluate the body structure, travel, and form-in-motion of each horse, particularly of breeding animals. Each entrant is compared to the established ideal for the relevant breed.

Equitation classes are those in which the skill of the rider is primarily what is being evaluated. Performance classes encompass a number of events designed to highlight the horse's performing abilities, which are judged according to the particular requirements of the show or sponsoring association.
➢ *See also English riding; individual classes and events*

TERMINOLOGY

Flat class. Walk, trot, canter in front of a judge.
In the hole. Refers to the rider waiting after the rider who is on deck.
On deck. Refers to the rider waiting to enter the course after the current rider finishes.
Rider up. Refers to a rider who is mounted and ready to go.
Trip. One completion of a course over fences.

ENGLISH COMPETITION CLASSES

Showmanship. Skill of handler in presenting a halter horse and turnout of horse are judged.

Conformation (Halter). Body structure of breeding and/or performance animals is judged.

Performance. Horse's performing abilities are judged. Includes events such as English Pleasure, Park Horse, Show Horse, Hunter, Jumper, Dressage, and Combined Training.

English equitation classes

English riding competitions in which the skill of the rider is evaluated, with an emphasis on the ability to use natural aids in controlling the horse. The standard English equitation classes are hunt seat equitation, which may include group rail work, individual patterns, and fence jumping; and saddle seat equitation, which is a similar but somewhat more animated and showy display of horsemanship. Each category has its own requirements concerning appropriate tack and attire.

➢ *See also Equitation*

Hunt seat equitation

English performance classes

English riding competitions in which judges evaluate the performing ability of the horses. Events in these classes can include English pleasure, park horse, show horse, hunter, jumper, dressage, and combined training.

Saddle seat equitation

English pleasure horse

A type of horse (not confined to one breed) that is structurally well-suited for providing the smooth, comfortable travel desirable in pleasure riding.

English riding

English riding has several roots. Horses have been used in Europe from ancient times to the early 20th century to get from one place to another before the motorcar, to pull heavy loads, to hunt, to play polo, to race, and to carry soldiers and weapons in war. The ancient Celts of Germany, France, and later the British Isles raced horses as sport and to prove whose horse was better. All of these activities have resulted in a body of knowledge of how best to sit on a horse in an English saddle and direct its progress, generally known as English riding.

Disciplines

Because of its genesis in military and aristocratic life, English riding tends to be more formal than other forms of riding. Much of the emphasis on performance mimics the ways horses were ridden in former, some would say more elegant, times. Several specific disciplines are gathered under the general heading English riding. These are:

- Hunt seat, a descendant of the old way of fox hunting; anyone who rides in a close-contact or all-purpose English saddle is riding some form of hunt seat
- Balanced seat, a new, improved form of hunt seat that borrows from dressage
- Forward seat, a form of hunt seat invented by Italian cavalry officer Federico Caprilli in the early 20th century after his study of the psychology and body mechanics of the horse; this type of riding is most often used for jumping

- Dressage, the French word for training; originated with the training done with cavalry horses
- Polo, invented in India and played on small horses called ponies (though they may be taller than an actual pony) in modified English tack
- Steeplechase, invented by two Irishmen in the middle of the 18th century when they placed a bet about who could ride faster from one church steeple to another, jumping anything in their path; today, both amateurs and professionals ride in this sport, against each other.

Both amateur and professional riders practicing hunt seat, balanced seat, and forward seat generally include jumping in their training; dressage riders do not, preferring to perfect and improve the horse's movement without obstacles. Dressage riders almost "dance" with the horse in choreographed patterns, using a variety of gaits, tempos, bends, and extensions, just as a ballet dancer would.

Preferred Horse Breeds

For classic hunter riding, lighter, faster horses with sensitive sides, such as Thoroughbreds and Thoroughbred crosses, are still preferred. Quarter Horse crosses, and even registered Quarter Horses as long as they are longer and leaner than the foundation style of Quarter Horse, probably run a close second in jumper and hunter-jumper rings. Basically, any short-coupled horse that can jump can be used. Other common breeds are Morgans (although they can get rather long-backed), Appaloosas (although purist judges sometimes mark them down because of their Western appearance), and the warmbloods of Northern Europe. You might also see *Selle Francais* horses, a light Thoroughbred-type horse from France, or even a draft horse–Thoroughbred cross. No matter the breed, the horse should look, act, and move as if he might have been a mount in one of those long-ago steeplechases or fox hunts.

For dressage, warmbloods with good extension (like human dancers) are often favored, although any breed, including gaited horses, can be taught dressage. Polo ponies are small, sturdy, quick horses of any breed. Endurance horses may be of any breed, as long as they have stamina, tough feet, and the ability to move through alien territory without becoming upset or frightened. Akhal-Teke horses, the ancient horses of the Russian steppes, are particularly good at endurance, although they are fine jumpers, as well.

Attire

Attire for hunter riding mimics the fashion of the upper classes of a bygone era, including breeches and formal jackets for all forms of hunt seat. Men wear ties and women wear chokers (called ratcatchers), which are kept in place with a decorative, equine-motif pin or a hidden button. English riders wear a hunt cap, a brimmed, black-velvet helmet that looks like the caps gentlemen wore in days of old. (Women, back then, rode sidesaddle, in long skirts called habits and with veiled fashionable hats to match. Sidesaddle is a subdiscipline of hunt seat; even today there is a U.S. national champion in the sport.)

REQUIRED SKILLS FOR ENGLISH RIDING

Riding a horse English-style requires development of leg strength and a sensitive feel in the hands and in the seat. The horse is asked to use his body in ways that both serve the rider and keep the horse sound. The horse is suppled and encouraged to bend, to balance himself and not pull on the rider's hands, to go forward when asked with light leg pressure, and to stop or slow down at minimal direction (with legs and hands) by the rider.

For the lower levels of dressage, hunt attire is worn. In international-level competition, riders wear either black single-vented dress coats or cutaways. With the dress coat, a derby is worn; with the cutaway, a top hat is used. Tall boots are worn for both dressage and hunt seat, and they are usually black (although brown is permissible as long as the rest of the leather — belts, gloves, and so forth — matches). White gloves are worn for upper levels of dressage, as well as black leather boots. The outfit also includes breeches tucked into the boots. (For hacking or lessons, short boots may be worn or chaps over jeans. Neatness usually is preferred, though, so polo shirts with collars are preferable to T-shirts in most riding arenas.)

Competitions and Events

Riding in all English disciplines depends on the rider's skill, knowledge of technique, and balance while using minimal equipment: a flat saddle with no horn, relatively mild bits in the horse's mouth, and spurs without teeth. Some competitions are judged on the rider's position, knowledge, and appearance; others are judged on the quality of the horse and his movement. Still other competitions combine aspects of both rider and horse in the judging.

English-style riders sometimes participate in endurance rides of up to 100 miles; however, they ride in specially built endurance saddles for this event. Or they may take their horses on a trail ride, to see the country — without chasing a fox. In this case, they generally ride in their favorite close-contact, all-purpose, or even dressage saddle; English riders may prefer a dressage saddle for trail rides because its deeper seat offers more security over rough terrain than the close-contact or flat saddle.

WESTERN OR ENGLISH?

Bias in the English-riding world holds that Western riders depend on their equipment rather than their skill, and that their horses are less trained than English-trained horses. Bias in the Western-riding world holds that English riders make horses perform tasks that are not natural to them and are sometimes dangerous. But neither style of riding is really "better" than the other; it is all a matter of individual preference.

Because it is generally possible to ride a quiet, well-trained Western horse in a deep saddle without much instruction, and because there are so many trail rides offered to nonriders by dude ranches and Western trail companies, it is easy to fall into the trap of denigrating Western riding. However, Western riders who train and compete in events such as pole bending, roping, and pleasure classes work as hard at getting themselves and their horses fit and skilled as English riders do. Moreover, skilled riders in either the Western or English worlds can generally refocus their skills fairly easily to ride in the other style, especially with just a little bit of good instruction in the new discipline.

Whether one decides to ride English or Western depends on how you think you would enjoy the skills and events each offers. Go to shows in each discipline as an observer before making a decision. Know that some breed disciplines, however, such as Paso Fino and Saddlebred, often prefer riders to be English-trained (at least a little) because the tack is similar and they require the fine-motor skills of English riding. Some Western riders learn English basics first, just so they can more easily try other equine sports and forms of riding. But well-trained, experienced cutting horse riders or pole benders are also often able to make a transition from Western to English riding relatively easily.

Shows

Almost all English shows are open shows; that is, your horse need not be of a particular breed to show there. In dressage, warmbloods compete against Thoroughbreds, Paints, and any other horse that can do the work. The same is true in hunter-jumper and jumper shows, and in three-day events. Breed associations sometimes hold shows, however, in which any English discipline may be included, as may Western disciplines and events particular to that breed alone.

➢ *See also American Saddlebred; Bit; Dressage; Endurance riding; Fox hunting; Morgan; Paso Fino; Polo; Saddle; Sidesaddle, riding; Steeplechasing; Thoroughbred; Western riding*

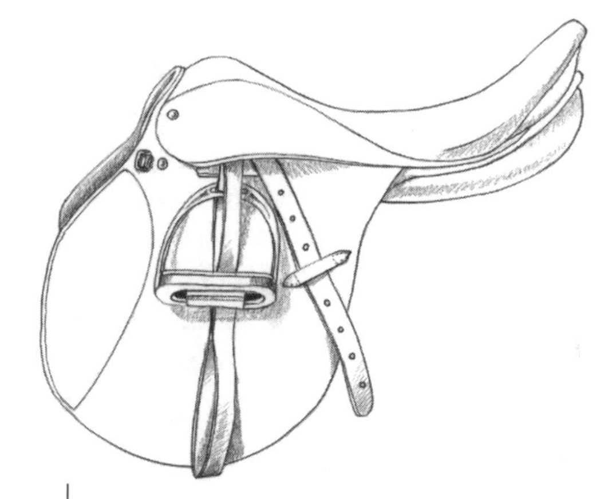

An English all-purpose saddle, shown with leathers and irons

English riding style

This style includes, but is not limited to, the disciplines of Hunt Seat, Dressage, and Saddle Seat riding. English riding style is usually clearly distinguished from Western riding by the saddle used and attire worn by riders. English riding often appears more formal because of the traditional attire required at competitions. The tack used in English riding is also different. The relatively lightweight saddles provide a shallower seat and do not feature a saddle horn.

More subtly, English riding encourages development of a secure balance in the novice rider since its saddle does not offer much to hold onto. Riders must rely on development of balance through proper posture and muscle development in order to feel safe while riding. This style also requires use of direct reining with both hands on the reins; in contrast, Western riders use neck reining with one hand on the reins. English riding may include jumping events, which are not part of Western horsemanship.

FITTING A SADDLE

- The saddle should rest on the horse's ribs, but not on his spine, and it must be as wide as the horse's withers.
- The *gullet* is the "spine" of the saddle, which keeps the weight of the saddle off the horse's spine. You should be able to see through the gullet when the horse's head is down.
- When the horse raises his head, the withers should not press against the pommel.
- The saddle should fit securely even before you fasten the girth.

English saddle

The kind of saddle used in English riding. Although there are several styles, English saddles are generally smaller than Western saddles and have no saddle horn. Within the English category, you'll find all-purpose or combination saddles for general riding; jumping saddles designed to support the forward-riding position required for jumping events; dressage saddles with extended flaps and stirrups; and flat saddles with shallow seats, which are used primarily in saddle seat riding.

➢ *See also Saddle*

Enteritis

The medical term for inflammation of the intestines. Enteritis is one of the serious conditions that falls under the broad umbrella of colic, and, as such, it merits the prompt attention of a veterinarian.

Your horse's intestines can become inflamed due to a variety of causes, including, but not limited to: parasitic infestation of the digestive tract; infection; or ingestion of sand, chemicals, or poisonous plants. Diarrhea is the primary observable symptom, but the horse may also exhibit behaviors identified with colic, such as rolling on the ground and kicking at his

abdomen in an attempt to relieve the pain. Because enteritis also prevents the absorption of fluids, it often results in dehydration as well. Your veterinarian can treat this by administering intravenous fluids.
➢ *See also Colic; Diarrhea*

Eohippus, primitive ancestor of the horse

Enteroliths

Stones that develop when layers of minerals accumulate around a foreign object in a horse's intestines. An enterolith can become quite large without causing any apparent problem, but it must be removed surgically if it is located in an area where it obstructs the digestive tract. Feeding your horse bran or bran mash on a regular basis may increase his likelihood of developing intestinal stones, so it is not generally recommended.

Entire

Term to describe an uncastrated horse.

Eohippus

The precursor of the modern horse. *Eohippus,* known by the proper scientific name *Hyracotherium,* was a dog- or deerlike animal, probably yellow, with four toes on each front foot and three toes on each back foot. According to some estimates, it was a mere 14 inches high at the shoulder and weighed as little as 12 pounds. Other estimates place the height at up to 20 inches and the weight at about 80 pounds. It appeared in North America some 60 million years ago but disappeared from the continent sometime after the Ice Age, crossing to Asia and Europe via land bridges, presumably in search of more abundant pasture.

Over millions of years and dozens of permutations, *Eohippus* evolved into a larger creature about the size of a pony. Its multiple toes gave way to single toes, or hooves. It survived the constant danger of predation by virtue of its speed, excellent vision, and quick reflexes. Later, during the Stone Age, four distinct types developed (the Caspian pony, the Celtic pony, Przewalski's horse, and the Spanish Barb) that are more immediate predecessors of the modern-day horse, known as *Equus caballus.*
➢ *See also Evolution of the horse*

His mane is like a
river flowing,
And his eyes like
embers glowing
In the darkness
of the night,
And his pace as swift
as light.

—Bryan Waller Procter,
The Blood Horse

Epinephrine

Also known as adrenaline, epinephrine is a hormone sometimes used by veterinarians to treat equine anaphylactic shock, a life-threatening condition.
➢ *See also Anaphylactic shock*

Epiphysitis

Inflammation at the ends of immature or still-growing leg bones. The condition can lead to leg deformities, contracted tendons, or the formation of abnormal growths. It often results from overfeeding a foal or young horse in an attempt to stimulate rapid growth. Allowing a horse to mature naturally in open pasture reduces the likelihood that he may develop epiphysitis, as does feeding him a balanced diet free of excess protein and grain.

EPM

➢ *See Equine protozoal myeloencephalitis (EPM)*

Equine dentist

➢ *See Dentist, equine; Teeth*

Equine encephalomyelitis

In both humans and horses, this disease of the central nervous system, also known as **sleeping sickness,** is contracted through viruses carried by mosquitoes. However, in North American horses the disease falls primarily into one of two strains — Eastern equine encephalomyelitis (EEE) or Western equine encephalomyelitis (WEE). The former has occurred in the eastern and southeastern parts of the continent; the latter is known throughout North America. In both cases, the mortality rate exceeds 80 percent. An additional strain known as Venezuelan encephalitis, which can be transmitted from horses to humans, has apparently been eradicated in North America. Some veterinarians still recommend vaccinating against it if your horse is exposed to horses from foreign countries.

Equine encephalomyelitis is a sporadic disease transmitted primarily during peak mosquito season. Symptoms include fever, loss of appetite, fatigue, poor muscular coordination or staggering, partial or complete blindness, brain swelling, and paralysis. Currently, there is no available cure, but close medical attention can serve to reduce mortality.

A low-risk vaccine that combines the antigens for EEE and WEE is available, and vaccinating your horse is an essential preventive measure. The vaccine is typically given in two initial doses, followed by an annual spring booster and, in areas with a prolonged mosquito season, a second booster in the fall. Mosquito control measures can also minimize the chances of a local outbreak.

Equine herpes virus (EHV)

A virus that causes the disease known as **rhinopneumonitis.** The EHV-1 strain of the virus is known to cause abortions in broodmares; the EHV-4 strain induces respiratory illness, particularly in young horses.

➢ *See also Rhinopneumonitis*

Equine infectious anemia (EIA)

An insidious, potentially lethal viral disease that ravaged the horse population in the 1960s, EIA is now far less common. The virus is transmitted by biting insects, but a significant contributing factor in the past was infection transmitted via unclean needles and syringes. Today's disposable medical equipment and heightened awareness of the need for proper sanitation have played a major role in reducing the incidence of EIA, which is also known as **swamp fever.** However, the disease has not been eliminated.

EIA is most commonly seen in horses that graze in wet pasture areas, where blood-sucking insects thrive and may transmit the virus. The disease can then spread from one horse to another, because the virus is

EQUESTRIAN CENTERS

Equestrian centers and riding stables are ideal places to begin riding, whether you are a child or an adult. In evaluating a riding school, investigate the following:

- ❑ Does the instructor have a reputation for being safe, attentive, and responsible?
- ❑ Does the instructor have a reputation for teaching ability and effectiveness?
- ❑ Are novice riders matched with calm, mature horses?

See also Instructor

present in the blood and all secretions of an infected horse, including saliva, semen, and mare's milk.

The symptoms of an acute attack include fever, loss of appetite, anemia, depression, excessive sweating, and watery discharge from the nostrils. However, the chronic form of EIA is more common. An episode may involve intermittent low-grade fever, anemia, irregular heartbeat, muscular weakness, staggering, and unexplained weight loss. A horse with the chronic form of the disease may be symptom-free for extended periods but is subject to relapses and is always a carrier of the virus.

There is no effective treatment for EIA at this time, but a blood test known as the Coggins test has become a powerful tool in detecting and halting the spread of the disease. Following strict sanitation procedures and quarantining or destroying infected horses are also necessary preventive measures.

➢ *See also Coggins certificate, Coggins test*

Equine influenza

A highly infectious virus that affects a horse's upper respiratory tract. Symptoms can include fever, dullness and depression, coughing, weakness, lack of appetite, runny nose, swollen eyelids, and watery eyes. Equine influenza can become serious if a horse is subjected to heavy activity or significant stress during the early stages of the disease.

The recommended treatment regimen involves providing absolute rest, a nourishing diet, and good ventilation in the facility where the horse is recuperating. Antibiotics may be administered. These are not effective against the virus that causes influenza, but they help prevent secondary bacterial infections such as pneumonia and strangles.

A vaccine is available, and is particularly useful for horses that have been exposed to infected horses. However, the vaccine provides only short-term protection that lasts three to four months. By contrast, a horse that has recovered from equine influenza will develop built-in immunity that lasts for at least a year.

I will not change my horse with any that treads but on four pasterns. When I bestride him, I soar, I am a hawk. He trots the air. The earth sings when he touches it.

—William Shakespeare, *Henry V*

Equine protozoal myeloencephalitis (EPM)

A neurological disease that affects a horse's spinal cord and nervous system. EPM is caused by a protozoan, *Sarcocystis neurona,* that is carried by opossums and possibly by birds. In most cases, horses contract the disease by eating feed or drinking water containing opossum feces. Once inside the horse, the protozoan migrates to the spinal cord, resulting in inflammation and nerve damage that can be crippling.

Early diagnosis and treatment is key to minimizing the damage caused by EPM. Subtle symptoms that may indicate the presence of the disease include slight changes in the horse's stance, gait, movements, or coordination. Later symptoms include hind-end lameness, muscular weakness or atrophy, noisy breathing, seizures, leaning to one side, and drooping ears and lips on one side of the face. The diagnostic process should include a spinal tap, and treatment can be both lengthy and expensive. However, the majority of horses recover well enough to resume normal activities.

Logically enough, the most effective preventive measures are those that prevent feed and water from becoming contaminated by opossum feces or bird droppings. These include keeping your supply of grain covered at all times, buying heat-processed feed (heat destroys *Sarcocystis neurona*), preventing birds from living in your barn or stable, and removing opossums from your property by live trapping or other means.

Equine veterinarian

A specialist in the care and treatment of horses. Equine veterinarians are represented in the United States by the AAEP.

➢ *See also American Association of Equine Practitioners (AAEP); Veterinarian*

Equine viral arteritis (EVA)

A viral disease of special concern to horse breeders, as it causes abortion in broodmares when transmitted through the semen of an infected stallion. EVA is also conveyed through other means of direct, horse-to-horse contact, and is highly contagious. However, it is not generally life-threatening in nonpregnant horses.

Symptoms often resemble those of a mild attack of influenza, including fever, coughing, nasal congestion, and short-term respiratory infection. Swelling can occur in the muzzle, legs, and eyelids, and stallions may suffer swelling of the sheath and scrotum. In severe cases, a horse may become weak and unable to stand. These cases require intensive treatment, particularly in foals and older horses. A horse that recovers from EVA is no longer contagious and benefits from immunity to the disease that lasts for several years; however, most stallions continue to spread the virus in their semen.

A vaccine is available that effectively limits outbreaks and protects susceptible broodmares. Breeding stallions that are not infected can be vaccinated so they do not contract the disease from an infected mare. However, because the vaccine contains a modified live virus, it is not risk-free. Consult your veterinarian to determine if vaccination is appropriate.

Equipment

➢ *See Attire; Bathing horses; Blankets and blanketing; First aid; Grooming; Handling horses; Tack*

Equisetum (horsetail, scouring rush, jointfir) (*Equisetum* spp.)

Several species of plants known variously as horsetail, scouring rush, and jointfir. The plants grow primarily in areas with high water tables, and they sometimes turn up in meadow hay. When ingested in large amounts, equisetum is toxic to horses. Symptoms of poisoning include excitability, diarrhea, trembling, staggering, and muscular weakness. Treatment generally involves the injection of large doses of thiamine (vitamin B_1), which is destroyed in the horse's system by an enzyme contained in the plant. This can be a lifesaving measure.

➢ *See also Poisonous plants; Toxic substances*

EQUIPMENT TIPS

To save money, you can join forces with other horse owners to buy equipment that is not frequently used. These items might range from clippers and show preparation equipment to trailers and pasture maintenance equipment.

When purchasing equipment, seek the recommendations of veterinarians and other professionals. Remember, too, that where horses are concerned it is important to buy high-quality, durable tools and equipment and not to settle for cheap items that will wear out quickly.

POISONOUS PLANT

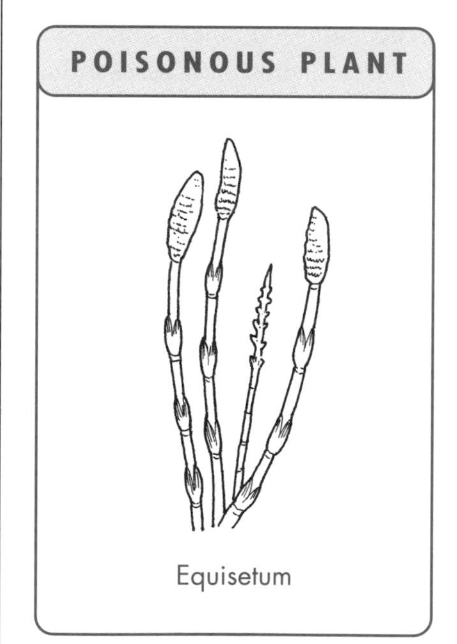

Equisetum

Equitation

The formal art and practice of riding horses. Skilled equitation involves maintaining the body position that is considered correct for the type of riding involved; making appropriate use of "body aids," meaning the hands, legs, seat, and upper torso; and demonstrating the mental composure necessary to control the horse and correct any mistakes he may make. It is an art that typically takes a number of years to master.

Stock seat equitation

Classes

Equitation classes are categories of performance in the ring or arena. In these classes the rider, not the horse, is being judged. However, the horse must be well trained and capable of performing the required routines.

The main equitation classes are hunt seat equitation, Western horsemanship (or stock seat equitation), and saddle seat equitation. Youth equitation events are offered in each class, as riders of similar ages compete with one another.

➢ *See also English equitation classes; Western riding; individual classes*

Equus caballus

The scientific name for the modern horse.

➢ *See Evolution of the horse*

Ergot (fungus)

A fungus that can grow on grasses including wheat, rye, barley, bluegrass, red top, brome, and reed canarygrass. A horse that consumes ergot regularly with his feed may develop circulatory problems that cut off the blood supply to extremities, such as the ears, tail, and feet. In severe cases, gangrene can develop and the horse may die.

Ergot (growth on fetlock)

Horny growths found at the rear point of the fetlock on all four legs. Some horses have virtually no ergots; others have large, readily visible growths. Ergots can generally be removed by peeling them off immediately after a bath, when they are soft and somewhat pliable.

Estrus

The period when a mare is in heat, which may last 6 to 10 days. The mare typically ovulates 24 to 48 hours before the end of this period.

Signs of estrus in a mare include increased interest in stallions and geldings, sometimes demonstrated by aggressive behavior such as kicking or nipping; rubbing against the male horse or against a fence; squatting, urinating frequently, elevating the tail, and "winking" the clitoris. However, some mares experience **silent heat,** meaning they are in estrus but do not exhibit the behavior or symptoms associated with heat.

➢ *See also Anestrus; Breeding; Diestrus; Foal heat, breeding at; Silent heat*

EUROPEAN PONIES

The ancient working ponies of Europe can still be found in some rural areas. Increasingly, however, they are being bred to produce riding ponies. The indigenous ponies are descended from the Asian Wild Horse and the Tarpan, which roamed Europe in enormous herds after the Ice Age. Their modern-day descendants include Bosnian, Fjord, Halflinger, Hucul, Iberian, Icelandic, Konik, Northlands, and Swedish Gotland ponies.

Euthanasia

The practice of humanely ending the life of a horse that is critically injured, terminally ill, in extreme chronic pain, or unable to function. The decision to "put down" a horse rather than prolonging his misery is likely to be the most emotionally difficult one a horsekeeper will face. To reduce the trauma, it makes sense to plan in advance how such a decision would be made and carried out.

EVA

➢ *See Equine viral arteritis (EVA)*

EVACUATING HORSES

The ability to evacuate your horses quickly, calmly, and safely can spell the difference between life and death in an emergency. Because you may need to transport your horses off the property in a trailer, it's essential to train them thoroughly in loading procedures. Practice loading them in a variety of circumstances — different times of day, different weather conditions, and so forth. Perform a periodic drill to reinforce the lesson, just as schools perform periodic fire drills. Keep your truck and trailer properly maintained and ready to roll. Pack essential supplies and identification in the trailer, or keep them in a handy place where you can grab them in a hurry. Make a list of things you will want to turn off, unplug, or lock before evacuating.

You'll also need to plan where to take the horses. It's a good idea to forge agreements with other horsekeepers in your area and to make a list of facilities willing to put your horses up on a short-term basis. When your plans are in place and well-rehearsed, stay abreast of weather reports and other information that can alert you to approaching emergencies.

See also Trailering a horse

A fly, Sir, may sting a stately horse and make him wince; but one is but an insect, and the other is a horse still.

—Samuel Johnson (1755)

Eventing

Combined training that incorporates three kinds of riding disciplines. The first discipline is dressage, in which a horse is judged on his ability to perform precise movements in an obedient and willing manner. The second discipline is cross-country riding, a riding event performed on a prescribed course that includes natural or natural-looking obstacles that must be jumped. The third discipline involved in eventing is stadium (or show) jumping, which involves jumping fences placed within a show ring. Eventing is an Olympic event, but competitions exist for riders of all skill levels.

➢ *See also Cross-country riding; Dressage; Stadium jumping (show jumping)*

Events

The individual competitive activities that take place at horse shows. Each class, or category of competition, features its own events.

Evolution of the horse

As currently understood, the history of the horse began in North America some 60 million years ago with an undersized ancestor commonly known as *Eohippus* or Dawn Horse. The proper scientific name for this 14-inch-tall, deerlike forest animal is *Hyracotherium. Eohippus* had four toes on each front foot, three toes on each hind foot, canine-like pads on the bottom of its feet, a small brain, and small teeth suitable for chewing only the tenderest of browse.

Over the next 50 million years, the horse evolved in literally dozens of directions, sometimes slowly, sometimes rather abruptly. Through the fossil record and other means, scientists have discovered evidence of many of the intermediate species — enough pieces of the puzzle to regard the horse as one of the best confirmations of classic evolutionary theory. However, equine evolution resembles a densely branched tree rather than a straight line. It's not at all clear that nature proceeded with the "intention" of producing modern horses — nor that the modern horse represents the crowning achievement of a logical progression.

The first recognizably "horselike" horses appeared about 15 million years ago, developing into four primary groups. *Pliohippus* began as a three-toed horse but gradually lost its side toes, becoming the first one-toed (hooved) horse. It resembled the modern horse, but it had curved teeth and a dished face with concave depressions on the sides. *Astrohippus,* a likely descendant of *Pliohippus,* arose some five million years later. It had one toe but, again, had marked depressions in its face.

Another one-toed horse, *Dinohippus,* appeared in North America about 12 million years ago. Its teeth, skull, and feet bore a striking resemblance to the modern horse, and there is evidence of a gradual transition over time from *Dinohippus* to *Equus,* the modern horse genus.

Modern Times

Two to three million years ago — recent times in geologic terms — a number of *Equus* species migrated from North America to Africa, Asia, the Mideast, and Europe. Those in Africa evolved into the modern zebra. In desert areas of Asia, the Mideast, and northern Africa, they developed into onagers and asses. The "true" horse, *Equus caballus*, inhabited Asia, the Mideast, and Europe. A million years ago, there were vast populations of *Equus* species throughout the world. But as the Ice Age began, many mammals became extinct, including all horse species within North and South America. Thus, the region where horses had originated was now barren of horses.

The early versions of *Equus* were not quite what we are accustomed to today. These horses had stocky, zebralike bodies, shortened heads, short, rigid tails, and manes that stood straight up.

However, the *Equus* species continued to evolve, developing into four primary types of primitive horses. The first, called the **Celtic pony,** inhabited the wet coastal regions of western Europe, developing a water-resistant coat in response to its cold, damp environment. This rugged pony is the forerunner of many contemporary pony breeds.

HOW THE HORSE EVOLVED

A simplified list of equine ancestors:

Eohippus	Eocene Epoch
Mesohippus	Oligocene Epoch
Merychippus	Miocene Epoch
Pliohippus	Pliocene Epoch
Dinohippus	Pliocene Epoch
Equus	Pleistocene Epoch (Ice Age)

In the late 19th century, in the even colder environs of the Mongolian steppes and portions of China, a species that was a relic of prehistoric times was discovered by Nikolai Przewalski, a Russian explorer. Named **Przewalski's horses,** these wild, primitive horses were extremely cold-resistant and had retained the yellow body and dorsal stripe of *Eohippus*. It is believed that the wild population of these horses became extinct in Mongolia during the 1960s, although more than a thousand Przewalski's horses now survive in captivity.

A third type of horse, known as the **Barb,** thrived under quite different conditions. It became adapted to the deserts of central and southern Asia, later migrating to North Africa and ultimately to Spain. The Barb is the ancestral father of the Andalusian horse, which is associated with Spain.

The fourth type of primitive horse also inhabited a desertlike environment. The **Caspian pony** lived in the early agricultural center of Mesopotamia and surrounding arid regions. This pony developed resistance to heat as well as great speed and endurance, allowing it to flee from predators. It is the predecessor of the Arabian horse.

Horses did not return to the Americas until 1494, when Christopher Columbus transported them on ships during his second journey to the New World. Spanish conquistadors brought greater numbers of horses to North and South America in the early 16th century. Today the horse population in the United States alone is estimated at more than 7 million.
➢ *See also Andalusian horse; Arabian; Caspian pony; Celtic pony; Eohippus; Przewalski's horse*

The hooves of a horse! Oh! Witching and sweet is the music earth steals from the iron-shod feet; no whisper of love, no trilling of bird can stir me as hooves of the horse have stirred.

—Will H. Ogilvie

Ewe neck

A conformation fault in which a horse's neck is thin, weak, and affords little or no flexion at the poll, characterized by a concave line from poll to withers. A ewe-necked horse may carry his head and neck quite high, which can limit his ability to see what is beneath him.

Exercise (for horses)

Daily exercise is often the difference between a healthy, happy horse and one that is beset with health or behavioral problems. Your horse gets exercise whenever you ride him; however, even if you ride him every day, exercise between rides is important. Horses evolved to be range animals, and their systems function best if they are allowed freedom to exercise.

Turnout areas — pens, runs, corrals, or paddocks — provide a place for horses to run, longe, or "kick up their heels." Be certain that these areas offer sure footing, safe fencing, and, in hot climates, protection from full sun. Grazing at pasture can also provide your horse with some activity, but it's no substitute for a sustained period of brisk exercise. For most horses, the recommendation is one hour of this type of exercise per day, although this can vary according to age and time of year.

The newborn foal and the newly unburdened mare both benefit from exercise.

Although a broodmare in late pregnancy may demonstrate little propensity to move, she will benefit from a daily walk. Mild exercise at this stage helps prevent or minimize edema, a painful condition in which tissues become swollen because of retained fluids.

Following birth, exercise allows the mare to regain her strength more quickly, and it also helps the foal gain strength and coordination. It's helpful to turn the mare and foal out in a safe enclosure that contains no other horses for the first few days after birth. This ensures that the bonding process will be completed without disruption.

Because stallions are inherently energetic, daily exercise is imperative. A bored, cooped-up stallion is likely to become difficult to handle and will develop vices, such as wood chewing. Riding, driving, or longeing is the best exercise; when this isn't possible, lead him on a long daily walk or provide him with the space to exercise on his own.

Consider your horse's age, his health, and the prevailing weather conditions when you provide exercise. In hot, humid weather, horses are prone to overheating, particularly older horses or those in less-than-peak physical condition. Avoid feeding your horse immediately after vigorous exercise, and limit his intake of water to small sips until he is thoroughly cooled down.

➢ *See also Cooling down (after exercise)*

Exercise (for riders)

A program of regular exercise can make the time you spend riding your horse much more rewarding and productive. Even a novice rider needs the muscle strength to stay mounted for an hour or more while learning. As many new riders have discovered, this is no insignificant feat. Skilled, experienced riders need additional muscle strength, as well as a high degree of flexibility, precise reflexes, an excellent sense of balance, and cardiopulmonary stamina.

The program that's right for you will depend on your age, your current physical condition, and your riding agenda. You should consult your riding instructor, a physician, or a fitness professional when developing your program. However, in general terms, virtually all riders benefit from stretching exercises to improve flexibility and strengthen muscles. Stretching immediately before every ride will help minimize the chance of muscle cramping, strain, or injury. Isometric exercises, in which certain muscles or sets of muscles are repeatedly contracted and released, are effective in building strength — without the need for elaborate fitness equipment.

Participation in certain compatible sports and activities is also helpful. Those that nicely complement riding include bicycling, cross-country skiing, walking, gymnastics, and various types of dancing, including ballet.

Exercise wrap

A bandage used to provide support for a horse's lower leg tendons during exercise. The bandage is wrapped around the legs, extending from the middle of the cannon bone downward to the fetlock.

WARM-UP STRETCHES

Warming up before a ride helps make the experience safer and more enjoyable. Start with an activity such as walking, jogging, or running in place for five minutes. Then place your feet directly beneath your hips, bend your knees slightly, tighten your abdominal muscles, and try these simple stretching exercises. Repeat each exercise four times, holding each move for a count of ten.

- ❑ Reach way above your head with both hands. Reach higher with the left hand, then to the right.
- ❑ Stretch your arms out on both sides. Twist gently to the left, then to the right.
- ❑ With your hands on your hips, bend sideways to the left, then the right.
- ❑ With your hands on your hips, bend forward from the waist, then backward as far as you can without discomfort.

Adapted from *Getting the Most from Riding Lessons* by Mike Smith

Exertional rhabdomyolosis

➢ *See Azoturia; Rhabdomyolosis*

Experience of rider

➢ *See Rider, levels of experience*

Eyes

Equine eyes have played a key role in the long-term survival of the horse. Keen vision is the horse's primary asset as he seeks to avoid a panoply of dangers, ranging from predators to prairie-dog holes. His distance vision is superior to human vision; his peripheral vision covers a remarkable radius of about 340 degrees, allowing him to see almost back to his tail on both sides.

There are limitations, however, to the horse's visual tools. He has only a small field of binocular vision in which he can clearly see an object with both eyes. This area is located some distance in front of his forehead (see diagram). The horse must position his head properly in order to focus. With his head elevated, he clearly discerns objects close at hand. With his head lowered, he sees objects that are far away. With his head in its normal or forward position, he sees objects at medium distance. When the horse looks outside of this field of binocular vision, he sees a separate image with each eye, and his brain receives two different messages.

Because his eyes are placed toward the sides of his head, the horse is subject to blind spots, areas where he cannot see things clearly unless he turns his head or moves his body. Primary blind spots include the area directly behind his tail and the area immediately in front of his forehead. For your own protection, and to avoid upsetting the horse, alert the horse to your presence whenever you are in or near one of these blind spots. Talking calmly to the horse is the best way to let him know that you are there. Countless injuries happen because a horse is startled by a handler.

In some cases, horses suffering from impaired vision are not safe to ride. These horses may be skittish in response to environmental stimuli or simply unable to see and avoid hazards. However, many visually impaired horses adapt successfully to impairment, including horses that function with only one eye. Some handlers believe that "pig-eyed" horses — horses with small eyes — are more likely to be high-strung or difficult to handle because of diminished peripheral vision.

EYE COLOR

Most horses have brown eyes, but a few have no pigment in their irises: these are called *walleyes.*

See Iris

Eye Problems

A horse's eye is protected from many potential injuries by a fatty cavity that lies behind it, allowing the eyeball to retract inward in response to a blow or an impact. However, this protection is not absolute, so injuries to parts of the eye are not uncommon. Torn eyelids result when a horse rubs his face against sharp objects, such as nails or barbed wire. Ulcers can form on the eye as the result of irritation from embedded slivers or seeds. An eye that is struck by a tree branch, whip, or the tail of another horse can develop a corneal injury that may require treatment with an antibiotic ointment.

The most common equine eye problem is **conjunctivitis,** an irritation of the lining of the eyelid. This is typically caused by eye contact with irritants such as flies or dust, and the symptoms include squinting, tearing, and a discharge of pus from the eyes. If treated early with eye ointment or eyedrops, conjunctivitis is a minor problem. However, untreated conjunctivitis can become a vision-threatening problem.

The most common cause of blindness among horses is known as **moon blindness.** This inflammation of the inner eye generally occurs when a preexisting infection triggers an abnormal immune response, which in turn damages the eyes. Early symptoms include watery eyes, constricted pupils, and sensitivity to light that causes the horse to squint or close its eyes.

Another cause of equine blindness is **night blindness,** an inherited condition seen with greatest frequency in Appaloosas. **Cataracts,** which are cloudy spots behind the pupil, also often result in blindness, but they can, in some cases, be successfully removed.

It's important to note that all eye problems are potentially serious. Even minor problems, if left untreated, can develop into conditions that could result in impairment or even blindness. Inspect your horse's eyes regularly, and contact your veterinarian at the first sign of problems.

➢ *See also Cataract; Conjunctivitis; Iridocyclitis; Night blindness; Periodic ophthalmia; Vision, of horse*

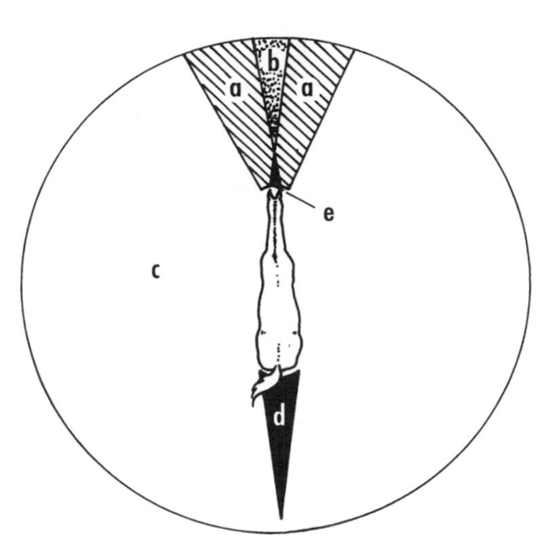

A horse can see either straight ahead (a and b) or to the sides and back (c). He has binocular vision only in a small area in front of his head (b). His blind spots are indicated by the black areas (d and e).

f

Face

The horse's head. Also, in Western riding, when a horse turns his body to present his head to a cow.

FACIAL SHAPES

The shape of a horse's face, or head, may be an identifying characteristic of his breed. For example, a dished face — one with a wide, flat area of cheek below the eye and a relatively small muzzle — is typical of an Arabian horse. The dished shape also allows the horse better frontal vision than the Roman nose, in which the area between the eyes that runs down to the muzzle between the nostrils is noticeably convex. The central ridge impairs frontal vision. For most breeds, a straight profile, with no ridge down the center of the face and only a mild, if any, "dishing," of the cheek, is preferable.

The eyes have it
Some horses have very large eyes, set high on wide foreheads. This placement allows them a great deal of lateral vision and the ability to see things approaching from the rear sooner than a horse with another face/eye shape would. It is believed that the high, wide-set eyes can lead to spooking at times, when horses with smaller eyes would not see anything to spook at.

Face brush

A small, very soft brush that allows dirt to be removed from the tender sides of the face, beneath the forelock, the backs of the ears, and along the front of the face down to the nostrils, as well as from the deep groove under the chin. Although this brush is soft, it should not be used near the eyes. Both the corners of the eyes and nostrils should be gently wiped with a soft, damp cloth.

➢ *See also Grooming, equipment*

Face flies

Flies that resemble houseflies and feed on the secretions of a horse's eyes and nostrils and on the blood from other fly bites. Use fly repellent on the horse's face to keep face flies at bay; horses that are badly affected might require a fly mask for protection when outdoors in summer.

➢ *See also Flies*

Face markings

Facial markings are used to identify horses on forms such as breed registry certificates and Coggins tests. The more different types of markings a horse's face has, the flashier he is considered to be. Markings are also referred to as "chrome" — the same word that describes the fancy detailing on cars. (A horse is not considered to have lots of chrome unless he also has markings on one or more of his lower legs.)

➢ *See also Leg markings*

Facial Shapes

Arabs are known for their dished profiles.

Most horses have fairly straight profiles.

A Roman profile is typical of some draft breeds.

Facial expression

Strictly speaking, a horse does not exhibit facial expressions. While humans can wrinkle their noses, squint their eyes, purse their lips, hollow their cheeks, and more to communicate emotion, horses do not have quite as much motor control of their facial features. They can expand their nostrils, widen their eyes (or make the lids droop), draw back or wiggle their lips, and, most importantly, move their ears in almost a 360 degree pattern, as well as flatten the ears against the back of their necks and incline the ears forward. Indeed, many consider the ears to be a horse's most expressive feature.

SPEECHLESS COMMUNICATION, EQUINE STYLE

Here are the most common equine expressions and what they mean:

- Flehmen (also known as Flehmen reaction) is an expression most often of stallions in which the horse curls his upper lip back, showing his teeth, and opens his mouth in order to draw a lot of air across an organ in the roof of his mouth that detects mares in estrus.
- A horse pins his ears, turns them backward and flattens them on the neck when he is being defensive. Often, ear pinning is followed by other protective or aggressive behavior, such as kicking or biting.
- When a horse moves each ear in a different direction, and then rotates them again like a propeller, the horse is probably alert and attentive to his surroundings and the rider.
- Perked ears. Ears that are perked forward may mean the horse is intently looking at and listening to an object or event, or it may be a sign that the horse mistrusts something in his environment and may spook.
- Rolling eyes usually indicate pain.
- Drooping eyelids may indicate merely that the horse is tired, or it may be a sign of illness if it is persistent.

See also Ears

"PROPELLER" EARS

PERKED EARS

DROOPING EYES

FANCY FACES

Some horse owners like a plain face, while some like lots of markings. Distinctive markings help identify horses that are not tattooed inside their lips (as are Thoroughbred racehorses), or branded (as are many of those belonging to warmblood registries). Here are drawings of each type of marking, although a horse may have any combination of markings.

STAR

SNIP

STRIP (OR STRIPE)

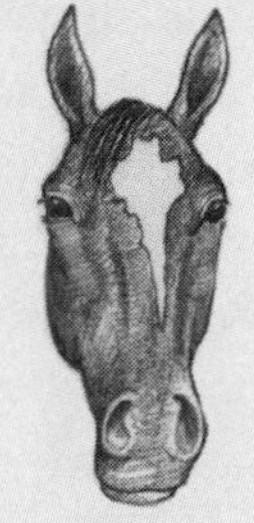
STAR AND STRIP (OR STRIPE)

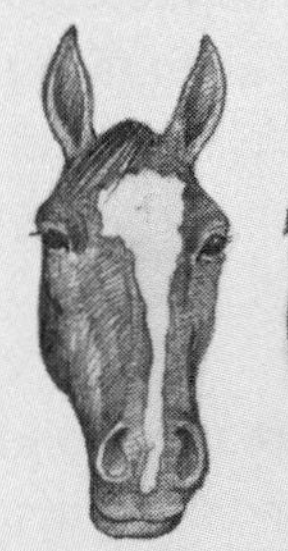
STAR AND NARROW BLAZE

BALD

CHIN SPOT

Facilities

Facilities for horses range from simple fenced pastures with run-in sheds, which provide shelter in hot and cold weather and precipitation, to riding and training facilities with multiple-stall stables, indoor and outdoor arenas, roundpens, hot walkers, and even swimming pools for rehabilitation of injured horses. However, the basic needs of every horse are:

- Fresh water in sufficient amounts
- Food
- Shelter for inclement (including very hot) weather
- Space for exercise outdoors
- Daily observation and, if needed, care by a knowledgeable person

Note that if a horse is recuperating from an injury or surgery, he should be rested in a stall large enough for him to walk around, lie down, and get up easily.

When you are designing a place to live for horses, there are several basic considerations that should go into your blueprint.

Home, Sweet Home

Each horse should have a stall that fits, so a barn is best built with more than one size stall available. A pony may live happily in a 10' x 10' stall. Most horses will do fine in a 12' x 12' stall, although large Thoroughbreds and warmbloods, and even the racing type of large Quarter Horse, are bulkier animals that would do best in a 14' x 14' enclosure. Mares with foals, too, need at least a 14' x 14' space to live in; better still, build one pair of stalls side by side with a removable partition, and remove the partition when your mare has her foal.

➢ *See also Barn*

Fields

Pasture is necessary for exercise, sunshine, and food. A good pasture will be large enough to support the number of horses assigned to it. It will offer them grasses, a mixture of clover (except alsike clover, which can be toxic) and other legumes. The pasture should be of sufficient size and variety to prevent overgrazing of the field. It should also be uncontaminated with plants poisonous to horses. It should be checked regularly to be sure it is free of holes (from burrowing animals and other causes), sharp objects, and slick spots. Naturally, the fields should have fences that are both safe for horses and sufficient to keep them contained.

The field, pasture, or paddock should also contain a water supply sufficient for the number of horses using it, unless the horses are turned out for no more than a couple of hours at a time and then let back into their stalls for fresh water. The outdoor water supply can be in troughs, either self-filled or manually filled, or fresh streams, ponds, lakes, or rivers on or adjacent to the property.

➢ *See also Fields, riding in; Pasture*

Run-in sheds

Some horses prefer to live most of their lives outside. Some farms acquire extra horses that cannot be accommodated in the barn. And some farms offer field board. In any case, provisions must be made for the horses to shelter themselves from the worst weather, and a run-in shed can be the answer.

LOCATION

The layout of your equine real estate is key to your horses' health and well-being. The barn itself, or the run-in sheds, should be built on clear ground higher than the rest of the facility so that water drains away from it. This location will keep the stalls dry and prevent various injuries and illnesses that can be caused by dampness and standing water. Your barn will also have fewer insects on high ground than in low-lying, wooded, or swampy places.

These three-sided sheds should be large enough to accommodate all the horses living or grazing in the field, about 140 square feet per horse, which will give them room to jostle for position. Like the barn, the shed should be built on high ground to allow for drainage, and the open side should be toward the south, to keep cold north winds out in winter. The roof should slope toward the closed back side so puddles do not form in front of the shed.
➢ *See also Fencing materials; Paddock*

A three-sided shed must face away from prevailing winds.

Water

Most horses will drink between 8 and 12 gallons of water a day, and more if they are in strenuous work, the weather is hot, or their feed is primarily dry hay. They may drink less in cold weather and when they have abundant, lush pasture available. In the stall, water should be provided in buckets kept full at all times, with refillings no less than two times a day in warm weather. Water in pastures and paddocks may be provided in troughs that are filled regularly, or by clean ponds, streams, and rivers. If you provide water in buckets and troughs, be sure to clean them frequently to prevent water quality from declining as horses drop food bits into them. These will disintegrate and cause bacteria to grow.

If you live in a cold climate, providing drinkable water year round is a big challenge. First, you'll have to bury water lines at least 3 to 5 feet deep to prevent their freezing. The hydrants, however, and the pipe leading out of the ground to them, might also freeze. Depending on climate, wrapping the pipe with insulating material might be enough; but in some places, an electric heating element will be needed. Remember, any wiring in a barn is a potential hazard, all the more so around water. Drain hoses after every use and/or store them in a heated area.
➢ *See also Barn; Fencing materials; Hazards; Paddock; Pasture; Poisonous plants; Water*

TEN COMMON STABLE HAZARDS

The main hazards to avoid around the stable and yard include:

1. Animal holes
2. Baling wire (loose pieces)
3. Broken posts
4. Dangerous or hazardous fences
5. Flammable liquids
6. Low overhangs
7. Poisons
8. Sharp projecting objects
9. Traffic
10. Trash

See Hazards

Falabella

The best-known breed of miniature horses, named after the Argentine family that developed it. Kept as pets and curiosities for centuries, miniature horses can be used to draw small carts, but they are not suitable for riding. The smallest known Falabella was a West Virginia–bred horse named Sugar Dumpling; she stood only 20 inches high. Conformationally, the best miniature horses mimic the Shetland pony; those with serious conformation faults may have weak hocks, crooked legs, or heavy heads.
➢ *See also Miniature horse*

False pregnancy

Your mare may have become pregnant, but up to 15 percent of mares with confirmed pregnancies lose the embryo before the 100th day. Losing the embryo doesn't make all mares return to their prepregnancy state. Many will continue to produce progesterone, the hormone that maintains pregnancy. False pregnancy also keeps the mare from coming back into heat, a situation that may continue for several months.

Famous Fictional Horses

Horses figure prominently in the mythology of many cultures and have been the subject of innumerable stories, poems, songs, tall tales, and novels. Some of these stories were so much a part of our childhoods that it seems hard to believe that they only exist in our imaginations. Herein is a short list of some of the best known and most beloved, with apologies for having omitted any favorites.

Black Beauty (Anna Sewall). Quite possibly the most famous fictional horse of them all, Black Beauty, writing in the first person, has given generations of readers some insight into what it must be like to be a horse. From his birthplace, where he is gently handled and carefully trained, he descends through a series of owners, some of them dreadfully brutal, until he at last finds a peaceful resting place and is reunited with Joe Green, the groom from his first home. A well-made horse with a shiny black coat, Beauty has one white foot, a perfect white star on his forehead, and one small patch of white on his back.

The Black Stallion (series by Walter Farley). Another black horse that has captured the imaginations of many readers, the Black Stallion fulfills the fantasy of so many children who long to tame a wild horse and be the only one able to ride him or her. Described as a giant of a horse "with a wonderful, physical perfection that matched his savage, ruthless spirit," the Black Stallion is marooned on an island with Alec Ramsey, who tames him and brings him back to New York where he becomes an extraordinary racehorse.

Blaze (series by C. W. Anderson). Lucky Billy, to have such a wonderful pony (and such wonderful parents!). Together, horse and boy have all sorts of adventures.

The Houyhnhnm (from *Gulliver's Travels* by Jonathan Swift). After leaving Lilliput and Brobdignag, Gulliver encounters an intelligent and civilized race of horses, the Houyhnhnms.

The Maltese Cat (Rudyard Kipling). The homely, flea-bitten hero of this well-known tale is a cart-horse in Malta who changes hands as partial payment of a betting debt and becomes a polo pony after being shipped to India with his new owner. Using his "wise little, plain little, grey little head," he leads a ragged bunch of polo ponies to victory against a sleek and pampered team, with native intelligence, a love of the game, and fierce determination. This wonderful story is unusual in its evocation of the heat of the game and its believable dialogue between the horses.

Misty of Chincoteague (Marguerite Henry). Although Misty actually existed, with her copper and silver coloring and the rough outline of the United States across her withers, her story is largely fictional. After the book was published, the pony, the island, and the Beebe family became celebrities, and Pony Penning Day now attracts thousands of tourists each year, drawn in part by the romance of the story of two children whose love and determination tame the wild Phantom and her small foal. The Phantom eventually returns to the wild life of Assateague, but Misty stays on Chincoteague and has several more books written about her.

My Friend Flicka (Mary O'Hara). Daydreamer Ken McLaughlin can choose any foal on the ranch, and he chooses Flicka, a beautiful sorrel,

He was a giant of a horse, glistening black — too big to be pure Arabian. His mane was like a crest, mounting, then falling low. His neck was long and slender, and arched to the small, savagely beautiful head. The head was that of the wildest of all wild creatures — a stallion born wild — and it was beautiful, savage, splendid. A stallion with a wonderful physical perfection that matched his savage, ruthless spirit.

—Walter Farley,
The Black Stallion

with a long, blond mane and tail. With bad blood on her dam's side, she is deemed "loco" by Ken's tough father, but love and patience win out when Flicka is badly injured. After Ken nurses her back to health, she becomes his faithful companion and an ideal saddle horse.

The Red Pony (John Steinbeck). Many tears have undoubtedly been shed over this story of Jody Tiflin and his red pony, who comes to him with rough, thick fur, a long, tangled mane, and bared teeth. With patience and love, he wins the colt's trust, but loses him to sickness before even being able to ride him.

Smokey the Cow Horse (Will James). From wild mustang to bucking bronc to outstanding cow pony, this mouse-colored horse has many adventures, narrated in a flavorful Western style.

Famous Horses of History

Here are some celebrated historic horses, in chronological order:

Bucephalus, Alexander the Great's horse, was known for his wildness, and only Alexander was able to ride him. After carrying his master to many victories, he was buried in an alabaster tomb and had a city named after him (Bucephala).

Inciatus was the favorite horse of the Roman Emperor Caligula, who stabled him in a marble stall and ordered he be serenaded by musicians.

The **Byerly Turk,** the **Darley Arabian,** and the **Goldolphin Arabian** were the foundation stallions of the Thoroughbred breed. These small, tough horses passed on the gifts of speed, courage, and heart, hallmarks of this versatile, athletic breed.

Justin Morgan's horse, Figure, is the progenitor of the only breed to be named after a single stallion. This little horse was renowned in his day for his speed, strength, and good nature, all of which he passed on to generations of offspring.

Traveller, Robert E. Lee's famous gray gelding (originally called Greenbriar), is buried under a monument on the campus of Washington and Lee University.

Comanche was ridden by Captain Miles Keogh in the Battle of Little Big Horn in 1868. At the end of the day, he was the only creature left alive on the battlefield.

Black Hawk and **Ethan Allan,** descendants of Justin Morgan's horse, were natural trotters that established lines of harness horses. Other famous trotting sires were Messenger and Hambletonian.

Marengo was a white stallion that carried Napoleon at Elba and for eight hours during the battle of Waterloo.

Baby Doll, one of the best bulldogging horses of the last century, carried dozens of riders to nearly a half a million dollars in prize money.

Dan Patch was for years the best pacer ever seen. His 1906 record of a mile in 1:55 ¼ stood for 33 years. The current world record, set in 1993 on a much improved track with greatly superior vehicles, is 1:46 ⅕.

Man o' War, Secretariat, Native Dancer, Kelso, Eclipse: See *Triple Crown* for a list of legendary racehorses.

A hurry of hoofs in a village street,
A shape in the moonlight, a bulk in the dark,
And beneath, from the pebbles, in passing, a spark
Struck out from a steed flying fearless and fleet;
That was all!
And yet, through the gloom and the light,
The fate of a nation was riding that night.

—Henry Wadsworth Longfellow, *Paul Revere's Ride*

Farrier

Farriers trim hooves and shoe horses, but they can do a lot more. In fact, many of them refer to themselves as equine podiatrists because they use their knowledge of the structures of a horse's leg and hoof to improve not only the horse's comfort, but also his **way of going** or quality of movement. They do this by trimming the hooves properly, fitting the shoes, and selecting the type of shoe and/or pad that will help keep your horse's feet most comfortable for the kinds of work you are asking him to do.

EVALUATING YOUR FARRIER

How can you tell if your farrier is doing a good job? There are a few main indicators of quality farrier care.

- ❑ Are you comfortable with how the farrier handles your horse?
- ❑ Does your horse move as well, or better, after being trimmed or shod as he did before?
- ❑ Do the pairs of feet appear to have the same angle? (If you are unfamiliar with hoof angle and how to determine it, ask your farrier to explain this to you.)
- ❑ Take a good look at the coronary band. Does it appear to be relatively parallel to the ground? The coronary band indicates the effects of balance or imbalance in hoof care.

See also Foot care; Hoof; Shoes, horse; Way of going

FAULT

In jumper competitions, faults are assessed for these mistakes:

- ❑ When a horse knocks down a component of a jump while crossing it;
- ❑ When his foot lands in the water during a water jump;
- ❑ When he completely knocks down the jump;
- ❑ When his trip around the course takes longer than allowed;
- ❑ In some events, when the horse touches a rail with his legs but the rail stays in the jump cups.

Fats

Both riders' and horses' diets should be balanced for the amount of fat needed. Fats provide stored energy, while carbohydrates provide instant energy. Body fat, which can be made from anything you eat, cushions and protects organs, insulates against the cold, and stores the fat-soluble vitamins A, D, E, and K. An active rider's diet should contain between 10 and 20 percent fat to be used as energy.

Horse feeds are generally rated according to the percentage of protein in the feed, ranging from 10 to 14 percent protein. All of them will contain some fat, derived from the grains — oats, millet, barley, and corn — generally found in them. The fat will usually range between 2 and 4 percent. However, if you'd like your horse to gain some weight or have a glossier coat, adding one or two ounces of high-quality cooking oil or flaxseed oil to his ration will produce both results. A healthy horse can tolerate up to 16 percent fat in his diet.

Fear (in horses)

Most equine fears arise from horses' ancient instincts. When they were preyed upon by coyotes, panthers, wolves, and so on, horses' only defense was rapid flight. So, when a horse believes a human action or construction will impair his ability to flee danger, he may exhibit fear.

The horse may also be afraid of anything that smells like a carnivore, from wild pigs to humans. He may fear any object that looks like a predator crouching to spring. He may be afraid of ropes or wire. In addition, he may fear wet spots, standing water or ponds, lakes, streams, and mud; in the wild, these areas could keep him from escaping predators. Most horses will also be startled by loud noises.

Expressions of fear

Horses used for pleasure or work purposes cannot flee as can a horse on the open plains. So, when he is not able to flee to avoid what he perceives as danger, he may resort to other means of protecting himself, including kicking with his hind feet, striking with his front feet, or biting. Or the horse may crowd you uncomfortably while seeking protection.

A frightened horse may rear or strike out in an effort to escape.

Fear (in humans)

Fear and/or a feeling of anxiety is a normal, expected response to being hurt by or thrown from a horse. Humans thrive on their sense of control of their surroundings. Being hurt by or thrown from a horse is the ultimate reminder that you are never fully in control of the animal you are riding or of the environment in which you are riding. Horses have minds of their own, and situations arise that you may never have imagined. Both are common factors in creating the situation that results in humans being hurt by horses.

It doesn't matter whether you are a novice or weathered horseman; nobody truly cherishes the idea of being hurt by or thrown from a horse. (Even professional cowboys would likely agree that they prefer to dismount at the time of their choosing rather than being thrown before the 8-second buzzer.) So don't ever blame yourself for what you feel. Feelings and emotions are real. They are also beyond our control. They happen as a natural response to our situations and relationships. Don't let anybody tell you to ignore your fears or "be strong."

The key to overcoming the fear and anxiety is to become aware of their presence. Once you allow yourself to face the fear, you then have to honor its presence in your mind and body. Then your focus turns to learning and applying new skills to help you control the effects such fear has on your ability to participate with horses at the level you desire.

Feathering protects the legs of draft horses from brambles, branches, and other hazards encountered while plowing or hauling lumber.

Feathers

Heavy horses — draft horses such as Clydesdales and Shires — and certain other breeds have long hairs at the back of the pastern and under the ergot. In winter, many breeds will develop some feathering, even if it is not typical or desirable of their breed, as it is for Clydesdales, Shires, and other draft breeds. For pleasure and show horses, the feathers are removed with electric clippers to show off the horse's conformation. Another reason to clip feathers is to keep dampness and mud from collecting there and causing diseases, such as scratches.

➢ *See also Anatomy of the horse; Clydesdale; Ergot (growth on fetlock); Fetlock; Pastern; Scratches (grease heel); Shire*

Fecal balls

The form a horse's manure takes; sometimes, because of their rounded shape and their prevalence on country roads used by riders, they are called "road apples." It is best to look at your horse's manure pile when you know he is well and has had a normal day — complete with normal activities and his usual feed — so that you will have a basis for comparison if his fecal balls change later on. Here are some guidelines for the appearance of fecal balls:

- Normal horse manure will have well-formed balls with enough moisture to allow a pile to stay heaped. Some undigested roughage will be observable.
- Loose manure may not form into identifiable fecal balls, but rather will be a softer mound. In this case, the material passed through the horse relatively quickly. It is likely that the horse's feed changed from grass hay to richer alfalfa, or he has had both extra salt and water. It might also indicate an irritation in the horse's gut that needs to be evaluated. The horse may be anxious or excited.
- Dry, hard fecal balls can be normal for some horses, or it may mean your horse has not had enough water. Typically, this change will be apparent during cold weather, when a horse drinks less because he is working less, or because he finds the cold water less palatable than warmer water.

FECAL IMPACTION

If a horse's water intake is inadequate, his manure may become too dry and hard to pass through the small colon and rectum, even though the horse may strain forcefully. The fecal matter must be removed manually (a tranquilizer is often advisable) and a soap enema administered to lubricate and stimulate the anus and colon. A soap enema consists of a piece of soap in several quarts of warm water, stirred until it dissolves. Three or four quarts of mineral oil given by nasogastric tube will help break up any impaction that has occurred in the large colon.

See also Impaction

Feces

Fecal balls or any form of stool excreted by a horse; also called manure.

➢ *See also Manure, as indicator of health*

Fédération Equestre Internationale (FEI)

A global organization that establishes rules and regulations for international equestrian sports, including the Olympic Games. It has also developed a set of standardized aids; if you ride a well-trained horse in the United States, Germany, or France, he should respond to the same aids, which makes it easy for instructors to work with horses and riders from many national backgrounds.

Feeder

More often called hay mangers, these fan-shaped metal bars are attached to the corner of a stall. Hay is dropped into the feeder and the horse can pull it out a few wisps at a time. Feeders should not be set too high; the natural position for a horse to reach fresh roughage is with his neck bent toward the ground. In fact, if you keep the stall very clean and there is little danger of the horse snorting up sand or sawdust as he eats his hay (for instance, you use wood shavings for bedding), then it might be best not to use a feeder. Instead, put the hay on the floor of the stall.

Feeding and nutrition

A horse's feed and nutrition have an enormous impact on his ability to do the job he is trained to do, on his mental attitude, and on his longevity.

Because a horse has only one stomach and a cecum (rather than the four stomachs of a cow, for example), a horse must be fed so that his simple digestive system is not overwhelmed and its processes interrupted, which can result in a dangerous colic.

The following are some considerations concerning equine feeding and nutrition.

Activity Level

A working horse needs more and higher quality feed than a nonworking or retired horse. An idle horse weighing 1,000 pounds needs about 25 pounds of hay daily to maintain his weight and normal bodily functions, plus minimal activity. (Of course, this may vary from individual to individual, with some horses being "easy keepers," and others needing more food for maintenance.) A working horse will need more hay, and he should be fed a sweet feed or pelleted feed as well. Some horses also require supplements of vitamins and minerals; if your horse loses condition or weight, ask your vet to recommend a diet and supplements to help.

Age

Older horses, even those in light work, may need supplements to maintain condition and weight. Very old horses may need special softer food, as their teeth are worn down. Older broodmares, particularly, may need special rations and supplements recommended by the vet to keep them and their fetuses in good condition.

Colic

A severe colic can kill a horse. In recent years, surgery to correct torsion has been developed, but it is expensive and recuperation is lengthy. It is best to avoid colic by practicing good horse husbandry. If you do suspect colic, however, call the vet immediately.

Minerals

Minerals are required in small amounts by all living things to maintain health. Horses obtain minerals through the grasses and grains they eat as well as through supplements. The essential minerals for horses are calcium, phosphorus, magnesium, sodium, chlorine, and potassium. Horses need trace amounts of iodine, cobalt, copper, iron, zinc, manganese, and selenium. Various soils provide, through hay and grasses, many of these minerals in varying quantities. In some regions, supplements are needed of one or another mineral; in other regions, supplementation would cause overdose. Veterinarians can help determine specific needs in your area.

Overfeeding

Overfeeding can be just as bad as underfeeding. A fat horse has more weight to carry and may not have the energy he needs for the work he does. Overfeeding of supplements can be deadly. Keep an eye on your horse's condition. If he loses weight during work, ask your veterinarian before adding foods or supplements to a normal horse diet of grain, hay, and pasture forage. If the horse gains weight during a lay up, cut his rations to the amount that will allow him to maintain his weight in minimal or no work.

Horses and poets should be fed, not overfed.

—Charles IX

COMMON SIGNS OF COLIC

- ❑ Restlessness (a mild colic)
- ❑ Tail swishing, looking at his belly, pawing, stamping, and lying down indicate a potentially more severe case
- ❑ Getting up and down repeatedly, staggering, throwing himself on the ground, and rolling point to severe colic
- ❑ Sweating or rolling of the eyes
- ❑ Ignoring food

SPECIAL FEEDS FOR SPECIAL NEEDS

At different phases of life, horses have different dietary needs. Here are some guidelines.

Weanlings
Weanlings can do very well on pasture and hay, sometimes with supplementary grain, especially if they spend a lot of time out playing with other weanlings. Overfeeding, especially on grain, can cause too-rapid growth, however, and result in problems, such as epiphysitis.

Yearlings
Yearlings are like young children. The growth plates at the ends of their long bones, particularly in the legs, may become deformed if overfeeding causes bone growth spurts that the soft tissues cannot keep up with. For yearlings, it is particularly dangerous to overfeed because the imbalances in the ration may lead to OCD, osteochondritis dessicans. This condition results in an abnormal turning of cartilage into bone, which can lead to pain, stiffness, and even arthritis later in life.

It is best to work with a knowledgeable veterinarian or breeder when formulating a ration for yearlings.

Stallions
The purpose of keeping a stallion is breeding; overfeeding will diminish his stamina, sex drive, and fertility, especially if he gets visibly fat. A balanced diet and regular exercise will keep him in top breeding condition.

Older Horses
The digestive processes of older horses may slow down, the rate at which they absorb minerals may decrease (zinc especially), and they may have trouble chewing. Keep their teeth in good condition, and feed three or four smaller grain meals, rather than the usual two most horses eat. If you can pasture the horse, he will be able to readily nibble grass, which is easier to chew and digest than hay. If not, alfalfa will give him both nutrients and roughage; add alfalfa to his ration.

Pasture

Probably the best food for horses is what grows fresh from the earth. However, because we generally keep a greater number of horses on pastures much smaller than they would roam naturally, pasture must be supplemented with prepared feeds (grain and hay), and the pasture itself must be kept in edible condition. Ask your county extension agent for the proper mix of grasses to plant for horses in your region. Then let the pasture grow to about five inches in height before turning horses out onto it. Allow them to nibble it down to two inches, then rotate them to a different pasture and let it grow again before reintroducing the horses. Careful pasture management will prevent the bald or muddy spots that allow weeds to take hold and destroy pasture.

Pelleted Feeds

There are three types of pellets; grain, hay, and a combination of hay and grain with vitamin and mineral supplements added. Generally, pellets are convenient because they are dry and not dusty or sticky. They are easy to transport and easier than sweet feed to keep vermin out of; they have less of an odor and are not sticky, so they store well in covered bins. The problem is that pelleted feeds don't provide much roughage; if you use them, supplement the rations with hay and/or pasture.

Protein

Protein is needed for growth and for building and repairing muscle and other tissues. Horses may need extra protein:

- When they are growing
- When they are pregnant, in the last trimester especially
- When they are nursing a foal
- In cold weather
- In spring, to help them shed out their winter coats

Because protein creates heat, horses working hard in the summer should be fed extra protein cautiously.

Salt

Salt, or sodium chloride, is essential for equine health. Grasses do not provide it; usually, it is added to commercial feeds. A horse doing minimal work will need about half a pound of salt per week, depending on how much he sweats. Because excess salt is simply flushed from the system, as long as the horse has sufficient water available, it is safe to provide free access to salt. Usually, blocks are hung in a rack in the stall or just set on the floor for the horse to lick at will. Large blocks can be used in pastures for field-boarded horses.

Supplements

Horses lived for eons without supplements; in fact, without anything but the grasses they found where they lived and roamed. However, keeping them in artificial conditions in which they cannot go and forage for themselves means we have to be sure to supply the things nature would have provided. We do this generally through scientifically prepared supplements. Technically, grain is a supplement to a wild horse's diet of grass, water, and the minerals he consumes in his natural food.

Trace Minerals

These are minerals needed in very small amounts for excellent health, and include iodine, cobalt, copper, iron, zinc, manganese, and selenium.

Vitamins

Vitamins are organic compounds needed in small amounts by every living creature for normal body function. Horses can obtain virtually all the vitamins they need if they are kept on good pasture and not worked beyond the limits of the energy found in the forage they eat there. If a horse is partially pastured and partially grain and hay fed (particularly if

STORAGE OF FEEDS

Keeping rodents and moisture out of hay and grain is the aim of good feed storage. Large amounts of hay should not be stored in the stall barn because of the fire hazard; rather, a roofed shed on well-drained ground, with the hay on pallets off the ground, is more desirable. Keep just a few bales for current needs in the barn itself.

Store grain and pellets in rodent-proof containers with tight fitting lids, or in metal-lined bins that close tightly. Wherever you store grain, keep the area safe from horses by using horse-proof latches. Should a free horse discover available grain, he would likely eat enough to colic or founder before you arrived at the barn.

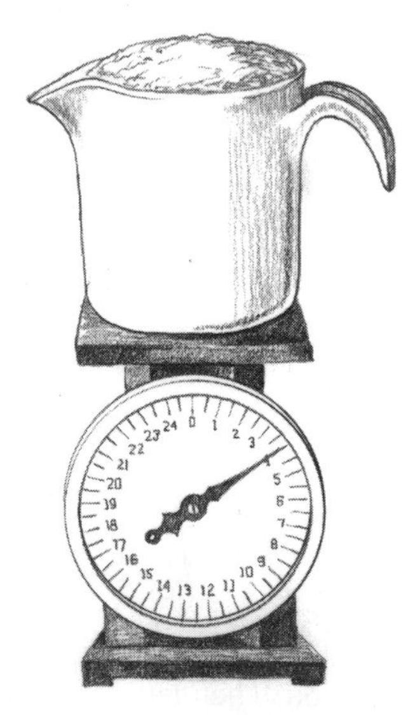

A scale is a useful element of feeding equipment to ensure exact measurement.

he does a substantial amount of work) it is necessary to analyze his rations and his condition to decide if any supplemental vitamins are needed and which ones.

Water

Water is just as essential for equine digestion and metabolism as it is for humans. Horses will generally drink water periodically while grazing. It is also essential to provide free access to water while they are eating food in their stalls, despite the old-fashioned belief that drinking during or just after eating will wash grain through the stomach too fast. In fact, water is more important for horses fed hay and grain. Insufficient water intake can lead to a variety of digestive disorders; some are fatal.

The only time to withhold water is when the horse is overheated; at this time, offer him just a small amount of warm water (to avoid cold-water colic). Offer free access to water once he has cooled down.

Weighing Feed

Rations should be weighed rather than measured or "eyeballed." Feeding flakes of hay is tricky because not all hay is created equal; some balers bale tightly, and their flakes weigh more than flakes from a loose baler. A minimally active horse needs about 25 pounds of hay, in two or three feedings, per day. Weigh a bale of each new batch of hay and divide to see how many flakes of *that hay* it takes to equal 25 pounds.

Not all grain is equal, either. A bushel of corn weighs nearly twice as much as a bushel of oats. Weigh each new feed mixture, whether home-made or commercial, to see how much your standard measure of *that feed* weighs and then calculate the volume of that feed needed to supply what your horse requires.

Each time you change your formula or change commercial rations, weigh again to determine how many scoops equal the pounds needed for good health.

➢ *See also Calcium; Digestion; Floating teeth; Grains; Hay; individual minerals; Pelleted feed; Phosphorus; Poisonous plants; Protein; Salt (sodium chloride); Scoop for grain; Supplements; Vitamins; Water; Weighing feed*

Feeding equipment

Stable-kept horses, and even those on paddock or pasture board, require various feeding equipment to be sure they get the benefit of the hay, grain, and supplements you provide. Here is a list of basic equipment for any stall:

- Two 10-gallon water buckets, wall-hung with snaps for removal and cleaning.
- A feed bucket, either corner-fitted and hung, or a feeding pan for the stall floor for horses that won't eat properly if their feed is above ground.
- A salt-block holder.
- A hay manger, corner-hung or flat-hung (optional).

HEATING UP COLD HORSES

Increase feed by about ten percent to cope with sudden or extreme cold temperatures. And consider adding a warm bran mash, perhaps with some cut-up carrots and apples and a couple of spoonfuls of molasses. It's the horse equivalent of a nice bowl of hot soup and can also maintain bowel health.

To two pounds of bran, add very hot water from the tap or even boiling water. Stir until crumbly but not soupy, like the consistency of oatmeal. Stir in cut carrots, apples, or a little molasses. More than a few horses have been known to enjoy raisins! Feed the mash when it is warm but not hot.

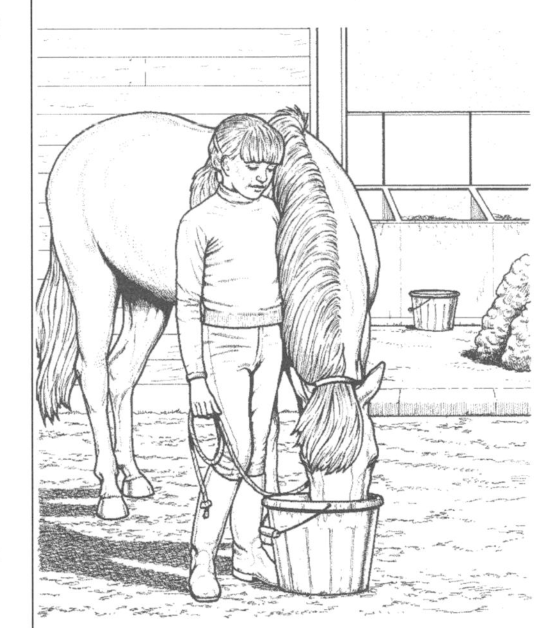

For stall and paddock boarders, provide:

- A large water trough, always at least one-third full.
- A feed bucket, firmly attached to the fence for each horse that gets grain; place them far enough apart to avoid fights at mealtime.
- A clean place to distribute flakes of hay, away from the mud and manure that can build up in some paddock areas.
- A container placed in the field to hold a large salt block, which horses may lick at will.

In an area with moist soil, you can use a single hot wire.

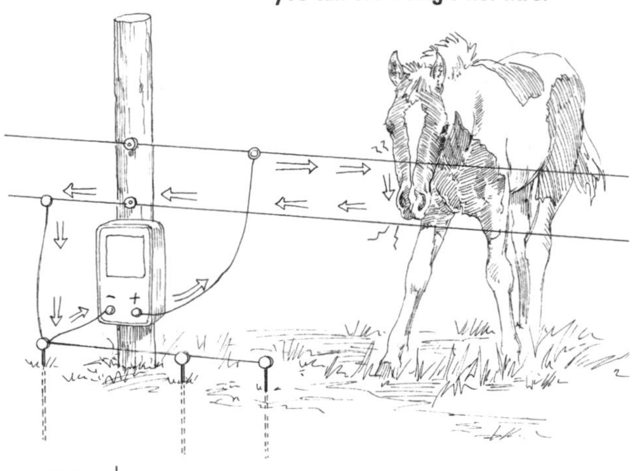

To complete the electrical circuit in dryer regions, it is necessary to use a hot wire (top wire) and a ground wire (bottom wire).

Feet

➢ *See Hoof*

FEI

➢ *See Fédération Equestre Internationale (FEI)*

Fences

The obstacles a horse jumps in hunter and jumper shows. Each discipline has its own variety of fences. In hunter classes or tests based on how horses might approach fences in open fields, the fences resemble actual fences, gates, and other obstacles found on farms and estates — painted board fences, natural wood fences, stone or brick walls, chicken coops, hedges, and brush.

In the show jumping ring, the obstacles are more varied, more colorful, higher, and wider. The fences may include jumps constructed of two sets of rails, one behind the other (oxer); the same type with the poles set unevenly (uneven oxer); or in a fan arrangement (fan jump). There may be three sets of rails, one behind the other, for a triple-bar jump. There may also be a set of three rails with the central set higher than those on the ends; this is called a hogback jump. There might also be solid planks (verticals), and some jumps might even have an artificial pool of water up to twelve feet wide beneath them. Any variation of a hunter fence may also be used.

➢ *See Jumps*

Fencing materials

The three requirements for fencing for horses are strength, safety, and cost-effectiveness. These characteristics bear evaluation by each farm owner or manager and will also take into account the type of horses to be fenced in: for example, energetic, high-strung young horses or older, retired draft horses; stallions, mares, or geldings.

The following is a list of types of fencing, along with notations about each material's uses, benefits, and disadvantages:

Barbed wire. This is effective for cattle, but can be disastrous for young horses. It can be damaging even for older horses who are used to looking out for themselves. It can cause permanent injury or death.

Electric fencing. This is a tape or smooth wire that can keep horses on one side of it — if they are taught to respect it. To teach a horse to obey electric fencing, entice him to touch it with his nose. Still, when in a gallop or when frightened, a horse may simply run through the fence. And if the power goes off, there's no protection. Electric fencing is best used in conjunction with other types of fencing to prevent over-the-fence battles, or to subdivide larger, more substantially fenced pastures into smaller paddocks; then, it would matter less if horses broke through it as they would still be contained in the larger area by solid fencing.

Net or diamond mesh wire. Safer than barbed wire, this material can be stretched and braced between poles; using "no climb" netting in small squares or diamond shapes will keep even small horses and foals from getting a foot stuck in it.

PVC (polyvinyl chloride) tubing can be used for fencing. Usually, it comes with a 15- or 20-year warranty. About twice the cost of wood fencing, it is generally lower maintenance; horses won't chew it and termites won't eat it.

Rubber or nylon tape. These strips are run though upright posts and then stretched. The tape may occasionally need restretching. The advantage is that when a horse runs into the fence, the tape is quite forgiving and inflicts less damage. The disadvantage? Horses might chew the tape and get indigestible fibers lodged in the gut. Under the right circumstances, tapes can be cost-effective.

Smooth wire. Twelve- or thirteen-gauge barbless wire can be stretched between poles to make an adequate boundary fence. Topping it with a rail or pipe, or running a strand of electric fencing inside it, makes smooth wire workable. Using smooth wire in smaller paddocks may cause horses to lean over and through it, however. To avoid life-threatening injuries, mark the fence in a manner visible to horses.

Steel pipe. More durable than wood, especially if coated with a rust-free paint, steel fencing needs fewer uprights than wood. But it has drawbacks. Steel pipe is unforgiving if a horse runs into or kicks it. And there have been a few cases of horses dying when they got their heads stuck between rails of pipe. Wherever you would like a fail-safe fence, though — at road boundaries or for stallion paddocks — steel pipe can be a good choice.

Wood fencing. Wood is a traditional type of fencing, particularly three- or four-board fences painted white, brown, or black. Wood is sturdy, durable, and aesthetically pleasing. Boards should be nailed to the posts on the inside of the post, facing into the paddock or pasture, making it harder for a horse to run through; the board would have to break rather than tear out by the nails. Horses can damage this fencing, though, by kicking, running into, leaning over, or chewing it. It is helpful to run an electric fence wire inside the top rail of wooden fencing to eliminate most of these possibilities. Wooden fences require a lot of upkeep, from replacing boards to repainting and weatherproofing. Occasionally, an upright post will snap and need labor-intensive replacement.

➢ *See also Chewing; Cribbing; Gates; Paddock*

TAKING THE STING OUT OF BARBED-WIRE FENCING

If you acquire a farm that already has barbed wire, until you can replace it, run an electric wire fence several feet inside of the barbed wire. This way, your horses won't be close enough to lean through the barbed wire and hurt themselves. And even while playing or fighting, a zap from the fence will probably send the horses back out of real harm's way. By running the electric strand inside the barbed wire fence, you contain the animals as well as creating a visual barrier that most horses — and people — will respect.

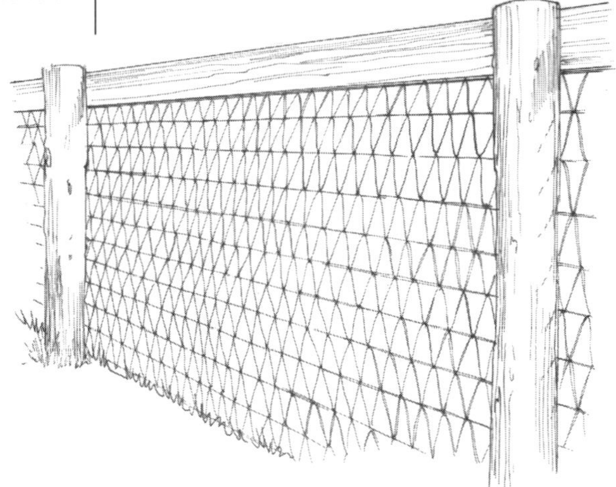

Net or diamond mesh wire fencing

GOOD FENCES MAKE GOOD NEIGHBORS

It's true for humans and equines alike: good boundaries make it easier to have peace in the neighborhood. Here are special circumstances calling for special fences to keep everyone concerned safe and happy:

Mares eat oats and foals eat oats . . .
And if these plants are in the neighboring field, the horses will find them. Fencing for mare and foal should be very safe and very durable, because foals are inquisitive and will try to get through a fence to something that intrigues them — edible or otherwise. Wood or net wire fencing are good choices.

Weaning worry-free
Weaning enclosures should be completely hazard-proof; well-stretched diamond mesh wire with good support comes closest.

Small spaces invite mayhem
Small paddocks seem to invite rowdiness in horses. Or maybe it's just that they're big and the paddock is small, so a lot of damage can be done. The best bet for chew-proof, kick-proof, lean-proof fencing is diamond mesh wire.

Stallion alert
Stallions can be rough on fencing material, or they can go over it. A stallion pen should be 5 feet high or more, made of wood or pipe, and equipped with a reinforcing electric strip inside.

Fender

The rectangular or triangular panel of leather that hangs down from a Western saddle. This piece holds the stirrup and protects the rider's leg.

➢ *See also Western riding; Western tack*

Feral horses

Free-running horses that have escaped from farms and ranches, or their offspring. The Mustang of the American West is a feral, rather than a wild, horse.

Fertility

Many factors affect the fertility of both mares and stallions. The most important is the level of skill and knowledge of the manager. Some other causes of low live foal rates are:

- Poor teasing practices; improper or inadequate teasing.
- Failure to breed the mare at the optimal time (close enough to ovulation for a viable egg to be present at the same time there is viable sperm)
- Spontaneous abortion. From 10 to 22 percent of all conceptions end in miscarriage within the first three months. Increased use of ultrasound indicates that these miscarriages are the result of Mother Nature aborting poorly formed or twin fetuses.
- Heat signs without ovulation; common in early spring and late summer.

- Stress due to the artificial nature of breeding. Some horses simply are stressed by the entire process, which reduces their fertility.

In equines, fertility is a variable concept. To start, semen from different stallions displays very different sperm life spans. The average sperm life in the mare's reproductive tract is 40 to 60 hours. Some stallions' sperm is viable only for 24 hours, although others can breed a mare as much as a week before her ovulation and still make her pregnant.

Sperm production is believed to be related to testicle size. As a stallion matures, he becomes more fertile. Testicles shrink in winter, producing only 50 to 70 percent of normal capacity. May and June are good breeding months, and the sex drive peaks at this time. Testicles also shrink with age.

Mares are usually **anestrus** (that is, they do not ovulate) during the winter. Mares that are too thin or too fat may not cycle, or they might have unpredictable cycles. If you've bred a mare several times, and she has not "caught" or "settled" (become pregnant), have her checked for thyroid problems if she is under 10 years old. Mares older than 10 who have not been bred before may be hard to get in foal.

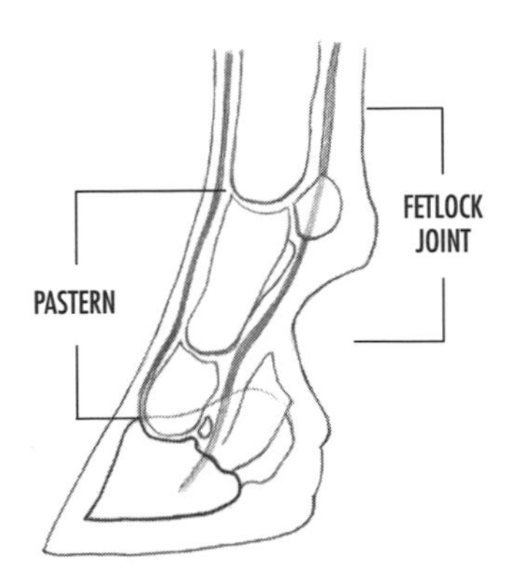

Anatomy of the pastern and fetlock

Fescue grass *(Festuca elatior* syn. *Festuca arundinacea)*

It's not the grass itself, tall fescue, that causes problems but rather *Neotyphodium coenophialum* (also known as *Acremonium coenophialum*), the fungus that grows on much of it. The fungus plays havoc with mares, causing retained placenta, failure to produce milk, and even abortion.

Fetlock

The fetlock is the rounded, bony, bulblike structure on a horse's leg just above the pastern; the pastern is the narrow, sloping area above the hoof. It corresponds to the joint attaching a human's fingers to the long bones of the palm, and the toes to the long bones of the foot (joining metacarpals and metatarsals to proximal phalanges). On some breeds, especially draft horses, long hairs, called **feathers,** grow from the back of the pastern. In winter, though, most horses will grow some long hair in this area. The extra hair can be clipped for grooming and showing off the horse's legs.

POISONOUS PLANT

Fiddleneck

Fiddleneck (fireweed, tar weed, yellow burr weed) (*Amsinckia* spp.)

While fiddleneck isn't a very palatable plant to horses and they probably won't eat enough by choice to poison them, it grows in wheat and grain fields and may be mixed in with carelessly harvested and prepared grain. An accumulation of small amounts of fiddleneck's pyrrolizidine alkaloids can cause liver failure and death.

➢ *See also Poisonous plants*

Fields, riding in

Horses associate fields with freedom; riding in fields requires more skill from the rider and some knowledge.

The horse will probably be more easily distracted in the field than in the enclosed riding ring. There's more to look at — birds, insects, maybe deer or larger animals somewhere nearby. Riders need to be both firm and gentle to get the horse's attention without being mean. On the other hand, your horse knows his feeding and sleeping quarters are back at the barn; if it's near feeding time or he's tired, he may want to head for the barn whether or not you do.

Here are some other concerns to be aware of when riding in the field:

- A horse's main line of defense is fleeing; in the field, he has lots of room to do it. From the moment you mount, be more alert than usual to things your horse may perceive as dangerous and reassure him that he is in a safe environment.
- Your horse may slow down leaving the barn area; use extra leg to move him on.
- Your horse likes to be with other horses, so let him. Keep him about one horse-length behind the horse in front of you, so that you can avoid sudden speed spurts while playing catch-up.
- If your horse tries to leave the group, correct him. If he is allowed to gallop away the other horses will want to follow, creating a dangerous situation for other riders.
- A change of pace that your horse chooses means something; perhaps fright. Keep his attention on you and correct him, bringing him back to the gait you have chosen.

➢ *See also Facilities; Mounting, technique; Pasture; Trail riding*

Figure eight

This pattern is used in both the education and the evaluation of riders. It is quite simple: ride two circles that join in the middle. It can be executed at a walk or trot/jog while riders learn how to maintain and change bend by using leg/hand/body coordination. However, at the canter or lope, it becomes a much more serious test of riding ability and horsemanship. Riders must be able not only to maintain proper bend and symmetry of both circles, but they must also plan for a proper change of lead in the middle of the two circles. Riders learn first how to perform a simple change of lead, which is done by asking the horse to come down to a trot, walk or halt prior to signaling for the new lead. Once this is mastered, riders learn how to perform a flying lead change, in which the change occurs midstride without any break of gait.

Filly

A female horse up to four years of age.

Finished horse

Fully trained for the job at hand. This could be considered the opposite of the "green" horse, although neither term is truly standardized. Essentially a "finished" horse is one that is fully trained for a particular purpose. Thus, the same horse could be "finished" as a riding horse, but be "green" as a driving horse.

MOUNTING IN THE FIELD

A rider is most vulnerable to injury during mounting and dismounting. Having to do either one, especially mounting, in the field is extra dangerous. To minimize the risk:

- ❑ Check the girth and be sure it's tight; without a mounting block to help you, you may put more pressure and weight in the stirrup iron, causing the saddle to slide unless it is secure.
- ❑ Use the landscape for help. Find a gentle slope and stand on the uphill side to mount; standing on the downhill side would make the horse even harder to mount.
- ❑ Point your horse away from the barn while mounting to reduce his inclination to head for home.
- ❑ Keep the reins a bit shorter than usual to keep his head up; there's probably grass around and he may want to nibble it while you mount.
- ❑ Keep rein on near side shorter than other rein. If horse bolts, you can maintain control.

It is somewhat of a golden rule that finished horses are the most suitable for novice or inexperienced people. These horses can be excellent teachers and partners in the process of learning.

Fireweed (tar weed, yellow burr weed) (*Amsinckia* spp.)

Similar to fiddleneck toxicity, a kind of poisoning that can also be caused by tar weed, buckthorn, and yellow burr weed.

➢ *See also Poisonous plants*

First aid

When an emergency occurs, call the veterinarian immediately. If you suspect colic, the vet will probably arrive while you are walking or calming your horse. Aside from colic, though, there are some equine ailments and injuries that do require significant first-aid attention from owners.

Abrasions

Scraped skin may bleed or ooze fluids and is sometimes slow to heal. Clean the area and apply antibiotic ointment to keep the wound soft, rather than crusty, as it heals.

Bleeding, Controlling

Immediately apply the cleanest pressure bandage that you can create quickly from towels, shirts, stable quilts, or other material. Wrap a piece of wood or even a rock padded with clean cloth against the bandage and bind it with bandage or strips of an old sheet. Continue wrapping until the bleeding stops, even if it takes many layers wrapped fairly tight. This will allow you time to call the veterinarian. Don't use a tourniquet; it can cause permanent damage by blocking blood flow to the area beneath it.

Bruises

This soft-tissue injury is caused by a bump or blow. The skin does not break, but there may be considerable swelling caused by broken blood vessels and lymph leaking within the damaged area. For severe bruises, a veterinarian may insert a large-diameter sterile needle to allow drainage. Cold water and ice can help reduce minor swellings. All bruises should be treated to avoid pressure-caused damage to the underlying tissue; such secondary problems might cause permanent lameness by interfering with tendons.

Burns

A severely burned horse may go into shock and die. First aid for serious burns includes applying ice packs or even cold water from a hose to reduce the depth of the burn and minimize tissue damage; the sooner you stop the heat from penetrating deeper into the tissues, the better. Call the veterinarian, who will administer painkillers and antibiotics and show you how to clean and dress open burn lesions.

BE PREPARED!

Post the veterinarian's main number and his emergency number in a conspicuous place. Someone other than you might need to attend to your horse, and even you might forget the numbers during the upset of an emergency.

WHEN IS IT AN EMERGENCY?

It saves lives and money to know the difference. It's an emergency

- ❑ When temp is <3°F above normal for your horse.
- ❑ When resting pulse or respiration is <20% above normal for your horse.
- ❑ When respiration is greater than pulse (indicates serious shock).
- ❑ When capillary refill time is <3 seconds.
- ❑ When normal gut sounds are lacking.
- ❑ When lameness is a grade 1; horse can't bear weight or hops while trying to move.
- ❑ When an injury spurts blood (indicates an artery is involved), is near a joint and amber-colored fluid is seeping from wound (indicates joint capsule is injured), or exposes bone, tendons, or ligaments.

Choking

Signs of choking include water or bits of food exiting from the nostrils, drooling, coughing accompanied by a spray of food or water, or nose- and head-slinging. Choking can be caused by too large a piece of apple, carrot (beware letting a horse eat an apple or carrot whole!), or other treat; dry alfalfa cubes eaten too rapidly; or from a too rapid ingestion of grain, pellets, or grass. Call the veterinarian immediately, and keep the horse calm by slowly walking him and gently massaging the throat, which may help the blockage to move.

Fractures

Most fractures occur below the knee or hock and are caused by a misstep or fall, thrashing against a stall wall as a result of the horse being cast on his back, or a kick from another horse.

Friction Burns

Friction burns happen when a horse's legs get caught in a lead rope, whether that lead is cotton, nylon, or a leather shank. The wound is a combination of a cut, where the line has sliced into the flesh, and a burn, where heat from the sliding rope has abraded the skin. More skin tissue is damaged than in a simple cut, and healing may take longer.

Minor friction burns may be treated like abrasions. Deeper friction burns, or those with rope particles embedded, require veterinary attention.

Head Injuries

Horses toss their heads in alarm, and they use their heads to investigate things on the other sides of fences. Horses also may fall or flip over in fright and hit their heads. A horse may be kicked in the face by another horse; there may even be lacerations or a break of a facial bone. If the injury is superficial, merely a minor cut, treat it with first aid.

However, if there is swelling, suspect underlying bone damage. If the horse bleeds from the nostrils or shows rapid eye movements, there is serious head injury. This is also the case if he cannot get up, or if he rises, then staggers or even falls again. Call the veterinarian immediately; prompt administration of dimethyl sulfoxide (DMSO) intravenously may keep brain tissue from swelling and causing permanent disability.

Hemorrhagic Shock

This is shock due to blood loss. Keep the horse warm with blankets and leg wraps while you work at stopping the bleeding. When the veterinarian arrives, he or she will administer fluids or blood transfusions and medication.

Incised Wounds

An incised wound is a clean cut. These cuts bleed profusely because of the shearing of numerous small blood vessels. If a horse loses a pint or two of blood, it's not serious. However, losing more than two gallons can be fatal.

Kick Wounds

Kick wounds don't all look alike. Small ones that break the skin but don't cause much swelling — those caused by the point of a shoe or toe, for example — may have chipped an underlying bone. Those caused by

FIRST AID FOR FRACTURES

As in all other injuries, first be sure to do whatever is necessary to stabilize the horse and prevent shock. Then call your vet. First aid treatment until the vet arrives largely depends on the type of fracture; ask the vet what you can do until he or she arrives.

In the absence of veterinary advice, use common sense to stabilize the broken bones as best as you can. Keep the horse as still and calm as possible. Many barns keep tranquilizer or herbal calming remedies on hand for such emergencies. If the break is in a leg, create a makeshift splint and attach to the leg with bandaging materials from your first aid kit, or use polo/exercise wraps. PVC pipe, cut in half lengthwise, makes an excellent splint; it's recommended to keep one on hand. Otherwise, a piece of wood or metal wrapped in cloth can do in a pinch. If the break is in the pelvis, shoulder, spinal column, or head, it is imperative that you receive advice from a veterinarian on how to administer first aid. The two priorities are to maintain calmness and stabilize the bones.

HEMORRHAGIC SHOCK AFTER FOALING

Keep the mare as quiet as possible, summon the veterinarian, and blanket her to conserve body heat; heat the space if you can do so safely. Keep the foal where she can see him, perhaps protected in a "cage" of hay bales just outside her door.

the landing of a whole hoof may exhibit a lot of swelling but no broken skin; they may do nothing more than cause temporary soreness and unsightliness. Assess not only the size and nature of the wound, but the horse's disability after the kick and the length of time it takes to resolve. Serious lameness for more than a short time, exposed tendons, or serious bleeding call for veterinary intervention.

Kick wounds can be minimized by removing hind shoes from horses living together in herds, separating horses at mealtime or when coming inside for meals, introducing new horses to herds gradually, and permanently separating horses that don't get along.

Lacerations

These wounds have jagged edges and often bleed less than incised wounds. Clean them with running water until all dirt and debris and dried blood are gone and pink flesh appears. Apply a nonirritating wound dressing, or none at all if you think the veterinarian might need to stitch the wound. Above all, avoid caustic disinfectants, such as peroxide and alcohol, which will slow healing and might even destroy so much tissue that the veterinarian can't apply stitches.

Porcupine Quills

Most porcupine quills on horses are found in the nose because the horse gets curious and sniffs the porcupine. Have someone hold the horse still with a halter and pull the quills out straight with needle-nosed pliers. Take care to avoid breaking them off; they'll be harder to remove if you must grapple at skin level for a hold. Don't clip off the ends; this, too, pushes the shaft deeper. If the horse is nervous or won't stand, obviously you cannot use a twitch applied to the nose. Try a Stableizer, a device that looks like an Indian war bridle and works by putting pressure on the upper gums and behind the ears. When the quills are out, apply a soothing antiseptic cream or gel.

➢ *See also Porcupine quills, first aid for*

Pressure Bandage

This is a tool used to control bleeding until the veterinarian arrives. You can make a pressure bandage from towels, shirts, stable quilts, or other material. Wrap the bandage over the wound, then wrap a piece of wood or even a rock padded with clean cloth against the cut. Bind it there with bandage or strips of an old sheet. Continue wrapping until the bleeding stops, even if it takes many layers wrapped fairly tight. This will allow you time to call the veterinarian. Don't use a tourniquet; it can cause permanent damage by blocking blood flow to the area beneath it.

Punctures

These wounds appear as small holes in the skin, bleed little, and may begin to heal from the outside first, trapping infection below the skin. Open the puncture hole a bit more, clean it with hydrogen peroxide or Betadine, and put antiseptic-soaked cotton or gauze into the hole. Change the dressing daily until the wound has healed. Avoid strong disinfectants, which may burn tissue and cause even more trouble below the skin's surface, leading to infection.

CLEANING WOUNDS

Whether the injury is fresh or old and caked with dirt and dried blood when you discover it, clean it with running water from a garden hose. If the horse will not stand, seek help and use a twitch. If an old wound opens and new blood flows, don't worry; it is important to clean the wound, and the flow of blood will help.

When it's clean, assess the injury. A deep cut may need stitches. If you can see tendons call the veterinarian. If the wound is over a joint and it is deep, or if it is below the pastern, call the veterinarian. Cuts that penetrate the coronary band at the top of the hoof may permanently affect hoof growth. Your veterinarian may suggest calling a farrier, as well, to take care of these cuts, or wounds on the bulb of the heel (shoeing might help protect the area so it can heal).

ARTERIES VERSUS VEINS

When a horse is bleeding copiously, it is essential to determine whether the blood is from an artery or a vein. Arterial blood is bright red and squirts or gushes out with each beat of the heart. Be aggressive in stopping blood that is spurting from a cut artery. Veins, when cut, ooze dark blood continually. A puncture to the vein can produce major bleeding, but it usually gives you more time to control it than a cut artery does. Still, you must take effective measures immediately to control the bleeding.

Rope Burns

➢ *See Friction Burns on page 140*

Secondary Shock

Triggered by circulatory failure of the blood vessels, secondary shock is brought on by pain and tissue damage. If mucous membranes become pale, call a vet; fluids, adrenaline, and steroids must be administered immediately to prevent irreversible damage caused by lack of oxygen to vital tissues.

Shock

Whether the cause of shock is blood loss (hemorrhagic shock) or another problem (secondary shock), a horse going into shock will exhibit pale mucous membranes, subnormal temperature, falling blood pressure, rapid but feeble pulse, and shallow respiration. He may be restless and anxious, or very dull. He will sweat, but has cold, clammy skin. Keep him warm and comfortable until the veterinarian arrives.

Snakebite

Because of his large size, a horse can handle a snakebite better than most mammals. However, a bite on the face or neck may cause enough swelling to interfere with breathing; suffocation could result. If you suspect snakebite because of puncture wounds, swelling, or behavioral changes, call the veterinarian. Depending on the location of the bite, the size of your horse, and the toxicity of the suspected venom, your horse may need a dose of antivenom.

RELAXING INJURED HORSES

Horses pick up on their handlers' fears, so control yours well around an injured horse. Be confident that all will work out well, and tell your horse so. Walk and talk calmly, and stroke him if he enjoys it. Consider an herbal calmative if you are certain its use will not interfere with veterinary care.

FIRST-AID SUPPLIES TO HAVE ON HAND

The following is a list of items to be kept in one place — clean, dry, and frequently checked on and replaced as needed — along with clean buckets and a supply of towels.

- ❏ Wound dressings (nitrofurazone or a similar herbal preparation), povidone iodine, and wound spray
- ❏ Chlorhexedrine ointment or zinc oxide preparations for fungal infections
- ❏ Povidone iodine solution in a spray bottle
- ❏ Tincture of iodine for thrush, quicked hoof, or hoof puncture
- ❏ Dexamethasone
- ❏ Hydrogen peroxide
- ❏ Phenylbutazone (also called Butazolidin or bute), in tablet form, to relieve pain and swelling
- ❏ A flushing solution mixed by your veterinarian for deep or puncture wound washing
- ❏ Disposable syringes and needles (if you have learned how to give injections properly)
- ❏ Gauze pads
- ❏ Strips of clean sheets rolled into bandages or extra clean white stable bandages, at least six or seven
- ❏ Nonstick sterile wound dressings
- ❏ Rolls of cotton, cotton batting, or cotton leg quilts
- ❏ Stretch bandage that sticks to itself, such as Vetrap
- ❏ Duct tape
- ❏ Adhesive tape in a variety of widths
- ❏ Masking tape
- ❏ Bandage scissors or bent trimmers
- ❏ Equine thermometer, a length of half-inch cloth ribbon (at least 3 feet long), and an alligator clip
- ❏ Flashlight and extra fresh batteries
- ❏ Stethoscope
- ❏ Pocketknife
- ❏ Humane twitch
- ❏ Petroleum jelly
- ❏ Mineral oil
- ❏ Protective boot for hoof
- ❏ Large bottle of rubbing alcohol
- ❏ A watch or clock with a second hand
- ❏ Disposable razors
- ❏ Extra clean squeeze bottles
- ❏ Well-washed worming syringe for administering dissolved tablets

Stitches

Some wounds heal better if they are stitched, but the wound must be relatively fresh — two hours old or less for a head wound, six hours old or less for a leg wound, and eight hours old or less for wounds of the body. Stitching cannot be done if the wound has been further damaged by application of caustic liquids, such as strong disinfectants. If you think a wound may benefit from stitching, clean it but do not apply antiseptics and call the veterinarian.

Tetanus (Lockjaw)

Tetanus is best prevented by keeping a horse's vaccinations for it up-to-date. If they are not, and the horse sustains a wound that breaks the skin, he should be vaccinated within 24 hours.

Tetanus is a disease with a high mortality rate among all mammals, but horses and humans are particularly vulnerable. It is caused by a spore-forming bacterium, *Clostridium tetani,* that lives in the digestive tracts of many animals and in soils rich in animal manure. Clean wounds rarely result in tetanus; punctures that contain foreign matter, especially soil, are dangerous.

The incubation period for tetanus in a horse is 1 to 3 weeks, although cases have occurred even later. The toxin produced by the bacterial spores travels along nerve trunks, causing muscle spasms. Eventually, the muscles involved in respiration virtually become paralyzed and suffocation results.

Wounds

While most wounds are not life-threatening (except if accompanied by great loss of blood or penetration to vital organs), they will need first-aid attention. The types of wounds are abrasions, bruises, incised wounds, lacerations, and punctures. A sixth — rope burns — combines a burn with an abrasive wound.

First cycles

Most mares in the Northern Hemisphere do not have estrus cycles in the winter. About the time of the vernal equinox, March 21, many mares begin to cycle again, but the exact season depends on the climate in the mare's home location. The first few cycles are less fertile than those later on, as the summer solstice, June 21, approaches. During first cycles, a mare may be bred, but probably won't become pregnant. If you must breed early, work closely with a knowledgeable veterinarian.

Fistula of the withers

This can be a life-threatening injury for a horse. It has many causes; most commonly an ill-fitted saddle causes bruising. If left untreated, infection might develop under the skin, draining painfully between the horse's shoulder blades and spine. Infection can eventually open up a hole through the skin and into the muscle below that is large enough to fit a man's fist. Cures, at this point, are difficult and slow; often, the horse must be destroyed.

Fistulous withers can also be caused by tack bumping the withers repeatedly and causing bruising, or by cuts in the area further aggravated by bumping or by unclean saddle pads.

LOCKJAW'S CLASSIC SYMPTOMS

Tetanus infection is commonly called lockjaw because of its characteristic progression. A horse's facial and chewing muscles will be affected first, making it hard for him to open his mouth to eat or drink. The third eyelid protrudes and the hind legs stiffen, causing an unsteady gate. His tail may be held out stiffly. The horse will be anxious with ears up, eyes wide, and nostrils flaring.

Treatment is often unsuccessful, so prevention through vaccination is key.

WEIGHT ESTIMATION FOR MEDICINE DOSAGE

To give the proper dosage of many medications, you need to know your horse's weight. A truck or livestock scale is best. To use a truck scale, first weigh your truck and trailer empty. Then load the horse and weigh again. The difference is your horse's weight.

If a scale is not available, measure the horse's girth at the heart. Multiply that number by itself, then by body length (measured from middle of chest to middle of tail). Divide the total by 330 and you will have an approximate weight for the horse. You may also purchase special weight tapes, which estimate the horse's weight through his girth measurement, at many tack shops.

Fitness

➢ *See Conditioning program; Interval training*

Five-gaited horse

Although there are numerous breeds of horses that perform more than the traditional gaits of walk, trot/jog, canter/lope, and gallop, the breed most commonly associated with the term "Five-Gaited" is the American Saddlebred. These horses are born with the natural ability to perform two additional gaits, the rack and slow gait.

➢ *See also Amble; American Saddlebred; Pace; Paso Fino; Peruvian Paso; Rack; Racking horse; Tolt*

Fjord Pony

A breed of sturdy ponies that looks remarkably like Przewalski's horse, with the same dun coloring, dorsal stripe, and erect mane. An extremely strong animal with a stocky build, the Fjord Pony has been bred in Norway since the time of the Vikings without alteration. A strong-willed but pleasant character and a willingness to work make these horses popular in Norway and Denmark.

➢ *See also Przewalski's horse*

Flagger

In Western roping events, a mounted rider who holds a flag with his outstretched arm until a competitor successfully ropes the cow. At that time, he drops his arm and the flag.

Flag race

In Western competition, riders must pull flags out of cans or buckets while riding a prescribed pattern designed by the association hosting the event.

Patterns for a Stake Race

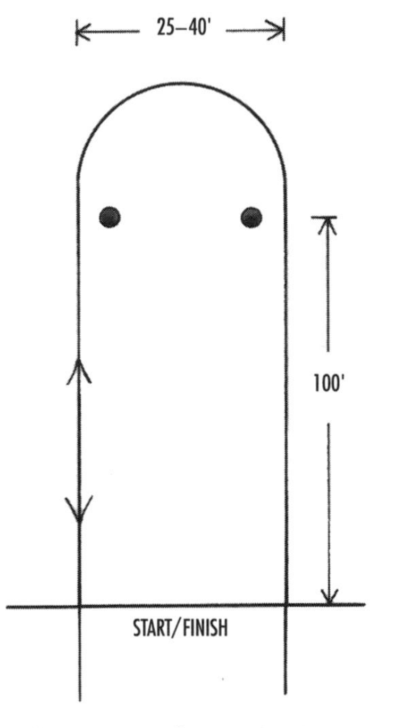

Flag races are also completed around set patterns, usually more complicated than the stake race patterns.

Flank

In roping, a Western event, flanking means holding a calf by its flank and leg and placing it on its side.

The word "flank" can also refer to the area on a horse's side between the thigh and the barrel.

Flat bone

The desirable condition that results when a horse's cannon bone seems wide and flat when viewed from the side. The tendon lies well behind the bone, with tissue intervening. In **round bone,** on the other hand, the tendon is tied in very closely to the bone. Flat boned legs tend to hold up better under stress, because the tendon is subject to less wear than those that continually slide along the bone.

➢ *See also Anatomy of the horse; Round bone*

Flat feet

An inherited or man-made condition (due to poor hoof care) in which the sole of the hoof is not concave, so that there is no natural way for the horse to avoid bruising the sole on rocks. Shoeing with pads added between hoof and shoe can help, as can toughening the sole by applying iodine to it. If this measure is taken, the hoof wall must be avoided (iodine dries it out), as must other tissues that iodine will burn.

Flat-foot walk

A true four-beat walk used by most horses in pleasure and competition.

Flat racing

From the ancient Greeks to the Romans to the Anglo-Saxons, horse racing and hunting on horseback have been popular sports. King Charles II, who ruled from 1660–85, is known as "the Father of the British Turf." He founded the Royal Stud and developed a small racecourse at Newmarket into the famous center for racing that it remains. In the United States, racing followed the colonists, whose informal races at short distances led to the development of the Quarter Horse, which can outrun a Thoroughbred in a sprint. The first American oval track was created on Long Island in 1665, and racing survived the disruptions of the Revolutionary and Civil Wars to become firmly entrenched in the American sport scene. The advent of the mechanized gate and the photo finish simplified the process of getting the race started and determining how it finished.

Though not permitted in all states, there are dozens of racetracks of various sizes and reputations around the country, with many famous and prestigious contests, including the races that form the coveted Triple Crown: the Kentucky Derby, the Belmont Stakes, and the Preakness. Lexington, Kentucky, is the undisputed center of American racing, though the industry is also well established in Florida, New York, and California.

➢ *See also Harness racing; Horse racing; Speed, of horse; Thoroughbred; Triple Crown*

Flatulent colic

➢ *See Colic; Feeding and nutrition*

RACING TERMS

Backstretch: The long section of the track on the far side of the racecourse.
Blinkers (blinders): Patches that are attached to the bridle to restrict the horse's vision and reduce his inclination to shy at objects coming up behind him.
Breeder's Cup: The annual racing championship, consisting of eight races held at a different track each year and offering a total of $13 million in prize money.
Claiming race: A race in which the runners may be purchased by preregistered buyers at previously set prices.
Classic: A traditionally important race. Also refers to the length of a race; an American classic is 1¼ miles, while a European classic is 1½ miles.
Cuppy: A dry, crumbling track surface.
Daily double: A wager in which the bettor must select the winners of two consecutive races (often the first and second of the day).
Daily Racing Form: A paper containing statistics about races, horses, and jockeys. Note that it is not "The" Daily Racing Form.
Exacta: A wager in which the bettor must select the first two finishers, in order.
Furlong: One eighth of a mile (220 yards).
Handicap: A race with weights assigned according to each horse's previous performances or the process of selecting horses (frequently for the purpose of placing bets) based on past performance.
Homestretch: The final section of the track leading up to the finish line.
In the money: A horse that places first, second, or third in a race.
Length: In a race, the distance of about 12 feet, which a Thoroughbred covers in ⅕ of a second.
Mudder: A horse that races well on a muddy track.
OTB: Off Track Betting
Place: To finish second.
Scratch: To withdraw a horse from a race before it starts.
Show: To finish third.
Silks: The brightly colored shirts and caps worn by the jockeys. Each owner or stable has its own colors and design.
Stakes: A race for which the owner must pay a fee to enter.

Flehmen response

A facial expression in which the horse curls his upper lip and raises his head in order to draw air across a specialized organ located in the roof of the mouth. It can be a reaction to odd tastes or smells in the environment, or it can be an effort to obtain the scent of a mare in heat.

Flex

Bending and flexing are directly related, but are not the same. When a horse flexes at the poll and in his jaw, it means he's yielding to pressure on the bit by relaxing his jaw and flexing forward/backward and side-to-side at the poll. Bending is reflected by movement within the spine and body. It is seen especially well when a horse is viewed from above while performing a circle. He bends from head through tail along the curve of the circle.

➢ *See also Bending; Lateral exercise*

During a flexion test, a veterinarian evaluates a horse's soundness.

Flexibility

In a horse, the ability to use the body athletically without undue stress on muscles and ligaments. It is both natural (some horses are more flexible than others, just like people) and can also be developed by working all of a horse's muscles while not overworking just a few.

Flexion

The vertical movement of the spine and limbs, including the joints of the horse's lower jaw, poll, neck at the withers (both up and down and sideways), back, and croup, as well as the knees, hocks, and stifles.

➢ *See also Anatomy of the horse; Flexion test; Hock; Stifle; Withers*

Flexion test

Part of a prepurchase exam performed by a veterinarian to determine whether the horse is sound for the purposes you intend to use him for. It can also be used to find lameness in horses. The veterinarian holds the hock or other leg joint in a flexed position for an extended period of time. When he releases it, a helper jogs the horse away and the veterinarian assesses whether or not the horse has regained normal use of the stressed joint within the appropriate length of time.

Flexor tendon

The tendon in the leg that allows the horse to pick up his hooves.

Flies

No matter how clean a stable or barn is, there will be flies annoying both people and horses. Barns are the perfect environment for flies, complete with plenty of manure in which to lay eggs and plenty of animals to bite (some fly varieties need a blood meal to complete the egg-laying cycle). In addition to inflicting often painful bites, flies carry a host of diseases and can cause skin allergies and hypersensitivity reactions. Be careful when grooming and handling horses during fly season, as they may stomp and kick at flies while you are within range.

Good stable and barn hygiene goes a long way toward controlling these pests. Clean stalls daily and cover manure piles with black plastic or spray them periodically with insecticide. Clean up spilled or leftover grain, and eradicate leaky faucets and other sources of dampness. You can protect your horse against flies by providing shelter during peak fly times (generally morning and early evening), using insecticidal wipes or sprays, installing screens on stall windows, using fly sheets (mesh blankets) and masks, and judiciously spraying the stable area.

➢ *See also Bot flies; Deer flies; Horse flies (tabanids); Parasites*

Stable fly

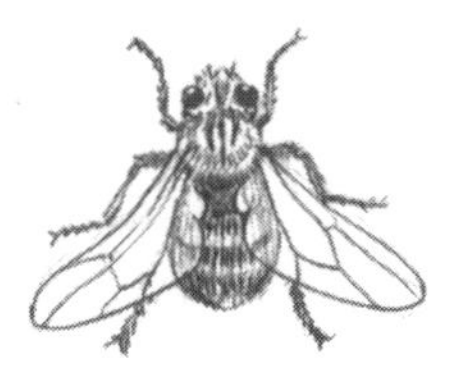

Horn fly

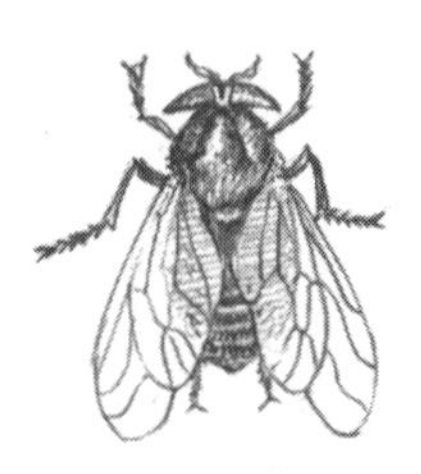

Horse fly

Flight instinct

Horses, having no claws and no ripping teeth, have only one choice when they perceive danger: flight. For the past 75 million years, horses have been fleeing danger; even in domesticated settings, horses will attempt flight if frightened beyond their faith in the human's ability to protect them.

Floating teeth

The equine term for rasping teeth to remove sharp points; the equine dental rasp is called a **float**. In their natural state, horses wear down the outer edges of their molars, but in domestic settings, they do not normally eat enough rough material to grind off those edges. Sharp points develop on the outer surfaces, not the grinding surfaces, and can be painful to the inside of a horse's cheek or to his tongue. Floating does for kept horses what foraging does for those in a natural environment.

After the age of two, horses should have their teeth checked at least once a year and floated when necessary.

➢ *See also Dental care*

God in His wisdom
made the fly
and then forgot
to tell us why.

— Ogden Nash

Flooring in barns

Although horses spend only a few hours a day lying down, they spend numerous hours supporting a half-ton of weight on four feet smaller than the average man's two feet. For this reason, flooring and floor quality are vitally important in barns.

Dirt floors are often preferred because they are easy to keep dry, providing some natural drainage into the earth, and they are not too firm or unforgiving if a horse lies down a lot or falls. Covered with rubber stall mats, they are even better.

Whether or not you use rubber mats, keeping the floor bedded with wood shavings or chips makes the horse more comfortable; helps keep the floor dry despite urine and fecal deposits; and, if they are pine chips or shavings, can help with odor control and cleanliness.

➢ *See also Barn; Bedding; Calluses; Facilities*

Flu (influenza)

An infectious disease of the upper respiratory tract, similar to the disease of the same name in humans. Flu causes fever and can become serious if the horse is worked hard or stressed in the early stages. The disease is very contagious and best treated with good nursing care; antibiotics are useless against the virus that causes it. However, it may be wise to give antibiotics to prevent secondary bacterial infections from taking hold in a flu-weakened horse.

➢ *See also Equine influenza*

Flukes

Parasitic flatworms that live in freshwater snails and spread Potomac Horse Fever (PHF).

➢ *See also Potomac horse fever (PHF)*

Fly mask

Fine-woven mesh protective gear that fits over a horse's eyes. Masks are usually held securely under the chin with Velcro strips, with a padded strap around the ears to secure them from above. The horse can see through the mesh, but the flies cannot get in to feed on the discharge from the horse's eyes. These masks are appropriate for any horse grazed outdoors in daylight during fly season (they are taken off when the horse returns to his stall), especially horses that are particularly bothered by flies.

Flying lead change

A way to change a horse from his left to his right canter or lope lead, or vice versa. A change of lead can be accomplished by bringing the horse back to a trot or jog; this is called a **simple change.** In a flying lead change, the horse never breaks out of the canter or lope, but rather changes lead midstride.

Uses of Lead Changes in Jumping

Horses approaching a fence to jump it — not to mention negotiating the corners and turns to reach each fence on the course — must be well balanced to clear the obstacle while carrying a rider's weight. The leading foreleg must hit the ground just prior to takeoff, or the horse will jump from a distance too far away from or too close to the jump to achieve the best trajectory, and the leading hind leg must be able to push the weight off the ground.

Uses of Lead Changes in Western Riding

Many Western horsemanship pleasure events use the same serpentine pattern requiring eight lead changes in the pattern. Pole bending competitions also demand horses that change their leads easily — the better to get close to and around each pole very quickly. In pole bending, riders often have their horses so attentive to lead changes that they will change on voice command alone. Reining horse competitions also require numerous well-executed lead changes.

BREED CLOSE-UP

A gaited horse that traces back to the Spanish conquistadors, the Florida Cracker Horse was later bred to herd and pen half-wild cattle. A strong, tough horse with great endurance and a well-developed herding instinct, the Florida Cracker naturally exhibits either a running walk or an amble. Free-roaming herds have been introduced into a couple of state preserves by the Florida Department of Agriculture and Consumer Services.

Horses bothered by flies may benefit from wearing a fly mask while outside during summer months. Horses can see through the mask well enough to graze and play; to prevent coat damage of the fine facial hair, it should be removed when the horse is brought inside.

THE FLYING LEAD CHANGE IN DRESSAGE

In some advanced dressage trials, the horse is required to change his lead at the canter every stride. This very precise motion makes the animal appear to be skipping.

Foal

The offspring of a mare and a stallion. A male foal is a colt; a female is a filly.

Foal care

Foal care means treading a fine line between performing the activities and procedures horse owners need to perform for horses in domestic settings and interfering with the foal's bonding with its mother and proper nursing. But foals are just plain cute, and people want to be around them and handle them. Now that imprint training has become fashionable, there is even greater danger of tactless horse owners supplanting the foal's dam.

Learning to care for foals comes through doing it. Still, there is a base of knowledge regarding foal care that would be hard to do without.

An easy way to keep a foal warm is to zip it into a large sweatshirt.

Antibodies

These are the components of a mammal's blood that enable it to fight off bacterial, viral, and fungal intruders in the system. Foals are born without antibodies and must quickly obtain them by nursing on the dam's colostrum.

Cold Weather and Foals

If a foal is born in cold weather, you may have to dry the baby with towels and even place a heat lamp in the corner of a stall where it cannot be bumped or put into contact with anything flammable, like hay or shavings. Or you can use a zippered sweatshirt, putting the foal's front legs in the sleeves and zipping it under his belly; the hood, if it has one, can help warm his neck. For greater convenience, buy a foal blanket prior to any birth you know will happen in winter — or even in early spring in cold climates.

Dehydration

Sick foals generally stop nursing. Because the fever and any diarrhea soon dehydrate the sick foal, veterinarians often give fluids by stomach tube if the gut will tolerate and absorb them. Otherwise, veterinarians may give oral medications to eliminate infection of the digestive tract, or they will soothe the gut lining with mineral oil and other gentle preparations (given by stomach tube).

Disease Prevention

Disease prevention consists of good nutrition — mare's colostrum and milk, unless the horse is orphaned and fed a substitute — and more active measures such as vaccination and cleanliness of the stable and paddocks. Vaccinations sometimes fail when a foal is still on his mother, or when he is already incubating the disease to be vaccinated against. Check with a veterinarian for the best times and circumstances under which to give all the preventive vaccines a foal will need.

Exercise

Exercise will help a foal become coordinated. It will stimulate his appetite and digestive tract, make him stronger, and decrease the risk of epiphysitis. Mare and foal should be turned out in a pasture or paddock after the foal has nursed several times, the mare has disgorged the placenta, the mare and foal have bonded, and all appears normal.

Facilities

Until weaning, a foal should be housed and turned out with his mother. When weaning time arrives, do not separate them completely, but rather house them in adjacent stalls or pens, and turn them out in adjacent paddocks, where they can see each other. This will minimize stress and separation anxiety in the foal. Safety of the foal is paramount (for example, the foal may try to get through fences to touch his dam), so materials must be chosen carefully.

➢ *See also Fencing materials*

Feeding

The best feed for a foal, if his dam milks well, is good green pasture. Otherwise, the foal can be started early on a little grain and good alfalfa hay. If you feed pellets, the foal may need help and encouragement to eat them at first; sweet feed, with its molasses content, will probably be accepted more easily.

Foals will usually learn to eat grain by watching their mothers and nibbling at any she leaves. When it is time to give the foal his own ration of grain, put it out separately from the mare's and prevent her from eating it. A foal creep feeder can be constructed by erecting an enclosure for his feed tub that the foal, but not the mare, can get reach into.

Handling

Handling a foal properly teaches him that humans are creatures he must live with and deal with, who provide pleasure, and who need not be feared.

Imprint Training

Handed down to us by Native American horsemen, this method of conditioning is used by breeders to improve their horses' compliance with different training methods. Imprinting can have huge benefits or do great harm. The results depend on the level of knowledge and expertise of the person doing the training. Seek counsel from someone who has had success with imprint training before attempting it on your own. (See page 211 for more information.)

Infections

Infections are more dangerous in foals than in mature horses, often causing death within a few hours. A foal's immune system is usually insufficiently established to fight pathogens effectively. And the organisms that can harm him are abundant in the environment. The foal ingests or breathes in the material, or it enters through a naval stump that is incompletely sealed off (or left unsealed for too long). Leg joints are particularly susceptible to infection because of the stress of weight bearing and other factors.

GROWTH AND WEIGHT

Foals should gain one to three pounds per day. Healthy, nursing foals will not usually have a problem gaining this much, but orphans might. If the foal doesn't gain or he has a rough coat, potbelly, or loose bowels, he needs a better milk replacer. If the foal looks healthy but is simply always hungry, just feed him more of what you're feeding, and begin him on alfalfa hay, grass, or foal pellets as soon as he seems interested.

Like small dogs, foals are sometimes better handled and led in a harness. Slip the noseband of a full-sized horse halter over the foal's head like a collar; position it around the base of the neck above the shoulders so that the jaw strap rests on the foal's withers. The crownpiece will be at the girth. Buckle the crownpiece around the foal's belly, just snugly enough so he can't catch a hind foot while playing or scratching.

Injections

Some veterinarians now believe that no injections are necessary for mare and foal shortly after birth, as used to be assumed. In fact, tetanus toxoid injections, once given as a matter of course, can do more harm than good by causing serum hepatitis in the foal especially. Keep the mare's vaccinations up-to-date; because of her antibody-rich colostrum, the foal will not need to be vaccinated until he is older and better able to tolerate the effects of the vaccines.

Leading Lessons

When the foal is several days old, "lead" him with your arms around his front and rear to get him used to movement controlled by someone else. After a few times, you can switch to a rope around his chest and buttocks with a figure eight over his back. When he has gone willingly with you in this restraint, begin putting a foal halter on him to take him and the mare from stall to pasture and back. Keep using the rope around his buttocks for a while, though, so he cannot pull back or get away.

For a while, it's best to teach foals to follow by controlling their head and rear with the foal halter and an extra-long lead rope around the back end.

Milk Replacer

Do not feed a foal straight cow's milk. Although foal milk replacers are based on cow's milk, they have added nutrients to match a foal's requirements. Some also have probiotics to help digestion. Different formulations offer either more carbohydrates or more fat; foals whose guts cannot handle carbohydrates often do well with the high-fat formulations.

Neonatal Isoerythrolysis (Rh Foals)

Rh foals have a different blood type from the mare, inherited from the stallion; the incompatibility causes antibodies in her colostrum to attack the foal's red blood cells. Nursing will be fatal, and substitute colostrum must be used.

➢ *See also Neonatal isoerythrolysis (Rh foal)*

Nursemares

A nursemare is used for orphaned foals, when possible. If you have a mare who has recently lost her own foal, she might be induced to accept the substitute, if the births are fairly close. She may be inclined to nurse the orphan, or you may have to apply some Vicks VapoRub to the mare's nostrils to prevent her smelling the strange foal. Rub some of the mare's fresh afterbirth over the orphan where she will sniff him naturally, so she comes to accept the foal as her own.

Nursing hobbles will prevent the adoptive mother from kicking at the second foal while he nurses.

Hobble the mare's hind legs so she can't kick, if she is at first reluctant to accept the foal. She can move around, but not hurt the foal. Tie her head for the first nursing, but not so tightly that she can't reach around to smell the foal.

Nursing

Nursing is not as automatic as most people think. Sometimes, mares are confused, especially if they are first-time mothers, or they may have had a prior bad experience. Some mares habitually ignore or reject their foals. Usually, once the foal nurses, though, the mare will accept him.

Foals, too, may be slow to seek the udder. Don't worry as long as he finds it within the first two hours or so after birth, when his intestinal lining is still ready to absorb the antibodies in the colostrum. If two hours have passed and the foal hasn't nursed, help him. Halter the mare if she's moving around too much, and keep her still. Talk to her soothingly. If the foal is vigorous, this may be all the help he needs. If he is clumsy, timid, or her kicking has scared him, you may need an assistant to guide him to the udder while you hold the mare.

Sometimes a mare can be persuaded to accept a second foal and raise it with her own.

Orphan Foals

Orphan foals are those who have lost their mothers during or shortly after childbirth, foals of mares who refuse to nurse their baby, or foals of mares who don't produce sufficient milk. Orphans can be fed either by a nursemare, if one is available, or by formula, as long as the foal has gotten several feedings of colostrum. Colostrum can be obtained either from the foal's mother while she is alive or expressed from her udder after her death, or from another nursing mare. Otherwise, the veterinarian will be needed to administer colostrum replacement if the foal is to survive being orphaned.

> **UDDERLY STARVING**
>
> Foals are the eating champions of the farm. While calves nurse only about four times in each 24-hour period, foals nurse 48 times — every half-hour on average! After a few weeks, the foal's hunger can be satisfied by hourly nursings. Later on, it will drop to a mere 18 times in 24 hours.

Premature Foals

Survival of premature foals depends on how well developed they are when they are born and good nursing care. Sometimes, those that would not survive even with nursing care can be saved in an intensive care unit at the veterinary clinic.

Stalls

Stalls must be big enough for mare and foal to walk around in and lie down in without harming each other. Their stalls must also be cleaner than normal. Removing manure and urine-soaked bedding may make a stall look clean, but odor is important, too. Ammonia develops in stalls at or near the floor, and the foal will breathe it in while napping. The foal spends more time napping than a mature horse, and chemicals are more toxic to his developing tissues, especially the brain. Lie down in the foal's bedding; if you can smell ammonia, so can he.

➢ *See also Barn; Facilities; Stall*

Standing Up

Often, people rush to help a foal stand, especially if he has fallen down a few times in the attempt. But it's not necessary — and it can be harmful. Sometimes, a foal has cracked a rib coming through the birth canal. These generally heal fine, except if you displace it picking him up. On his own, the foal would probably compensate and stand in his own way, in his own time.

Vaccinations

Colostrum is the first immune-system help a foal receives. When foals get their first set of vaccinations from the veterinarian, they often need a booster a few weeks or months later to produce full immunity. Sometimes, vaccination fails because of interference from maternal antibodies, if the foal is vaccinated too early.

Weaning

Weaning, removing the foal from its mother's udder and introducing him to a diet of pasture, hay, and grain, should occur between two and six months of age. Considerations about when to wean include:

- Growth rate. A heavy, fast-growing foal should be removed from the dam before her rich milk overfeeds him and causes leg-growth problems.
- Mental attitude. Many slower growing, less aggressive foals do better mentally by being left with their mother until five or six months of age. Foals can be left unweaned even beyond six months, but it's difficult for a mare if she is pregnant again.

➢ *See also Weaning foals*

Weanling

Weanlings are all foals under one year of age that have been removed from exclusive feeding at the mare's udder.

COUNTDOWN TO HEALTHY FOALING

To ensure a healthy mare and foal, there is a countdown that should be followed.

- ❑ Six to seven months into the pregnancy, give the mare a booster shot for flu and tetanus, unless she had them earlier
- ❑ At five, seven, and nine months of gestation, give her a rhinopneumonitis shot.
- ❑ Two to four weeks before foaling, repeat the flu and tetanus shots to create more antibodies and enhance the protection the colostrum gives the foal.

Yearling

Yearlings are all foals over one year old but under two. While the age of horses is calculated from January 1 each year, to determine if a foal is a yearling, use his actual date of birth.

Foal heat, breeding at

The dam may go into heat three to 12 days after foaling. This is not a good time to breed a mare, as most will have minor infections for a short while after foaling.

A new mother and her baby will benefit from a deep layer of clean straw for bedding.

Foaling

While horses have been giving birth (foaling) in the wild for thousands of years, when they give birth under the artificial conditions in which we keep and raise them, intervention is often called for. Human intervention improves the chances of getting the best result: that is, a healthy dam and a live, healthy foal.

Active Labor

This is also called second stage labor; it is the time when the foal is expelled from the uterus.

Bedding

Good, clean straw is the best choice for bedding in a foaling stall. It should be put down over a disinfected and completely stripped stall that has had lime sprinkled on it first.

Delayed Foaling

While the development of the foal triggers the delivery, the mare has a great deal of control over when it actually begins. If she is nervous or upset, she can withhold the voluntary contractions of her abdominal muscles that accompany the involuntary uterine contractions and prevent the labor from happening until she feels it is safe and/or private enough for her.

Early Labor

This may last a few hours or as long as 72 hours. During this time, mild uterine contractions position the foal for birth with his head and front legs pointing toward the birth canal. The mare may go to a faraway spot in the field, or stand with a distant look in her eye. She may be restless, pacing and nibbling a few bites of food. If you suspect the mare is in early labor, check her every 10 to 15 minutes.

Facilities for Foals

A mare needs room to foal, and mare and foal need ample room to live. The foaling stall should be at least 14 by 16 feet, although even larger is better. It is nice to also have a door to an attached paddock for mare and foal, until the foal is strong enough to be turned out with his dam and other horses, and to make it easier to introduce him to handling outside by humans.

Gestation

Gestation may vary between 305 and 395 days, averaging 11 months and one week or 340 days. However, light breeds tend to have longer gestation periods than do heavy breeds. Also, some stallions customarily sire foals that are carried a longer or shorter time than usual.

As a general rule, mares with first foals often have shorter gestation periods than those who have borne foals before.

Grooming a Mare before Foaling

Keep the mare clean and well-groomed for her comfort during the gestation period, with extra attention just prior to foaling. The cleaner she is, the less chance of infection when her genital tract is exposed during foaling. When labor is imminent, wrap her tail in a clean stable bandage to prevent dirt from getting into the open tract as she gives birth. A wrap will also keep the tail free of blood and mucus.

In Pasture

If a mare has had easy births and healthy foals before, a dry, level, grassy pasture where she can be by herself is often the best place for her to foal. There's less risk of infection than in a stall or corral, and it's a more natural environment.

In Stall

Foaling may take place in a good pasture if the mare is healthy and foals easily; it may also take place in a stall. Any birth expected to be difficult should take place in a large, clean stall with appropriate help available. It is also easier to predict foaling when a mare can be observed frequently in a foaling stall.

FOALING SUPPLIES

To attend the birth of a foal, more than hot water and towels are needed. Here's the basic list of foaling supplies:

- ❑ Clean, new five-gallon bucket
- ❑ Clean container — a wide pan about four or five inches deep — to milk into if needed
- ❑ Tail wrap for mare's tail (use clean nylon stocking)
- ❑ A quart of disinfectant for washing mare's hindquarters
- ❑ Small bottle of tincture of iodine; also a wide-mouth small bottle or jar for dipping the navel stump
- ❑ Shoulder-length sterile obstetric gloves
- ❑ Three or four clean bath towels
- ❑ Paper towels
- ❑ Roll of sterile cotton
- ❑ Large, sturdy plastic garbage bag to hold placenta
- ❑ Foal enema kit or prepackaged adult human enemas
- ❑ Flashlight and new batteries
- ❑ Heat lamp and safe means to secure it away from hay and flammables
- ❑ Obstetric chains or straps
- ❑ Sterile syringes and needles
- ❑ Suction bulb
- ❑ Fresh injectable antibiotic (consult with your veterinarian regarding best type)
- ❑ Obstetric lubricant
- ❑ Oxytocin to help mare shed placenta — use only if advised by your veterinarian

Prediction

Foaling can be predicted by purchasing test kits, which check the calcium content of the mare's milk (levels increases as foaling approaches). Foaling can also be predicted by alarms fitted around the mare's girth, which sound off if the mare lies down prone. Prediction can be made by transmitters fitted to the vaginal lips, which issue signals when the vulva begins to open. Finally, foaling can be predicted by observation through television cameras installed in the foaling stall. This footage should be monitored by knowledgeable breeders, some of whom may be in remote locations.

Presentation

A foal should emerge from the dam front feet and head first. Anything other than this position signifies a malpresentation, which must be corrected immediately if the foal is to be born live.

FOALING PRESENTATIONS

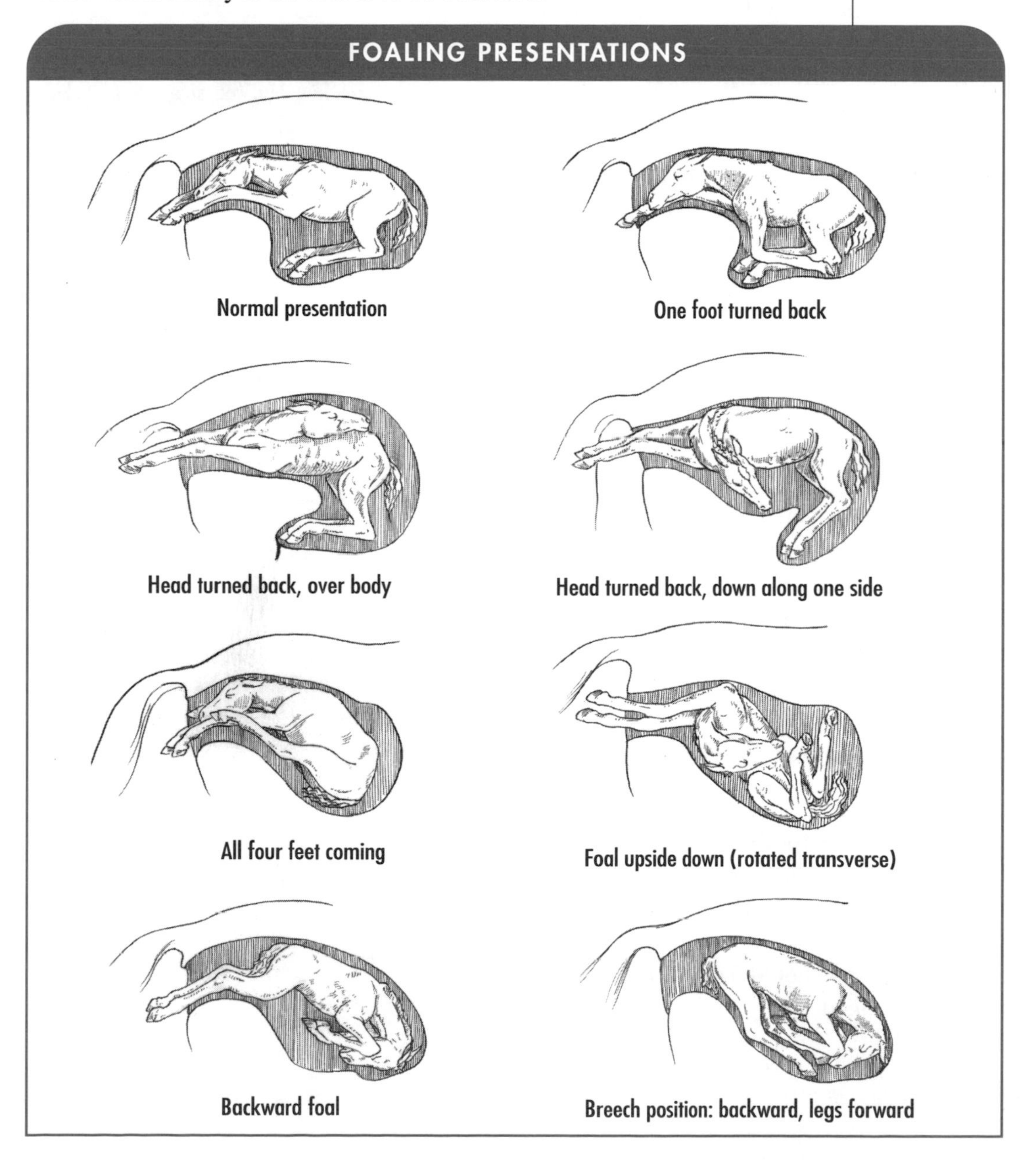

Privacy for Foaling

Mares instinctively leave the herd and go off by themselves to foal. Providing privacy for foaling will help the birth happen on time, and it will relax the mare.

Signs of Foaling

A month before foaling, the mare's udder will look fuller and larger, especially when she is at rest. About two weeks before foaling, the udder will fill with milk and look shiny. A few days before, muscles on each side of the tail droop away from the root of the tail and the vulva may appear relaxed and swollen. There may also be drops of secretions from the teats. Some mares "wax," or produce a congealed secretion at the end of the teats, anywhere from ten days to 24 hours before labor begins.

Straw Bedding

Straw is the preferred bedding material for foaling because it is less dusty than sawdust, does not harbor *Klebsiella* bacteria, and will not get into the foal's mucous membranes (as sawdust might).

Foot care

Enormous animals, called "huge, elemental beasts" by writer Dick Francis, run on four very small feet. To prevent pain and unsoundness, taking proper care of those feet is essential. The following are the basics of foot care.

Brittle Feet

The outer wall of a horse's hoof is made of a substance similar to fingernails. Due to dry conditions, poor nutrition, or genetics, that material may be brittle and susceptible to cracks.

Cleaning Feet

A horse's feet, especially if they're shod, must be cleaned often to remove stones that might bruise the sensitive sole and mud and debris that could set the stage for fungal infections. Cleaning is done with a **hoof pick,** basically a handle with a rounded, metal shaft attached. The shaft is not too sharp or pointy, to avoid piercing the sole. Hoof picks often have a stiff brush attached for complete cleaning of sand and dirt and for brushing off the frog, which should not be scraped with the pick itself.

Conformation

How a foot is built plays a role in the foot's ability to hold up under strain and to allow the horse to do the sorts of work it is intended for. Hooves should be wide at the heels. All four feet should be slightly concave at the sole, the hind feet even more so than the front. There should be a thick, resilient hoof wall. The frog should be large and centered; if it is not, the feet or legs might be crooked.

Anatomy of the foot

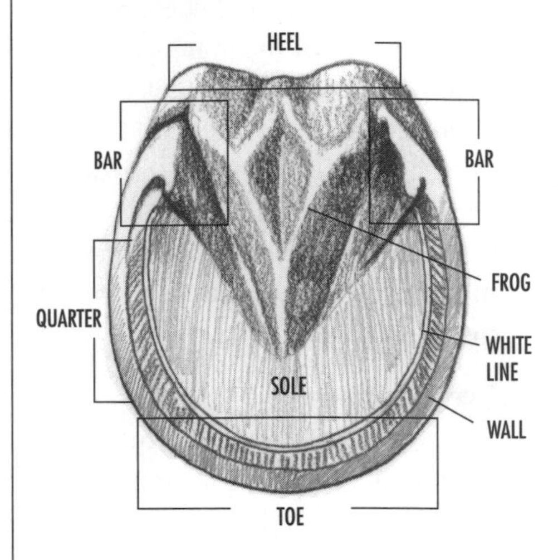

The foot of a horse, from the bottom, contains these structures: quarter, toe, heel, frog, bar, wall, sole, and white line.

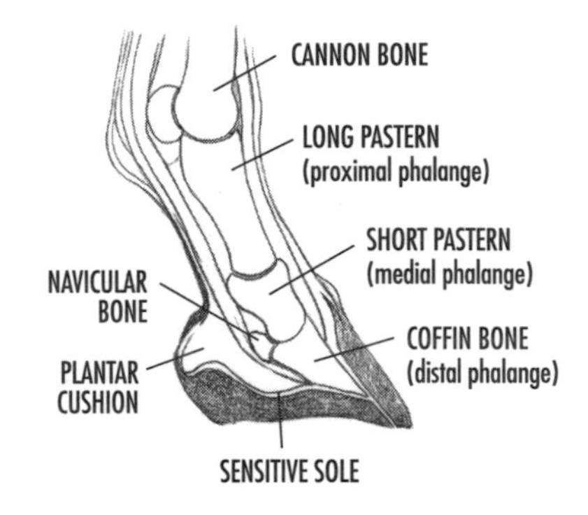

From the side, the following structures may be identified: cannon bone, coffin bone, sensitive sole, navicular bone, and plantar cushion (sensitive frog).

Contracted Heels

Improper or unnecessary shoeing, or shoes left on too long without trimming, can cause the heel to contract.

➢ *See also Contracted foot, contracted heel; Lameness; Proud flesh*

Drainage Hole, Protecting

When a foot is abscessed or has a puncture wound, it is necessary to open a drainage hole to prevent infection from being sealed up in the foot and damaging the delicate structures. Protecting the drainage hole can best be done by bandaging it between soakings — the usual treatment, which draws out infection and fluids. Usually, cotton or gauze is loosely packed into the bottom of the hoof. Then a self-stick veterinary bandage is wound securely around the bottom and sides of the foot. Duct tape may be applied over this layer or a hard rubber protective boot may be used.

Farrier

A professional who specializes in the care of horses' hooves. These days, some farriers call themselves equine podiatrists. A good farrier is knowledgeable not only about horses' feet and legs, but also about their locomotion, about illnesses that affect the feet, and about the wide range of corrective and therapeutic shoeing required by equine athletes. They also know how to shoe horses to correct deficiencies of conformation, to protect injured or delicate structures, and to keep the horse pain-free and sound through a lifetime of work. Farriers may fit shoes in concert with a veterinarian's recommendations, as well.

➢ *See also Farrier*

Flat Feet

Feet that are not as concave as they should be. It is an inherited condition or can result from poor hoof care. When traveling over rough terrain, the feet can be protected with Easyboots, or the farrier can apply shock-absorbent pads between the shoe and sole. Some soles can be toughened up by applying iodine on days when the horse will travel on rocks or gravel.

Frog of Foot

This V-shaped structure in the middle of the horse's hoof helps absorb concussion and regulates hoof moisture. It also assists the circulatory system by aiding venous flow of blood through a "pumping" action that occurs with each step.

Growth of Hoof

The hoof grows downward from the fleshy coronary band (coronet) at the top of the hoof. Average hoof growth is about ¼ to ⅜ inch per month.

Handling

Handling a horse's feet often can accustom him to the process. It allows you to clean out any debris that would injure the hoof or cause infection, or find any infection that is already present. Handling the feet helps you assess the condition of the horse's lower legs by feeling for heat, which may signal injury or disease.

SUMMER DRYNESS

Unless your horse has a stream in his pasture to walk through and lots of moisture in the grass, his hooves are likely to become dry and brittle in the heat of summer. Stomping the feet to get rid of flies helps dry feet become even more brittle. The answer is a good hoof dressing applied according to the farrier's or veterinarian's directions.

PICKING UP THE FEET

Proximity and leverage are keys to picking up a horse's feet properly and safely for cleaning or the application of medications.

Run your hand down the back of the horse's front leg and gently squeeze just above the fetlock joint to encourage him to pick up his foot. Cup it in your hand while working.

Lean against the horse to help him shift his weight before asking for a hind foot. Run a hand down the back of the leg, as with the front leg, and gently pinch the tendon at the back of the leg to encourage him to pick up his foot.

Rest a hind foot on your thigh to work with it, if you and the horse are comfortable with the process. If the horse tries to take his foot away, grab the toe and bend the fetlock joint firmly; this gives you more leverage, and it is more difficult for a horse to remove the foot with the joint flexed. This procedure is not recommended with arthritic or injured horses.

Thrush

A disease of the hoof caused by bacteria, or sometimes fungi, exacerbated by allowing horses to stand in muddy pens, dirty stalls, and boggy pastures with infrequent hoof cleaning. Thrush is characterized by a foul odor; there will also be black secretions at the edge of the frog. Later, dark spots may develop on the sole near the white line. In early stages, the condition is easily curable with daily applications of iodine, bleach, or commercial thrush remedies. Left untreated, thrush can cause lameness and damage the sensitive structures inside the foot.

Thrush can be particularly difficult in winter because the horse may stand in water or mud more often, especially at the end of winter. If he picks up ice balls in his feet, the ice can melt when he comes inside and make his soles wet. If a horse is turned out for the winter, it may be best to remove his shoes to minimize the chances of the hoof filling with the mud or ice that a shoe holds.

Footing safety in winter

If a horse is turned out in winter, he will probably be safer without shoes. He will have less chance of building up snowballs, which are held by shoes. He will also have better traction and feel for his footing if left unshod. If a horse must cross paths and driveways, spread gravel, sand, or rock salt on icy places. If a horse falls or sprawls, he could damage soft tissues, or worse yet, fracture his pelvis.

Footwear

The main footwear for a horse is the shoe; shoes may be plain or orthopedic. Traction may be improved for specific purposes, such as galloping cross country or performing on grass, with **caulks.** Caulks are pointed metal studs of various shapes that screw into holes in the shoe. **Bell boots,** made of plastic, rubber, or other materail, are sturdy, flexible, protective devices that cover the horse's hoof from just above the coronet to the lower part of the hoof. These boots prevent a horse from hitting the bulb of his front heels with his hind feet if he overreaches.

➢ *See also Boots, horse; Shoes, horse*

Forage

Roughage, required by horses. Forage is usually thought of as hay, but fresh grass is also a forage, as is silage, which should not be fed to horses.

➢ *See also Hay*

Forage poisoning (botulism)

The disease-causing agent in forage poisoning is a bacterium, *Clostridium botulinum*, that produces spores. It is found in soil or in feed or water contaminated by the bacterium's toxin. The toxin can also be found in spoiled grain. Rodents or small animals caught in hay as it is baled are also a source of the bacterium. Forage poisoning is almost always fatal; horses should be vaccinated against the bacterium. Foals retain some immunity from colostrum, but are more susceptible than older horses. Foal vaccinations against the disease should begin at between two and four weeks, with two additional inoculations at two-week intervals.

Forehand

The horse's head, neck, shoulder, and front legs; the part of the horse in front of the rider's hands.

Forehand, turn on the

A turn in which the horse moves his hindquarters around his forequarters; the horse's forefeet remain virtually in the same spot while his hindquarters move in a semicircle of 180 degrees so that he faces the opposite direction. It is a basic exercise taught in the initial stages of training for horses from nearly every discipline. It helps teach the horse how to move away from leg pressure, and encourages him to use his body, to use his hindquarters, and to help him learn what your leg aids mean.

➢ *See also Hindquarters; Legs, use of when riding*

Forest Pony

➢ *See New Forest Pony*

Forging

Hitting the toe of a forefoot with the toe of a hind foot. It can be caused by conformation problems, such as sickle hocks or a square frame (ideal conformation is a longer underline than top line). Young, tired, or poorly

HANDLING AND USING FORCE

Force is never a good approach to working with a horse. It is liable to trigger a defensive reaction that could cause serious harm to the horse and/or handler. It is best to consider a horse's natural behavior and instincts and learn how to work with, rather than against, those natural tendencies. Demonstrate leadership and firmness, as if you were the dominant member of the herd.

conditioned horses may do it, as may those whose toes have grown too long and need trimming.

➢ *See also Overreaching*

Forward seat

A style of hunt-seat riding. While balanced seat asks for the rider's torso to be upright with weight on the seatbones and heels dropped down through a slightly bent knee, forward seat asks for the rider to incline her torso a few degrees in front of the vertical during all work above a walk. Irons (stirrups) are usually one to two holes shorter than in balanced seat, creating a deeper bend in the knee.

This seat is also balanced — that is, the rider's weight is distributed behind the withers of the horse and in line with the horse's center of gravity — but the balance is achieved through an S-curve rather than the more static position of balanced seat. Balanced seat riders most often shorten their stirrup leathers one or two holes when jumping big fences to allow their buttocks to clear the saddle as the horse jumps up and rounds his back. (Horses rarely round their backs over low fences.) Forward seat riders generally carry the same stirrup length over fences and on the flat.

➢ *See also Balanced seat; Center of gravity*

Foundations, for barns

➢ *See Barn*

Founder

➢ *See Laminitis*

"Four-beater"

A lope (Western) or a canter (English) is a three-beat gait, with first one hind foot striking the ground, then the opposite hind and forefoot of the same side as the first beat, and last, the front foot opposite the leading hind hitting the ground. In a four-beat gait (also called a "trashy lope" by Western riders), the horse will set his front hoof of the second beat down before he has completely pushed off the hindquarters. The gait makes for an uncomfortable ride, at best, and is frowned upon by show judges.

4-H

In the United States and Canada, 4-H is a program for young people run by states or provinces. These programs have existed since the early 1900s, and their aim is to teach young people about agriculture and livestock. Now more than six million youths ages five to 19 are involved in 4-H with projects ranging from horses and livestock to Communications and Expressive Arts. The programs are administered primarily through extension offices of land grant universities within the United States and Canada. The Horse Project focuses on horsemanship education rather than just riding skills. So members learn the wide range of skills required to become responsible horse owners and/or managers.

A dog starv'd at his master's gate
Predicts the ruin of the State,
A horse misus'd upon the road
Calls to Heaven for human blood.
Each outcry of the hunted hare
A fibre from the brain does tear,
A skylark wounded in the wing,
A cherubim does cease to sing.

—William Blake, *Auguries of Innocence*

These include:

- The skills needed in horseback riding
- Safe riding and horse-handling practices
- How to have fun with horses through games, shows, and clinics

Fox hunting

A sport that was developed in England for a bona fide agricultural purpose: ridding farms of foxes, which killed small livestock and were a general nuisance. When the activity was no longer necessary for that purpose, it became a sport. Live foxes may be hunted, or riders may chase after a "drag," a fox-scented object drawn over the land before the hunt. The sport is enjoyed in Great Britain, Ireland, and the United States. Although it is now more of a social event and sport than an agricultural practice, the old traditions remain.

➢ *See also Hunting*

Fox-trot

One of the accepted slow gaits for a five-gaited horse. The defining trait of the fox-trot is its rhythm. The sound of this rhythm is produced by one front foot touching the ground a split second before the diagonal rear foot, then a pause, followed by the other front foot and finally the other rear foot a split second later. This rhythm has been described as having the same cadence as the phrase "a chunk of meat and two potatoes."

➢ *See also Missouri Fox Trotter*

Fractures, first aid for

➢ *See First aid*

Frame

The side view of a horse's body.

Freedom of movement

A horse's ability to use his muscles in a strong and supple manner to perform the movements or gaits being asked for by a rider. Such a horse appears to be floating, without restriction in any of its muscles or joints, since movement is so effortless for him.

Restriction is usually most obvious when watching a horse trot. A freely moving horse should flex the joints in both fore and hind legs, so the movement flows freely throughout the limbs. Horses with conformation faults of the hips, shoulders, or legs, or those experiencing pain, will often appear stilted in their movement. An obvious example of restricted movement is when a horse with early stages of laminitis or navicular disease appears to be tiptoeing around the arena, trying to avoid bearing normal weight on the front feet, which are in pain.

The dusky night rides
down the sky,
And ushers in the morn;
The hounds all join in
glorious cry,
The huntsman winds
his horn,
And a-hunting
we will go.

—Henry Fielding,
A-Hunting We Will Go

French Trotter

A breed of horse developed from the Thoroughbred, the Norfolk Roadster, and the old utilitarian Norman horse. It is a tough competitor in races featuring the diagonal gait. The French Trotter has immensely powerful quarters, strong legs and hooves, and good size (about 16.2 hands high).

➢ *See also Missouri Fox Trotter; Thoroughbred*

Friesian

A breed influenced by the Spanish Barb and the Forest Horse. The Friesian can be ridden, but the massive animal excels as a harness horse. It has luxuriant feathering, a kind and alert face with short ears, and strong shoulders and neck. It may be as small as 15 hands high.

The Friesian's strength comes from the heavy, primitive Forest Horse. Similar horses carried knights to the Crusades. Today, the Friesian is ridden by those who prize its strength and kindness; it is also popular as a circus horse and, because of its black coloring, it is also still used for the funeral business.

➢ *See also Barb*

Frog of foot

A V-shaped structure beginning at the heel and extending into the sole about two-thirds of the length of the foot. The frog is the shock absorber for the hoof.

➢ *See also Foot care*

Full pass

A movement (also known as **side pass**) in which the horse, responding to the rider's requests through aids, moves to the side without bending, crossing both front and hind legs instead.

➢ *See also Aids; Bending; Leg, use of while riding*

Fungal infections

Fungal infections can cause difficult skin conditions in horses. In the winter, with its dampness and lack of sunlight, the horse is especially likely to catch a variety of fungal infections. Some infections may be spread from horse to horse by grooming tools.

➢ *See Girth itch; Mold (fungi) in feed; Rainrot (rain scald); Ringworm; Scratches (grease heel); Warts*

Fungi (mold) in feed

Dampness causes mold in feed, whether hay or grain. Rich alfalfa hay is especially susceptible to mold and must be kept dry. Feeding a horse moldy hay may poison the horse, cause colic, or even cause abortion in pregnant mares.

➢ *See also Mold (fungi) in feed*

I have seen flowers
come in stony places
And kind things done by
men with ugly faces,
And the gold cup won
by the worst horse at
the races,
So I trust, too.

—John Masefield

Furioso

A heavily built horse with a fine head indicating its Thoroughbred ancestry, the Furioso is a good all-around mount. Bred in Hungary, a country long known for fine horses and excellent riders, the Furioso (named after one of the founding stallions) is popular with the Csikos herders, whose trick-riding skills are admired worldwide. The breed standard calls for a height of about 16 hands, solid color, a strong back, good shoulders, and well-made legs and feet.

Furlong

One-eighth of a mile (220 yards or about 201 meters).

Fusarium moniliforme

An organism that produces deadly toxins and is the factor in moldy corn poisoning. It lives in soil and is commonly found on moldy corn stalks or corn that has been rained on or stored wet. Small doses make horses lose weight and stop eating. They may appear listless. Large doses are usually fatal.

➢ *See Mold (fungi) in feed; Toxic substances*

Futurities

Competitive events sponsored by a variety of breeds and disciplines. In futurities breeders are confident enough in the offspring of certain crosses that they are willing to gamble nomination fees to the futurity event that begin to be paid while the foal is still *in utero*. The most famous futurity in the United States is the Kentucky Derby. However, nearly every major breed in this country sponsors futurities that range from halter competition with a few hundred dollars paid to the winner, to international performance competitions with tens of thousands of dollars paid to the winner.

Horses must be nominated for futurities based on their conformation and abilities. Often there are large prizes involved; they are meant to influence the conformation and performance standards of a breed or discipline.

The horse, the horse! The symbol of surging potency and power of movement, of action, in man.

—D. H. Lawrence, *Apocalypse*

g

Gait

The way a horse travels across ground. Numerous natural gaits are inborn in horses, depending on the breed; humans also alter the natural gaits to produce artificial gaits, most often for competition. The most commonly known natural gaits are walk, trot or jog, canter or lope, and gallop. There are a number of less well-known gaits exhibited by **gaited horses.** These include the pace, the rack, the slow gait, the largo, and the running walk.

Artificial gaits are not inborn but are developed by humans through special training methods.

➢ *See also Artificial gaits; Canter; Five-gaited horse; Flat-foot walk; Fox-trot; Gallop; Jog; Lope; Pace; Pasi-trote; Paso llano; Running walk; Sobreandando; Tack; Three-gaited horse; Tolt; Transition; Trot; Walk*

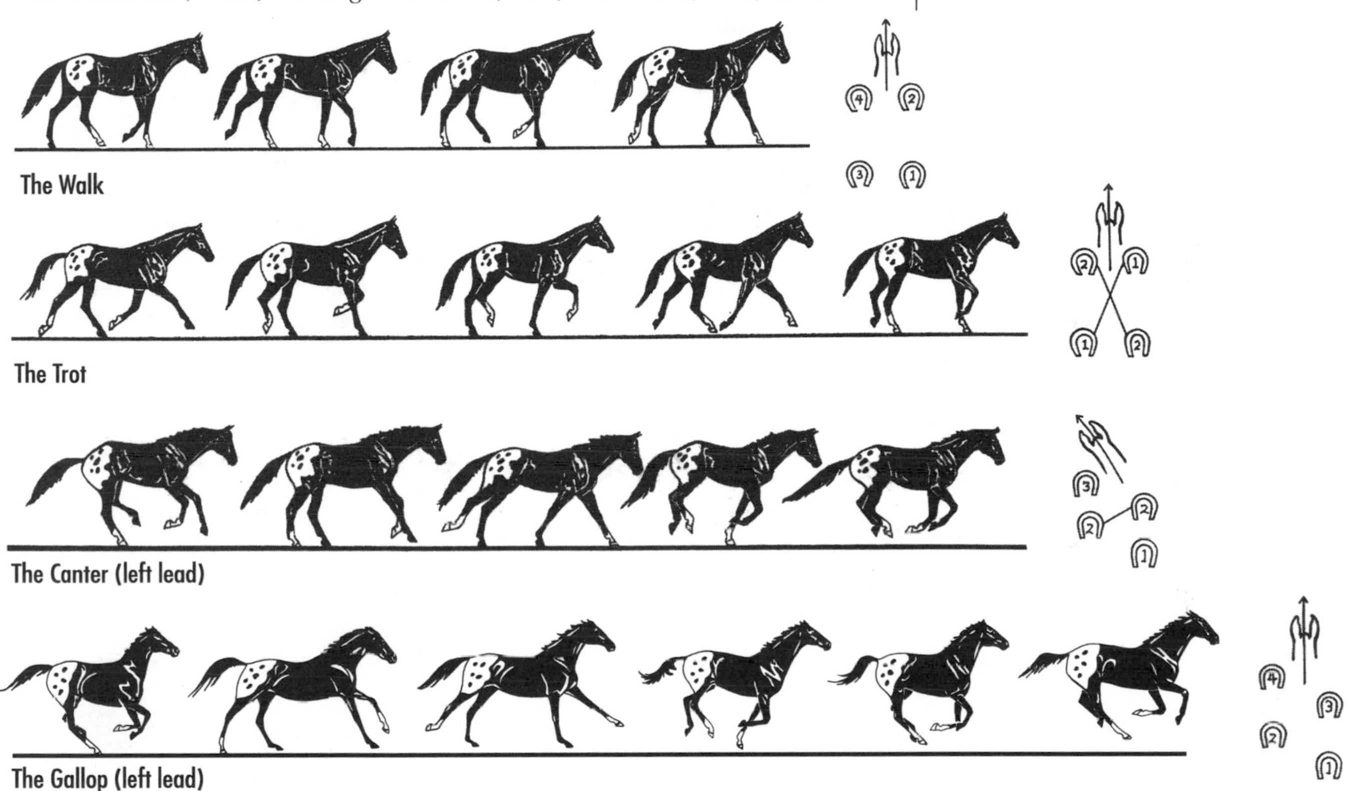

HOW HORSES MOVE

Horses move in a greater variety of ways than any other quadruped. They can move two legs at once, on the same side or opposite sides. They can move one leg, then two, then one. They can move one leg at a time. They have two-, three-, and four-beat gaits.

The slowest natural gait is the walk, a four-beat gait in which the horse moves one foot at a time forward, in this sequence: left hind, left fore, right hind, right fore. Next is the two-beat gait known as the trot, jog, or pace. In trot and jog, paired diagonal feet leave the ground simultaneously and, before the first pair touches down, there is a moment of suspension. In the pace, lateral pairs move forward in turn, also with a period of suspension with all four feet off the ground. In the canter or lope, a three-beat gait, one hind leg pushes off, then the opposite hind and its diagonal foreleg create the second beat; and the final beat occurs when the opposite foreleg (diagonal to the first beat) reaches out and strikes the ground. Again, there is a moment of suspension, which accounts for the sensation of flying that many people feel when they ride. They *are* flying!

Gaited horse

A horse that performs in gaits other than walk, trot/jog, and canter/lope. The gaited horse is esteemed for the smoothness of its gait, whether it is a four-beat lateral gait or a four-beat diagonal pattern. Among the breeds of horses that are trained in these additional gaits are the Tennessee Walker and the Peruvian Paso.

The Tennessee Walker performs at the flat-foot walk (actually its natural walk), the running walk, and the rack. It also has a smooth, rocking-chair canter, and a smooth movement known as the slow gait.

Although the Peruvian Paso is capable of cantering, it prefers its natural, ambling gait, which it may have inherited from the Spanish Jennet. Unique to this gait (called *paso,* meaning step) is the way the forelegs arc out to the side at each step, as in a swimmer's stroke. Like other four-beat gaits, paso gaits are exceptionally smooth, fast, and comfortable.

The predisposition to gait is genetically inherited. Generations of breeding will strengthen this tendency.

➢ *See also Amble; American Saddlebred; American Standardbred; Andalusian; Pace; Paso Fino; Peruvian Paso; Rack; Rocky Mountain Horse; Running walk; Slow gait; Tennessee Walker*

A GALLOPING HORSE

To an observer, it may seem that a galloping horse is pitching forward. However, to a rider, the canter and gallop are smooth gaits, because there is less concussion, and three to four sliding beats are easier to sit to than two concussive ones.

Gallop

The gallop is the horse's fastest gait, with speeds that can average 18 miles per hour and more. The gallop is a four-beat gait. The sequence of footfalls, if the horse is beginning with the left lead, is as follows: right hind, left hind, right front, left front, all hitting the ground separately in a quick rhythm. The right lead starts with the left hind, then the right hind, the left front, and the right front. A moment of suspension follows the four beats, when all four feet are off the ground.

Galvayne's groove

A groove that usually appears on a horse's upper corner incisors when he is about 10 years old.

➢ *See also Age*

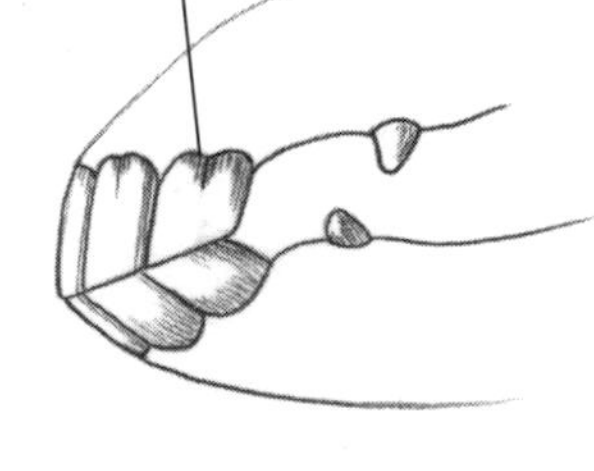

Garden plants, toxic

Although horses usually avoid eating toxic plants, hunger, curiosity, or boredom may lead them to eat things that are not good for them. Landscaping of yards, gardens, and barnyards of fairgrounds can offer many poisonous nibbles. Worst of all are yews, which are poisonous to people and all livestock — particularly horses. In horses, death from yew poisoning may be so sudden that there is some of the plant left in the mouth. Among homegrown vegetables, the leaves of potato plants and even potato skins are toxic to horses. The following is a list of other garden plants horses should not be allowed to eat:

- Buttercup
- Daffodil (narcissus)
- Lily of the valley
- Larkspur (wild delphinium)
- Rhubarb leaves
- Castor beans
- Lantana

Gas colic

➢ *See Colic*

Gaskin

The muscle in the horse's upper hind leg, between the hock and the stifle, comparable to the human calf muscle. The tibia and fibula are the bones associated with the gaskin. Weak or underdeveloped gaskins can lead to problems with drive or impulsion.

➢ *See also Anatomy of the horse*

Gasterophilus (bot fly)

Also known as the **bot fly,** this is a brown hairy fly the size of a small bee. It lays its eggs on the horse's hairs, one to a hair. The eggs, once swallowed by the horse, attach to the stomach, feeding and growing for eight months to a year. They pass out in the manure, emerging as adult flies in three to five weeks to start the process again.

➢ *See also Bot flies*

Gastric ulcer in a foal

A raw spot in the digestive tract. In foals, gastric ulcers can be caused by overfeeding the foal or by keeping the mother on a heavy grain ration, making her milk extra-rich during nursing. Ulcers can also be caused by prolonged stress, such as being confined in a small stall with no outlet for youthful energy. Certain drugs, such as phenylbutazone (Butazolidin or bute) or other painkillers, can cause ulcers, as can a lack of selenium in the diet. The condition can also result from providing feeds that contain anabolic substances, such as rice oil. These feeds are commonly fed to foals to be shown at halter.

Foals with gastric ulcers usually have pain in the gut, which they reveal by grinding their teeth, showing colic symptoms, or losing weight. The signs may be subtle, however; indeed, the foal may continue to eat because lining the stomach with food temporarily relieves the caustic action.

Gates

On a horse farm, gates take a lot of abuse and require extra thought and care. Horses see the gate as a way to get back to their feed and hay and may lean against the gate while waiting for it to be opened. The gate should be as tall as the fence itself, and hung on extra-sturdy posts. It needs to be at least 4 feet wide; wider is better, although wide gates can be difficult to shut while handling a horse. Gates should not be made of netting; horses are likely to paw at the gate and can catch their feet or pull off shoes.

A gate used with an electric fence needs an insulated handle so you can unhook the gate without getting shocked. It must also be wired to become dead when unhooked, so it won't shock you or a horse or spark and start a fire in dry grass. You can bypass the gate by burying the electric wire or

GATE GUIDELINES

Keep in mind these tips for best gates:

- ❑ Gates must swing freely and be light enough to handle.
- ❑ Gates must be wide enough for a horse and human (4 feet wide).
- ❑ Avoid gates with narrow walk-throughs for humans; horses can get hung up in these.
- ❑ If your gate is heavy, rest it on a block at both its closed and open positions, to keep it from sagging and take the weight off the hinges.
- ❑ Avoid flat metal gates with sharp corners and edges.

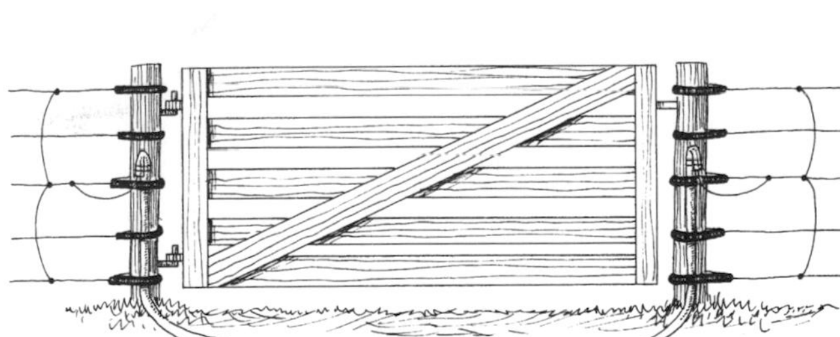

An underground connection is important both on an electrified gate *(top)* and a non-electrified gate *(bottom)*. The cable should be designated for burial (12½-gauge or larger) and run through ½-inch PVC pipe or garden hose.

by erecting a tall lintel over a standard gate, high enough for the tallest horse to pass under and then some, and running the electric wire over that spot. However, you will then have a permanently shock-free area at the gate, which may partially defeat the purpose of the electric fencing. On the other hand, once a horse is shocked going through a gate, he may never want to go through it again.
➢ *See also Fencing materials*

Gelding

A castrated male horse. The advantages to owning a gelding over a stallion, especially for sport purposes, are that geldings are easier to care for, less prone to injury, usually have a consistent attitude (making them easier to show and haul), and may be put out to socialize with other geldings. Stallions, on the other hand, must be kept in paddocks by themselves because of the risk of serious fighting for dominance, which is potentially life-
threatening.
➢ *See also Castration; Stud colt*

Gender

A baby horse is a **foal,** a female foal is a **filly,** and a male foal is a **colt.** The father of a foal is called a **sire;** the mother is called a **dam.** An adult female is a **mare,** while an adult unneutered male is a **stallion.** A neutered stallion is called a **gelding.**

Genetics

The science of heredity; the scientific study of the differences and similarities among related individuals. An understanding of genetics is a vital key to breeding success, regardless of the size of the program. Skilled breeders can produce animals with traits desirable to the breed ideal, marketplace, and/or show arena and can even change the direction of an entire breed. Unfortunately, many breeds have been negatively influenced by uneducated breeders who crossed individuals with highly heritable faults.

Two primary concepts to understand in equine genetics are **phenotype** and **genotype.** Phenotype is the expression of genes that can be measured or observed by the human senses: what you see when you look at a horse (color, conformation, movement, and so on). Genotype is the actual genetic makeup of an individual as determined by its genes: what you cannot see when you look at a horse, but is inherited from its ancestors and passed along to its descendants.

You can view photos or even videos of a horse's ancestors, and determine what characteristics are consistently passed on from one generation to the next. Choosing the mare should also be done according to characteristics that are observable in her ancestry. Nothing in breeding is foolproof, however, and both parents may carry a recessive gene for an undesirable trait that simply hadn't been paired up before. One recessive gene will not result in the trait being present in the foal.

This foal was unlike its dam, unlike its sire, unlike any horse on the Goose Bar ranch. It resembled only one — the Albino. It was almost like having the Albino right there in the stall! Was the power and ferocity of the great outlaw enclosed within that mottled baby hide of pink and white? The thought made shivers go through Ken.

—Mary O'Hara, *Thunderhead*

Color

Color is determined by recessive and dominant genes. If a foal inherits the same dominant gene from each parent and two different recessive genes from them, his coat will be of the dominant color. However, if both parents pass on the same recessive gene for color, the foal will be that color, no matter what color the parents are. For example, a bay stallion and a gray mare might each carry a recessive chestnut gene, so their foal will be a chestnut.

In the three major horse colors — bay, gray, and chestnut — gray is dominant over bay, and bay is dominant over chestnut. To be gray, a horse must always have one gray parent; the trait cannot skip a generation, as chestnut can. A chestnut bred to a chestnut can only produce a chestnut, because the two recessive chestnut genes are present. However, a chestnut can result from mating two bays or two grays, or one of each, if each parent carries the recessive chestnut gene.

Other colors — black, Palomino, dun, paint, or Appaloosa — are more complicated to predict. Some colors happen because a recessive factor "dilutes" a more dominant gene. In such cases, a bay becomes a buckskin, brown becomes red dun, and chestnut becomes Palomino. When there are two recessive dilution factors, the bay becomes **perlino,** or almost white with rust-colored mane and tail. A chestnut becomes a **cremello,** creamy white with light mane and tail and blue eyes.

In order to breed for color, you need to know if that color is dominant, recessive, or heterozygous (mixed). Knowing the genotypes of each parent is also important, although you can get clues by seeing the colors of their previous foals and checking the colors of ancestors; this will help indicate the presence of recessive genes.

Dominant colors are bay, gray, tobiano paint, blanket Appaloosa, and line-backed dun. Recessive colors are chestnut, perlino, and cremello. Mixed colors are white, roan, buckskin, and Palomino. For black, there must be multiple traits present, and the color can occur even if the parents aren't black. A dilution factor will produce grulla from black or brown coats.

DEADLY COLORS

White and roan horses are very rare for the simple reason that if dominant genes for those colors are present, it is lethal to the embryo. For a live foal of those colors to result, the colors must be produced by the pairing of a recessive and a dominant gene.

Combined Immune Deficiency (CID)

➢ *See Combined Immune Deficiency (CID)*

Crossbreeding

Breeding individuals of different breeds to combine the best or most desirable traits of each. Partly, this combination is made from choosing traits and parents carefully; it is also partly due to a desire for hybrid vigor (heterosis).

Defects

Defects often run in families, although mutations and poor pre- and postnatal care can be responsible for many problems. Inbreeding or mating closely related animals is risky unless you are certain there are no bad recessive genes in the family, genes that will cause a defect when doubled. Parents of a defective foal may have at least one ancestor in common; thus, the foal could have inherited from both dam and sire the recessive defect genes of the common ancestor.

Dilution Factors

The recessive genes that produce nothing alone, but affect other color genes to create interesting variations in coat color. Examples include buckskin, red dun, Palomino, perlino, cremello, and grulla.

Dominant Traits

A trait that is produced by a dominant gene. A dominant trait will always be present in a foal instead of a trait controlled by a recessive gene.

Genes

Chemical codes that transmit various traits of a living creature. Genes are located inside chromosomes.

Heterozygous

Referring to the inheritance of a mixed pair of genes, one recessive, one dominant. The foal will show the dominant gene, but it can pass on either gene to its own offspring.

Homozygous Dominant

Two dominant genes together create an animal that is homozygous dominant for the trait in question. The horse both expresses that trait and passes it on to his offspring; indeed, a homozygous dominant horse can pass on no other trait to its offspring but the homozygous dominant one.

Homozygous Recessive

Two recessive genes come together in the foal to create a homozygous recessive animal. This horse expresses the trait and passes on the recessive gene to his offspring.

Hyperkalemic Periodic Paralysis (HYPP)

A muscular disease that affects both horses and humans. It is caused by a hereditary genetic defect that disrupts a protein called a sodium ion channel. It can result in uncontrolled muscle twitching, profound muscle weakness, and even collapse and sudden death from cardiac arrest or respiratory failure. Severity of the disease varies, with homozygous horses affected more severely than heterozygous ones. The original genetic defect that causes HYPP was a natural mutation that occurred as part of the evolutionary process, not from inbreeding.

➢ *See also Appaloosa; Paint and Pinto; Quarter Horse*

Inbreeding

The mating of two closely related animals, which should be avoided. Inbreeding, in general, increases the probability of undesirable recessive traits being expressed in the offspring. Inbred horses are also usually less hardy than other horses. Over time, inbreeding dramatically decreases the genetic variation within a bloodline, concentrating both good and bad genes. If undesirable genes are present, inbreeding offers a greater chance of birth defects.

➢ *See also Linebreeding*

GESTATION AND GENETICS

The genetics of the foal are thought to have an effect on the length of gestation; not even veterinarians can agree as to what the average gestation period is. Determining the length of gestation of foals with similar genetics (same breed, same line, length of gestation of each parent, and so on) may help you to estimate gestation time. The accepted average is 340 days or 11 months, one week.

Lethal Genes

When some genes are doubled up as an expressed trait, they result in a fetus whose organs are improperly formed. Some of these are apparent, such as the lack of an anus or eye sockets, or the less readily observable but deadly water on the brain. However, the most common lethal gene is that for white coloring. It may produce roans and whites that die soon after conception; it may also produce a white overo pinto that dies soon after birth because of a faulty intestinal tract, which accompanies the white gene. The lethal white situation occurs when two horses carrying the gene (two true roans, for example) are mated.

Linebreeding

The mating of the descendants of a particularly outstanding horse who are distant relatives. This type of breeding is meant to fix the traits of the superior ancestor. *Caution:* The undesirable traits of the ancestor and the breeding partner should also be assessed to be sure undesirable traits are not being fixed, as well.

Outcrossing (Outbreeding)

The mating of unrelated individuals within a breed; often, because their genetic makeup is more varied, the offspring are of superior quality.

Pedigree

The pedigree is the lineage of the horse, the family tree showing all his ancestors. It is useful to check the pedigree for desirable traits that have been passed down from one generation to another before selecting a horse to mate with your horse. It is also a reflection of a horse's potential as proven by ancestors.

> **SELECTIVE SERENDIPITY**
>
> When two individuals produce an offspring superior to what would be expected by their mating, it is called *nicking*. This is indeed a lucky phenomenon that occurs through an inexplicable combination of genes.

Gestation

The length of time it takes for a fertilized egg to become a foal. Gestation varies widely, between 305 and 395 days, with an average of 340 days, or 11 months and one week.

Gingering

A process used to make a horse consistently hold his tail in a high position, primarily for the show arena. Ginger or ginger extract is inserted into the rectum, causing a burning sensation that makes it uncomfortable for the horse to lower his tail. The procedure has become less common since animal protection groups began policing competitive horse events.

Gingivitis

An irritation of the gums that may be caused by rough feed, taps between the teeth, injuries and bruises, ill-fitting bits, improperly cared-for teeth, twitching and lip chains, and weakness in the tissues of older horses. Gingivitis can lead to other dental conditions, including problems in the underlying bone.

Girth

In English riding, the leather, cotton, or synthetic strap that buckles around a horse's belly and holds the saddle in place on his back. In Western riding, a **cinch** holds the saddle in place, is buckled at each end, and often is made of fleece or webbing.

Girth itch

A fungal disease that usually begins where the girth or cinch has rubbed the skin raw behind the elbow. If left untreated, the infection may become bad enough that a saddle cannot be placed on the horse, and the skin peels, leaving widening, sore, inflamed spots. The spores of the fungus are durable and can persist in brushes and tack, often spreading the problem throughout the barn. Disinfecting and using different tack and equipment for each horse helps diminish the problem.

➢ *See also Fungal infections*

Glass eye

A term for the blue eye of an Appaloosa, Paint, or mixed breed.

Gloves

Gloves serve a multitude of purposes for horse enthusiasts, ranging from sturdy work gloves that can literally save your skin during barn work to the elegant white gloves worn by Olympic dressage competitors. Although the rules of each competitive discipline set standards for the use of gloves, it is considered good horsemanship to wear gloves while showing your horse. Some horsemen even believe gloves are an essential part of a safe dress code for handling horses.

In the hunter show ring, gloves are worn with show attire. These are black (to match formal black boots) and are made of either fine leather or a knit with rubber pimples on the palm side for grip. Brown boots are rarely worn, but if they are the color of the gloves should match. In the dressage ring, gloves are white, either fine leather or knit, with pimple palms. For field hunting, it is permissible to wear any neat leather glove, or leather string-backed gloves. (Leather string-backs are also popular for training and schooling in warmer weather, as they allow air to circulate.) Wool knitted gloves in bright colors are also available with pimple palms; these are popular in cold climates, especially with young riders, for lessons, trail rides, and hacks. Cotton roping gloves are worn to protect the roping hand in some Western events. Leather gloves matching other leather items are also appropriate for Western pleasure show attire.

Gnats, protection from

Using an equine fly spray may help protect the horse from gnats. Providing a run-in shed will also help, as will keeping marshy ground and standing water to a minimum. In very humid weather, especially when the sun emerges from the clouds, additional applications of fly spray may be called for.

A poorly fitting girth or one that pinches can make a horse cranky.

TYPES OF GIRTHS

Different disciplines and different types of saddles call for different types of girths. A racing girth, for example, is usually made of elastic to allow the horse room for maximum ribcage expansion at a gallop. These are often used in conjunction with *racing surcingles,* which cross over the top of the saddle for extra safety if the girth breaks. A dressage girth is shorter than the average girth, because the dressage saddle has more depth.

Leather girths have the advantage of durability, if well maintained. Cotton and synthetic groups cost less and are often cooler, absorbing less sweat.

Goats, as companion animals

Horses are herd animals and like to have another of their species nearby. If it is not possible to keep two horses, a goat will often make an acceptable companion for the horse. Particularly sociable horses may even benefit from having a goat share not only their paddocks, but also their stalls. Goats also make ideal companions for orphan foals.

I heard a neigh, Oh, such a brisk and melodious neigh it was. My very heart leaped with the sound.

—Nathaniel Hawthorne

Goat's milk substitute

Goat's milk makes a good substitute for mare's milk if the mare has died, is ill, or is not producing enough milk. It is also usually digested well by premature or sick foals who cannot tolerate mare's milk.

Goat tying

A timed Western event in which the goat, with a collar around its neck, is staked at one end of the arena. The competitor starts at the other end and gallops her horse toward the goat, halting as close to it as possible. The competitor then dismounts, flips the goat on its side, and ties any three legs with a leather thong or braided nylon goat string. At the completion of the tie, the clock is stopped.

➢ *See also Rodeo*

A gooseneck trailer hitch is permanently attached to the truck bed.

Goatweed (St.-John's-wort, Klamath weed) *(Hypericum perforatum)*

Also called Klamath weed, St.-John's-wort causes a reaction to sunlight (photosensitization) on unpigmented areas (white markings) and light-skinned horses. The reaction includes severe red, swollen lesions that are intensely itchy and sore. The skin also peels or comes off in sheets. Affected horses must be moved out of the sun and their lesions must be treated.

➢ *See also Photosensitization; Poisonous plants*

Gooseneck trailer

A trailer that attaches to the hauling vehicle, generally a sizable pickup truck, via an adjustable metal stanchion permanently affixed to the bed and underlying framework of the truck. Some regard these trailers as more stable than ball hitches, even when part of a heavy-duty tow package. The downside is that the permanent fixture makes it difficult to use the truck for other purposes.

Grade horse

An unregistered horse of uncertain parentage. Even a horse with known parentage is a grade horse if you do not have his papers. While that will not prevent you from participating in open shows in your discipline, you will not be able to show your horse in breed shows hosted by the breed registry.

Grains

The four major grains fed to horses are oats, barley, corn, and milo (sorghum). Each has its advantages, but all are considered high-density energy sources that are added to a horse's natural diet of grasses.

Barley

Barley is comparable to oats as a horse feed, except that it has a lower fiber content. It is considered a heavy feed, with a great deal of energy density. The kernel is also harder than an oat kernel; barley feed needs to be rolled or crushed. However, because it is so heavy, especially after processing, barley should always be mixed with other, bulkier feeds, such as wheat bran, to avoid the possibility of colic.

Bran Mash

Most wheat grown in the United States is used for human consumption, with just the husks left over. However, those husks are useful as a high-fiber horse feed in the form of bran. Twice as bulky as oats, bran is poorly digested, offering very little energy. But bran is good as intestinal filler, and it helps increase the amount of manure passed. Bran is also used as a tonic for a sick or tired horse, to provide more moisture in the bowels, or for a laxative after foaling. If bran is fed dry, it should be no more than ten percent of the total ration.

Corn

Corn is a high-energy, low-bulk feed that can make a horse too fat unless fed sparingly. It must also be checked thoroughly for mold; moldy corn can kill a horse. If corn is fed, it should be mixed with other feeds — preferably high-protein and high bulk. It takes 15 percent less corn than oats to keep a horse conditioned, provided the ration is balanced with additional protein.

Crimped, Steamed, or Rolled Grains

Crimping, rolling, or steaming cracks the outer covering of the grain, making it easier to chew and digest. The food/energy value often rises when grains are processed in these ways. The proof can been seen in manure: When whole-kernel grains are fed, many kernels pass through in their entirety, without having released full nutritional value.

Nutrition

For horses, nutrition is derived from three categories of feed: roughages, concentrates, and supplements. **Roughages** offer relatively little energy, but are high in fiber and necessary for proper functioning of the intestines. This category includes pasture, hay, forage crops, and silage. **Concentrates** are high in energy and low in fiber; they include grains and grain by-products. Supplements are used to balance rations and make up for deficiencies in protein, minerals, or vitamins.

Oats

Oats are the most popular feed for horses. This grain is palatable to horses and appreciated by handlers for its fiber content, which is higher than either barley or corn.

Grain is higher in energy and lower in fiber than hay.

A COLD WINTER'S NIGHT

Most horses enjoy a hot bran mash on a cold night. Mix up a couple pounds of bran with enough hot water to make it look like oatmeal. You can add chopped apple, carrots, or any other treat your horse enjoys.

Toxicity

Grains can become toxic if infected with fungi or molds. **Ergot,** a fungus, sometimes infects wheat, rye, and barley seed heads. An affected horse may show circulation problems if small amounts of the toxin are ingested daily. Blood vessels to the feet, ears, and tail may constrict, and the horse may even lose his ear tips or ears. He may also develop gangrene in his feet and legs, leading to death. **Moldy corn poisoning** can occur if the horse eats feed or even pellets containing *Fusarium moniliforme,* a soil-borne fungus. The affected horse may become depressed, lose weight and his appetite, and be subject to bleeding disorders. A large dose is usually fatal; it may make the horse run into fences, press his head against a fence or wall, cross his legs or walk sideways, circle aimlessly, or lose other motor control as his brain deteriorates. He may die within a few hours or may linger for several days.

Wheat and Wheat Bran

Wheat is usually not fed because of its high cost; if it is, it should be rolled or crushed to make chewing and digestion easier. Wheat should be fed as less than 20 percent of the grain ration, and it should be mixed with bulky grains, such as oats, or with bran to avoid colic. Wheat is a high-energy, high-protein feed, but it is low in fiber. Wheat bran, on the other hand, is high in fiber and should only be added to the ration as 10 to 15 percent of a dry ration. Wheat bran can also be fed as a bran mash to improve intestinal functioning, or it can be added as a treat to a warm mash.

Winter Nighttime Feeding

In the winter, a horse's minimum nighttime feeding should be larger than the daytime feeding (or feedings) to help him maintain energy and body heat through the cold night. He should be fed as much hay as he will eat in the overnight period; feeding less may induce him to nibble posts, bedding, or anything else within reach because he is cold, bored, and craves more roughage. If he is bedded on straw he may even eat that, leading to possible impaction or other digestive upset.

ENERGY-TO-PROTEIN RATIO

Getting the proper energy-to-protein ratio in a horse feed is essential to the horse's health. Commercial feeds are balanced for the purposes described on the packaging. Some horsemen, however, add grains or supplements to commercial rations to increase growth or enhance performance. *Caution:* Adding grains or supplements without good knowledge of the changing protein-energy ratios or mineral balance may harm the horse rather than enhance his performance.

Gran Prix, Grand Prix

A competition at the highest levels of dressage or jumping. Commonly, the term used in dressage is Gran Prix; for jumping, it is generally written as Grand Prix. Grand Prix jumping competitions often carry substantial prize money for at least the top six finishers. These events are judged on speed and accuracy: that is, not knocking down fences or landing in water jumps. Gran Prix dressage is an international sport with the contents of its competitions decided by the Fédération Equestre Internationale (FEI) rather than the American Horse Shows Association (AHSA), which determines the content of tests up to Fourth Level.

➢ *See also American Horse Shows Association (AHSA); Dressage; Fédération Equestre Internationale (FEI); Stadium jumping (show jumping)*

A Gran Prix dressage competitor.

Granulosa cell tumor

A tumor in a granulosa cell, which is part of the ovary and produces the female hormone estrogen. A granulosa cell tumor in the mare causes reproductive problems but is usually not malignant, although it can rupture. If it has become cancerous, signs may include personality change, irritability, or aggressive or vicious behavior. The mare may also exhibit stallionlike behavior around other mares, especially those in heat, because the tumor alters the mare's hormonal balance.

Grass clippings for feed

Grass clippings should not be used as horse feed. Because they are soft and easily swallowed, the horse may overeat and experience an impaction. Clippings also spoil and ferment relatively quickly, so a horse should not be grazed on lawns that have been recently cut. If a horse does get into grass clippings and exhibits signs of colic, call the veterinarian.

➢ *See also Colic*

Grass hay

Grass hay is often less dusty than legume hays, such as alfalfa. When making hay, good choices are timothy, smooth brome, and intermediate wheatgrass. (Downy brome, or cheatgrass, should not be used as hay.) In general, dry-land grasses are preferable to swamp grasses, as the latter are usually coarse. Hay should be bright green and clean-smelling.

➢ *See also Hay*

Grass toxicity

Although most grasses are natural feeds for horses, some can cause problems. The following is a list of problem grasses and their effects:

- Buckwheat, when dried, can cause photosensitization.
- Perennial rye grass can cause photosensitization.
- Sorghum and Sudan grass, common in Southern pastures, are dangerous particularly after a frost or drought, which changes glycosides in the plant into cyanide. An affected horse, starved for oxygen, breathes heavily, grows weak, falls into a coma, and dies within a few hours.
- Tall fescue causes abortion, retained placenta, and failure to produce milk in mares. It is not the grass itself, but the fungus *Neotyphodium coenophialum* (also known as *Acrimonium coenophialum*), that infests much of this grass and causes the problems.
- Torpedo grass, abundant in the South and in Florida, causes severe and often lethal anemia.

Gravel

An abscess of the foot, usually formed between the sensitive sole, frog, or laminae and the insensitive or bony structures nearby. The abscess is often the result of a bruise; the blood, which is not sterile, pools and forms an abscess. When the pus works its way up and out of the hoof through

THE GREAT HORSE

A forebear of today's draft horse, in particular the Shire, the Great Horse was developed in England in the 12th century by crossing French and Fresian heavy horses with lighter native stock. Initially used to carry armored soldiers, they eventually became more widely used for farming and hauling. These massive horses stood 16 to 17 hands high and weighed over half a ton.

the coronary band, it is called a *gravel*. Sometimes, the gravel does not exit fully, or it may be reabsorbed. In that case, the pressure often forms a horizontal crack in the hoof wall; this happens three to four months after the initial abscess, and probably causes some degree of lameness.
➢ *See also Abscess*

A short horse is soon curried.

—John Heywood, *Be Merry Friends*

Gravel, in turnout area

If you use gravel as a surface in pens and turnout areas, be sure it is ⅜ of an inch in diameter or smaller. This is small enough that it will not catch in horses' shoes and will not be scooped up when shoveling manure.

Grease heel, Greasy heel (scratches)

A condition caused by either a fungus or a bacterium, which occurs when a horse stands in a muddy or dusty paddock. The infection enters through a cut or scratch on the skin. The infected area becomes crusty, scabby, and thickened. It may ooze and the leg itself may swell, even to the point of causing lameness. Grease heel affects unpigmented (white) areas more than darker ones.

The first treatment is to thoroughly wash and dry the area. Then apply medication; there are many commercial preparations available, from zinc oxide–based ointments to mixtures of dimethyl sulfoxide (DMSO) and nitrofurazone and dewormer paste containing benzimidazole (one part each). There are also several new herbal and natural remedies on the market. (*Hint:* DMSO reduces inflammation and helps the fungicide in the dewormer penetrate more deeply into the tissues.)
➢ *See also Dimethyl sulfoxide (DMSO); First aid; Scratches (grease heel)*

Green horse

A horse lacking sufficient experience and training to be considered safe for an inexperienced person to handle and ride. Some consider a horse with *no* training to be green, while others consider a horse that has only had basic training under saddle to be green. Bottom line? A green horse requires an experienced rider *and* a confidant human to work with him until he has reached a more finished level.

Grooming

Old horsemen say grooming is as good as feeding for putting condition on thin, weak, or aged horses. In fact, the massage afforded by grooming does increase circulation and relaxes the horse's mind. It gives the unkempt horse's coat a shine that will improve his appearance and builds trust and confidence between horse and handler.

For any horse, grooming removes dirt, sweat, glandular secretions, dead skin cells, and loose hair. At the same time, it brings natural oils, which protect the horse from weather, to the surface of the skin. Grooming gives you an opportunity to closely inspect the skin, head, mane, tail, legs, and hooves so you can monitor the conditions of those parts. During

grooming, you can, for example, find ticks and bots adhering to the hair or skin, and you can discover small nicks and cuts before they become bigger problems.

➢ *See also Clipping; Mane, care of; Tail, care of*

Equipment

The basic grooming tools are:

- Curry comb (hard rubber or plastic)
- Grooming gloves
- Small terry cloth towels
- Stiff-bristled (dandy) brush
- Soft body brushes
- Soft face brush
- Hoof pick
- Hoof brush (often part of a hoof pick, but any small, very stiff brush will work)
- Hoof dressing brush (a large brush for applying liquids and semiliquids to the hoof, if needed)
- Spray bottles
- Comb
- Sweat scraper (for removing water after baths or spraying)
- Sizable tote to keep equipment neat
- Bottle of alcohol
- Bottle of liniment
- Fly spray (in summer)

Using a Vacuum

Horses can be vacuumed to remove heavy deposits of mud if they have rolled, or to help remove hair when they are shedding. You can vacuum the horse as part of the grooming process any time, as long as he tolerates the noise and sensation. *Never* vacuum a wet horse, or you could seriously shock him or yourself. Vacuuming contains flying dander and dirt, allowing less of it to rise into the barn air — unlike ordinary grooming. *Note:* Exhaust air does exit from the nonbusiness end of the vacuum. If the barn floor has not been swept, the exhaust will lift all of that material into the air.

Using a Wisp

This is both a grooming and conditioning technique. Use the wisp as a curry or for **stropping;** to do this, lightly swat the wisp against the flat musculature, moving it briskly off again in the direction of hair growth. This causes the horse's muscles to contract, then relax. The contraction and relaxation of the muscles acts as a massage does, carrying waste products from the muscles and nutrients to them.

➢ *See also Wisps, making and use of*

GROOM AND GLOW

1. A good grooming begins by rubbing your horse firmly in a circular motion with a rubber curry comb; move counter to hair growth to dislodge dirt. Curry only the body, neck, and upper half of the legs where there is fleshy tissue.

2. Then use a stiff-bristled brush in the direction of hair growth, with the brush flat on the coat, to remove more hair and dirt. End each stroke with a flick into the air.

3. Next use a softer body brush, going all the way down the legs, in the same direction as hair growth to clean the coat further. Many horses also enjoy having their faces brushed with a small, soft brush especially for that purpose.

4. End with a spray of water or dilute skin bracer on a terry cloth towel; wipe the coat with the towel in the direction of hair growth, removing the last bits and particles of dirt and pressing the hair down again on the skin.

5. Use plain water on a clean sponge or towel to remove dirt from the nostrils and around the eyes.

6. Carefully brush, comb, or pick through the mane and tail.

7. Pick out the hooves.

A basic grooming kit

Ground poles (ground lines)

Poles placed just in front of the vertical elements of a jump in order to help the horse judge the takeoff point. The term can also refer to a pole placed on the ground to teach a horse to go over an obstacle or to teach a rider how to ride over jumps without height. They can be used, also, for flexibility exercises for the horse. In this case, three or more poles are placed a stride apart, and the horse is asked to trot over them. Ground poles can be used similarly to improve the rider's balance through the slight change in the horse's elevation or rhythm as he goes over them. If a ground pole is placed a canter stride away from a jump, it is called a **timing pole,** and it is used to help novice jumpers learn when to prepare the horse and themselves for the jump.

➢ *See also Cavalletti, cavalettie*

Ground poles help teach a horse to pay attention to where he's putting his feet.

Ground tie

A method of keeping the horse still by dropping the lead rope on the ground. It is not actually tied to anything in or on the ground, but seems to work by mentally tying the horse to the earth.

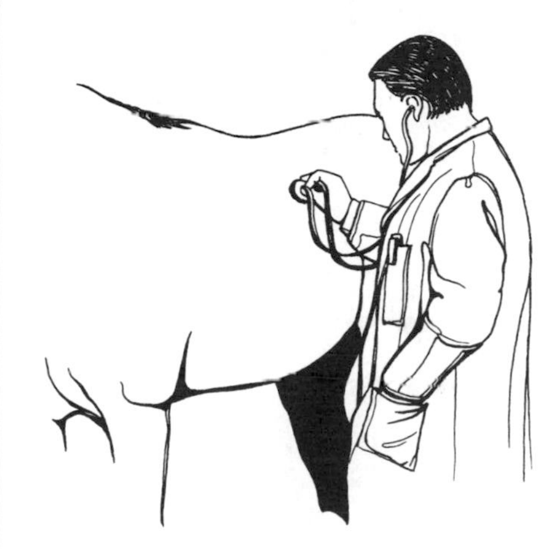

Ask your veterinarian to teach you how to recognize healthy gut sounds. You may save a horse's life with what you learn.

Grulla coloring

A variant coat color of black or brown caused by a genetic dilution factor. The actual color will range from bluish gray to brownish gray.

➢ *See also Genetics*

Gum problems

➢ *See Dental care; Gingivitis; Teeth*

Gut sounds and illness

A healthy horse has a variety of gut sounds, with short spans of silence between them. You can listen for sounds with a stethoscope, or with your ear pressed directly to his side. If you hear a constant rumbling, the horse's gastrointestinal tract may be overactive. If you hear none, there could be blockage or a shut-down gut. In either case, call your veterinarian. Listen to your horse's gut frequently when you are sure he is healthy and eating normally so you will get a sense of what sounds are normal for him. This way, you can act if you detect significant variation.

Gymkhana

A program of competitive mounted games, usually with a timed element. Events may be for individuals or teams.

h

Habronemiasis (summer sores)

A condition caused by habronema roundworms, which can produce summer sores on the horses skin. It also may result in gastric tumors.

Hackamore

A specialty bridle that is best used by very experienced riders with good hands. The hackamore has no bit; instead, it puts pressure on the sensitive areas of the nose, poll, and chin groove. This bridle is sometimes used briefly for a horse who has a damaged or overly sensitive mouth. Prolonged use of the hackamore can permanently injure areas of the face.

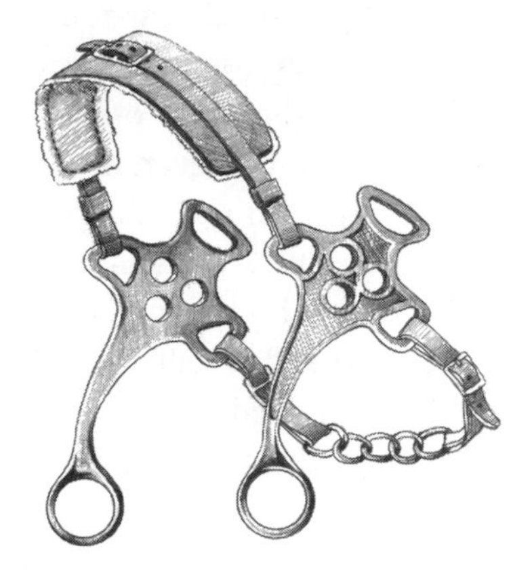

A mechanical hackamore; there are other, less severe styles.

Hackney

A showring harness horse of British origin. At one time, Hackneys were popular horses for pulling carriages in everyday life. The Hackney Horse Society was founded in 1883 to promote this breed, which is derived from the Yorkshire and Norfolk Roadsters (old English riding horse breeds), although there is Darley Arabian blood in its background as well. The Hackney, a warmblood, stands about 15.3 hands high. He may be any solid color, although some white markings — particularly socks — are found. He has brilliant action that appears effortless.

The Hackney pony shares the studbook of the Hackney horse, although the pony was created by breeder Christopher Wilson of Cumbria, England. He based his pony, which he developed in the 1880s, on the local fell pony, a hardy, small horse, and Welsh crosses. A champion Yorkshire Trotter and a great Norfolk horse also figured in the pony's background. The pony, which measures 14 hands or less, is courageous, has great stamina, and also exhibits the high-stepping action and brilliance of the Hackney horse. It also is found in solid colors with some white primarily on the legs.

God forbid that I should go to any heaven in which there are no horses.

—Robert Bontine Cunninghame-Graham, letter to Theodore Roosevelt, 1890

Haflinger

Also sometimes called the Edelweiss pony, this is the horse of the Austrian Tyrol. Technically a coldblood, it has an Arabian foundation sire, El Badavi XXII. The base stock was the extinct Alpine Heavy Horse. The Haflinger works easily on steep slopes, can be ridden or driven with a sleigh or cart, and is a popular show pony with children and small adults. Haflingers are not usually worked until they are four years old, but often continue, active and healthy, working at 40 years of age. Registered members of the breed have a brand with the letter H at the center.

➢ *See also Coldblood; Warmblood*

Hair

Horses have hair, rather than fur, and benefit — like humans — from a good deal of brushing. In winter, for warmth, the hair may grow quite long, and a second coat also may come in, depending on the breed. The hair that keeps the horse warm in winter also makes him harder to cool down after strenuous exercise; often, horses and ponies worked in the cold months are trace clipped so that they will cool off more easily. Show horses and ponies may be body clipped year-round, or at least during show season, whatever the weather, for cooling purposes and for the sheen and definition the uniformly short coat of hair gives to the horse's body. Such clipping can be avoided, however, if you follow a careful daily grooming routine.

A horse should not be clipped in cold weather unless you will blanket him when he is stalled or turned out to play. It will also be necessary to cover a body-clipped horse with a quarter sheet while you begin his warm-up at the walk; once he begins to warm up, the quarter sheet should be removed. When you're finished, cover him with a cooler so that his body doesn't cool down too fast.

Many horses, but especially body-clipped horses, are clipped in some spots, most often the muzzle and the inside of the ears, to create a cleaner appearance. Understand, though, that the muzzle hairs are nature's way of helping the horse find where things are; they are feelers. Similar hairs are located around the eyes, and they serve the same purpose. These feelers help protect the eyes from injury by signaling when his head is too close to an object. The hair inside his ears helps keep out the cold in winter and flies and gnats in summer. If you must clip the ears for showing, try to leave at least a tuft inside the ear protecting the canal itself.

➢ *See also Clippers; Clipping; Cooler; Feathers; Grooming; Hunt clip; Trace clip*

Curly Hair

Some Appaloosas have kinky tail hair. It is thought that the curliness is caused by a recessive genotype. The Buckskin registry lists horses with curly coats all over their bodies, and the American Curly (formerly Bashkir Curly) also has curly body hair and a kinky mane and tail. This breed is native to Russia, and was developed from the primitive Tarpan and extinct (except for experimental herd) Przewalski's Horse. The thick, curly winter coat hair can be spun to make cloth.

➢ *See also Przewalski's horse; Tarpan*

Half-breed

A horse produced by breeding a purebred with a horse of different breeding. A half-breed registry is an organization that registers, transfers, and maintains official records of horses produced by breeding a purebred from a specific breed with a horse of different breeding (a grade horse, a purebred of another breed). Many such registries determine and uphold policies to ensure the integrity of the half-breed.

Half-halt

A signal of preparation that can be used to help the horse maintain balance during changes of gait, direction, or speed. The rider essentially calls the horse to attention by increasing contact with the lower legs while simultaneously increasing the rein contact ever so slightly. This creates a contained form of energy that can then be put to a variety of uses, such as transitioning from one gait to another.

HALF-CIRCLES

In a half-circle, the rider presses into the horse with the leg that will be on the inside of the half-circle. The supporting rein is kept on the outside of the circle. The aids are the same regardless of the discipline: guide the horse by means of a coordinated use of the inside rein and the outside leg. The outside rein and inside leg simply help determine the degree of the turn, so they are less active.

Half-hitch

Technically not a knot but a **hitch,** used primarily to make other knots. Although easy and popular, it is not the safest knot to use with horses. When the strain is constant, this hitch is fairly reliable, but it can be difficult to loosen if a horse has thrown its full weight onto the line.

> **MAKING A HALF-HITCH**
>
> 1. Grab the two ends of the rope.
> 2. Make a turn around a post, bringing one end of the rope around under the other.
> 3. Bring the end up from under, then down through the loop, and tighten.

Half-pass

A lateral movement used by riders in a variety of disciplines. In dressage, this movement is called a half-pass; in Western riding it is known as a **two-track.** It is a movement created by leg and rein aids that asks the horse to travel sideways as well as forward with his bend in the direction of travel; in the similar **leg yield,** the bend is away from the direction of travel. The half-pass is more demanding on the horse physically than the leg yield and requires a greater degree of skill from the rider.

Developed as part of cavalry training, the half-pass is used both as a suppling exercise for the horse and as a foundation for the rider to coordinate aids controlling both forequarters and hindquarters. The horse should be slightly bent around the inside leg of the rider in order to give more freedom and mobility to the shoulders, adding ease and grace to the movement. The outside legs pass and cross in front of the inside legs. The horse looks in the direction in which he is moving. He should maintain the same cadence, balance, and impulsion throughout the entire movement.

When roping, a rider will often rise out of the saddle in a half-seat to get a better angle from which to throw the rope.

Half-seat

In Western riding, this means rising up out of the saddle with the weight in the stirrups. In English riding, the term may mean either:

- Lightening the seat by pushing more weight down into the heel and slightly inclining the body forward (but not so far that the rider is in a two-point, or jumping, position).
- A position used for the hand-gallop, a speed faster than canter. It allows the horse's back to move more freely under the rider. *Note:* All the weight should not be shifted to the irons, but distributed along the inside thigh and the calf.

Some instructors use the term half-seat interchangeably with two-point or jumping position. However, a true two-point requires the rider's upper body to come forward, more or less mimicking the angle of the horse's neck.

➢ *See also Jumping; Two-point seat (hunt seat, light seat)*

Halt

Perhaps the most important skill for horse and human to master during foundation training, the halt is both the request by the rider for the horse to cease movement and the technique for making the horse stop. Whether on the lead or the longe-line, in harness or under saddle, safety depends on the ability of a horse to halt on command. A wide variety of methods are taught for achieving a halt. The bottom line is that both horse and human must consistently adhere to the method used.

➢ *See also Rearing*

Halter

The basic headgear a horse wears to enable humans to handle and control him. *Haltering* is the method of putting on a halter.

A halter alone is insufficient for handling a horse; it must always have a lead rope attached. Trying to lead or handle a horse by the halter alone, even for a short distance, may result in all sorts of difficulties, including injury to the handler.

Attaching a lead rope at least 6 (but preferably 8) feet long will give the handler leverage and the opportunity to get out of the horse's way without setting him loose. Be sure the snaps on lead ropes are in good order, though; most lead ropes are thick enough and made of such durable materials that their only weak spot is at the snap.

If you leave halters on pastured horses in the field to make it easier to bring them in, make sure those halters are either leather or breakaway nylon with a leather crownpiece; these halters will break and let the horse free if he catches it on something. Never leave a halter on a foal when it is unsupervised.

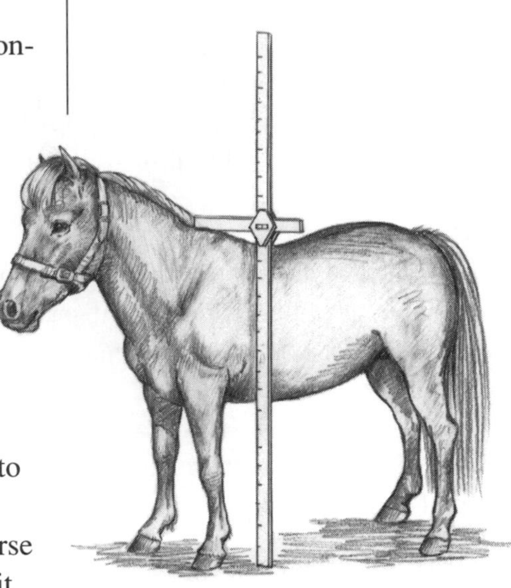

Horses are measured with a *stick*, the common name for a measuring device consisting of an upright pole with a sliding arm, one end of which rests on the horse's withers, the other end attached to the pole so the measurement can be read. The arm slides up the pole and crosses to reach the correct measurement. The horse should be standing on a level surface with a handler keeping him still while he is measured.

Halter classes

Show events that may be entered by nonriders. These classes are judged on the horse's conformation, quality, and breed type.

➢ *See also Showmanship classes*

Hand, as unit of measurement

The standard unit of measurement applied of the height of a horse at the withers. One hand equals four inches or 10.2 centimeters. The measurement is derived from the ancient habit of using the width of the palm of a man's hand to determine how tall a horse was. If your horse measures 61 inches at the withers, the highest point on his back where it meets his neck, he is 15 hands and 1 inch tall. This measurement is written as 15.1 hands high (hh). It is spoken as "fifteen-one."

Hand-breeding

➢ *See Breeding*

Handling horses

No matter what a human does with a horse, he or she must handle that horse in some way, compelling him to move or behave in the manner we desire without causing injury or undue stress to either party. Over the eons, humans have figured out what is needed for a horse to submit willingly (usually) to their requests. Below are tactics to use in performing the essential task of handling humankind's equine partners.

Body Contact

When being handled, the horse must always know where the handler is if he is not to become frightened and spook. The handler must also define for the horse the limits of his movements that will be accepted. Be consistent in setting and maintaining these limits.

Bribery

Using bribery, such as treats, to handle a horse may have unfortunate consequences: If you feed by hand, you may inadvertently create a horse that bites or aggressively seeks treats. In short, your horse will be spoiled and may become dangerous.

Chain Restraint

Some horses will ignore a tug on the noseband of a halter but *will* respond to a chain over the nose or under the chin. The chain gives much greater control by allowing the handler to exert pressure on various parts of the horse's head. If you choose to use a chain restraint, be compassionate and learn proper use and adjustment techniques from a professional.

Consistency

Horsepeople will tell you a horse's mind is like a two-year-old human's. This means that consistency in how you treat a horse — in what you reprimand him for and when you reward him — is paramount. A horse can learn bad habits just as easily as good ones. Be aware that every interaction you have with your horse is an opportunity for instruction.

Distance, Maintaining

Your personal space, when handling a horse, should be determined by the amount of space that makes you and your horse comfortable, on average about 18 to 24 inches around your body will suffice. If the horse runs into that space, quickly and firmly correct him with methods effective for you and your horse.

Dominance

In herds, horses fight for dominance. In a domestic setting, the rider/handler must be the dominant "horse," preferably by practicing benevolent dominance rather than dominating by brute force. This means being consistent and firm in your expectations of the horse, firm in your corrections when he missteps, and consistent in how you correct him.

Endorphins

Brain chemicals that make a horse calmer and easier to handle. One way to prompt the release of endorphins in a horse's system is through massage. Twitching is a form of restraint that releases endorphins.

Farriers

Prepare your horse to accept interactions with the farrier *before* they are needed. Accustom your horse to having his feet handled and held in the positions required for trimming and shoeing. Handle your horse's feet often; clean stones and debris out of them at least once a day. The farrier can't do good quality work if your horse struggles against him.

When the farrier arrives, work with him to find methods that keep your horse calm. Focus your attention on your horse and the work the farrier is doing. (Keep in mind, however, that some farriers may cause discomfort to the horse by overextending his legs or raising them too high. If you observe this, mention the possible problem to the farrier.)

HANDLING HORSES THAT PULL

Horses that pull backward when you are tying them can be dangerous; it is best to teach young horses to tie properly to avoid problems later. However, a horse may acquire an aversion to being tied, due to mishandling or trauma. In that case, you can work at retraining him.

To avoid injury to the horse, tie an old inner tube around a tying post, and then tie the horse's lead to that. There will be considerable give when he pulls back, and this may help him get over his fear. Alternatively, you may cross-tie him, attaching leads hung high on either side of a barn aisle to either side of his halter at the rings where the noseband and cheek pieces meet. Even that is too much for some horses, though.

Most horses are willing to "ground tie," if you simply drop the lead rope to the ground and let it hang under their chin. However, it will be hard to trailer this horse unless you can get him to accept an actual tie, either to a ring, post, or cross ties.

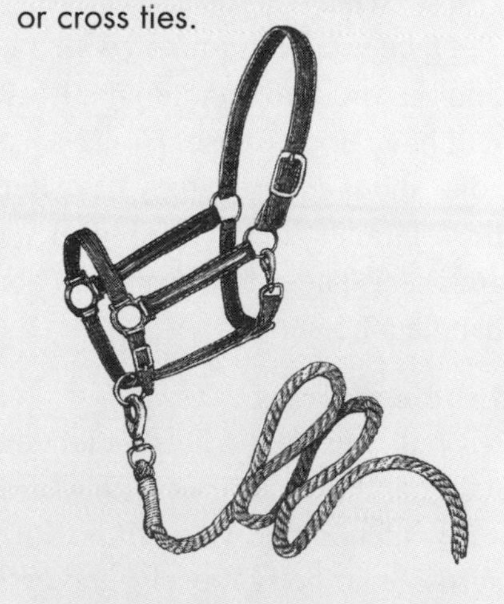

Always lead your horse with a halter and a lead rope.

Force and Handling

Using force when handling horses will have an unpredictable effect. It is impossible to force a horse to do anything unless he chooses to comply.

Hard-to-Catch Horses

There are a few types of hard-to-catch horses, each of which is best handled in a way that gets him over his reluctance to be caught. The hard-to catch horse may succumb to treats and grain taken to the field, but then he will only come if there are treats. A better method is to catch him by waiting for him to come to you out of curiosity, no matter how long it takes, and then rewarding him with a treat when he is where you want him to be. This approach is tedious, but it's the only way to get over the problem for good.

A timid, young horse may also be hard to catch. Gain his trust by catching him every day and making the result pleasant for him. The spoiled old horse can also be hard to catch. You may have to turn him out in a very small paddock or pen for awhile. Then catch him for everything — feed, water, grooming — until he understands that he is dependent on you for those good things and is eager to come to you to get them.

Holding a Horse

Begin with a properly fitted halter, or add a chain over the nose or a lip chain for more control; a few horses may require a bridle with a bit if they are particularly difficult. Choose the location to hold the horse according to what's needed. If you need room to work around the horse, an open location with nothing nearby for him to bump into is best. If he must be perfectly still, holding him next to a wall or fence is preferable. If he is prone to moving backwards to get away from things he doesn't like, back him up against a fence or wall so he can't rush backward when the object of his scorn — the veterinarian, the saddle-fitter, or anything and anyone else — arrives. Be aware that a confined horse may become aggressive due to his defensive instincts.

Horse Sense

Once you're "in tune" with a horse's nature and no longer have to think in order to handle your horse appropriately in various situations, you will have horse sense. Horse sense consists of two basic pieces of knowledge about equines: they have better memories than we do, and they are prey animals (that is, he will flee from anything he perceives as an attacker). These facts can help you decide when, where, and how to handle a horse.

Leading Horses

Walk beside the horse's left shoulder, whether you are leading him on a halter and lead rope or with a bridle and reins. Grasp the lead rope or reins a few inches below the chin with your right hand. Pass the lead or reins across your body loosely, and grasp the end, or a coil of the remaining rope or reins, in your left fist. Some horsemen will grab the halter/bridle

HOW HORSES THINK

When handling horses, keep in mind that their brains work very differently from ours and that their senses are far keener. What we may interpret as misbehavior or willfulness is often genuine fear of the unknown or confusion about what is expected. The urge to flee is strongly ingrained in the equine brain. Punishing a horse in a situation where he is already fearful will only increase his stress level and make him less likely to learn what you are trying to teach him. If you punish him consistently, he will probably begin to associate the object of fear (a trailer, a bridge, a particular tree on the trail) with the punishment and feel even more afraid and stressed when approaching it. In most cases, your first course of action should be to let your horse become familiar with the object of his fear and to make sure that your signals to him are clear and consistent. Once he does something correctly, praise him and move on to something else rather than overschooling him.

itself for better control, but this method is dangerous to human hands and fingers. A better way to achieve more control is to use a nose or lip chain. Never walk in front of a horse you are leading. At best he may bump into you or clip your heels; at worst, he will disrespect your space — or simply panic at something — and knock you down if something incites his urge to flee.

Punishment

Beware that punishment applied inconsistently or aggressively can trigger a dangerous, defensive response from the horse. Punishment should be viewed more as assertive discipline. It needs to be consistent and applied immediately following the act it's used to correct.

Relaxation

A handler who is tense will transmit that tension to the horse and make him generally harder to handle. It is essential, when handling horses, to be both firm in decision and intent and relaxed about carrying out whatever needs to be done.

Rubbing, Discouraging

The easiest way to discourage a horse from rubbing on you is *not* to rub on him. If you do, he will think rubbing is all right; further, he will use rubbing to challenge you for dominance in the relationship. If you simply must rub, then insist your horse stay in his space and not rub against you.

Safety Tips

Be alert when handling horses, no matter how well you think you know them. Pay attention to their body language and to what's going on around them. Then, if they spook, you will have a better idea of what might frighten them, which way they might move, and when — and you can arrange to either move with them or move out of their way.

➢ *See Safety*

Stallions

Stallions must be handled with firmness and respect. Because they are prouder, stronger, and bolder than geldings and mares, and also highly exuberant and sensitive, you invite serious trouble if you abuse a stallion. Handling him with respect and firmness, especially if he has a good temperament to begin with, will result in a tractable horse — as long as he gets sufficient exercise as an outlet for his abundant energy. Here is a list of stallion-handling dos and don'ts:

- Halt any disobedience — charging, rearing, striking, biting — while the horse is young and experimenting with his limits.
- Keep your temper and never punish a stallion in anger.
- When you have to reprimand and correct his behavior, cease the punishment the second the horse responds and behaves.
- When an incident is over, allow the relationship to return to its friendly and relaxed state as soon as possible.

GROUND RULES

Establishing clear ground rules for handling defines your expectations for your horse and for anyone else who is allowed to handle him. The rules will vary, depending on where you learned horsemanship, but might include the following:

- ❑ The horse must respect your space. He's not allowed to crowd you, walk into you, or rub against you at any time.
- ❑ The horse is not allowed to nip or bite, even if it might be interpreted as a playful gesture.
- ❑ The horse must respond to basic commands such as "ho," "go" (cluck, kiss, verbal commands for forward motion), and "turn" (again, based on various commands), without hesitation.
- ❑ The horse must stand quietly when asked, whether being led or ridden, without excessive fidgeting or fussing.

Temperament of Horses

Work with the horse's temperament when handling him. If he is bold, be sure you don't allow him to dominate you. Make sure lazy horses don't get away with cheating and avoiding what you've asked. Don't be so overbearing, however, that timid horses become more afraid and skittish.

Turnout Terrors

These horses bolt into the pasture, often bucking and kicking, as soon as the lead is unsnapped or halter removed. Some handlers simply try to let the horse loose as soon as possible, but that simply makes the behavior worse. Make such a horse relax before you let him go. Loop the lead rope around his neck so he can't go rushing off as soon as he feels the lead drop away or halter come off. Vary the interval between pasture or paddock entry and when you actually let him go until he understands that *you* will leave him, rather than him leaving you.

Veterinarians

Owners act as handlers for their horses during most veterinary examinations and procedures. As such, it is vital that even the most inexperienced handler become aware of how to do so safely. Methods may vary, but handlers should become familiar with the wide range of concerns that veterinarians might have for the safety of all participants: self, handler, and horse. The best way to learn more about this is to communicate with your veterinarian.

➢ *See also Restraints*

Voice Tone

A soothing voice can calm a nervous horse. For many horses, a disapproving voice is sufficient punishment for misbehavior and gets the desired results: a change in the behavior. An approving tone of voice, on the other hand, can be a powerful reward for most horses.

Hands, rider's

A rider's hands can be either a subtle or a brutal way of communicating with the horse. An old rider's adage says that to have good hands, a rider must first attain a good seat. It's true; the hands must be able not only to deliver varying amounts of pressure to the reins (and, in turn, to feel the horse's mouth and weight and movement through them), but they also must be able to be used independently of the rest of the rider's body. For that to happen, the **seat** — the way the rider sits in the saddle or posts (rises) from it — must be developed to a point at which the rider doesn't need to use her hands for balance and staying on.

➢ *See also Aids; Reins; Seat*

Hand twitching

If a twitch is not available when you need to calm a horse, try grasping his nose and squeezing; you may be able to apply enough pressure to release endorphins and get him to calm down and stand still for a few minutes.

TYING HORSES

- ❑ Use strong halters and ropes that are in good repair.
- ❑ Tie to something solid; horses have been known to pull all manner of boards and posts after them as they break free, a danger to themselves and anything they encounter.
- ❑ Avoid tying near a wire fence; feet can get caught and hooves ruined, and large lacerations can occur, perhaps even fatal ones.
- ❑ Tie short, but with enough room for some freedom of movement for head and neck, and keep the tie even with the head, not below or above it. The horse should only be able to lower his head to his chest, no lower.
- ❑ Never tie with reins; if the horse pulls or bolts, the reins will break. Worse, the bit may do significant damage to the horse's mouth.

Having good hands is important in both English and Western riding.

Hanoverian

A German warmblood breed founded in 1735 when England's King George II founded the Celle Stud with 14 Holsteiner stallions and local mares. (George, being German, was also Elector of Hanover.) The intent was to create a better draft horse. Later, Thoroughbred blood was introduced; after WWII, both more Thoroughbred and some Trakehner blood was introduced to further refine the horse. Today, the large, powerful Hanoverian excels at both show jumping and dressage. Hanoverians stand about 16.2 hands high, and are all solid colors.

➢ *See also Dressage; Stadium jumping (show jumping); Trakehner; Thoroughbred*

Set the cart before the horse.

—John Heywood, *Proverbs*

Harness

The equipment of a horse that is driven, as opposed to one that is ridden. Harnesses are made in a wide variety of styles and sizes, suited for the many styles of driving and the various breeds of horses used. The most common harness types are the single-hitch draft or carriage harness, the fine harness, and the team/draft harness. Although each style differs, most harnesses include:

- A **headstall** that holds one or two bits
- A breast collar or neck collar, some type of **surcingle** or bellyband, and **traces** to aid in securing the shafts of the vehicle to the horse
- Breeching or tugs to aid in stopping the vehicle
- A **crupper** to help prevent the harness from being pulled too far forward on the horse's body

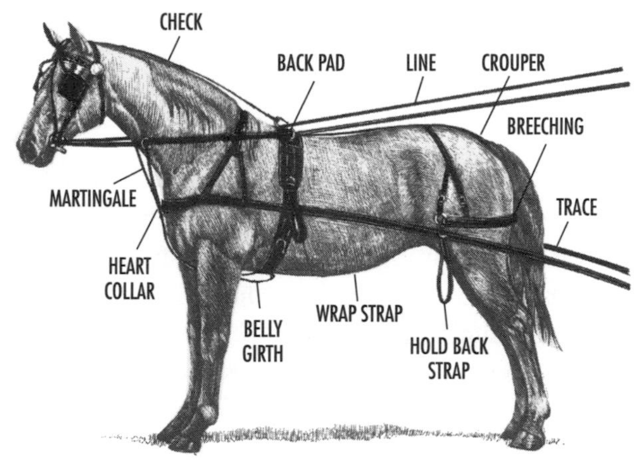

Single light horse harness with breast collar

Harness, foal

Harnesses for foals are often safer and more humane than halters. They place pressure on the sturdy shoulders and chest rather than on the fragile head. You can purchase a foal harness. Or you can use an adult halter, put on upside down, as a harness for young foals.

➢ *See Foal*

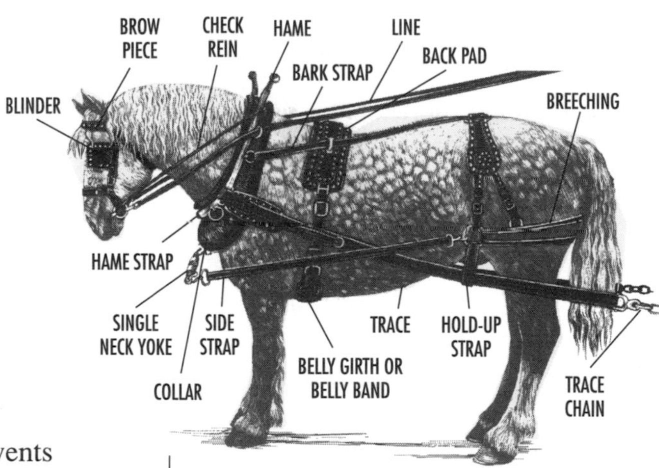

Single draft harness with neck collar

Harness racing

Although chariot racing was one of the original Olympic events (in around 1000 B.C.), harness racing as a sport is a peculiarly American tradition. From its origins on country roads and at county fairs, harness racing garnered legions of new fans when Roosevelt Raceway opened on Long Island in 1940. Today many cities in the Northeast and Midwest have racetracks, and the sport is also popular in Florida, California, and Canada.

Both trotters and pacers vie for wins in their own Triple Crown races. Pacers race in the Cane Pace (Yonkers Raceway), The Messenger Stakes (Ladbroke at the Meadows), and the Little Brown Jug (Delaware, Ohio).

The trotters compete in the Hambletonian (the Meadowlands), the Yonkers Trot (Yonkers Raceway), and the Kentucky Futurity (Lexington's Red Mile).
➢ *See also Horse racing; Pace; Speed, of horse; Standardbred*

Hat

Vital to every horseman's safety, regardless of the level of skill or style of riding, is the protective hat known as a helmet. Almost every horse show association now endorses rules that make protective headgear acceptable in every class of competition. It is best to learn the rules for apparel in the competitions you plan to attend, so you become familiar with acceptable styles of hats for your class.
➢ *See Attire; Helmet, safety*

Hay

Horses by nature are grazing animals. This means that they naturally need forages as the mainstay of a healthy diet. Hay provides a perfect replacement for the grasses they may not have access to as stabled animals turned out in dirt paddocks or pastures that lack sufficient forage to maintain health.

The average mature horse generally consumes 2 to 2.5 percent of its body weight in feed each day. Ideally, at least half of that should come from hay or pasture for optimum growth and development. Mature horses that are not working or being used as breeding animals can actually thrive on high-quality forages alone. However, those horses that are still growing, that are being used as breeding animals, or that are in athletic training programs usually require supplementation with grains and other concentrates to maintain optimum health.

Nutritional Value

Hay provides many nutrients. The exact nutrient value of any given sample of hay can only be determined through forage testing, which is available through your local extension office. Generally, hay is high in calcium and low in phosphorus. It often contains high levels of potassium and vitamins A, E, and K. If it is dried in the sun rather than via chemicals, it likely will have high levels of vitamin D. The protein content of hay varies almost too much to generalize. But some guidelines include the following:

- Legume hays (such as alfalfa or clover) can provide as much as 20 percent crude protein.
- Grass hays (such as timothy or bromegrass) average about 10 to 15 percent protein and can dip as low as 3 percent.

Production Methods and Protein Content

Hay production techniques largely determine protein content. Hay cut early has a higher protein content. Hay cut much past midbloom stage may offer inadequate levels of protein, so the horse must be fed protein supplements to maintain optimum health.

HEALTH CONCERNS

There are several health concerns to keep in mind when feeding hay.

- Horses are highly sensitive to molds, mildews, and fungi. Hay that has been baled too early in the day or that has been rained on will often be wet enough to be a perfect breeding ground for these toxins.
- Sudan grass and sorghum grasses can lead to life-threatening complications; *never* feed them to horses.
- Some varieties of tall fescue are known to contain a fungus that can cause life-threatening complications. Make sure your pasture is free of this plant.
- Blister beetles, common in alfalfa hay grown primarily in the southern United States, can cause irritation of the digestive tract, which often leads to death. Beware of these critters!

Note: Horses usually won't eat noxious grasses. However, if these are the only grasses available in a poorly managed pasture, horses may eat plants that their instincts warn them about.

High-quality hay is the result of careful hay production methods. It is cut early and should appear leafy (if legume) or feel soft in texture (if grassy hay), and be free of mold, dust, weeds, and other undesirable matter. Its smell should be appealing, not musty. If you are unfamiliar with hay selection, it would be worth your time to visit with your local extension agent or agricultural education instructor to learn more.

Types of Hay

There are two main types of hay with which most horsemen are familiar. These can be fed singularly or in a mix known simply as "mixed hay."

Legumes. A family of plants having root nodules that produce nitrogen; they also have stems that leaf out into leaflets as is typical of the clover plant. Legumes are the highest in protein of all hays; averages show they can contain two to three times the amount found in grass hays. They also are higher in calcium, beta-carotene, and vitamin E. Some believe legumes are also the most palatable of grass hays. This is why many horsemen prefer alfalfa or alfalfa mix hays over grass hay for growing, breeding, or highly athletic horses. Still, their high protein content makes legumes potentially dangerous to feed to mature, idle horses. Be careful, and become knowledgeable about horse nutrition.

Alfalfa is the most commonly known and grown legume. Some estimates list alfalfa or alfalfa-mix as constituting more than half of all hay grown in the United States. Other legumes include clovers, bird's-foot trefoil, and lespedeza.

Grass Hay. Grass hays come in a much wider variety than legumes, since numerous types of grass are native to each region of the United States. Grass plants grow with tall stalks and long, slender leaves that often wrap around the stalk, rather than branching out the way legumes do. This makes grass hay dry faster; as a result, mold is less of a problem. Grasses also don't have problems with blister beetles like alfalfa can. This means grass hay is often a better choice for horses with compromised breathing. Their lower protein and nutrient density actually makes grass hay more desirable as a feed for mature and idle horses that aren't used for breeding.

There are far too many varieties of grass hay to include here. But a few of the most common ones fed to horses are timothy, Bermuda grass, bromegrass, bluegrass, fescue, ryegrass, and orchard grass. Timothy is perhaps the most common and widely grown since it is easy to grow in a variety of soils and cold climates.

Mixed Hay. A combination of legume and grass varieties into one crop. Mixed hay provides the best of both types of hay. It is more appealing to horses because they often prefer the smaller leaves of legumes. The mix of nutrient values (high-protein legume and lower protein grass hay) makes this hay the choice of many professional stable managers because one type of hay can be fed to all types of horses.

➢ *See also Feeding and nutrition; Feeding equipment*

HEAT ENERGY FROM ROUGHAGE

More body heat is created in the horse by digestion of roughage than grain. If it is extremely cold outside, increase the ration of hay before you increase the amount of grain.

Hay net, hay bag

Hay nets or bags provide a method of feeding hay in a variety of situations. The primary use is at times when the horse needs to be tied for extended periods. The hay net can be tied in a place close to where the horse is tied so he can eat. Just be sure to tie the net or bag high enough that the horse can't get a foot caught if he paws. Tie with a quick-release knot in case the horse gets hung up in the net or bag.

Hay and grain feeder

Hay storage

Hay must be kept dry in a well-ventilated area. It can be a serious fire hazard: Newly baled hay can combust if put up at a high moisture content level. It also can provide fuel for fires that start elsewhere, making it nearly impossible to control the blaze. So it is best to store hay in a place away from your stalls and arena.

Hazards

There is no end to the potential hazards around a barn, including anything that could catch a hoof and hold it; cause a horse to slip and fall; cause a horse to rear and bang its poll; and anything flammable, poisonous, sharp, or uneven (such as a hole). Train yourself always to be on the lookout for things that could harm your horse or a rider.

Following are some of the main hazards and the simple cures.

Baling Wire

A horse may eat small pieces of wire with his hay, so be careful how you cut bales open. Long pieces lying in the barn may trip a horse or poke him in the eye. Discard all baling wire, and twine, too, which could be picked up and eaten if left in the hay.

Broken Posts or Fence Uprights

A horse can step on or trip over these objects and damage a hoof or leg. Be sure any old structures are completely removed from horse lands.

Dangerous Fences

This category includes fences constructed of materials hazardous to horses and otherwise safe fences that are in a state of disrepair.

Horse fencing made of large-mesh wire can snag a hoof and cause immense damage if the horse is frightened and tries to pull free. Safe fencing with flapping boards and exposed nails is also dangerous. Remove all dangerous fencing from your property and keep other fencing in good repair. Good maintenance requires daily observation of every fence in every field to ascertain what damage the horses might have done during their turnout, or what natural forces (weather as well as humans, beavers, birds, and dogs) might have done.

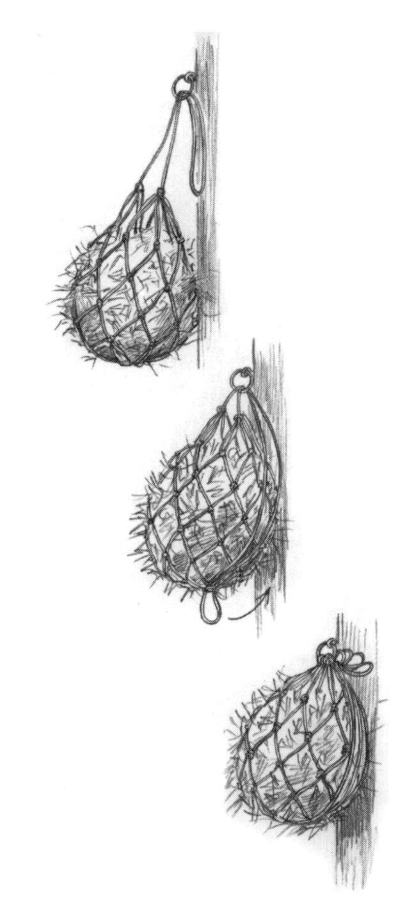

A hay net must be hung high enough so that a horse cannot catch his foot in it.

Hazardous Materials

Never for an instant leave any flammable liquid in the barn where a person or a horse (or a cat or mouse) could knock it over. Never leave a poisonous substance where drops or pieces of it could get into a horse's feed, hay, or water.

Low Overhangs

Your run-in shed should be a minimum of 7 feet high at the open side; higher is even better. A playing or fighting horse could rear and run into a corner, doing considerable damage to his head, neck, or eyes.

Public Rights of Way

If your pastures or paddocks abut a road or other human thoroughfare, stay alert always for things thrown onto your equine land, including broken bottles, plastic bags (which, if they contain food, might be swallowed by a horse) food wrappers, household junk, and discarded small machinery.

Survey frequently; if possible, install double fencing at road borders, with one horse-safe fence at the property's edge, and another run of fence several feet inside the first. This will leave you a "no man's land" you can easily survey and clean up when needed, and it will make it that much harder for passersby to discard trash on your land. It will also protect your horses from breaking a fence board, jumping or stepping out, and getting into traffic.

Rodent Holes

These will be made by different "varmints," depending on where you live. In the East, they are often called chuck holes, because woodchucks make them. Whatever makes holes in your area — woodchucks, moles, prairie dogs, snakes — be alert for them and fill them in. When riding in fields, survey well ahead before you take up a gallop; a horse that steps suddenly into these holes can pull a tendon or break a leg.

Sharp Projections

These include old nails from which something has fallen, broken hooks, separating sections of wood, and more. A horse could catch his skin, which would result in a painful and possibly laming tear. He could also catch an eye or eyelid with even worse consequences. Completely remove broken gate latches or any other fastener or holder; check frequently for nails coming out of boards, for nails left behind by workers, and for dangerous wooden structures.

Head

A horse's head is important for reasons beyond the obvious. First, combined with the neck, it aids the horse in maintaining balance. Second, it becomes a point of identification between breeds and is often prominently described within breed ideals. Finally, depending on its attachment to the neck, it can either help or hinder a horse's training and performance while being ridden.

➢ *See also Conformation; specific breed*

Head bumper

A flat, padded, leather or neoprene cap that attaches to a halter and protects the horse's head while he is being trailered. If the horse rears or is bounced up to the top of the trailer, the bumper will protect his poll.

CONSTANT VIGILANCE!

It's impossible to avert all disasters, but be aware that horses, with their highly developed sense of curiosity, excel at getting into peculiar situations. If it's possible for them to get into trouble, they probably will. Every horse owner has a story about an injury or a near-miss that defied common sense and preventive measures but happened anyway. Think like a horse to imagine what kind of trouble he could get into. Then take steps to block off hazardous areas, lock feed storage bins, dispose of all trash, and keep stalls danger-free.

Even if a horse is a good traveler, a head bumper is a wise precaution; what happens on the road is often unpredictable, both in terms of horse behavior and driver/trailer incidents.

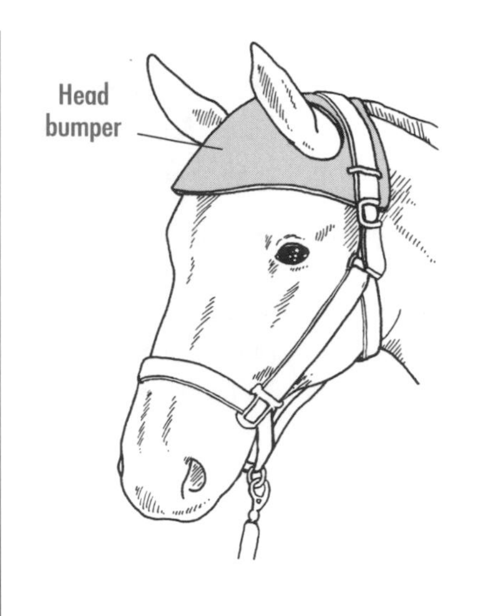

Header

In Western riding, a roper who brings down calves by settling the catch rope over the animal's head. In driving, a person who stands by the head of a horse during idle times to help keep the horse calm.

Headshy

A horse who will not let you handle or groom his head, face, or ears. Often, the behavior is a result of trauma, either purposeful or accidental. Most headshyness can be cured through repetitive gentle retraining and by gaining the horse's trust.

Headstall

Another term for bridle.

Head-turned-back presentation

➢ *See Foaling, foaling presentations [box]*

Healing with alternative therapies

Alternative therapies include chiropractic, massage therapy, acupuncture, acupressure, and herbal therapies. Many horse owners believe using one or more of these helps keep their horses sound and happy. More and more veterinarians are recognizing the value of some alternative therapies, and many veterinarians are becoming skilled at chiropractic and acupuncture in addition to their conventional medicine practices.

➢ *See also Acupuncture; Chiropractic treatment; Herbal therapy; Massage therapy*

Part of caring for your horse is regular grooming and checking of feet.

Health care

Health care of the horse includes all the same systems as health care of any other mammal. Because horses spend so much of their time, especially if they are pasture-boarded, beyond the eyesight of their human handlers, it is especially important to bring them in, handle them, and assess their physical condition frequently — at least once in each 24-hour period. Remember, an ounce of prevention is worth a pound of cure.

➢ *See also Dental care; Digestive problems; Diseases, prevention; Ears; Eyes; First aid; Infectious diseases; Parasites; Signs of illness; Skin problems; Toxic substances*

Heat cycles

Mares' heat cycles vary widely, from one to ten days. Ovulation occurs about 24 to 48 hours before the end of heat. Also, mares are not consistent over their lifetimes. One year, a mare may have 5-day cycles; the next, 10-day cycles.

Heated barns

While they may be more pleasant for humans, heated barns are a very unnatural environment for horses and can cause them to become prone to respiratory problems and viruses. Plus, heating a barn is expensive and difficult. It is better to heat only those areas you need warmed when the situation demands it, such as when nursing an aged, sick, or injured horse, or for foaling in very cold weather.

Heat stroke

Also called **hyperthermia,** heat stroke is caused by being confined in hot, stuffy stalls or trailers, or a paddock in extreme heat and humidity with no shade. The condition can even be prompted by working too hard on a hot day. Unless it is treated immediately, with the aim of bringing down the horse's body temperature, he may go into shock — and it may be fatal. Call your veterinarian immediately, then hose the horse down with cool water, starting with the lower legs.

➢ *See also First aid*

Heaves (chronic obstructive pulmonary disease, or COPD)

A disease characterized by loss of ability to perform, constant or intermittent cough, and a watery discharge from the nostrils. As the condition worsens, the horse may lose weight because he is using so much energy to breathe. COPD is not curable by antibiotics because it is caused by dust and mold; after the first attack, there will probably be relapses of varying severity. This often causes infertility in mares.

Hedges

Beware of hedges around horses; they are often formed of plants, such as yew, that are highly toxic. Other toxic plants that tend to be grown as hedges are boxwood, azalea, privet, rhododendron, chokecherry, and cherry laurel.

➢ *See also Poisonous plants*

Heeler

In Western riding, a roper who brings down steers by roping the steer or calf's hind leg.

Height, measurement of

In English riding, events are designed for ponies or horses, and some are even divided into divisions for small, medium, and large ponies. Height determines what division these equines are eligible to show in, so they must first be measured. The measurement will be in hands and inches.

➢ *See also Hand, as unit of measurement*

TRAILERING IN HOT WEATHER

Trailering in the heat of summer and remaining stopped for more than a few minutes can be extremely stressful for a horse. While driving, keep the trailer's windows and optional doors open to create a cooling breeze. When you arrive at your destination, take the horses out of the trailer and into a stall or shaded paddock immediately. If you must tie them to the trailer — at a show, for example — tie to the shady side, return to hand-walk them frequently, and be sure clean water is available within their reach.

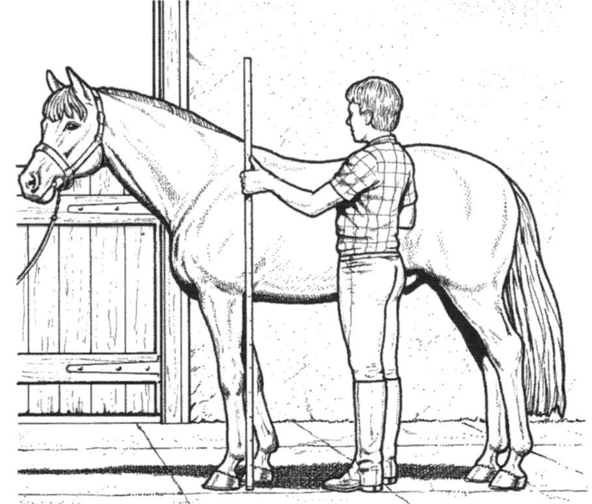

Measuring a horse at the withers to determine its height

Helmet, safety

A protective riding helmet is the first item of equipment a rider should buy. Be sure the helmet is designed specifically for horseback riding and meets the standards established by the American Standard for Testing Materials (ASTM) and the Safety Equipment Institute (SEI). These standards do not guarantee that a helmet will protect a rider from head injury; however, if properly fitted and secured, a helmet should reduce the risk of injury.

Approved helmets are constructed of expanded polystyrene, which crushes on impact and absorbs the energy of the fall, cushioning the head. Bicycle helmets are not recommended for horseback riding, because they are too shallow and do not cover the head adequately.

Almost every horse show association now endorses rules that make protective headgear acceptable in every class of competition, so riders no longer have to risk their safety for the sake of fashion.

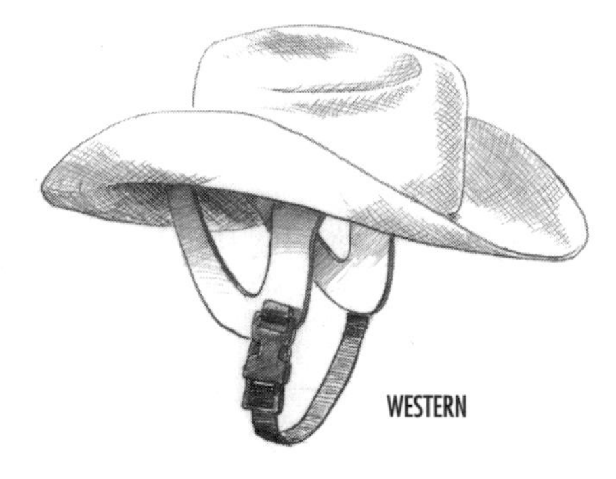

Safety helmets come in standard equestrian and Western styles.

Hemorrhage after foaling

➢ *See First aid, hemorrhagic shock; Foaling*

Hepatitis

There are two kinds of hepatitis, infectious hepatitis and noninfectious hepatitis.

Infectious Hepatitis

This disease is caused by using contaminated needles and syringes. Onset of symptoms is sudden and the horse is almost always violent, deranged, and impossible to catch. He will usually run into walls or fences. Ninety percent of horses with infectious hepatitis die, despite treatment with antibiotics, glucose electrolyte solutions delivered intravenously, and B-complex vitamins.

Hepatitis and/or death of liver tissue is often caused by vaccination, especially if the serum has been derived from equine tissues. The disease may occur 30 to 90 days after the injection; a delay of 6 months is not unheard of. Vaccines that have caused hepatitis include encephalomyelitis serum, tetanus antitoxin, pregnant mare serum, and anthrax antiserum.

Noninfectious Hepatitis

This disease is caused by poisons in the body and can be acute or chronic. The acute form comes on suddenly and is often fatal because liver damage is so severe. Causes can be toxins from poisonous plants or contaminated feed, or from an unrelated bacterial infection that produces liver-damaging toxins. The horse may have stomach pain mistaken for colic. He may stagger or drag his feet, and he may deteriorate mentally. He may stand with feet wide apart and head drooping, or he may walk into trees or fences. He may have muscle tremors, and he may become photosensitive. In some cases, the horse becomes violent and unmanageable. If no irreparable damage has been done, removing the source of the toxin and nursing care can help the horse recover.

HEMIONIDS

While hemionid means "half-ass" in zoological terms, these animals are not crosses between an ass and another equid. They have features of both horse and ass, as well as distinct characteristics of their own, most notably extremely long lower leg bones. Examples of hemionids include the Mongolian Kulan of Central Asia and the onager (also known as the "wild ass" of Biblical times), which is still found in the Middle East and Asia.

Herbal therapy

For equines, many herbal preparations are marketed. Like conventional medicines, they are aimed at treating specific ailments, from wounds to coughs. Unlike conventional medicines, which may be derived from organic or inorganic substances, herbal therapies all depend on natural substances. As is the case with herbal remedies for humans, herbal therapies for horses are not federally regulated: Proceed with caution.

Herd-bound

Term to describe a horse that craves the companionship of other horses. A herd-bound horse will engage in behaviors including running a fence line to try to see other horses, pawing or weaving in his stall, or making shrill calls whenever he hears other horses. If a herd-bound horse goes out with others, he may be difficult to catch and bring in. In lessons, he may call to other horses and dash to the gate.

Herd instinct

In nature, horses are members of distinct groups, on which they depend for companionship, protection, and reproduction. Most horses become upset if they are without the company of other horses for extended periods, although a few can adjust to an isolated existence. In group situations, such as riding lessons, horses naturally try to get close to each other, as they would in the wild.

Hernia

A rupture of the lining of the abdomen that can cause pain and a range of problems from infection, strangulation of the intestine, or even potentially fatal peritonitis. Call your veterinarian if your horse develops any swelling in his abdomen.

Herpes virus

➢ *See Equine herpes virus (EHV); Rhinopneumonitis*

Heterozygous genes

A mixed pair of genes — one dominant and one recessive. A foal with heterozygous genes will show the dominant trait, but he can pass on either gene to his future offspring.

➢ *See also Genetics*

Hindquarters

The hindquarters of a horse is the "engine" of that horse. It is therefore vital that the hindquarters be suited for the function expected of each individual. For example, a racehorse needs hindquarters that can easily propel him over moderate distances in a short amount of time, whereas a Clydesdale needs hindquarters that can propel him forward with shorter strides that allow him to pull heavy loads over long distances without causing exhaustion and fatigue.

➢ *See also Conformation*

Histamine

A chemical that is produced when the body recognizes that an invader — such as pollen or vaccine, or even chemicals, such as fly spray — has entered the system. A "histamine reaction" is synonymous with an allergic reaction. Horses are as prone to allergies as humans, so it's important to understand how an allergic reaction occurs. The body produces antibodies to fight the invader and the antibodies produce the chemical histamine.

Call the veterinarian if your horse has an allergic reaction; if he has a history of such reactions, keep epinephrine or antihistamines on hand.

History of the horse

The horse, as we know it, began its development about 60 million years ago in North America and Europe. The earliest ancestor is called *Eohippus,* or Dawn Horse. At the time, he was a dog-sized animal with fast reflexes, three-toed feet, great vision, and superior speed. He ate soft leaves. After the Ice Age, the animal developed sharp teeth to clip off tough grasses and molars to grind the grasses. *Eohippus* developed a long jaw and teeth that grow continuously so he could never wear them down. He eventually got bigger. By the time humankind came along, *Eohippus* had grown to pony-size and his feet had only one toe, a hoof. Stone-age humans hunted this creature.

As humans settled down and began to cultivate land, they found — at some point, and possibly with help from the horses hanging around for grain leftovers — that the horse could help them with their farm work. Not too long after, humans began selective breeding. Today, there is a horse perfectly suited for every use, from pulling huge wagons or logs, to competing in sports patterned after the vermin-elimination work (fox hunting) of earlier times, to being household pets in their miniature versions.

Hitches

There are three types of trailer hitches: fifth-wheel, gooseneck, and straight-pull. Each has advantages and disadvantages.

Fifth-wheel

This is the same type of hitch used to hook semitrailers to the big rigs used to pull them. It is the safest hitch available for hauling heavy trailers. Some education and considerable experience are required before an amateur driver is safe on the road.

Gooseneck

These hitches are welded to the bed of the tow vehicle. They are more maneuverable but are minimally acceptable for hauling larger trailers.

Straight-pull or Tagalong

This is the common "ball" hitch. Less expensive than gooseneck hitches and a good choice for two-horse trailers, straight-pull hitches can be hard to maneuver and hook up. They also may put more stress on rear end of the tow vehicle. This is the only hitch available for cars and vans.

➢ *See also Trailer*

How the horse dominated the mind of the early races, especially of the Mediterranean! You were a lord if you had a horse. Far back, far back in our dark soul the horse prances.

—D. H. Lawrence, *Apocalypse*

Hitching post

A sturdy vertical beam or round post of good diameter that is held up by substantial posts anchored in the ground. The hitching post is useful for tying horses while they are waiting to work.

Hobbles are used to fasten two of the horse's legs together to inhibit it from moving too far or too fast. Pictured are breeding hobbles that keep the mare from kicking at the stallion and hold her tail out of his way.

Hives (wheals)

Low bumps on the skin caused that might be caused by allergens that are either in the air or come into contact with the coat or skin. Hives can also be caused by the bite of an insect or the puncture of an injection. Usually, hives subside by themselves in a few hours or a day. Sometimes they are accompanied by diarrhea, fever, or respiratory problems. In these cases, call a veterinarian.

Hobbles

A form of restraint.

➢ *See Restraint*

Hock

The tarsal bones, corresponding to the human ankle. Along with the **stifle,** which correlates to the human knee, the hock is one of the hardest working joints in a horse's body. So it is vital that it be relatively large in circumference and height, and that the bones above and below it are in alignment. Stress injuries and lameness are common in horses with poorly conformed hocks.

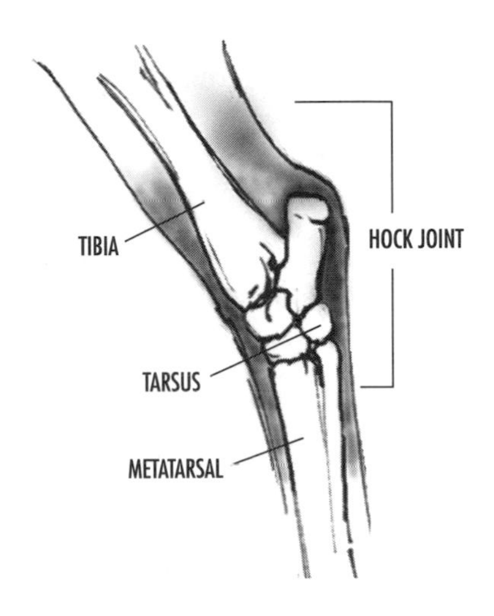

The hock is one of the hardest-working joints in a horse's body. It should line up directly under the hindquarters.

Hog's back

A type of jump.

➢ *See also Spread*

Hog tie

To tie three legs of a roped calf with a narrow rope.

➢ *See also Hooey*

Holding a horse

Here are the basic guidelines for holding a horse:

- Use a properly fitted halter
- Choose a safe place to hold him
- Do not wrap the lead rope around your hand or arm; rather, coil it and hold it on the outside of the coils
- Don't stand in front of the horse; stand beside his near shoulder.
- Be alert to the horse's attitude so you can predict and control possible movement — or get out of the way

Hollow back

A dip in the back between the withers and point of hip. Hollowing of the back indicates resistance within the horse's body. It is felt by sensitive riders as stiffness; less experienced riders will often just feel as if their horse is uncomfortable to ride. It has numerous possible causes including, but not limited to, an injury to the horse, poorly developed riding skills, mismatched horse/rider combination (for example, a rider who is too large for the horse), or an ill-fitting saddle.

Holsteiner

During the 17th century, the Holsteiner was recognized as an elegant, although heavy, coach horse. In recent years, the breed has been refined with the introduction of some Thoroughbred blood, and today the Holsteiner excels at show jumping and dressage. A large horse, he stands between 16 and 17 hands high. He is intelligent and comes in all solid colors.

A hondo is an eye or loop through which a rope slides when it is thrown.

Homeopathy

Invented in 1779 by German physician Christian Freidrich Samuel Hahnemann, the idea behind this medical system is that specific substances, even poisons, can chase away in a sick person the very symptoms the substance would cause in a healthy person. The short way of saying it is "like cures like." This approach is thought to work through the stimulation of the body's immune system. Commercial homeopathic remedies are available for horses.

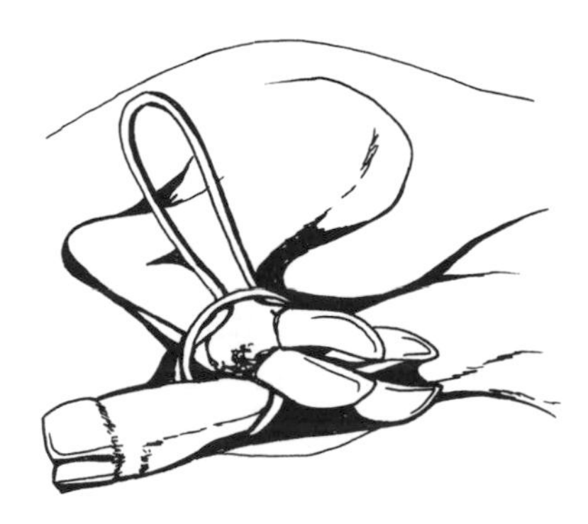

The hooey is used to quickly tie the hind legs of a calf. Wrap the rope twice around three legs, and then make a half-hitch or hooey around the hind legs.

Homozygous dominant

In this horse, two dominant genes for a specific trait come together, producing a horse who is homozygous dominant for that trait. The horse can pass on no other trait to his offspring.

➢ *See also Genetics*

Homozygous recessive

When two recessive genes come together, they produce a homozygous recessive individual, which will express that trait and pass only that trait on to its offspring.

➢ *See also Genetics*

Hondo

In Western riding, a catch rope — one with which cattle are roped — is a length of twisted fiber strands that ends in a loop. The hondo is the small, tied end of the rope through which the rest of it slides when the rope is thrown. Attached to the hondo is a **burner,** a strip of leather that helps prevent wear as the rope slides through the hondo while tightening.

Honest horse

A horse you can trust.

Hooey

A half-hitch that is tied around the hind legs of a calf that has been roped and is being hog-tied.

Hoof

The hard structure at the bottom of a horse's leg. The hoof is the remnant, after eons of evolution, of the original toes of *Eohippus,* the primeval horse. The ergot and chestnut are also such remnants and may likely disappear with further evolution. However, the original toes still exist as bony elements within the horse's hoof.

Although a horse's feet look solid and tough, they are actually full of delicate structures and are protected only partially by the hard outer wall. The sole of the foot is softer than the wall; the frog, a wedge-shaped area of soft tissue beginning at the heel and reaching two-thirds of the way toward the toe, is quite soft.

The shapes of front and hind feet are slightly different, but both grow outer hoof wall that extends beyond the softer portions, keeping them out of contact with most small pebbles on the ground; a large stone, however, may bruise even the artificially raised sole of a shod foot. Keeping a horse's feet in good condition is essential for soundness.

Abscess

An area of collected pus within a horse's hoof. Abscesses will make the horse lame. They may be caused by a puncture of the sole of the hoof, an infected stone bruise to the sole, or a horseshoe nail that entered (or "quicked") the sensitive tissues within the hoof. Treatment by the veterinarian involves piercing the sole of the hoof to reach the abscess and allow for drainage. Follow-up nursing care involves soaking the foot in Epsom salts. Often, the veterinarian will also prescribe topical and sometimes even injectable antibiotics.

Bruised Sole or Corn

Because the sole of the foot is not as hard as the load-bearing outer hoof wall, bruises can occur on hard or rocky ground. Letting shoes stay on too long can also cause bruising; the bulb of the heel may press on the metal, causing bruising beneath the sole, as a sharp rock might. A bruise or corn on the sole of the feet appears as a reddish discoloration on the sole; lameness might be the first clue, however.

Conformation

Feet should be large but in proportion to the horse's body. Small feet will suffer extra concussion and increased wear and tear; few horses have feet that are too big, but some horses (and some breeds, such as the Quarter Horse) may have feet that are too small, and can develop unsoundness later in life.

Cracks

Unshod horses often experience cracks and chips of the lower part of the hoof, but these are rarely serious. A shod horse may also develop cracks, usually starting in the hoof wall at ground level, traveling upward. A crack may also start in the heel or quarter area and travel horizontally around the food because of weakness of the wall in that area, perhaps as the result of a blow or injury. Minor cracks are generally dealt

CLEANING HOOVES

Clean a horse's feet every day. Use a hoof pick to dislodge dirt and debris by running the pick against the hard sole from back to front; avoid scraping the frog or puncturing it, but clean the channels beside it.

Cleaning the feet can prevent lameness from stone bruises; it can also prevent thrush, a bacterial infection caused by an anaerobic organism present in manure that thrives when it is trapped by muddy debris against the soft tissues of the frog.

See also Frog of foot; Thrush

It's extremely important to check and clean your horse's feet regularly.

with as routine matter by the farrier each time he visits. Some cracks, however, call for veterinary care or corrective shoeing by the farrier. Deep cracks may make the horse lame, as may those caused by injury to the coronary band.

Founder

Also called **laminitis,** founder is inflammation of the sensitive laminae of the foot. It can be caused by:

- Overeating, especially grain
- Drinking large amounts of cold water when overheated
- Uterine infections in mares that have recently given birth, often caused by retained placenta
- Grazing on lush pasture; this is more likely to affect ponies than horses, although it may affect either
- Various other causes, including viral respiratory infections, some drugs, and the overeating of beet tops

➢ *See also Laminitis*

Growth

Hoof growth is essential for soundness. Growth is enhanced by exercise, proper nutrition, and proper trimming and shoeing. A hoof grows about ½ inch per month in weanlings and about ⅓ inch per month in mature horses.

Navicular Disease (Navicular Syndrome)

This common condition is caused by inflammation of the tissues surrounding the small navicular bone in the front foot (or feet), followed by physiologic changes to the navicular bone. It begins as an irritation of the navicular bone and the deep flexor tendon, and causes intermittent lameness of one or both front feet. The condition often goes undetected until it is quite severe. Predisposing conditions include:

- Hard work, on unforgiving surface
- Disproportionately small feet
- Upright pasterns

➢ *See also Navicular disease (navicular syndrome)*

Shoeing

Shoeing must be done properly for the foot to function normally and for the horse to stay sound. The farrier must take into account the natural shape and size of the horse's foot and leg, the work the horse is doing, and any injuries to the foot or other areas the horse has suffered. Proper shoeing can protect the hoof wall from excessive wear and damage, increase traction, correct gaits, and reduce discomfort in horses with various physical problems.

INDICATIONS OF NAVICULAR DISEASE

It is tough to sell a horse afflicted with navicular disease; they are in pain and lame more often than those without navicular problems. Sometimes, the digital nerve can be cut, leaving the hoof without sensation, and, hence, no pain. Look for two small scars, ½ to 1 inch long, above the bulbs of the front feet. Do not buy such a horse for hard work; the wear on the tendon, especially in the absence of pain to tell the horse when to stop, may cause the tendon to rupture.

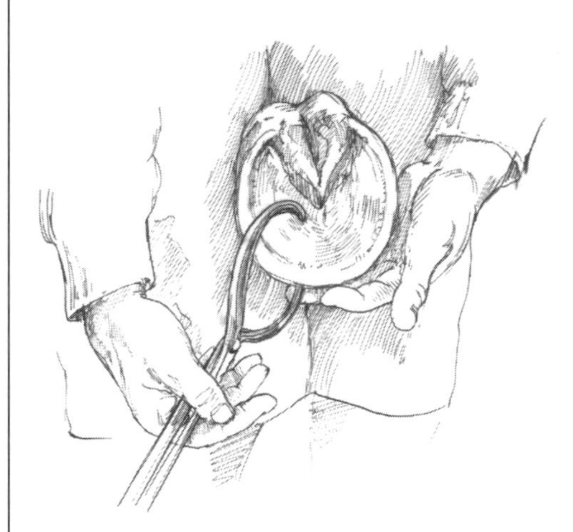

A hoof tester is a big caliper with which the veterinarian or farrier can put pressure on selected spots on the sole of a horse's foot to determine where there is soreness.

Trimming

Most shod horses grow enough new hoof to need trimming every six to eight weeks, and sometimes in as little as five weeks during hot weather or substantially increased work. Horses that are unshod may wear down hoof naturally and need trimming about every 8 weeks.

Wall

The hoof wall is the outer layer of the foot, made of horny material that encases and protects the sensitive structures within the foot. The hoof wall also provides a firm foundation on which to bear the weight of the horse.
➢ *See also Farrier; Foot care; Picking out feet; Picking up feet; Shoes, horse*

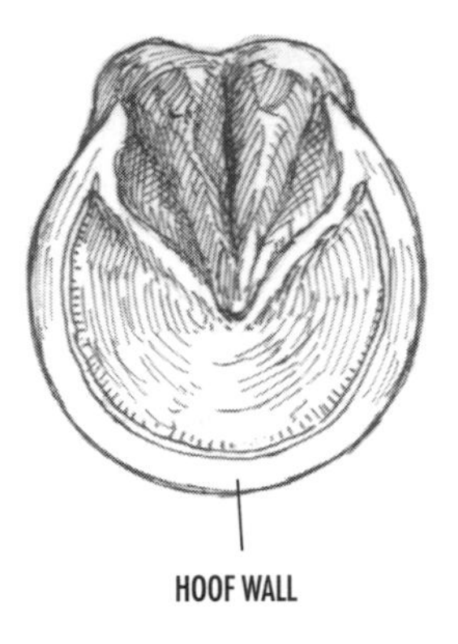

"Hoof bound"

Improper shoeing or injury can cause contraction of the foot; the two parts of the heel then get closer together and are not sufficiently separated by the frog. The position alone can be a problem, but it often leads to a condition in which the hoof wall itself may start to press against the coffin bone inside the foot. This condition, called "hoof bound," makes the horse lame and unsound.

Hoof pick

An instrument for removing dirt, stones, and debris from a horse's hooves. Picks come in several varieties, but all have either a sharp wedge or screwdriver point for digging out the material. Some have an attached brush for removing the remaining sand or grains, and some even include a small comb for pulling manes.
➢ *See also Hoof*

Hormones

Horses, like humans, have numerous hormones that regulate the systems of the body. Imbalances in hormones can wreak havoc in horses. It is vital therefore to become familiar with the function of the most important hormones. Synthetic hormones can be used to improve fertility and athletic performance.
➢ *See also Cushing's disease/Cushing's syndrome*

Horn flies

➢ *See Flies*

Horse

Properly, a horse is a full-grown male horse — gelding or stallion. A female, on the other hand, is a mare. The species, however, that is generally called horse is *Equus caballus,* a large, strong mammal with four legs, solid hooves, and flowing mane and tail. The horse is capable of being domesticated and trained to perform tasks for humans, such as drawing loads or carrying riders, as well as performing more artistic or sporting feats.

HORSEBALL

Invented in the 1970s by French riding instructor Jean-Paul Depons, horseball is most aptly described as rugby on horseback, with a little basketball thrown in. As fast and furious as polo, it is a more egalitarian sport, as each rider only needs one horse per game and it can be played at different levels of riding experience. Teams of six shoot at a basketball-type goal with a soccer-sized ball with leather handles that is passed or thrown from player to player and scooped from the ground when necessary. Players must get rid of the ball within 10 seconds and may steal it from another player. Two 10-minute halves are played. At the Pony Club level, riders play bareback, using only saddle pads and a surcingle with handholds.

Horse flies (tabanids)

These are part of a group of biting flies that attack horses. The flies may be black, brown, yellow, or gray, and they range in length from ¾ inch to 1 inch. The female must have a blood meal to reproduce; her bite slashes the skin so that the blood flows freely and she can drink it with her proboscis. This causes an allergic reaction, and a bump, in some horses. Horse flies are prevalent where there are areas of permanently wet ground because females lay their eggs on or near water.

Because they are such voracious feeders, often going from one horse to another, horse flies spread blood-borne diseases such as equine infectious anemia and encephalomyelitis.

Horsekeeping

➢ *See Facilities; Feeding and nutrition; Foot care; Handling horses*

Horsemanship

A term often used to describe stock seat equitation classes at horse shows; also, the body of knowledge and lore that is involved with working with horses. Horsemanship includes learning a wide range of knowledge and skills, including horse care (nutrition, health and hoof care, grooming, exercise needs), stable management, horse psychology, humane training, and effective riding, the mastery of which leads to one's being known as a "true horseman."

An example of horsemanship education would be a program that teaches the hows and whys of horse care, training, and riding. A horsemanship program begins with safety in ground handling, grooming, and tacking up; riding instruction begins in the arena, with the horse already prepared for the rider. After the lesson, horsemanship students untack, groom, and care for their horse and equipment before leaving the barn; in contrast, riding students typically walk their horse to cool him out before returning him to his stall or picket line, still tacked up.

Horsemeat

The topic of horsemeat and the slaughter of horses raises many moral and ethical concerns among horsemen. The 1999 IGHA/Horse Aid Projected Estimates are: 60,000 equines slaughtered in the United States; 29,000 equines shipped live to Canada for slaughter; and $58 million in projected revenue for horsemeat exported from the United States.

Most people are opposed even to the idea of consuming horsemeat, according to polls done in many states like California, where sale or slaughter of horses for human consumption (even for export) is illegal. Proponents for regulating the horse slaughter industry argue that the sale of horsemeat for human and nonhuman consumption creates a viable outlet for unwanted horses and horses that would often starve to death or die from neglect.

HORSE INDUSTRY, STATISTICS OF

- About 3.2 million people own an estimated 8.3 million horses; 80 percent are used for recreation, with the remainder used for profit-making activities, including agriculture, logging, racing, and exhibitions.
- More than 250,000 young people are working at equine 4-H projects — more than the combined number engaged in cattle and swine projects.
- More than 100 million spectators attend horse races every year; it has been the number one spectator sport in the United States for more than 30 years.
- A United States Department of the Interior study concluded that 27 million Americans ride at least once a year. More than half of the 27 million, however, ride on a regular basis.

To many, the words love, hope, and dreams are synonymous with horses.

—Oliver Wendell Holmes

Horse racing

From the Greeks to Roman charioteers to kids on bareback ponies, people have always loved seeing how fast their horses could go. One of the earliest recorded horse races was at the Olympic Games of 642 B.C.

In Great Britain, horse racing has long been known as "the Sport of Kings," given the keen interest that generations of royalty have taken in all matters equine. Horse racing in the United States takes two primary forms: flat racing (with a rider) and harness racing (with a driver). Steeplechasing, which is flat racing over fences, is popular in some parts of the country but attracts far fewer spectators overall, though in England it is much bigger business.

While any horse can be made to run around a track, and many races are held at local fairs and rodeos, the premier horse for flat racing is the Thoroughbred. In harness racing, the breed of choice is the Standardbred, with horses that either trot or pace.

➢ *See also Flat racing; Gaits; Harness racing; Pace; Speed, of horse; Standardbred; Steeplechasing; Thoroughbred; Triple Crown*

Horse sales

Horse sales may be as prestigious as the Keeneland Thoroughbred sales each spring — where young bloodstock is sold to international buyers for millions of dollars, with the prices based on the racing performance of the horses' sires and dams — or as down-home as a sale offering horses of various types and quality from the general region. Horses are also sold at agricultural sales where cattle, swine, and sheep are sold.

Horseshoes

➢ *See Hoof*

Horse shows

➢ *See Show Associations*

Horsetail (jointfir, scouring rush) (*Equisetum* spp.)

Horsetail, also called scouring rush or jointfir, grows in meadows and floodplains, or wherever there is a high water table. This plant may end up in meadow hay, where it can begin a cumulative poisoning effect. Symptoms include excitement, muscle weakness, trembling, staggering, diarrhea, and loss of condition. With early diagnosis, treatment with massive thiamin injections can save the horse.

➢ *See also Poisonous plants*

Hot walker

A mechanical device on which several arms extend out from a rotating center. A horse can be tied by the halter to each arm. When the hot walker is turned on, the horse must walk in circles until he is unhooked. This is a labor-saving device for trainers who lack the barn help to cool out hot horses by hand.

HORSE SENSE

Horse sense means practical common sense and also describes the quality of being in tune with a horse's nature, instinctively handling him appropriately in various situations. Horse sense consists of two basic pieces of knowledge about equines: They have better memories than humans do, and they are prey animals (that is, they will flee from anything they perceive as an attacker). These two facts can help guide you in deciding how to handle a horse.

Humidity concerns

Horses cannot pant as dogs can to cool down. Their only methods for cooling down are through air exchange in the lungs and by radiating heat away through the skin (which then evaporates as sweat). On humid days, sweat does not evaporate well, negating this cooling mechanism. Therefore, horses should be worked less on hot humid days to cause less heat buildup. They should be cooled down by additional means, such as walking in a shady place — especially if there's a breeze. You may also spray or bathe a hot horse with warm to cool water.

The hunt clip is used to increase a horse's comfort in cold weather.

Hunt clip

A clipping pattern for horses that do heavy work and sweat a lot in cold weather. The hunt clip removes most of the hair, leaving the leg hair and a saddle-shaped patch of hair on the back to prevent soreness.

➢ *See also Blanket clip; Clipping; Trace clip*

Hunter

A horse of any breed that is suitable to working as a hunt horse over fences. Hunters must be graceful, well-conformed, and be able to work in a long frame with ground-covering strides since show hunters are judged primarily on their pace, style, and manners, and classic fox hunt mounts are intended to withstand the rigors of the hunt while maintaining gracefulness and style. Today's use of hunters ranges from those shown only on the flat in hunter pleasure classes to show hunters that never see an open field of fences, and culminates with those horses that remain true to tradition in the rare fox hunts held primarily on the East Coast and in England.

Hunter pace

A game that may be played during horse camp or at lesson barns. Hunter pace is also a competitive event, often sponsored by a hunt or a hunter-jumper association. In the game/event a trail — sometimes a few miles long — is marked. There are often two choices: a flat trail and one with obstacles to be jumped. Competitors may choose the trail they think will get their horse, or more usually their team of horses and riders, to the finish line the fastest. Hunter pace is not ridden like a race, with teams of riders on the trail neck-and-neck, trying to win. Rather, each team sets off by itself at a given time. Each team's total time is recorded and the team that negotiates the course fastest wins.

Hunting

Humans have been hunting on horseback for thousands of years, though fox hunting as we know it today is a more recent development. The first mounted hunters were after large prey animals such as deer and boar (animals that could actually be eaten once they were caught). In medieval France and England, the hunt evolved primarily as a sport of the nobility. The French influence is still felt in some hunting terms, such as "Tallyho!" (from the old French *Ty a hillaut*, meaning that the quarry is in view and running).

The first mention in print of chasing foxes for sport appears during the reign of King Henry VIII. By the 17th century, fox hunting had become a popular pastime for gentleman, merchants, and farmers, as well as lords and ladies. Hunting came to America with the first settlers and became firmly established, especially in the eastern United States. There are many historic hunts in England and America that have been running for hundreds of years.

Legend has it that the famous "pink" jackets commonly worn by hunters (which are actually red) are named for the tailor who designed them, rather than the color itself. Evidence for the existence of such a tailor is scarce, but the term persists.

➢ *See also Hunter*

The English country gentleman galloping after a fox — The unspeakable in full pursuit of the uneatable.

—Oscar Wilde, *A Woman of No Importance*

Hyperkalemic periodic paralysis (HYPP)

➢ *See Genetics*

Hyperthermia

An abnormally high body temperature, often caused by exertion on a hot or humid day.

HYPP (hyperkalemic periodic paralysis)

➢ *See Genetics*

i

Icelandic horse

The size of a pony, between 12.3 and 13.2 hands high, this equine is nonetheless considered a horse. There has been no crossing with other breeds for more than 1,000 years, so its inheritance, from the hardy semi-feral Tarpan and the tough Fjord, is intact. Icelandic horses can live outside year-round, even in severe conditions. They have five gaits: walk, trot, and canter, plus the skeid (a pace) and the tolt (a fast running walk).
➢ *See also Canter; Fjord Pony; Pace; Skeid; Tarpan [box]; Tolt*

It is not best to swap horses while crossing the river.

—Abraham Lincoln (1864)

Idle horse

A horse on forced stall rest. The term is sometimes used when referring to the care of such horses. Idle horses require special care because the horse was not intended to be a stall-bound animal. Numerous problems can result from improper care of idle horses; however, overfeeding and feeding inadequate roughage are the main sources of serious complications.

Immune system

Healthy animals have a high degree of natural immunity to infectious diseases, especially to those that have been present in their environments for many years. The weaker animals have already died of those; the strong might have gotten a mild case and survived, carrying on with some gained immunity. Natural selection, then, ensures that primarily the most disease-resistant horses will be the ones to live and breed. Humankind, however, has selectively bred for its own purposes and may well have overridden the immune factors in some cases.

Fortunately, we have also developed vaccinations, providing introduced immunity to many diseases both old and new. Unfortunately, new diseases occasionally crop up. The immune system of any horse is based upon the health of his blood and other bodily serums, the antibodies in the blood, and the condition of his organs, musculaturc, connective tissues, and mucous membranes.

Disease Prevention

The immune system prevents disease when the antibodies in the white blood cells recognize an invader that is harmful and kill it. The immune system can be "taught" to recognize invaders through vaccination.

Warts

Warts are caused by a virus, often spread by contact directly between animals or by handlers who carry the virus from an infected animal to one not yet infected via their hands or grooming implements. Young horses are particularly susceptible to warts because they have not yet developed immunity. They occur in older horses who had not previously been exposed to the virus.

Worms

A horse's immune system fights some intestinal parasites as well as bacteria and viruses; the immune system recognizes the worms as invaders and produces antigens. The horse's counterattack depends primarily on damaging the egg-laying capacity of the female and the male's fertility. Still, parasites can damage or weaken the immune system, thus making the horse more susceptible to disease.

➢ *See also Vaccinations*

Impacted teeth

Teeth that are trapped below and behind teeth that have already emerged. In horses, impacted teeth almost always come through on their own. Some may, however, cause a tooth bump, a bony protrusion on the lower jaw, as they emerge. The bony protrusions eventually smooth out, usually disappearing by the time the horse is between 5 and 7 years old. If there are numerous tooth bumps or if they become rapidly larger and are sensitive to the touch, the horse's dentition might not be sorting itself out. The veterinarian should be called to x-ray the mouth and assess whether removal of any teeth is warranted.

Impaction

Impaction of a horse's intestine is life-threatening, leading to dehydration, colic, and death unless the blockage can be removed or dislodged. Impaction can happen in either the cecum or large colon, or both. When impactions happen in both, they are very difficult to alleviate and can often be corrected only by surgery. Other serious impactions may take several days' treatment and medication to lubricate the impacted material and keep it moving. This is often done at the veterinary clinic, with intravenous administration of medications, liquids, and nutrients to the horse. Here is a list of possible reasons a horse may become impacted:

- Pasture grazing by horses that are unaccustomed to it; such horses returned to grass must be carefully monitored to prevent overeating and possible impaction.
- Indigestible material, such as hay twine, rubber fencing, fibers from rubber tubs or old tires, and hay nets.
- Enteroliths, or intestinal stones.
- Certain feeds, such as beet pulp, which tend to make the manure too firm, leading to impaction.
- Bolting the food without sufficiently chewing it.
- Tooth problems, such as the worn-down teeth of old age, which let food pass into the intestines insufficiently chewed.
- Insufficient drinking water, especially in conjunction with dry feed. This problem must be guarded against in winter, when horses themselves may cut down their water ration because they have not gotten warm enough to be thirsty.

ABOUT IMPACTION

Impaction is the term for the situation when there is no movement of food through a horse's intestine. This causes steady and increasing discomfort, which the horse may show by pawing or lying on his side and not moving. Although the horse may have initial bowel movements emptying the lower part of the intestine, he will soon cease to defecate. Contact your veterinarian immediately if you suspect impaction.

See also Fecal impaction [box]

If your horse lies on his side for an extended period, he may be suffering from impaction, a serious condition that requires prompt medical attention.

Imperforate hymen

A condition where a membrane partition obstructs the opening of the vagina. This condition is common in maiden mares, those who have not been bred before. It is best to have a veterinarian surgically correct this before breeding; if it is uncorrected, the membranes will be torn at breeding, possibly leading to infection.

Imprint training of foals

The concept of imprint training was handed down to us by Native American horsemen, who discovered that foals spoken to *in utero* and handled within the first 24 hours of birth were often easiest to train as mature horses. Today, this technique is used by many breeders, both professionals and amateurs. Imprinting can only occur within the first 24 to 36 hours after birth; beyond that, any changes in temperament are caused by conditioning.

Imprinting can provide a tremendously positive foundation for future work with a horse, or it can literally destroy a horse's future. The impact depends on the education and experience of the person doing the training. Be sure to learn everything you can about imprinting before you try it on your own. Even then, on your first foal it is best to work side-by-side with someone who has had success with imprinting methods. And a crucial point is to be certain to allow a first-time mother to bond with her foal before you attempt to bond with him, to keep her from rejecting him.

When Allah created the horse, he said to the wind, 'I will that a creature proceed from thee. Condense thyself.' And the wind condensed itself, and the result was the horse.

—*King of the Wind* by Marguerite Henry, 1948

Impulsion

Impulsion is the driving force, or thrust, of a horse's forward movement. Most horses have it naturally, as part of their flight response for survival. However, poor training and riding habits can essentially cause a horse to lose his impulsion. Poor conformation can make it difficult for a horse to utilize his drive effectively. Without impulsion, it is nearly impossible to get a horse to do much, in ground training or under saddle. High levels of forward impulsion, when restrained, result in the high-level movements of dressage, such as the **piaffe,** which is basically an in-place high-stepping trot.

➢ *See also Piaffe*

Inbreeding

This is the mating of two closely related animals, and it should be avoided. Inbreeding, in general, increases the probability of undesirable recessive traits being expressed in the offspring. Inbred horses are also usually less hardy. Over time, inbreeding dramatically decreases the genetic variation within a bloodline, concentrating both good and bad genes. If undesirable genes are present, inbreeding offers a greater chance of birth defects.

➢ *See also Genetics*

Incised wounds

➢ *See First aid*

Incisors

A horse's front teeth, which are used for biting off grass.

➢ *See also Dental care; Teeth*

Indirect rein

In Western riding, laying a rein against a horse's neck, as in traditional neck reining. In English riding, the type of pressure placed on the rein used to maintain a bend in a horse's body.

Indoor riding ring

Very popular in cold climates, indoor rings are also useful for extremely hot and sunny days or when nighttime lessons or riding are necessary or desired. Prefabricated buildings are readily available in a variety of sizes, from the size of a small dressage arena to arenas that can accommodate a line of jumps.

Infection, in broodmares

Broodmares should be monitored to determine whether there's an infection; this can be done by an intrauterine culture during the heat period. Breeding the mare while an infection is raging could cause abortion later, even if breeding is successful. Genital tract infections should be treated as soon as they are detected, or they could cause uterine damage and permanent infertility. The mare should also be checked for signs of infection at foaling time, when bacteria can be introduced through torn and abraded tissues.

Mastitis

An infection of the udder relatively common in female mammals, ranging from cats to dogs to cows to humans. It can be brought on by bacteria and by lowered defenses resulting from stress, exhaustion, and cracked nipples; it may be accompanied by fever, fatigue, and colic. Many mares with mastitis will show a sudden unwillingness to have their foals nurse, often resorting to kicking or biting them if the pain is severe enough. Any of these symptoms should be reported to your veterinarian, who may recommend treatment with antibiotics. Warm, wet towels applied to the udder may help alleviate the mare's discomfort and enable her to again accept her foal's nursing.

Infection, in ear

Hearing impairment is rare in horses, but it can be caused by a middle ear infection that punctures the eardrum. Two other ear-area infections, however, may cause great discomfort to the horse, although they don't generally threaten his hearing. One is infection of the guttural pouch below the ear. This sac opens into the Eustachian tube inside the ear, and the pouch may fill with pus if infected, causing chronic nasal discharge. The salivary gland below the ear might also become inflamed from infection, causing pain and swelling and making the horse shy away from having his ear handled. He may also carry his head in an odd position.

Infection, in foals

Because of their immature immune systems, foals are susceptible to a variety of infections. When caught early, bacterial infections are generally treated successfully with antibiotics. Viral infections, which do not

SEEK PROFESSIONAL HELP

If you suspect that your horse is in pain, dedicate yourself to identifying the cause and dealing with it, whether it is illness, infection, impaction, or other condition. Contact your veterinarian sooner, rather than later, if your horse shows signs of pain.

respond to antibiotics, require excellent nursing care to help the foal fight off the infection.

➢ *See also Combined immune deficiency (CID)*

Infectious diseases

Infectious diseases are caused by pathogens: bacteria, fungi, viruses, protozoa, and parasites. Many of the most serious diseases can be prevented by vaccines. The following is a list of major equine infectious diseases (see individual entries for each):

- Botulism
- Equine encephalomyelitis
- Equine infectious anemia (EIA); swamp fever
- Equine influenza
- Equine protozoal myeloencephalitis (EPM)
- Equine viral arteritis (EVA)
- Leptospirosis
- Lyme disease
- Piroplasmosis
- Pneumonia
- Potomac horse fever (PHF)
- Rabies
- Rhinopneumonitis
- Salmonellosis
- Strangles
- Tetanus (lockjaw)
- Vesicular stomatitis (VS)

➢ *See also Diseases, prevention; Parasites; Signs of illness; Vaccinations*

Infertility

The inability to breed or reproduce. Infertility has different causes and treatments in mares and stallions.

Mares

Infertility in mares has many causes but usually is caused by recent infections of the genital tract or by a systemic infection. Body condition also plays a role with mare fertility; if a mare is too fat, she may have unpredictable heat cycles or none at all. If she is too thin, she may have abnormal cycles, or none at all. A mare with hormone problems may also fail to cycle or may cycle irregularly. High-level athletic mares may stop ovulating.

Stallions

Infertility in stallions has many causes but is typically caused by fever or inflammation. Anything that raises the temperature of the testicle for a length of time can interfere with sperm production. The decrease in viable sperm production may not be noticeable right away, as some sperm produced before the injury or illness could still be viable (it takes between 60 and 70 days to produce mature sperm). After any injury or illness, periodic semen checks will reveal whether there is a period of lower fertility or infertility, and breeding can be scheduled accordingly or postponed until sperm production is back to normal.

In front of the leg

Riding with your leg no farther forward than the girth or cinch. As the horse moves, you feel him shift power from behind, under your seat, out in front of you and into your hands. If your horse is not in front of your leg, however, the instant he slackens or quickens his pace, you are likely to become unbalanced in your seat. If he is in front of your leg, variations and even bobbles and trips will be less likely to unseat or upset you; you will be with or just behind his center of gravity, rather than in front of it, where you could tip forward or be whiplashed backward.

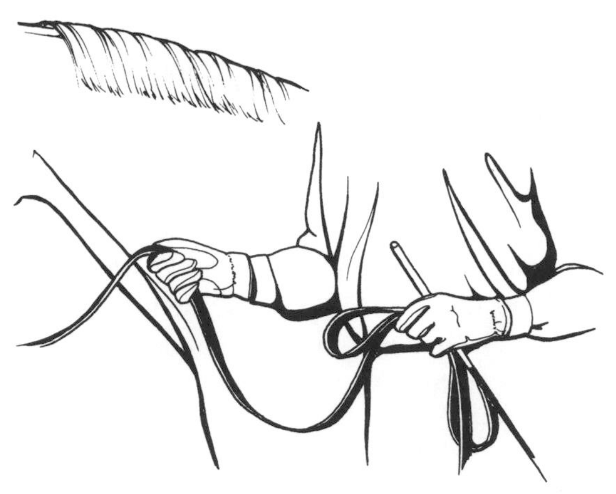

Using the in-hand position while leading gives you more control, especially if you happen to drop the lead rope from one of them.

Infundibulum

An indentation that shows up on the biting surface of the horse's incisors, usually when the horse is around six years old.

In-hand position

Whenever you lead a horse, whether on a lead rope or by the reins, stand on his left side between his head and shoulder and grasp the rope or reins with your right hand 8 to 12 inches below the horse's chin. Hold the excess with your left hand about 18 to 24 inches below that, allowing it to cross your body comfortably with a little slack. If there is extra length of rope or rein, do not wrap it around your hand; instead, loosely coil it first and grasp the whole coil with your fingers always on the outside of all the rope or rein. This position gives you the best control of the horse but also allows you to drop the rope if necessary. It is also easier to catch the horse than to be dragged across a field or arena because your hand was hopelessly attached to the rope or rein of a fleeing horse.

RIGHT OR LEFT?

It is customary to walk on the left side of a horse, because people are more often right-handed and thus have their stronger arm next to the animal. It is prudent, however, to accustom your horse to being led from both sides, in case it is ever necessary to lead from the right.

When leading a horse along a road, always walk in the same direction as the traffic, on the right side of the road, keeping yourself between the vehicles and the animal.

In-hand training

Training done while the handler is guiding the horse from the ground. In-hand training is a concept with definitions as diverse as the horsemen who train horses. Essentially the common ground to all in-hand training is the fact it is training done "in-hand" versus under saddle or while hitched. Most horsemen would agree that all in-hand training essentially begins the day you first handle a foal. It progresses as you train that foal to accept a halter and lead readily at your side. Such training encompasses all of the various good habits we expect our horses to develop for behavior during grooming and tacking up. From that point, the methods used by horsemen vary widely. Some horses are trained on the longe line while others move right into harness or saddle training.

➢ *See also Backing up; Loading horses into trailers; Side step (sidepass); Trailering a horse; Turn on the forehand; Tying*

Injectable supplement

Unless your veterinarian has determined that your horse needs injectable supplements, avoid them. You can cause serious harm or even kill the horse if the supplements are not needed and upset his nutritional balance.

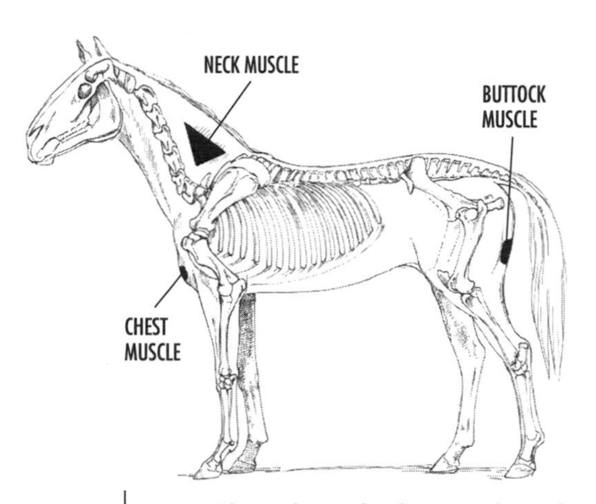

The neck muscle, chest muscle, and buttock muscle are the three primary sites for intramuscular (IM) injections.

Injection

A method of administering medication to horses, identical to the process used in human medicine. The medicine is drawn into a syringe via a hypodermic needle, then injected into the horse. It is a highly efficient mode of administering medicine since you can be confident the full dose was received. The primary types of injections are:

- **Intramuscular (IM),** given in the muscle, usually in the neck, chest, or buttock.
- **Intravenous (IV),** given into the vein, usually in the jugular vein of the neck.
- **Subcutaneous (subcu),** given between the muscle and skin.

➢ *See also Abscess*

ADMINISTERING VACCINATIONS

Many horsemen give their own shots, except a few, such as rabies, that are usually required to be given by a veterinarian. It is more economical to give common injections, such as antibiotics (some of which require multiple doses each day), yourself than to have the veterinarian visit each time a shot is needed, especially if you have a large herd. Be sure to learn under a veterinarian's guidance.

Have your veterinarian instruct you in the proper techniques for intramuscular shots; intravenous shots, which carry greater risk, should be given only by the veterinarian.

In-line trailer

No longer in production, these trailers were made for one or two horses, with the horses facing front. In a two-horse in-line trailer, the first horse had a hay net hung in front of him; the second didn't. The trailers were thought to stay on the road better than wider models, and, in the event of an accident, horses could be led off one after the other, or more easily freed from a trailer on its side by cutting roof or side panels away without a tangle of side-by-side horses struggling with each other to get free. A few of these trailers are still available, used.

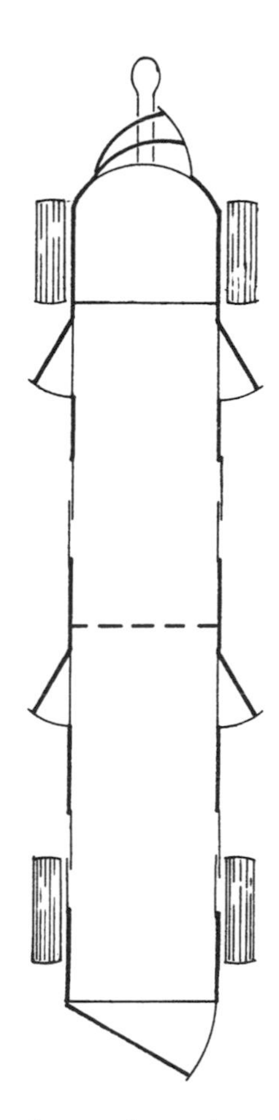

A two-horse in-line trailer may have four escape doors.

Inoculation

➢ *See Infectious diseases; Vaccinations*

Insect

The insects that bother horses most are flies, horseflies, mosquitoes, and gnats. These can be kept at bay, at least partially, by fly repellent sprays.
➢ *See also Flies; Lyme disease; Parasites; Ticks*

Insect control

There are several elements of good insect control in horse barns.

- Clean stalls every day and dispose of manure properly; some farms spread it on the pasture, especially a fallow pasture, to improve it.
- Use insecticidal sprays on horses to keep them comfortable and to encourage the insects to go elsewhere. It is also possible to get whole-barn insect spray systems. Set on timers, the systems release a fine mist of insecticidal spray into the aisles and/or stalls.
- Purchase additional fly attractant devices, if need be, and place them near entrances where the flies will enter the device and be trapped and killed.
- Strategically place "bug zappers," electronic flying insect attractant/killers, if helpful, especially in mosquito- and gnat-infested areas. In such locations, keep boggy land well-drained.
- In any barn, avoid having standing water, whether in natural depressions or unused troughs or equipment.

THE ROLE OF INSTINCT

A horse is born with certain instincts that have helped it survive in the wild over millions of years. It is important to understand and work with, not against, these instincts. These include:

- The instinct to eat, grazing for most of the day
- The instinct to live in a herd
- The instinct to reproduce
- The instinct to flee danger
- The instinct to strike out, if escape is impossible

Inside (leg or rein)

The leg or rein to the inside of the space one is working in, whether an arena, paddock, or contained field. The inside leg or rein would be seen by a person standing in the center of that space. In the open, the inside rein or leg is the one you desire the horse to bend around or move away from.

Instructor

Almost anyone may claim to be a riding instructor. It is not necessary to pass any tests to become a riding instructor in any discipline, although various forms of certification are offered. Hunt-seat instructors who have achieved any level of British Horse Society certification often promote themselves based on that achievement. Dressage instructors who have studied with the great masters of Europe or the United States often advertise that. Riders who were in Pony Club as children and passed some of the club's increasingly difficult tests may list that prominently on their résumés. Still, a person who is a great rider may not be a great instructor; there is a world of difference between knowing how to do something and being able to educate others. When an instructor's students have competed in important regional or national shows or they have their own valuable horses in training with the instructor, the title **trainer** or coach is often used. However, because riding is a sport and most teachers also coach the barn's riding teams, any instructor might use the term trainer.

HOW TO FIND AND EVALUATE AN INSTRUCTOR

Unfortunately, there is no standard licensing requirement for instructors of riding or horsemanship. Many instructors have absolutely no formal training in teaching or horsemanship, so it is in the best interest of novice horsemen and their parents to do some research before hiring an instructor.

- ❏ Ask your friends for referrals, and then observe the recommended instructors during lessons.
- ❏ If you don't know anyone involved with horses, then check with your local 4-H.
- ❏ Attend local equestrian events, such as clinics and horse shows or fairs. Observe how people interact with their horses, then go talk with those you admire. Ask where they learned about horsemanship.
- ❏ Interview the instructor just as you would any prospective employee. Ask for a résumé, credentials, certifications, honors, and so on.
- ❏ Ask about the instructor's safety record. Verify what you are told with the instructor's insurance company.
- ❏ Ask for references from students who are involved at the level you hope to achieve in your first year of riding and beyond (for example, pleasure riding, competitive riding, driving, ground work).
- ❏ Inquire whether the instructor offers a curriculum for horsemanship, in addition to riding, if you'd like to learn more than just how to ride a horse (for example, riding, training, ground work, horse care).
- ❏ Observe the instructor while teaching the type and level of lesson you plan to take.
- ❏ Is the instructor's style of teaching consistent with your preferred style of learning?
- ❏ Does the instructor seem capable of using a variety of teaching methods?
- ❏ Is he or she professional and worthy of your trust?

Once you've located an instructor, evaluate the following:

- ❏ Is the facility a place where you will feel comfortable spending time?
- ❏ Does safety seem to be important? For example, is the facility safe for horses and riders? Are basic safety rules posted and adhered to? Do riders wear the basics of safety dress (safety helmet, boots, form-fitting clothes that won't get caught on a saddle in case of a fall)?
- ❏ Is the equipment safe and in safe condition? Specifically, is the leather relatively clean and well-kept? Signs of cracking or wear at stress points such as the stirrup leathers or billet/cinch straps are cause for concern. Are the stirrups designed to allow your foot to be released easily in case of a fall?

The answers to these questions will help guide you as you select an instructor. Remember that there are many ways to teach horsemanship and riding and many ways to learn it. Always look for a good fit between an instructor's teaching style and your preferred method of learning.

Becoming an Instructor

Despite the fact that many instructors start their careers simply by teaching friends and family, the process of becoming a true teacher is long and complex. Instructors hold the futures of their students in their hands. Students' safety and their lifelong interest in learning depend on an instructor's ability to teach in a professional and effective manner.

Although anyone can "hang out a shingle" and claim to be an instructor, consumers are becoming more educated about what to expect from their riding or horsemanship education. Therefore, increasing numbers of horsemen interested in developing successful primary or sideline careers as instructors are pursuing formal education and certification as instructors.

The main advantages to instructors who attain such education and certification are:

- Network of mentorship and support from colleagues
- More diverse teaching methods
- Improved understanding of individual differences in teaching and learning relationships
- Increased credibility and marketability
- Decreased risk of liability from injuries to students caused by ignorance or negligence

Insurance

There are numerous types of insurance for horses and horsemen; several are described below. Speak with your insurance agent about your specific needs.

- **Liability insurance.** Similar to other forms of liability insurance, this protects you from lawsuits resulting from injury to other persons caused by you or your horse. A wide range of coverage is available; the most vital differentiation is between the personal and professional limits of such policies. If you receive any remuneration for your horse-related activities, you likely need professional liability. Personal liability can often be added to your renter's or homeowner's policy if you are not a professional.
- **Life insurance (for horse).** Known as mortality insurance, this comes in a variety of types ranging from specific perils coverage (covers death caused by limited number of causes) to full mortality coverage (covers death by most natural causes). Most policies require that surgery be performed if it's a lifesaving option, regardless of the cost of surgery. So many horsemen now purchase surgical coverage.
- **Surgical or major medical.** Covers the cost of most medically necessary surgeries and medical treatments.
- **Care, custody, and control.** Insurance for people who care for, or have temporary custody or control of, horses owned by others. Examples of those who benefit from this type of insurance include, but are not limited to, boarding stable operators, trainers, coaches, and people who receive money for hauling horses, whether professionally or as amateurs.
- **Professional errors and omissions.** Protects professionals from malpractice types of lawsuits.

The best horse doesn't always win the race.

—Irish proverb

Intercollegiate Horse Show Association (IHSA)

An organization founded in 1967 at Fairleigh Dickinson University. Its purpose is to promote competition for riders of any skill level, regardless of financial status, in both English and Western equine sport. Students compete individually and as team members at both the regional and national levels. In the year 2000, 285 colleges had teams; some competed in English events, some in Western, and some in both. To eliminate unfair competition by riders who own horses, the only horses used in IHSA events are those owned by the host school; horses and riders are paired by random drawing. The IHSA is a recognized member of the American Horse Shows Association (AHSA); all IHSA horse shows are judged by AHSA-recognized officials.

Interfering

A description of when, during travel forward, a horse hits his foreleg with a hind leg. Interfering can be caused by toed-out conformation or by **base-narrow** conformation — that is, the legs are set too far inside the horse's torso. The problem may result in tripping and can cause injury to the horse.

International Arabian Horse Association (IAHA)

An organization, based in Aurora, Colorado, that registers Arabian, half-Arabian, and Anglo-Arabian horses. It also conducts seminars about the breed, conducts shows and competitions, and offers the chance for owners of Arabians to get involved in the breed through 270 local clubs throughout the United States and Canada. The IAHA offers special programs for young riders, and publishes a magazine devoted to the interests of people who ride, train, or simply enjoy Arabian, half-Arabian, or Anglo-Arabian horses.

➢ *See also Arabian; Anglo-Arab; appendix*

International Equestrian Federation

➢ *See* Fédération Equestre Internationale (FEI)

Interval training

Used for endurance horses, distance horses, and race horses, this is much like human interval training. The training itself consists of anaerobic work, usually a short sprint at a gallop, on the flat or up a grade. Gradually step up the duration and speed of the workouts until your horse reaches to the peak of conditioning that you're looking for.

Intestinal obstructions

➢ *See Impaction*

Intussusception, as cause of colic

Telescoping of the gut; the intestine slides inside itself, much as a telescope would be closed, with the farthest section squeezing into the one behind it, and then both of those into the one behind it, and so on. When this happens to sections of intestine, the gut-inside-a-gut-inside-a-gut may happen enough times to completely close the space that creates a passage for digesting material and manure. In any case, the surface area for absorption of nutrients is reduced. The horse may recover from acute colic with minor treatment but then lose weight rapidly.

➢ *See Colic*

HORSES OF IRELAND

Ireland has been the source of many distinguished horses.

- The Irish Draft was the 19th-century forebear of the famous Irish hunter. The breed, developed over centuries from Anglo-Norman and Andalusian stock, was extremely versatile, capable of pulling a plow, trotting in harness, galloping under saddle, and in particular, jumping with courage and style.
- Breeders introduced Thoroughbred blood to Irish Draft stock to produce the Irish hunter.

After the Irish Famine of 1847, the Irish Draft declined, and unsuccessful efforts were made to improve the stock with imported heavy breeds.

In the 20th century, the Irish Draft made a comeback. They now stand taller, at 16 to 17 hands high, than did their forebears, who stood no more than 15.3 hands high, but they still exhibit the breed's athleticism and amiable disposition.

Iodine, antiseptic

A useful antiseptic to have in the barn, for use on both human and equine cuts and abrasions. It is also used, in mild solution, as a navel dip for foals.

Iodine can be applied to the bottom of the hooves of flat-footed horses so they don't bruise as easily; iodine toughens the tissue somewhat. Apply it only to the sole of the foot; avoiding spilling it down the hoof wall (it will dry out the hoof wall) or onto tender tissues above the hoof, where it may burn skin.

Iodine, supplement

Iodine is necessary in the diet for growth and to regulate metabolism. Only a small amount (as little as 1 mg daily) is needed. Sufficient iodine is included in most commercial feeds. If you still think your horse lacks iodine, a little iodized salt or mineral salt (such as is afforded by an equine salt or mineral block) can supply it. Kelp products also contain iodine. But the danger is usually of supplying too much iodine rather than too little. It's best to check with your veterinarian before adding iodine to a horse's ration.

Iridocyclitis (moon blindness; periodic ophthalmia)

Inflammation of the inside of the eye caused by an abnormal immune response. The triggering disease is widely debated. Early symptoms are watering of the eye, constricted pupils, and photosensitivity. The membranes become red and swollen, and the cornea becomes cloudy and opaque. Pus may form in the lower half of the eye. After a few days or weeks, the eye seems better. Attacks recur, however, and each one causes further damage. Untreated horses eventually become totally blind.

Moon blindness itself is not contagious, although the causative agent might be.

➢ *See also Periodic ophthalmia*

Iris

The pigmented area of the eye surrounding the pupil. The iris is capable of contracting or expanding to regulate the amount of light entering the eye.

Most horses have brown eyes (brown-pigmented irises). A **walleye,** however, is a horse with unpigmented or blue-pigmented irises. Some horses, just like people, have one brown and one blue eye. The color of the iris does not affect the horse's vision.

Iron

A trace mineral necessary to a horse's health. It is usually supplied sufficiently in the diet, although some horses seem to need more than the established requirement. Consult your veterinarian for more information.

➢ *See also Feeding and nutrition; Trace minerals*

HORSES OF ITALY

Italy was famous in the 17th century as a center of fine horse breeding. Today, several breeds are distinctly Italian, including the Maremmano, the Murgase, and the Italian Heavy Draft. The first two types vary somewhat in conformation, with the former being a rather coarse riding horse and the latter a light draft type. Both are honest and hard working, measuring 15 to 16 hands high and of solid colors. Both have been used for military and police purposes, as well as farm work and hauling. The Italian Heavy Draft is a heavy horse, with short legs, a well-muscled body, and a thick, upright neck. Generally chestnut with flaxen mane and tail, this horse is known for its rapid pace, even when pulling a considerable load.

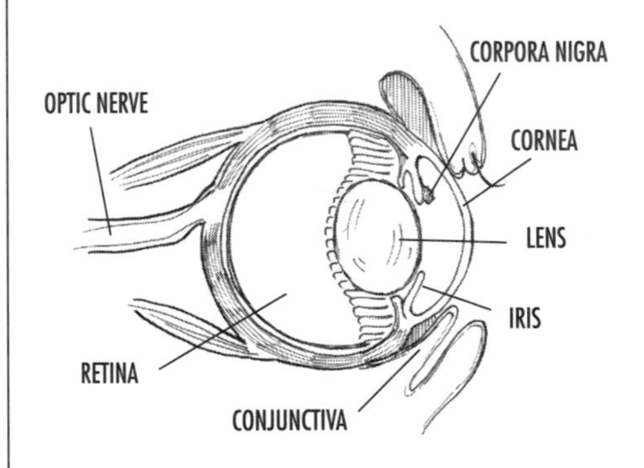

The iris regulates how much light enters the eye.

Irons

Stirrups.

➢ *See also Stirrups*

Isolation and stallions

It is unwise to isolate stallions from other horses completely. Isolation of *any* horse creates stress and unpredictable behavior. They are herd animals by nature. Although stallions cannot be turned out with geldings, whom they will likely injure or kill, or with mares with whom they might breed, stallions need to have other horses within eyesight for their emotional health. You can put a stallion in a pen or paddock next to other horses as long as the fencing is sturdy, and preferably double, between them. A stallion can be housed in the barn next to another horse, but be sure the wall between them is very well-built. With care, stallions need not be isolated, and can be housed in facilities with other stallions, mares, and geldings.

➢ *See also Stallion*

Itching dermatitis

A condition caused by mites, which lay eggs in tunnels in the skin. Infestation can spread from horse to horse through bedding, saddle pads, grooming tools, and clothing. The condition is also known as **barn itch, red mange,** or **sarcoptic mange.**

➢ *See also Mange*

Ivermectin

A drug (which goes by many brand names) that is effective against all internal parasites except tapeworms. It does not poison the parasites like other preparations, but rather paralyzes them by reacting with the d gamma-aminobutyric acid (GABA) in their nervous systems. The only worm ivermectin cannot kill is the tapeworm, as tapeworms do not have gamma-aminobutyric acid in their nervous systems.

IVs for foals

➢ *See Injection*

Half the failures in life arise from pulling in one's horse as he is leaping.

—Julius Hare, *Guesses at Truth* (1827)

Jaw

The jaw is a vital part of the horse's body. Proper conformation allows for healthy eating and efficient training. Horses with jaws that deviate too far from the ideal can have problems, especially with eating. If the upper jaw is long, giving the horse the appearance of a parrot (parrot beaked), he has an overshot jaw. If the upper jaw is too short, it is undershot (monkey mouth). Both conditions might cause trouble eating and, because the traits are inherited, it would be best not to breed these horses.

➢ *See also Overbite; Pulse and respiration; Teeth*

Jennet

A small horse of Spanish breeding.

Jenny

A female donkey.

Jibbah

Term for the characteristic bulge in the forehead of an Arabian horse.

Jimmyweed (rayless goldenrod) (*Isocoma* spp.)

A bushy, unbranched, yellow-flowered perennial common in the southwestern United States. It contains tremetol, a substance that causes muscle tremors, depression, stiff gait, and weakness. It can poison nursing foals if mares eat the shrub.

➢ *See also Poisonous plants*

Jimsonweed *(Datura stramonium)*

One of the many weeds that can poison horses. A close relative of the nightshades, it should be removed from all pastures.

➢ *See also Poisonous plants*

Jockey

A professional rider of a horse in a race; or an older term for any rider or driver of a horse.

Jodhpurs

Pants used in English riding. Jodhpurs come in two styles: hunt seat and saddle seat. Traditionally, those used in hunt seat were cut full above the knee and tight below, a pattern that originated in Jodhpur, a state in northwest India where polo and other equine sports held sway. These days, jodhpurs are usually not much fuller above the knee, if at all, because they are made of stretch fabric. This allows for the movement of the rider, which the extra fabric once accommodated. While these breeches were originally uncuffed for tucking into short or even tall boots, today they are usually finished in a cuff and are worn outside short boots (over which they fit tightly).

Adults may wear jodhpurs for hacking and lessons; young children (usually under twelve) may wear them in shows. The breeches are held

JOCKEY CLUB

The Jockey Club, founded in 1894, registers names of Thoroughbreds produced in the United States. Its mission is improving the breed and performing services for portions of the equine industry that use Thoroughbred horses, especially racing. The Jockey Club publishes the Stud Book, in which the ancestry of all registered Thoroughbreds born in the United States can be found. The organization also works closely with other countries that have significant racing and Thoroughbred industries, including Canada, Mexico, the Caribbean, and some nations in South America.

See appendix

POISONOUS PLANT

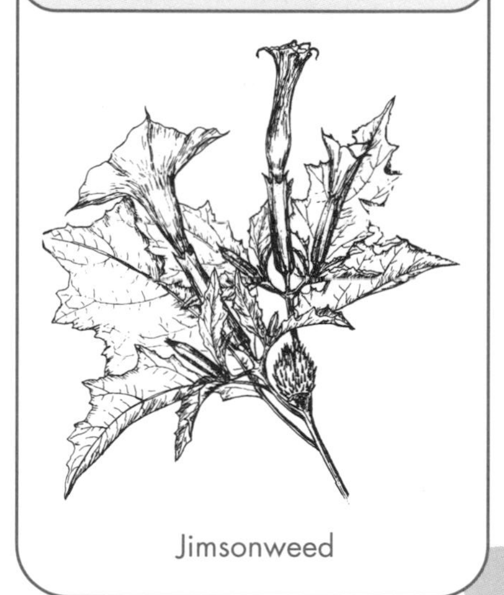

Jimsonweed

down by elastic "stirrup" straps that attach to either side of the cuff and pass below the sole of the shoe at the heel.

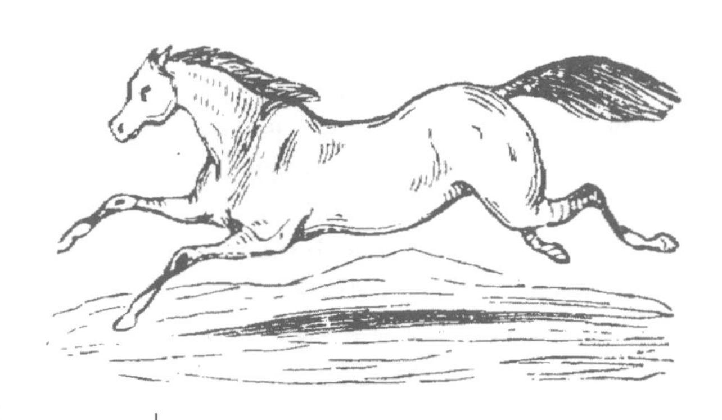

Saddle seat–style jodhpurs are similar to the older style of hunt jodhpurs. They are relatively loose fitting in the hips and thighs to allow freedom of movement to mount and ride. They are more form-fitting around the knee to prevent chafing, but then they flare out at the ankle to allow a smooth fit over the top of the traditional, short jodhpur boots. Saddle seat riders often have jodhpurs for both schooling and competition; the latter are made of much finer fabric that matches or complements the coat and vest worn for showing.

➢ *See also Attire; Breeches*

Jog

The Western term used to describe the two-beat, diagonal gait known as a trot in English disciplines. The jog is considered somewhat slower and softer than the trot because Western horses move with slightly shorter strides and a softer form of impulsion. Riders usually sit the jog rather than posting it the way English riders do while riding the trot.

➢ *See also Trot*

Now the great winds shoreward blow;
Now the salt tides seaward flow;
Now the wild white horses play,
Champ and chafe and toss in the spray.

—Matthew Arnold,
The Forsaken Merman (1852)

Joint capsule

Where two bones meet, there is a joint. The joint contains smooth, slippery cartilage that reduces friction between bones. Tough connective tissue surrounds the two joints and the space between them; it holds synovial fluid, a lubricant that further protects the ends of the bones. Ligaments, which hold these bones together, run into this space. The joint capsule is the entire structure: the ends of the meeting bones, the cartilage, the fluid, and the surrounding tough tissues.

Jointfir (horsetail, scouring rush) (*Equisetum* spp.)

➢ *See Equisetum; Poisonous plants*

Joint ill (navel ill)

An acute disease that settles in joints of foals, damaging the joints or creating permanent lameness if not caught early. The organism involved, streptococcus, can rapidly destroy joint cartilage and bone. There are two causes: entry of bacteria into the navel soon after birth, before it dries; and septicemia, an infection of the blood transmitted from the dam's diseased uterus. Prevention begins with health checks and treatment of uterine infections for the mare, maintaining a clean birthing area, and making sure the foal nurses soon enough to obtain antibodies that would help him fight the bacterial invasion. But the primary prevention strategy is to treat the umbilical stump with mild iodine as soon after birth as possible.

Jousting

A medieval sport patterned on the combat techniques of knights on war-horses. It involves two riders in armor and padding, grasping long poles and riding horses that also wear head, neck, and chest armor. Each combatant gallops toward the opposing rider and attempts to strike him on the chest and knock him off the horse to the ground. These days, jousting is performed in festivals with historic themes, in theme arenas, and at theme parks. It is the state sport of Maryland.

Judging

A rather simple term to describe the very complex process of evaluating classes of competition. Essentially, judges evaluate the classes in front of them according to a set of rules; those horses, handlers, or riders that come closest to meeting the ideals set forth in the rules win honors in the class.

Judges range from unlicensed novices with little experience in center ring to highly honored individuals who hold licenses with numerous organizations and have decades of experience officiating competitions around the world. The licensing process is not uniform throughout the industry, so each licensing organization sets its own standards. Perhaps the most prestigious and respected is the licensing system within the American Horse Shows Association (AHSA). The AHSA educates and licenses judges in more than twenty-five divisions.

Just as the judges in our court system must know the Constitution and case law, horse show judges must know the rules of the organization hosting the competition, as well as current trends. Trends should not dictate the judge's decision; however, it's important for judges to be aware of them.

Jumper classes

Classes that are sometimes part of a hunter-jumper show and sometimes held separately. Jumper classes are judged on time and faults. The fastest clean round wins. Rider and horse must be brave, strong, fit, agile, intelligent, experienced, and aggressive.

Jumping

Jumping is one of the most thrilling activities shared by horse and rider. Here is some basic information.

- Most horses, unless injured, are able to jump.
- Some horses like to jump; some do not.
- Most humans are able to learn to jump horses.
- Jumping itself is simple; the horse does all the work.
- Getting to the jump so that the horse can jump it well while carrying his rider is difficult. This is the rider's job.
- There are five parts to a jump: the approach, the take-off, suspension, the landing, and the get-away.
- The rider manages all but the suspension portion, although she may use that airborne time to encourage the horse to land in a particular way and/or head in a new direction.
- Equipment counts. Jumping is best done in a saddle built for the purpose and with the proper length of stirrup.

JUDGING JUMPING

Hunter-jumpers are judged on pace, style, and manners as they travel around a course of jumps; jumpers are judged on how many elements of the jumps they knock down and how much faster than the predetermined time they traverse the course.

- While horses naturally jump, they must be taught to do it on request in a way that keeps rider and horse happily together; this is not a task for a beginner.

Jumps, types of

Similar jumps are used for hunter-jumper and jumper contests, although there are some that may be used only for jumpers.

Hog's Back

This consists of three elements; front, middle, and back. The center part is higher than the two on the outside, which are of equal height.

Jumps for Jumpers

Only jumpers include diagonal jumps (verticals with one side higher than the other), uneven oxers (one end of the front element is high; the opposite end of the back element is high), fan jumps (obstacles with a spread between oxer poles that is wide at one end and very narrow with all poles gathered together at the others), water jumps (verticals with real

TYPES OF JUMPS

The **vertical** looks easiest for the rider, but it is tricky for the horse's depth perception. On the **hog's back**, the horse may miss the rear element; the rider needs to rock the horse back so he jumps bigger. **Spreads** are tricky, especially for new riders, because you have to encourage the horse to jump both height and width. The rider has to judge and balance between setting up for a vertical or an oxer. Should you jump the horse low and long, or high and with lots of arc? All but the tiniest spreads are best left for intermediate or higher level riders. **Roll tops** are intimidating for both horses and riders. The **oxer** is the easiest jump for the horse to see and negotiate, and should not be intimidating for new riders.

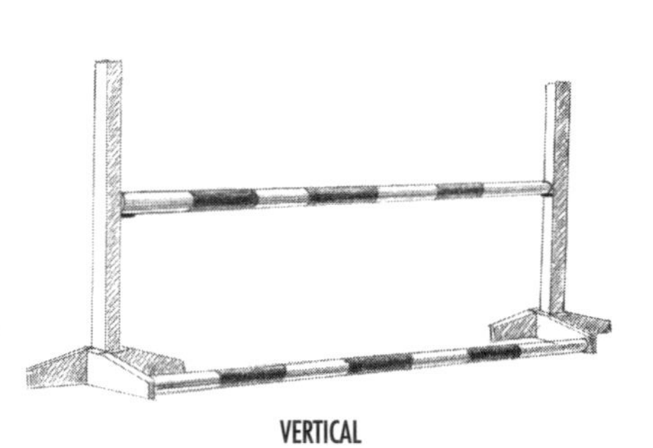
VERTICAL

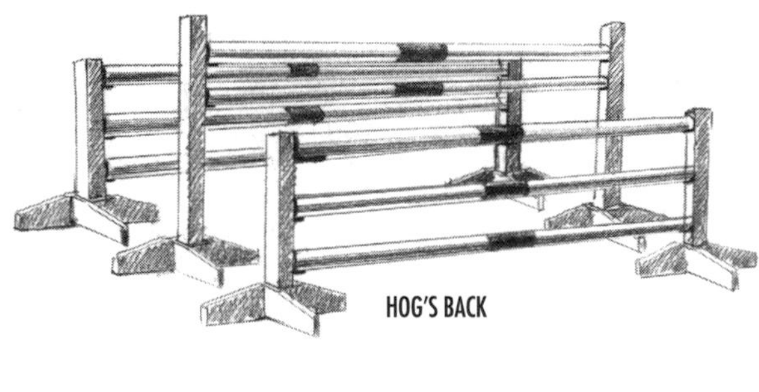
HOG'S BACK

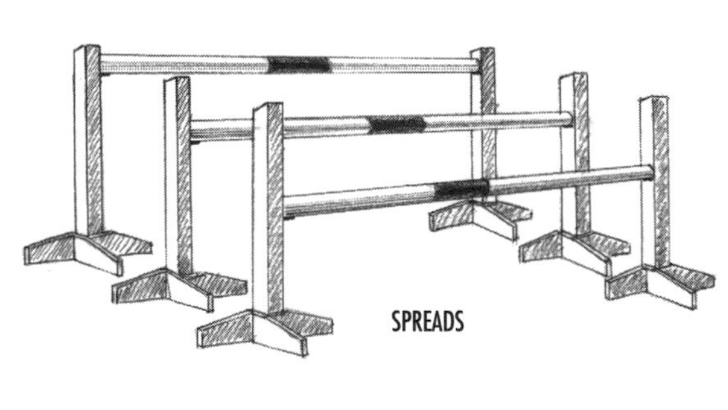
SPREADS

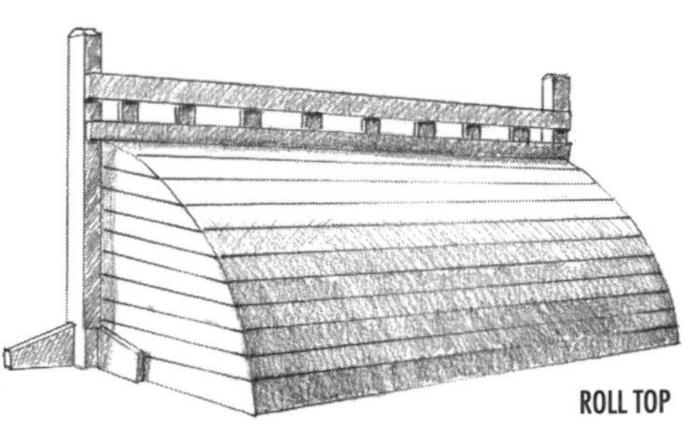
ROLL TOP

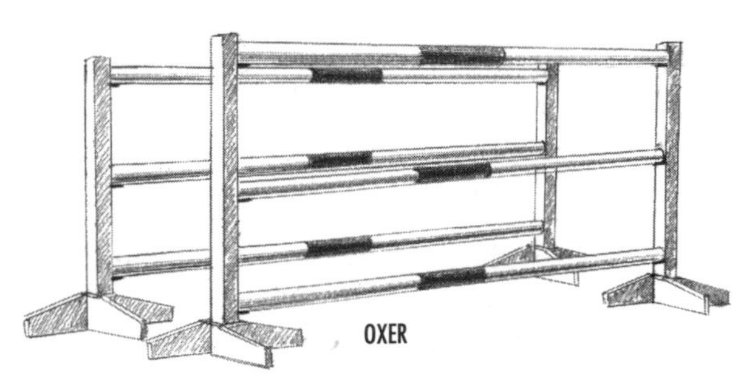
OXER

or simulated pool of water up to 12 feet across on the far side), and narrows (jumps with as little as 5 feet between the uprights).

Oxer

This is usually two verticals set close together, one behind the other, with the height closely matching the depth. Horses seem to like oxers, causing them to give a good, smooth arc over the top.

Roll Top

This solid, curved jump, a few feet in width, is often topped with a pole to raise the height. Sometimes, the jump is bare wood, painted wood, or wooden slats; more often, it is covered with "grass" carpeting and looks like a little hillock. It is usually intimidating to horses, if not to riders.

Spreads

These have greater width than verticals — perhaps as much as two elements the width of their standards apart. Sometimes, spreads are even made of three elements, like a hog's back, but the height ascends from front to back. A spread may also be created by putting two elements that have a predetermined width between two uprights. Often, 55-gallon drums are used, or the jump is made to look like a flat-topped stone or brick wall that is 2 or 3 feet across.

Verticals

A single set of poles or panels of the height being jumped. Verticals are usually the least intimidating for new riders, but they are more difficult for the horse to judge because of their lack of depth. Variations include: panels that look like gates, usually straight across the top (although the riviera may have a downward curve in the middle); brush boxes (narrow containers, not much wider than a jump pole, into which greenery can be stuffed to look like a fence with a hedge); and flower boxes (boxes not much wider than a pole with flowers sticking out; they're usually very low, to look like a garden border).

Junior horse

In showing, a horse that is four or five years old or younger, depending on the breed's definition. The term may also mean a suitable horse for a junior rider: one that is sensible, calm, well-trained for the discipline, kind (that is, will take action to save his rider), and suitable for the child.

Junior rider

A rider who is 18 years old (or 17 years old by some rules) or younger.

Jutland

A breed of heavy horse developed in Denmark, descended from medieval war horses. Jutlands are 15 to 16 hands high, mostly chestnut with flaxen mane and tail and feathered legs.

JUMPING SAFELY

- Riders should always wear safety helmets when jumping.
- For both educational and safety reasons, novice jumpers should jump only very experienced jumping horses.
- Good intermediate and advanced riders may want to ride some relatively inexperienced jumping horses to refine their own skills and increase their knowledge of how horses operate.

Note: Practice should be done on calm horses with a qualified instructor present.

HOW HIGH, HOW FAR?

The record for a high jump has been held for over 50 years by Chilean Thoroughbred Huaso, ridden by Captain Alberto Larraguibel Morales. In an official FEI competition, the pair cleared 8 feet, 1¼ inches (2.47 m) on their third attempt.

The longest official jump, 27 feet, 6¾ inches (8.4 m), was made by Something and André Ferreira at the 1975 Rand Show in Johannesburg.

(from *The Guiness Book of World Records*)

k

Kabardin

A breed of horses native to the northern Caucasus, valued for toughness, sturdiness, agility, and endurance. Kabardin horses make excellent jumpers. They are generally bay, dark bay, and black, standing about 15 hands high.

Kaimanawa horse

A breed of feral horse living in New Zealand.

Kentucky Saddler

Former name for the American Saddlebred.

➢ *See American Saddlebred*

Kentucky wire

Another name for diamond wire mesh fencing material. Horses can't chew it, it's relatively maintenance free, and best of all it's nearly impossible for a horse to get a leg caught in it. Common in Kentucky on big Thoroughbred breeding farms, hence the name.

Kevlar

This material, used in bulletproof vests, is also used in high-quality shipping boots. It will withstand bumps and knocks from other hooves without the need for additional padding.

➢ *See also Leg wrap; Shipping boots and shipping wraps*

Keyhole race

Both a game/exercise for your horse and a timed event at many Western shows. A keyhole race requires a horse to gallop down a narrow lane marked with lime (ground limestone), turn around in a circle at the end of it, and gallop back. If he puts a foot outside the boundaries, he is disqualified.

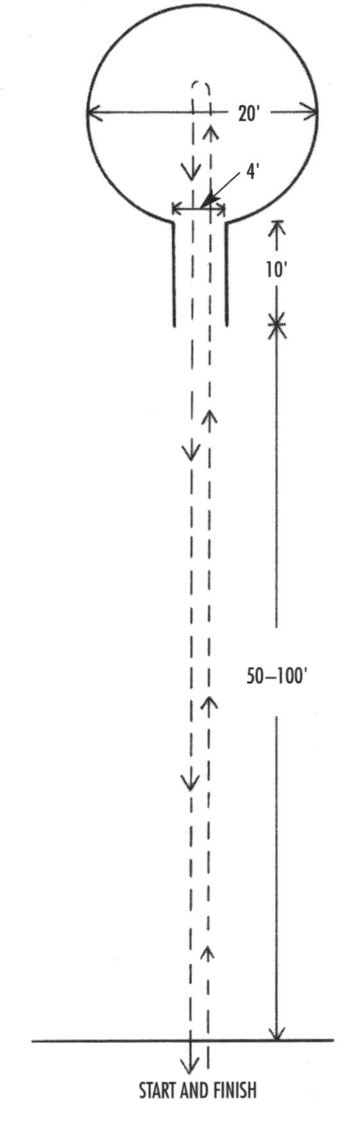

Riding between the lines, and quickly, is the key to winning a keyhole race; figures such as this one are marked out on the ground with lime.

Ki

➢ *See Chi*

Kicking

Kicking is one of the horse's only natural defenses; it is an inborn survival instinct. For human safety, therefore, it is vital that we understand and accept kicking as a fact of life with horses. If we are to coexist peacefully with our equine friends, we must learn about their instincts so that we can handle them in a manner that establishes dominance without inciting a defensive reaction such as kicking.

Kick wall

A horse's kick is powerful enough to go through many common building materials. As horses are unpredictable, it pays to construct walls between stalls of materials that will withstand hoof pressure. Make the walls at least 4 feet high. Some good materials are:

- 1½-inch tongue-and-groove lumber
- Full thickness (2-inch) rough-sawn lumber
- ¾-inch plywood covered with sheet metal

Whatever barrier you choose, ¾-inch rubber mats make excellent lining for stall walls. Besides cushioning the blow to the wall, they cushion and protect the horse's legs and feet, as well.

Kidney colic

Common wisdom holds that in horses there's no such thing. When a horse is experiencing intestinal-tract distress severe enough to become colic, he may act as if he is trying to urinate but can't. That stance gave rise to the term "kidney colic" among horsemen. Veterinarians went along with the phrase for years and simply treated the horse for what really was wrong with him.

Kimberwicke, Kimblewick bit

A bit designed to provide more control than a snaffle through the application of leverage. It can have either a solid or jointed mouthpiece but has two slots for rein placement as well as hooks for a curb chain. Placement of the rein in the upper slot causes the bit to work like a snaffle, with direct pressure only on the bars and tongue. Placement of the rein in the lower slot creates leverage that tightens the curb chain and adds pressure to the chin groove and poll. Horses worked in these bits often develop the bad habit of working behind the bit since there is no true release of pressure when using the leverage slot. It is also known as the *Spanish jumping bit.*

➢ *See also Double bridle; Pelham bit*

Kinesthetic reception

This is a cousin to kinesthetic intelligence, the capacity to use your whole body to express ideas and feelings (as actors, athletes, dancers, and mimes do), including the facility to use your hands to create or transform things (as painters, mechanics, and surgeons do). These same abilities help some people learn to ride. An instructor will notice that a student who learns through kinesthetic reception will be very coordinated; will have good balance, dexterity, flexibility, muscle strength, and speed; and will have a sensitive touch.

Klamath weed (St.-John's-wort, goatweed) *(Hypericum perforatum)*

Also called goatweed, this plant causes photosensitization and severe lesions in light-colored horses or unpigmented areas of a horse's body.

➢ *See also Goatweed; Photosensitization; Poisonous plants*

A horse! a horse! my kingdom for a horse!

—William Shakespeare, *King Richard III*, V, 4

POISONOUS PLANT

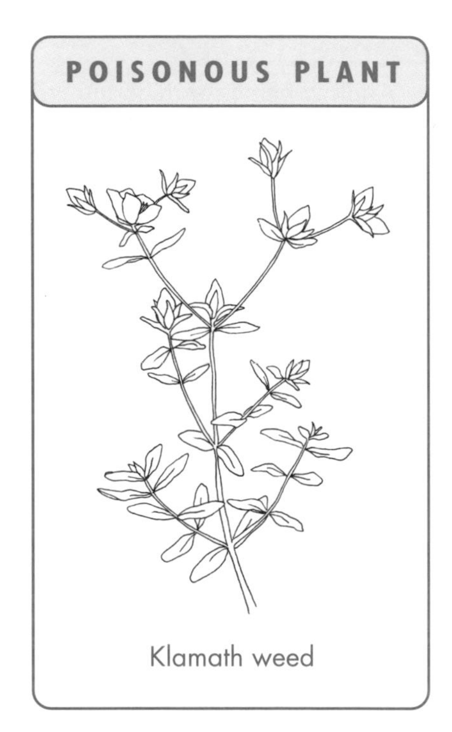

Klamath weed

Klebsiella bacterium

A bacterium harbored in wood products, such as wood shavings and chips. It can be very harmful to foals; use other forms of bedding in the stall of a mare ready to foal.

➢ *See also Foaling*

Knabstrup

A breed of spotted horse, developed in Denmark but of Spanish ancestry. Because of their showy markings, toughness, intelligence, and gentleness, Knabstrups have often been featured in circus performances. They are white with dark spots all over the body, sparse mane and tail, and mottled lips and muzzle.

Knee

Comparable to the human wrist; carpal bones compose this joint.

➢ *See also Anatomy of the horse*

Knee roll

The padded front section of the underflap on an English saddle, meant to cushion the rider's knee.

Knock-kneed

Conformation fault in which each front knee is closer to the other than the upper and lower parts of the leg are; puts strain on tendons in the leg and on the knee itself. It often causes mobility problems. Knock-kneed horses should not be bred.

➢ *See also Back at the knee; Bench-kneed; Over at the knee (buck knees); Tied-in knees or hocks*

Knots

Use only a quick-release knot when tying a horse. If the horse cannot break free from an object when panicked, he might take it with him, doing irreparable damage to himself in the process. *Caution:* Never tie a horse to anything by the reins attached to a bridle and bit; if he panicked it is likely the damage to his mouth would be severe or irreparable. Most people already know a slipknot, which is used by many horsemen. Below is an illustration of a quick-release knot you might choose.

➢ *See also Tying*

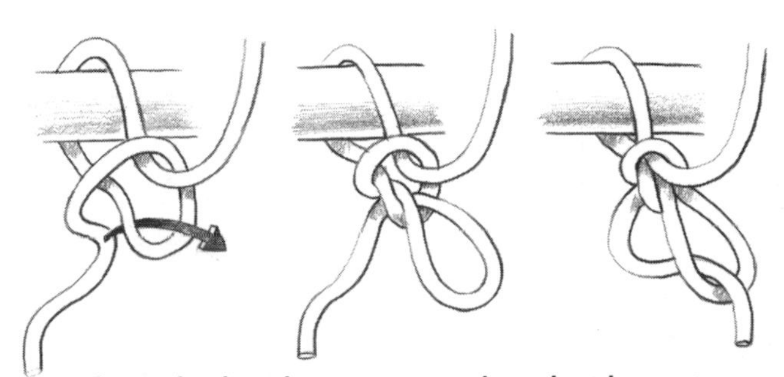

This quick-release knot is easy to make and quick to untie.

1

Laceration

A cut with torn, jagged edges. Healing may be slow because the irregular edges may not come together well, and there is often a great deal of soft-tissue damage at the site of the wound. Clean lacerations with cold, running water, which will help remove dirt and debris so you can see the extent of the injury. The cold water may also help numb any pain. A washcloth or paper towel used gently can also help clean the wound of foreign matter; clean until pink flesh appears. If you apply antiseptic or wound dressing, use a gentle, noncaustic preparation to avoid damaging the exposed tender tissues.

➢ *See also First aid*

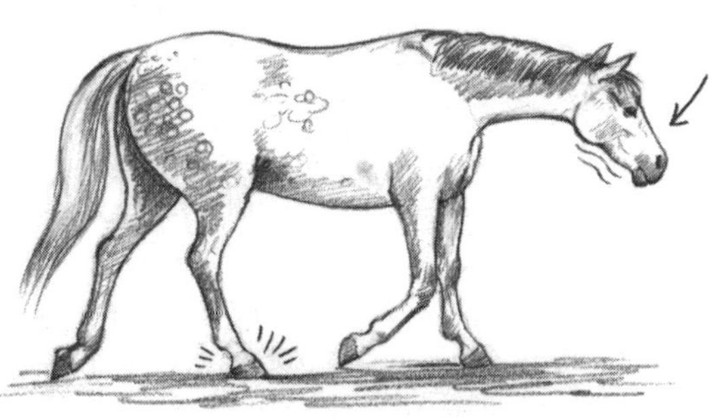

A lame front leg or foot causes the horse to lift his head *(top)*, trying to take the weight off the painful leg. A lame hind leg or foot causes the horse to put more weight on his front legs *(bottom)*, lowering his head.

Lameness

Pain that impedes locomotion. The condition may be caused by pain in a foot or leg that is caused by injury (a kick, for example) or infection (such as an abscess in the foot). It may also be caused by muscle soreness, a pulled tendon or strained ligament in the leg or even above the leg in the large muscles of the chest and hindquarters. Lameness can be caused by a bruise to the sole of the foot, or by a stone caught against the sole; in the latter case, removing the stone before it can do further damage may cure the lameness. A few of the many other causes might be arthritis, navicular disease, or inflammation of the coffin bone after a long ride over rocky ground.

To diagnose lameness, watch the horse as he walks or trots. If he has pain in a front leg, he will bob his head more than normal as he lifts his weight off the painful foot and places more weight on the good one. A painful hind foot or leg will create an uneven stride as he favors the problem side. This is often referred to as a hitch in the gait, a sort of stutter in motion before a leg is put on the ground.

A lame horse cannot be ridden. If you are riding a horse that suddenly exhibits signs of lameness, immediately dismount. Examine each foot for a stone, stick, or other foreign object. If you spot a nail in the foot, a veterinarian or farrier should remove it so that he or she can know exactly where the puncture wound is.

A veterinarian should treat a horse that is chronically lame. Medication or special shoes, such as egg-bar shoes, which are designed to take the weight off the heel, may be recommended to treat the condition.

Regular care from a farrier will keep your horse's feet and hooves in top condition. To minimize chances of a puncture wound, avoid riding in areas where your horse might be exposed to loose nails or sharp objects. Most importantly, if you observe signs of lameness, have your veterinarian or farrier examine your horse's feet as soon as possible.

➢ *See also Egg-bar shoe; Farrier; Hoof; Navicular disease (navicular syndrome); Shoes, horse*

Laminitis

An inflammation of the sensitive inner tissues of the hoof wall. The precondition of founder, this serious and potentially fatal condition is usually caused when an overheated horse eats too much grain or drinks too much water. Call your veterinarian immediately if you suspect laminitis. Prompt treatment can relieve the condition.

Other causes of laminitis include too much lush pasture in spring, sudden excessive weight gain, infections, complications arising after foaling, and the trauma caused by the hooves pounding against a hard surface.

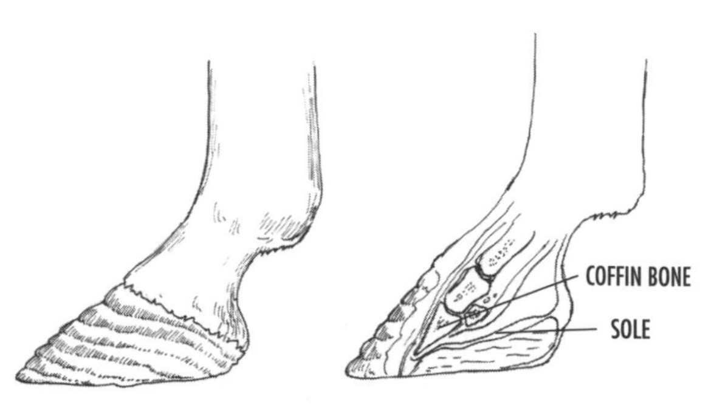

A foundered hoof with characteristic rings and ridges and a dropped coffin bone.

Laminitis begins with interference of the flow of blood to the feet. This results in a lack of oxygen and swelling of the **laminae,** the soft and sensitive vascular plates lining the inside of the hoof wall. The front feet are more likely to be affected than the hind feet.

If your horse has laminitis, he may stand with his front legs extended farther out than usual to take the weight off. If all four feet are affected, lying down may be the only way for him to relieve the pain. Fever, sweating and/or shivering, rapid pulse and breathing, and diarrhea are other symptoms. The feet may also feel hot to the touch.

Without treatment, the coffin bone of the foot will rotate downward, which means the horse has **foundered.** Founder is the term for the structural change that occurs in the foot as a result of laminitis. This stage is also known as **chronic laminitis.** Many horsepeople, however, use the terms founder and laminitis interchangeably.

Proper management will help prevent founder. New feeds should be introduced gradually over the period of a few weeks. A horse that drinks too much water or eats grain before he has completely cooled down after exercise is especially at risk of developing laminitis. Allow your horse to drink only a few sips of water until he has stopped sweating and his chest is not hot. He should be completely cool at least one hour before eating grain, although he can graze or eat a small amount of hay before then. Never allow a horse to become exhausted during a workout.

➢ *See also Cooling down (after exercise); Foaling; Foot care; Hoof*

Lampas

A swelling of the hard palate just behind the horse's top front teeth. This condition may occur in a young horse that has bruised the roof of his mouth after eating hard grain or pellets for the first time. Because it is painful, it may cause the horse to stop eating temporarily. The best solution is to allow the horse to eat pasture grass or to soak the grain in water to soften it until the condition is resolved.

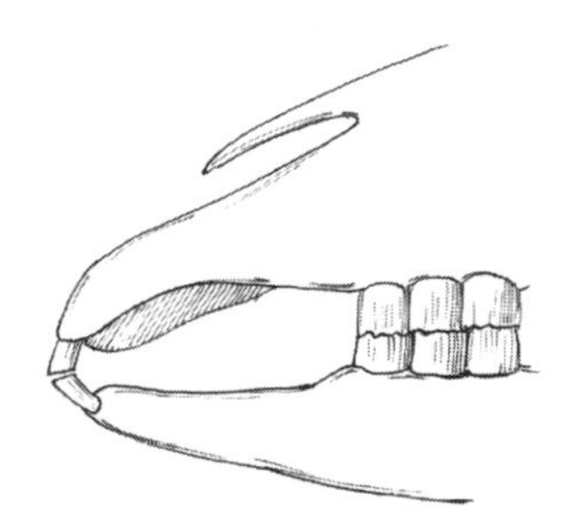

Lampas, a swelling of the roof of the mouth, can interfere with a horse's eating.

Lantana (*Lantana* spp.)

An ornamental flower, common in the Southwest, that is toxic to horses.

➢ *See also Poisonous plants*

BREED CLOSE-UP

The Landais Pony is an ancient breed, probably descended from the Tarpan, but with Arabian and Welsh blood mixed in. It is one of the indigenous ponies native to France, the others being the Ariègeois and the Pottok. The pony stands between 11.3 and 13.1 hands high, is ideal for riding, and is adaptable, hardy, and intelligent.

Large intestine

The portion of the alimentary canal, in particular the large colon and cecum, where intestinal microbes break down roughage into energy-producing fatty acids.

➢ *See also Digestion*

Large strongyles

Bloodworms, or blood-sucking parasites that lodge in blood vessels and cause anemia and debility. There are some 54 varieties of strongyles (worms), both large and small, that affect horses. Three of the large strongyle species do the most damage, causing loss of appetite, anemia, emaciation, rough hair coat, sunken eyes, digestive disturbances and a "tucked up" appearance. Often, the horse will exhibit an uncoordinated hind end. This level of damage happens with severe infestations; major infestations are most likely in horses grazed on permanent pasture where harrowing does not break up the manure piles and expose the light-sensitive eggs, interrupting the parasite's breeding cycle.

➢ *See also Bloodworms; Deworming; Parasites*

Lariat

A long leather or hemp rope used with a noose, particularly from horseback, to catch cattle and other horses.

Larkspur (*Delphinium* spp.)

A species of spired, brightly colored annual flowers (both cultivated and wild) that are mildly toxic to horses.

➢ *See also Poisonous plants*

Latches and locks

Horse-proof latches are a crucial element of a safe stable or a reliable fencing system. One of the best types for a turnout area is mounted on top of the gate. A chain secured on a peg or hook will also outwit a horse.

Inside the barn, the most popular type of stable door latch is a sliding bolt. If your horse has figured this one out, try a hook with a flange that you must compress before lifting it out of the eye. And be sure to latch your grain bin and/or grain room door. A horse that wanders in and discovers a supply of grain is capable of overeating to the point of death.

Lateral exercise

Exercise in which the horse moves on two tracks.

➢ *See also Half-pass; Leg yield; Shoulder-in; Side step (sidepass); Two-track (half-pass)*

Latigo

A long leather strip on the saddletree of a Western saddle, used to adjust the cinch. The latigo should be checked frequently for signs of wear and cleaned regularly with a wax-based leather conditioner.

➢ *See also Saddle; Western tack*

LASSO OR LARIAT?

In parts of Texas, the lariat was traditionally made of manila and called a *riata* or a *lasso.* Before the saddle horn was invented in the 1800s the end of the rope was sometimes tied to the horse's tail. The Californian lariat was twice as long (up to 60 feet), made of rawhide, and thrown in a loop.

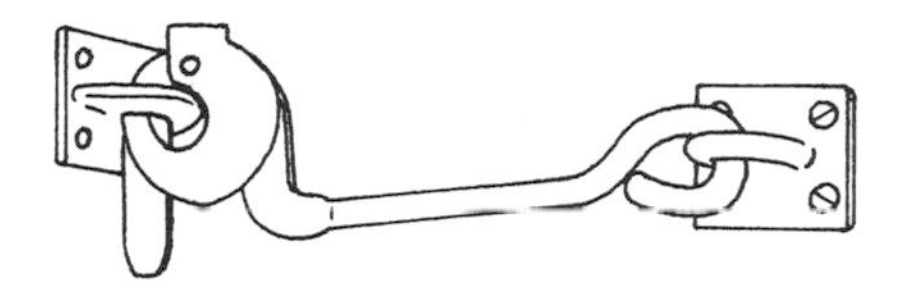

A hook with a flange will keep your stable door securely closed, as long as you remember to latch it.

Lawn clippings

Lawn clippings should never be fed to horses because the horse is likely to overeat and can develop colic. In addition, lawn clippings that remain in a pile will ferment within a few hours and can develop mold and other toxins.

Lead, when cantering

When a horse canters, one front leg leads the stride: that is, it strikes the ground farther ahead of the other front leg and the opposite rear leg. Horses are trained to lead with the inside leg, which helps them stay balanced as they canter in a circle. It is possible for a horse to be on one lead in front and another behind. This is called **cross-cantering** and is usually fairly obvious to the rider because she is jolted all over the place.

A horse can also (and often does) pick up the wrong lead, even when cantering in a circle. He might do this because he has a favorite lead (most horses do) or because he is feeling stiff or sore on one side. The rider may not have given sufficiently clear aids. To the beginning rider, simply getting the horse to canter is an achievement, but as the rider improves, she learns to obtain the correct lead. Most beginners must actually lean forward slightly to see which front leg is hitting the ground first, but eventually, they learn to tell by glancing at the shoulder. Advanced riders are able to tell by the way their hips move with the horse. If he's on the correct lead, the rider's outside hip will turn slightly to the inside. On the incorrect lead, the rider's inside hip will be pushed to the outside.

➢ *See also Canter; Gait; Flying lead change; Simple lead change*

Leading

One of the first tasks a foal learns is to accept a halter and follow a lead rope. All horses should know how to walk next to a handler without crowding or forging ahead. Nearly all horses are trained with the handler

CHECKING FOR THE CORRECT LEAD

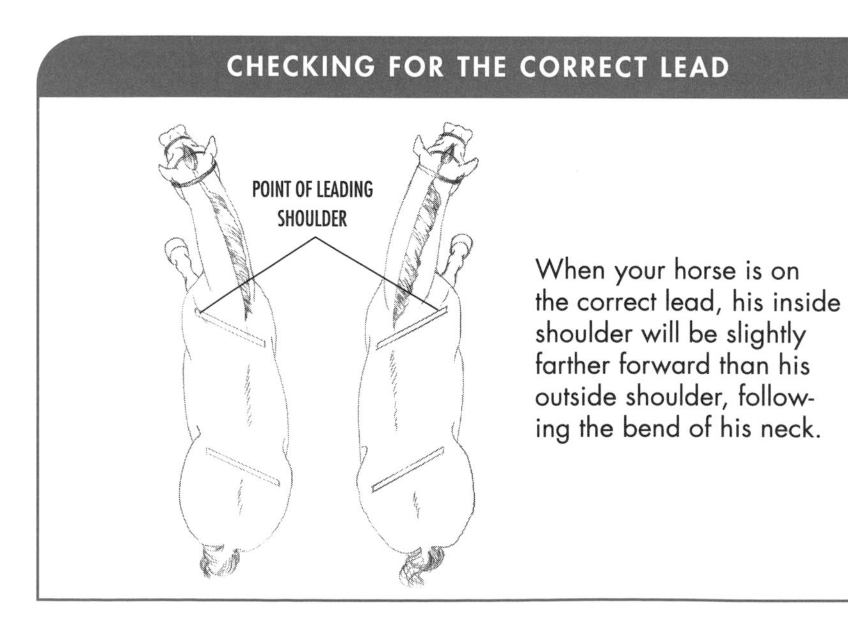

When your horse is on the correct lead, his inside shoulder will be slightly farther forward than his outside shoulder, following the bend of his neck.

I will not change my horse with any that treads but on four pasterns. When I bestride him, I soar, I am a hawk. He trots the air. The earth sings when he touches it.

—William Shakespeare, *Henry V*

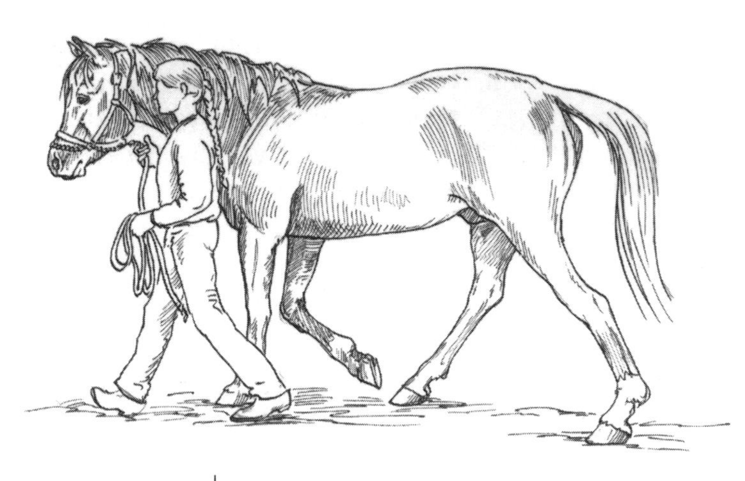
Proper leading technique

standing on the horse's left side. The handler holds the lead rope or reins near the snap in her right hand, with the slack gathered in her left hand. However, you should accustom your horse to being led from the right side as well, since situations may arise in which you'll need to be on his other side.

To lead a horse, stand near his head where he can see you. Step out briskly, saying "walk on" or whatever other vocal cue you use when you want him to move forward. He should immediately imitate your action. If he doesn't, don't get in front of him and start pulling on his head. Horses react to pressure by moving away from it, so leaning on the lead rope is counterproductive. If the horse doesn't step off right away, try shifting his head slightly toward or away from you so that he's not moving in a straight line. If he is always stubborn about stepping out, you might want to have a few leading lessons with a crop to reinforce the concept. A smart tap on the shoulder after you've gotten no response should get the idea across.

You might also need to carry a crop with a horse that crowds, or pushes into your space when you're leading him. Letting him walk right into the butt end of the crop a few times should discourage that behavior. You can also use your elbow. If he drags you along behind him, however, holding the whip in front of him might make him slow down. In sending any signals, however, be sure to do so without putting a lot of pressure on the horse's head or shoulder. Instead of moving away, he'll be more likely to just lean back.

TEAM LEADERS

In a four-horse team, the two horses in front are called the leaders. The one on the left is called the near leader, and the one on the right is the off leader.

FOLLOW THE LEADER

Keep these guidelines in mind when leading your horse:

- *Never, ever* wrap a lead rope around any part of your body. Hold the slack folded together in one hand, not coiled. You could be seriously injured or even dragged if the horse should bolt and you become caught in the rope.
- Always stay to the left side of the horse. If you're in front of him you can't see what he's about to do, which might include running over you because he just saw something scary.
- Always use a lead rope; it gives you a lot more control. If you're just holding your horse by the halter and he suddenly yanks away from you, he's gone. If you've got a lead rope, he should be no more than six or eight feet away, because you're still grasping the end of the rope.
- If you are leading the horse by the reins, bring them over his head and use them like a lead rope. You'll be much more likely to hang on to them if he pulls away than if they were still around his neck.

Your goal should be to have the horse respond to your voice commands more than to any signals from the lead rope or whip. This is especially important when it comes to stopping; you're obviously much safer with a horse that will stop when you say "whoa" than with one that just plows ahead. Practice with your horse in a safely enclosed space until you can get him to walk on, halt, and even trot on a long line (moving straight ahead, as well as longeing).

➢ *See also Handling horses; Safety; Shying, spooking*

IN AN EMERGENCY

You may find yourself needing to lead your horse somewhere without a lead rope handy. In a pinch, a length of baling twine, a leash, or a belt will do. If the horse doesn't have a halter, looping any kind of rope around his neck and holding it close under his throat should do the trick. Never tie anything around his neck, however. And unless your horse is trained to follow when you tug his mane, this isn't a good idea; he'll probably resist and learn that he doesn't have to do what you want.

To make a right turn with a leading rein, this rider moves her hand to the right. A leading rein is generally used as a schooling aid; it is not often seen in the show ring.

Leading rein

A rein that tells the horse which way to go by gently moving his head in the correct direction. The rider applies a leading rein by moving her hand laterally away from the horse's neck in the direction she wants him to turn; the rein should not be pulled upward or downward. The opposite hand continues to follow the horse's mouth without adding more tension.

➢ *See also Aids; Direct rein; Indirect rein*

Lead-line class

Class in which a parent, instructor, or older friend leads a pony or horse while the young rider follows the judge's directions. Many youngsters enter their first shows in lead-line classes.

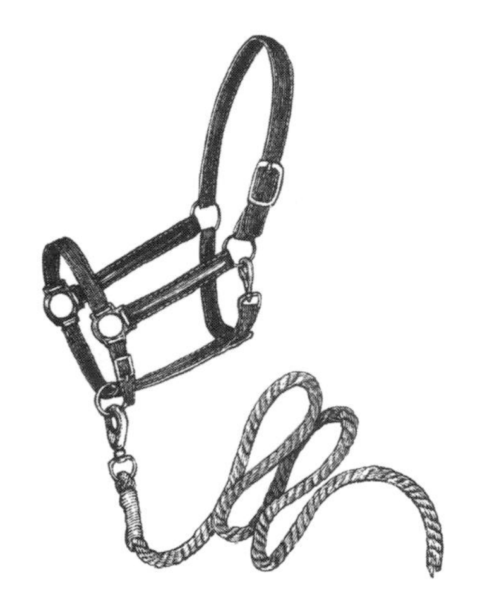

Halter and lead rope

Lead poisoning

Preventing lead poisoning is far easier than treating it. Lead accumulates in the tissues over time and a fatal dose can be ingested from licking at peeling paint (lead has a sweet taste that animals and children may find irresistible). Old batteries are another common source of lead poisoning. Keep your facilities clean and free of clutter, and make sure your horse never has access to dumps or other unsafe places.

Lead rope

A length of cotton or nylon rope with a snap at one end that fastens to the halter.

➢ *See also Bull snap; Halter; Panic snap*

Lead shank

A length of nylon or leather with a chain at one end that can be adjusted over, under, or around the nose for greater control. The chain is commonly used when handling stallions or other fractious horses.

➢ *See also Bull snap; Halter; Panic snap*

Lead shank

Learning, by horses

Two concepts are paramount in order for horses to learn: first, they must be able to trust their teacher/rider; and second, they learn by being rewarded for responding to a request. They also learn not to do things by being corrected swiftly, but the fewer missteps one takes with a horse, the less this negative reinforcement will come into play. Negative reinforcement, if overused, can easily make a horse sour or make his behavior worse.

Horses also must be able to connect correction to the misdeed; if they are punished for biting more than a few seconds after the bite, they have no idea why they are being punished. This leads to mistrust of the trainer. On the other hand, if the horse is asked to move forward and he moves, but the leg pressure asking him to do so is not removed, the horse assumes that there is no reward for good behavior. Without rewards, he will become resistant to the trainer or rider's requests very quickly.

A LEAN LOOK

"Lean" is an old horseman's term for a fine-skinned quality where veins and muscles show clearly.

Learning, by riders

Some riders learn best by seeing, others by hearing, still others by doing. Some people use a combination of all three methods. Students will generally seek an instructor whose main teaching style accords with their own learning style. Good teachers use all three styles about equally so that all students in a class will be served.

Students' ability to learn and the ways they learn are also affected by age, physical condition, social expectations and conditioning, fear and nervousness, learning disabilities and behavioral problems, preconceptions, ill-fitting clothing or improper riding gear, fatigue, illness, stress, and prior experiences with horses.

Leasing vs. buying

When riders decide they need a horse of their own, they often think immediately of buying a horse. There is also the possibility of leasing, however; for beginning buyers, leasing can be a good idea.

The pros of buying are:

- You don't have to worry about reporting to the horse's owner and riding it in the ways specified in your lease contract.
- You have an asset, if you'd like to resell later when you want a different sort of horse.
- You can bond with the animal to your heart's content.

The cons of buying are:

- After the purchase price, you still have to deal with lifelong feed costs or boarding fees, veterinary fees, and tack and equipment fees.

- It is your responsibility to care for the animal for the rest of his life.
- Your horse may develop a physical problem that keeps him from being resold, or you may discover that you don't like him after all.
- You may have to sell for one reason or another and you could take both a financial and emotional loss.
- If you end up with an aged horse that can no longer be ridden, you will have to continue to care for him.

The pros of leasing are:

- You can get the horse you need or want without the initial purchase expense.
- Often, the owner will include tack and pay for ordinary veterinary care, such as shots.
- In some cases, the owner will share the board fees.
- By making renewable short-term lease agreements, you can move on easily to a different horse.
- You do not have the responsibility of housing and caring for an aging horse.

The cons of leasing are:

- The owner may impose restrictions on your use of the horse.
- The owner may decide to withdraw the horse from leasing just when you are getting the most out of the arrangement.
- Your bonding with the horse may be limited.
- The pride of leasing is not the same as the pride of ownership.
- You may, depending on the lease, have to pay some veterinary bills you didn't expect for a horse you don't even own.

A CAROUSEL OF HORSES

In some areas, barns offer horse-sharing programs. For a set fee and number of hours per month, riders can use any one horse from a pool of horses suitable for their riding level. The barn keeps track of how many hours each horse in the pool has been ridden; a horse that has logged his allotted hours is pulled from the pool temporarily, but the others may be used.

One advantage of this system is that riders can experience a number of different horses without having to take a lesson to do it. Another advantage is knowing how many hours a week you'll be able to ride and exactly how much that will cost. Another plus is that a privately owned horse may get sick or go lame, and then the owner has nothing to ride. If you are a member of a horse pool, it is very likely that if one horse has a problem, another will be available for your allotted riding time.

Legal aspects of horse ownership

Horse owners have legal obligations to their horses, neighbors, other horse owners, visitors to their property, and others within sight or sound of their property. Owners also have legal obligations concerning confining their livestock to their land. Personal liability is a factor that applies to horse owners, whether they house their animals on their own farm or ranch or board them at a commercial stable or friend's farm or ranch. Liability for injuries a horse might cause to riders and nonriders is often available as part of a homeowner's policy that covers accidents caused by pets. It may also be available as a separate policy from the same agencies that issue mortality insurance for the horse.

➢ *See also Insurance*

Leg, of horse

A horse's legs should be straight. If a line were dropped from the point of the shoulder, it should be straight down the center of the leg and hoof. Toes should point forward and the hooves should be exactly the same width apart as are the forearms where they meet the chest. This is proper conformation.

Improper conformation could include bowlegs (knees that curve outside the imaginary line), **knock-knees** (knees that curve inside the line),

WHAT A HORSE'S STANCE REVEALS

If your horse is unwell, the position of his legs may indicate what is wrong with him. If a horse stands with one front leg in front of the other — a position known as pointing — he is probably trying to relive pain in that leg by keeping weight off. He may have a simple bruise, or he may have a more serious injury to tendons, ligaments, or deep muscles. This position may even be a sign of navicular disease. Pointing always bears investigation; if you are not sure whether the problem is due to overwork and sore muscles, it is wise to ask your veterinarian to assess the problem.

If a horse stands three-legged or hobbles when he puts any weight on the leg held up, get a professional assessment (unless the problem resolves itself in a very short time). It could mean he has a foot abscess, a strained or sprained fetlock joint, a nerve or muscle injury, or even a fracture. (If the horse begins to walk normally within a few minutes, he may have just banged his leg during play, or landed the wrong way while running; still, keep an eye on it to see that it doesn't recur.)

These problems are most often seen in front legs. However, hind legs may also be involved. If a horse is standing with one hip dropped and a hind leg dangling, or if the fetlock joint and pastern are knuckled under with the front of the pastern resting on the ground, suspect serious injury (very few horses rest their legs this way). A peculiar hind-leg stance usually means serious injury, a broken leg, or ruptured ligament or muscle high in the leg. If the horse is unwilling to move or bear weight on the leg, don't force the issue. Seek veterinary help right away.

Rigid hind legs are usually a signal of tying up, a condition also known as azoturia or Monday morning disease. The horse may be feeling a great deal of discomfort in the large muscles of the hind end, and he may break out in a sweat.

Pawing the ground with front feet may signal colic; look for the other signs before you decide to call the vet, as it may also signify boredom. Similarly, standing with front legs forward and hind legs back may signal gas pains. Standing with legs bunched together might mean pain in the back or chest, or it could mean serious systemic illness. If the horse is standing with most of the weight on his hindquarters, he could be suffering from neck pain or laminitis. Standing with the mid-back humped up can mean severe body pain from an injured back or ribs, chest pain, or even peritonitis (inflammation and infection in the abdominal cavity) or serious digestive tract pain.

Pointing may indicate pain in a front leg.

Standing with legs bunched together may indicate chest or back pain or illness.

and **bench knees** (cannon bones set outside the line). A horse may also have front-leg problems that are apparent when viewed from the side. These include being **back at the knee** (in which the front leg appears to be concave toward the rear), **over at the knee** (in which the knee protrudes a good distance in front of the rest of the leg), **tied-in knee** (in which the knee appears as a knot between the upper and lower halves of the leg), and cut in at the knee (in which the knee is set forward at the top and angles back substantially toward the cannon bone).

A horse's legs may also be referred to as **base-wide** or **base-narrow**. A base-narrow horse will have his feet set too close together, and will generally paddle them outward as he moves. A base-wide horse will have his feet set too far apart, and will generally "wing" inward as he moves.

A horse's hind legs determine how the horse moves, that is, whether he is fast or slow, clumsy or agile. The hock and stifle are the hardest worked parts of a horse's frame. The hock should be large and sturdy, flat on the outside and only slightly rounded on the inside edge. The angle of the hock and stifle are also very important. A hock joint that is too straight (post-legged) may cause dislocation. A hock too angled, which is known as **sickle-hocked,** may lead to bone spavin or curb.

Foals sometimes display crooked legs that sag or bend at abnormal angles. This may be caused by nutritional problems in the mare, or possibly because of the way the foal was positioned in the uterus. The condition often straightens itself out after a few days of weight-bearing and nursing. If it doesn't, a veterinarian or farrier may advise splinting or bracing the leg until it straightens out. This condition is rarely hereditary.
➢ *See also Anatomy of the horse; Azoturia; Curb (of hock); Foot care; Hindquarters; Ringbone*

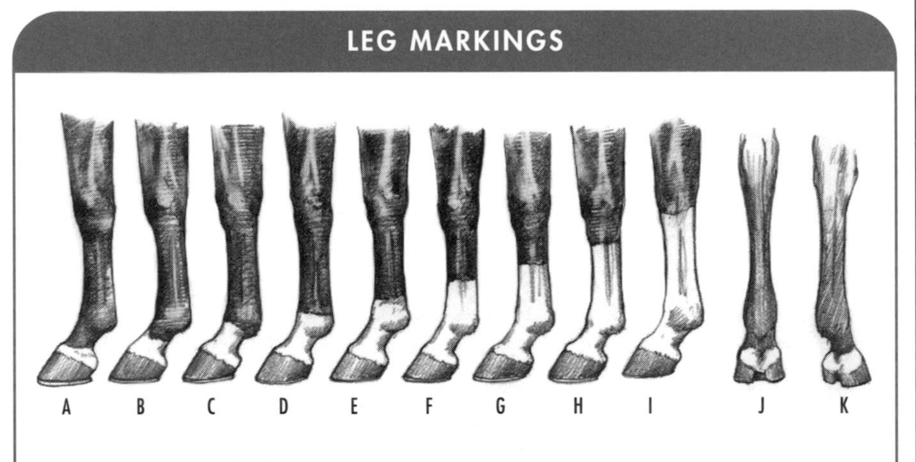

A, coronet; B, half-pastern; C, pastern; D, high pastern; E, ankle or sock; F, sock; G, half-stocking; H, three-quarter stocking; I, stocking; J, heel; K, half-heel.

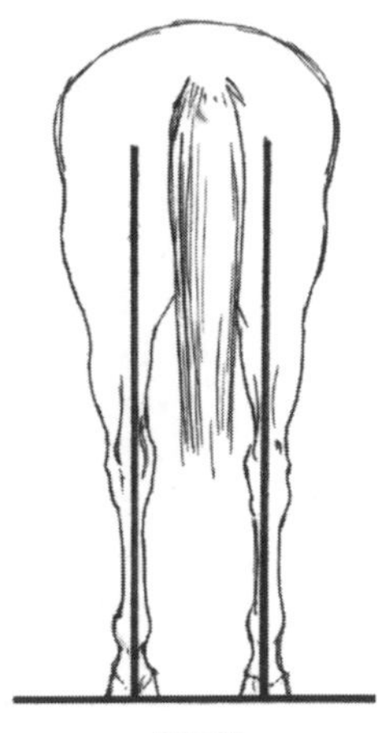

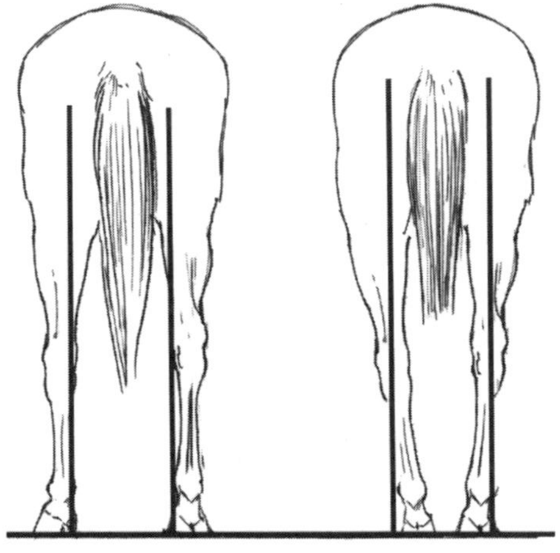

A base-wide horse seems to inscribe concave arcs in the air as he moves each foot forward; a base-narrow horse seems to inscribe outward, convex arcs.

Legs, use of when riding

A well-educated rider uses the legs and seat far more often than the hands in directing the horse. A rider's command of the leg aids, whether she rides English, Western, or any other discipline, will determine how effective she is. Resist grabbing with your hands when trying to balance your seat; you can't balance a 1,000-pound animal by pulling on his head.

Ahead of the Girth

When an English rider uses her leg ahead of the girth (with the ankle bone at the girth and toe at the shoulder of the horse), she is trying to influence the forehand; however, the rider's leg is then loose and unstable, so it is not kept there for long. In Western riding, the same position can cause the rider to brace his or her body against the cantle; this is also not a good position.

At the Girth

In English riding, many maneuvers are initiated by applying leg pressure at the girth rather than behind it. In fact, it is really the toe that appears to be at the girth; the rider's leg is slightly behind the girth, which is often referred to as the middle position. The position is similar in Western riding. It is usual for the rider's leg to be in continuous light contact with the horse's sides in this position; applying leg at the girth simply means tensing the calf muscle and/or squeezing it inward.

Bareback

Riding bareback allows the leg to lie more closely against the horse and may help riders learn to feel the movement of the horse and to more subtly use leg pressure to influence that movement.

Behind the Girth

Also called behind the middle position, this leg position is used in both English and Western riding to activate the hindquarters to move sideways.

Between the Legs and Hands

Horses should be ridden "between the legs and hands;" that is, the rider uses the legs to move the horse forward and the hands to help direct the movement and control its expression.

For Cantering

Position your inside leg at the girth (in the middle or normal position) and your outside leg behind the girth. Squeeze with your outside leg. Your inside leg will also squeeze, but the effort should mostly stem from the outside leg.

For Half Halts

To create a half halt, you must sit down and not follow the horse with your back, set your hands, and close both legs. These are the same steps you take at the halt, only with less intensity.

Inside Leg

This term refers to the leg on the inside of any circle, arena, or riding space you're working in. Out in the field? Even if you're going straight, eventually you'll have to turn. The leg to the inside of that turn is the inside leg.

Legs and Balance

A rider must be taught to control the position of her legs against the horse's sides before she can achieve balance. How the rider's leg, from hip to heel, lays on the horse is called "the rider's leg position."

Leg and Hand Coordination

Learning to properly coordinate the actions of your legs and hands is essential to proper communication with a horse. Never use a hand aid

My horses aren't hungry,
They won't eat
your hay. So fare thee
well, darling,
I'm going away.

—Traditional

without also using a leg aid. You cannot use a leg aid without automatically using your back. So, if you always use the appropriate leg aid any time you use your hands, you will also automatically begin using your whole body, which will eventually lead to achieving a team of horse and rider that seems to be one and the same.

Legs and Training

In training a horse, the legs are used to drive the horse forward, elevate him, elongate his frame or stride, move him laterally, or bend him.

Outside Leg

The leg to the outside of any space or circle you're working in.

PROPER LEG POSITION FOR BASIC SEAT

- The weight should be down in the heel.
- The inside of the calf should press against the horse's side.
- The lower leg should stay close to the girth, except when turning or controlling the horse's hindquarters.
- The knee and toe can be turned slightly outward.

Legume hay

Hay made entirely from alfalfa, clover, or other legumes; it should be fed sparingly because horses like it and tend to overeat it. Fine-stemmed alfalfa, useful for cattle, is too rich for horses and does not offer enough roughage to assist their digestion.

Leg up

Helping a rider mount is often called "giving a leg up," or even "giving a leg." To give a leg up, stand on the left side of the horse and tighten the girth. When the rider has taken the reins in her left hand and placed her right hand on the skirt of the saddle, stand just to the left of her. She should face the saddle and lift her left lower leg backward from the knee. Place your left hand under her left knee and the right hand around and under her left ankle. At an agreed signal, usually the count of three, the rider jumps up off her right foot, while you raise the left leg straight upward. As soon as the rider is high enough to pass her right leg over the horse's back, she mounts and turns the body forward, lowering into the saddle.

Leg wrap

Leg wraps come in several varieties and are used for different purposes. One type of leg wrap is the shipping wrap, which protects the legs during travel and if the horse scrambles or hits the lower leg on anything when getting into or out of a trailer. These wraps consist of fleece, quilted cotton, or spongy towel-shaped cloths that are the length of the horse's lower leg. They are usually held on by 9-foot lengths of 4- or 5-inch-wide fabric, which are stretchy and often come with hook-and-loop fasteners. The same kind of wrap may be used for soreness (with or without a bracer) or to hold medication on injuries.

Another type of leg wrap is the polo wrap. These 4-inch-wide lengths of fabric are usually soft and somewhat stretchy. Polo wraps lend support and protection to a horse's lower leg while he is being trained or worked; they act in much the way as elasticized athletic bandages do for humans.

Leg yield

A motion in which the horse moves both forward and sideways, away from the pressure applied by the rider's leg. The horse's body bends away from the direction of travel, with the sideways, or lateral, movement created by one pair of legs moving toward the other under the horse's body.

To perform a leg yield, bend the horse around the leg you wish to move away from. Use a leading rein on the side you want to move toward; use a direct rein, toward your hip, on the side you're going away from. Keep the inside leg at the girth. Move the outside leg — the one you're moving toward — behind the girth to keep the horse's hindquarters from swinging out. Use leg pressure to create forward movement, which you then turn into the lateral movement with your body weight and hands.

Leg Yield

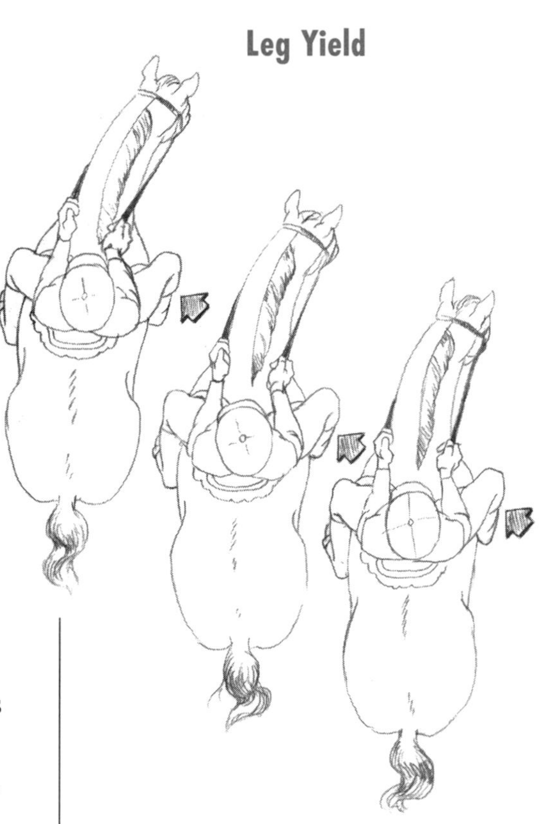

If you were to look through the horse and rider at the horse's legs, you would see the right legs moving both forward and toward — and even across — the left legs.

Leptospirosis

A disease of mammals and human beings caused by spiral-shaped bacteria. Leptospirosis can cause illness and abortions, and it is often spread by a carrier, an animal that recovers from the disease but harbors and then sheds bacteria in urine, saliva, and other secretions and excretions. Common carriers of the bacteria are rats, mice, and other rodents, as well as pigs, cattle, and dogs. Horses may be exposed particularly through rats that live in barns and contaminate feed and water.

In horses, leptospirosis is usually not a serious disease. It may cause fever, loss of appetite, dullness, and jaundice. However, pregnant mares may abort their fetuses, and there may be problems with the kidneys, joints, and eyes in some horses. Leptospirosis is one main cause of moon blindness, or periodic ophthalmia.

➢ *See also Periodic ophthalmia*

Lesson barn

A barn that keeps school horses, which are used to teach riders of all levels. These barns may also have boarders at various levels, and both lesson students and horse owners may show their horses. Some lesson barns host shows; others do not. Finding the proper lesson barn for your riding goals will help you meet them safely and pleasurably.

Discuss Your Goals

If your goal is to become an Olympic competitor, be sure the school is prepared to teach you all they can and prepare you for a national-caliber coach at the appropriate time. If you want to ride well and compete in shows, either rated or local, be sure the barn will offer that opportunity when you are ready. If you want only to learn to ride and enjoy classes and trail rides, be sure the barn is willing to provide that level of teaching; some barns are "show barns" and are not interested in students who don't care to compete. Some barns, on the other hand, are purely riding schools and have, at most, a few in-house shows each year with only barn students competing against each other.

Facilities

Don't judge a lesson barn by how fancy the establishment is; some of the best teaching is done with no more than a basic outdoor arena. If you live in a severe climate or you can only get to lessons after nightfall, however, you may want to find the best barn that offers an indoor, lighted arena. Also look for a barn that houses its school horses well and offers them adequate turnout; at many barns, the hard-working lesson horses take a back seat to the more lightly worked — but more expensive — privately owned horses or boarders. A good lesson barn will take good care of its school horses. These establishments will usually take good care of their students, as well.

Observe Lessons

If you are new to riding, you may not know whether a teacher is well-educated in the skills he or she is teaching. But you can tell by watching whether the students are enjoying themselves, whether they seem to understand the instructions, whether they seem to enjoy a good deal of success at the activities presented, and whether the instructor verbally abuses them. Teachers should have an encouraging attitude, and they should treat the horses and students respectfully; good teachers will not tolerate students who abuse horses by yanking on the bit, punching or kicking the horse, or bouncing carelessly up and down in the saddle while waiting to perform an activity.

Style of Riding

If you know you want to learn English riding, or hunt seat, look for a hunter, hunt seat, or English barn. If you want to learn Western, look for

LESSON BARN CHECKLIST

When you are selecting a lesson barn, find out if the barn:

- ❑ Welcomes beginners, or adult beginners, when appropriate
- ❑ Offers the style of riding you are interested in learning
- ❑ Has healthy, content, well-kept horses
- ❑ Is convenient; bear in mind, though, that the closest barn is not always the best
- ❑ Has an outdoor riding arena
- ❑ Has an indoor riding arena
- ❑ Has access to trails or at least expanses of open land where you can ride on uneven terrain
- ❑ Has well-kept facilities

For some people, riding in any type of weather is not important so they may not want to make the indoor arena a hard-and-fast requirement. Other people have no desire to ride on trails, wanting only to master technical skills taught in arenas . . . and vice versa. Keep these issues in mind when selecting a lesson barn.

a Western barn. Dressage is an English discipline offered at dedicated dressage barns, and at some hunter barns. A few Western barns also offer dressage. Other types of riding that have dedicated lesson barns are saddle seat and Paso Fino.
➢ *See also Dressage; English riding; Instructor; Paso Fino; Saddle seat equitation; Western riding*

Lessons, riding

These may be private, semiprivate, or group. They may be a half-hour (private) or an hour (private, semiprivate, and group.) At some barns, students must arrive at least a half-hour before a lesson to tack up their horses; they usually must remain at least that long afterward to cool down and groom the horse and put away tack. Adults, and even some busy junior riders, who cannot commit to a lesson each week at the same time are usually best served by scheduling private lessons individually, as time permits — but this approach will generally be more expensive per lesson. Missing too many group lessons (a different number at each barn, depending on how their program is set up) usually results in forfeiting them. In any case, it will certainly mean that the student lags behind the class in the skills they are developing.

If you are taking group lessons and miss one once in a while, taking a private lesson can help you catch up. If you want to progress faster than the class and can leap-frog into a more advanced class, you may decide that a few private lessons will aid you. Many students also take a private lesson or two before a show so that they can get individual, targeted help with the skills and procedures they will need at the show, in addition to what has been taught in class.

Dressage and reining lessons are almost always private. Few of these lessons are given on school horses because the aim of both disciplines is to demonstrate the capability, suppleness, athleticism, and obedience of a horse as the rider works to improve him. Some instructors teach very effective lesson by using semiretired "school masters," but most are given on the student's own horse.

HOW TO CHOOSE AN INSTRUCTOR

You are going to build a relationship with your riding instructor, so it is important that you get good "vibes" from that person. If you think you've found a good personality match, then observe the instructor during lessons. He or she should:

- ❑ Have several years' worth of teaching experience
- ❑ Have well-organized classes
- ❑ Pay attention to safety as well as to learning
- ❑ Be pleasant even when frustrated
- ❑ Be able to serve your riding goals
- ❑ Keep talk targeted on the subject matter
- ❑ Welcome questions
- ❑ Treat students and horses with respect
- ❑ Demonstrate when appropriate
- ❑ Keep the class interesting and fun

See also Instructor

Lethal genes

Genes that, when received from both parents, cause death before or soon after the foal's birth.
➢ *See also Genetics*

Levade

Performed by Lipizzaners of the Spanish Riding School, the levade is a test of strength and balance. Rising on his hind legs, the horse crouches at and maintains a 45-degree angle. The strongest stallions can hold this position for only 10 or 15 seconds.
➢ *See also Lipizzaner*

Levels of riding experience

➢ *See Rider, levels of experience*

Lice

Two kinds of lice infest horses, but they cannot be transferred to humans; lice are species-specific. *Sucking lice* feed on blood. A severe infestation can cause anemia. *Biting lice* are more active, causing skin irritation and itching. Lice are found primarily under the mane or on the buttocks, rump, and tail. With close observation, you can see the lice or the eggs if you pull out a few tufts of hair where the lice have been feeding (the hair comes out easily in these spots). Lice are most problematic in winter, when the hair is long.

Affected horses are treated with insecticidal sprays and powders. Usually two treatments are necessary, because ready-to-hatch eggs won't be affected the first time. Treat brushes, blankets, saddle pads, and other equipment used on the affected horse, as well, or you may reinfect him or infect other horses.

Light-driving harness

A harness used for driving horses that pull light carts and carriages, rather than heavy wagons (such as those pulled by teams of draft horses).
➢ *See Harness*

Light horse

A horse that is light in body weight and style, not coat color. (Heavy horses, such as Percherons and Clydesdales, are heavy in weight and body type.) The conformation of a light horse's back allows the saddle to fit easily and securely, and his way of going tends to be comfortable to ride or drive. Light horses include:

- Sport horses, such as Selle Francais, Irish Sport Horses, and Akhal-Tekes
- Stock horses, such as Quarter Horses and Appaloosas
- Hunters, such as the Irish Hunter, Thoroughbred, Thoroughbred cross, and Connemara
- Pleasure horses, such as Arabians and Pasos
- Animated horses, such as Tennessee Walkers, Saddlebreds, and Missouri Fox Trotters

➢ *See also American Saddlebred; Appaloosa; Arabian; Conformation; Missouri Fox Trotter; Paso Fino; Peruvian Paso; Quarter Horse; Tennessee Walker; Thoroughbred; Way of going*

Lighting, of barns

Barns should include unfiltered natural light to encourage vitamin D production in the horse. Sunlight is also a good sanitizing agent, and it helps keep barns from developing rank odors. Large sliding doors can admit this light and windows that open can be installed so that some light enters without being filtered through panes of glass. Of course, windows need to be closed for winter months in cold climates.

AT LIBERTY

Liberty horses are ridden without saddle or bridle, but with leg, weight, and voice aids alone.

LIGAMENTS

Ligaments attach bones to bones, unlike tendons, which attach muscles to bones. Ligaments help horses lock their legs to stand for long periods with little fatigue, even sleeping on their feet. They also help the horse absorb the shock of the concussion of hooves against the ground.

Injuries to ligaments are called *sprains*.

For stalls, work areas, and tack rooms incandescent bulbs are best; be sure they are caged anyplace a horse might come into contact with them, such as in his stall (where he might toss his head and hit anything installed in a low ceiling) and around the wash stall. Fluorescent lights diffuse easily and, thus, are only appropriate where you can install a large enough fixture to provide the concentration of light you need. Mercury vapor lights, which are several times brighter than fluorescent, are best for indoor riding arenas. They must be installed on high ceilings as they require 16 feet of space to diffuse.

Lightning

Lightning is a danger to horses, both in the barn and out. Field-boarded horses often seek shelter among trees, unless you have a lean-to or run-in shed for them. On any outside structure you should install lightning rods; horses standing under run-in sheds have been killed when their rodless sheds were struck. All horse and storage barns should also be fitted with lightning rods.

MINIMIZING LIGHTNING RISK

Make sure your pasture has a sheltered area where a horse can find refuge during a storm. Horses may leave exposed hilltop and seek out low ground. If your horse has a favorite spot (under a tree or by a building) check to see if puddles accumulate there after a rainfall. Such puddles can conduct electricity. You can fill this area with gravel to reduce the chance of a lightning strike.

Light seat

A seat that is halfway between a full balanced seat (in which most of the rider's weight is on the seatbones), and a two-point seat (in which the rider's entire seat is off the saddle, with only two points — the inner thighs — in contact with the saddle). A light seat allows the inner thighs to support more of the rider's weight than they would in a full seat; in fact, the rider's seat appears to be just touching the saddle. A light seat allows the rider to change from a driving seat in full contact to a two-point seat easily while jumping a course of fences, and it allows the horse to move forward naturally, as well.

➢ *See also Balance; Balanced seat; Forward seat; Two-point seat (hunt seat, light seat)*

In a light seat (which some instructors also may call a half-seat), the rider's seat touches the saddle but without putting the full weight on the seat bones. Some of the weight is absorbed by the inner thigh.

Lime, use of

Granulated or pelleted limestone, also called lime, is available at feed stores and farm supply stores in 50-pound bags. This product is used to deodorize and help dry stalls. It is sprinkled on the floor before the stall is completely rebedded, especially before receiving a new horse. Lime can be sprinkled on wet spots during daily cleaning as long as the horse will then be turned out and the stall windows and doors can be left open. New bedding must be installed over the lime before the horse is readmitted to the stall. Avoid powdered lime as it may cause respiratory problems in horses.

Limit class

In shows, this term describes a competition in which the horses entered may not have won more than a certain number of ribbons or cash prizes in previous competitions at the same level.

➢ *See also Ribbons; Show associations*

Linebreeding
A breeding technique that attempts to fix certain traits by mating distant relatives descended from a particular outstanding horse.
➢ *See also Genetics*

Lineup, in a show
When a judge has finished assessing the horses in a group class, the riders will be asked to line up with their numbers (on their backs) facing the judge before the winners are announced.

Lipizzaner
A compact and powerfully built horse, the Lipizzaner (often called Lippizan in the United States) is one of the great athletes of the equine world. Originally bred (in the 1500s) as light war-horses by Austrian emperor Maximilian II and his brother, the Archduke Charles, they have been trained for centuries at the Spanish Riding School in Vienna to perform advanced dressage movements based on military exercises. For more than 400 years, the Spanish Riding School, named for the Spanish horses that established the Lipizzaner breed, has been dedicated to the art of classical equitation.

During World War II the Spanish Riding School was nearly destroyed by the Nazis, who removed the breeding stock from Austria. The Allied army, led by General George Patton, rescued the mares and foals and returned them to their home.

Lipizzaners mature slowly and are not broken to saddle until they are five years old. Born dark, their coats lighten as they age to become almost pure white by the time the horse is 10. Although an occasional bay adult is produced, this has become increasingly rare. Lippizaners have strong, arched necks, short, powerful legs, and sturdy barrels. Their large, dark eyes and small ears add to their appeal.

Only stallions are ridden, and it takes many years of careful and incremental training to perfect the technique of both riders and horses. As in all dressage work, the rider, who rides without stirrups, makes no visible movements while guiding the horse through a routine of highly controlled movements and breathtaking leaps, sometimes referred to as "airs above the ground."

Some precision drills are set to music and performed in unison. Not all horses can manage the more difficult maneuvers, such as the **capriole,** an amazing leap from a standstill position, or the **courbette,** a series of hops on the hind legs.
➢ *See also Ballotade; Capriole; Courbette; Levade; Passage; Piaffe*

Lipizzaner Leaps

BALLOTADE

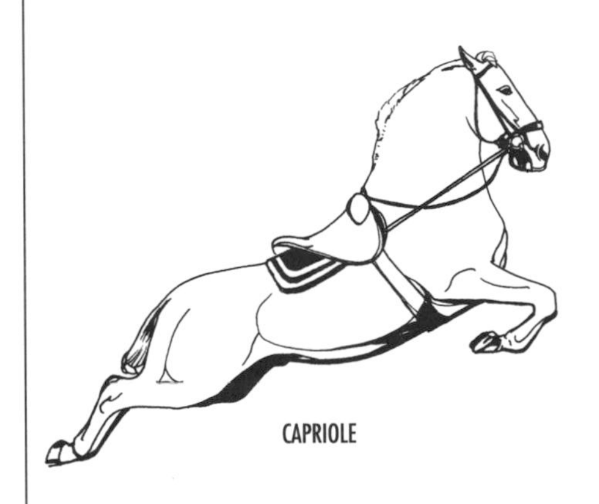
CAPRIOLE

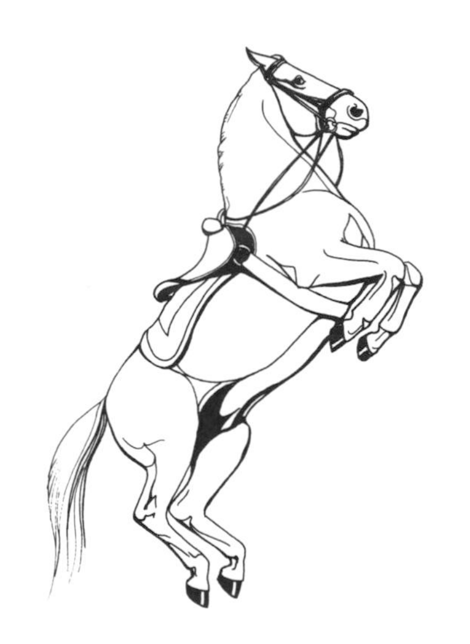
COURBETTE

Lipomas
Usually benign tumors that consist of slow-growing balls of fat on thin stalks. They occur in the abdomen of older horses, especially fat ones. Unless a stalk winds around the intestine, which will cause colic and a fatal blockage if not surgically corrected, lipomas usually do not cause problems — even if an older horse has a number of them.

Litigation

Litigation has been the prime motivator for increased safety in riding schools. Many schools are certified by the American Association for Horsemanship Safety, Inc., which may, in addition to lowering a barn's insurance rates, also promote stringent safety measures in order to avoid litigation for negligence. The philosophy behind safety measures is succinctly put by internationally known trainer John Lyons: "Human safety is first. Horse safety is second. Everything else is third."

➢ *See also Insurance; Legal aspects of horse ownership*

Live-foal guarantee

Part of a breeding contract that specifies the mare must produce a live foal (commonly defined as a foal that stands and nurses) after breeding to the stallion involved.

Liver chestnut coloring

A very dark red (chestnut), with the horse's mane and tail being the same color as the body.

➢ *See also Coloring*

Livestock sales

Usually auctions, livestock sales may involve any domestic livestock, including horses. Many of the horses sent to these sales are intended for sale as meat for the food market in Europe and Asia.

➢ *See also Horse sales*

Loading horses into trailers

Horses should be introduced to all sorts of trailers as part of their early training; nothing is more frustrating than missing a trail ride, show, or lesson because your horse won't load.

When arranging horses on the trailer there are several considerations. One of these concerns is balancing the load by estimating the weight of the horses, and putting them on in the correct order for the type of trailer you are using. Another concern is the personalities of the horses; some don't mind being leaned against, for example, when in a stock trailer with others. Some horses have trouble stepping down backwards when unloading from a step-up trailer. Putting these horses on first will allow you to later unload the rest so that there is room to turn the last horses around so that they can come out facing forward.

➢ *See also Trailering a horse*

Lockjaw

The common name for tetanus. It is called lockjaw because a horse suffering from the disease is unable to eat. The horse with lockjaw may die of nerve damage or other injury, but he can also die of starvation.

➢ *See also Tetanus*

LOADING LOGIC

- ❑ If trailering only one horse in a straight two-horse trailer, load him on the left side so that the trailer won't pull to the outside of the road when you are driving.
- ❑ If trailering only one horse in a slant-load trailer, load him in front to keep the weight close to the towing vehicle.

Locks

In some states it is illegal to apply a lock to a horse's stall because, in the event of fire, a person needs to be able to open the stall door and let or lead the horses out. Latches, however, are fine, and some are more horse-proof than others. They are easily removed in a hurry by a human yet relatively impossible for a horse to open.

Locoweed (crazyweed, milkvetch, poisonvetch) (*Astragalus* and *Ozytropis* spp.)

A member of the legume family that grows in arid and semiarid land.
➢ *See also Poisonous plants*

POISONOUS PLANT

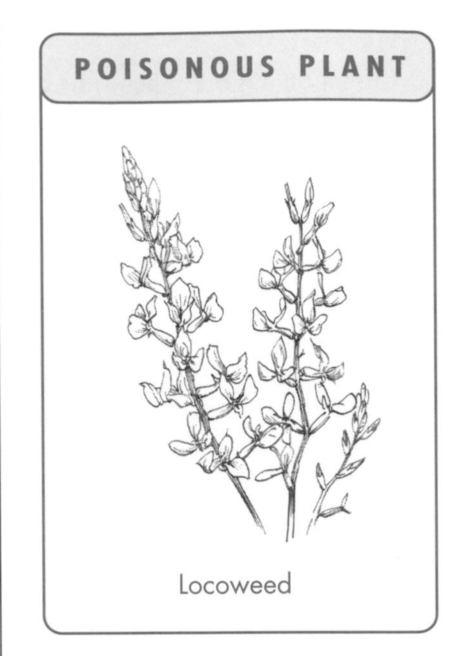

Locoweed

Longeing

Working a horse on a long rope in a circle about 60 feet in diameter. A long rope is attached to the halter or bridle or to a special noseband called a **cavesson.** The handler holds the line in one hand and a long whip in the other.

Many young horses are trained on the longe line before learning to accept a bit and saddle. Working on the longe teaches them to balance themselves as they move around a circle and to respond primarily to vocal commands. The whip is never used to strike the horse, just to keep him from moving in toward the handler and to encourage him to move forward.

Longeing is a good way to work off excess energy if a horse has not been ridden for a while or is known to act up under saddle. Lots of instructors use longeing lessons for their students, who can then concentrate on their seat and leg positions without having to worry about the reins.

Longe line

A long rope that is attached to the horse's halter or bridle while
the horse works in a circle at a set distance from the handler.
➢ *See also Longeing*

Long trot

The Western term for an extended or brisk trot.

Lope

The Western term for canter. A good lope is smooth and slower than the canter that's expected in English riding.
➢ *See also Canter; Gait; Lead, when cantering; Trashy loper*

Lope

Low-octane horse

A horse that requires little or no supplemental grain to maintain its health and body weight. Also known as an "easy keeper."

Lusitano

The Portuguese version of the Andalusian horse. Like the Andalusian, the Lusitano is able to learn and perform Haut Ecole movements with natural, quick, elevated action. He is now a very desirable saddle horse for any English discipline, and he also makes a showy carriage horse. The Lusitano was once the mount of the Portuguese cavalry, and he was developed from the fiery, agile Barb and the primitive Sorraia, which contributed hardiness and endurance.

Lying down

It can be a bit of a shock to see an adult horse lying flat on the ground, but horses often lie down to rest — although they are often just as comfortable standing up. You may find your horse in a recumbent position with his legs tucked under him and his head nodding, or he might be stretched out in the sun for a snooze. In either case, he can get to his feet remarkably fast, so don't get close to him until he's safely on four feet.

Scrambling up and down is not the average horse's most graceful moment; it is possible for him to injure himself, especially if he's prone to lying in his stall, or to get cast under a fence and become unable to get up. If your horse begins to lie down more frequently than usual, make sure he's not trying to relieve the pain of colic or taking weight off a sore leg.

➢ *See also Rolling; Sleep habits [box]*

Lyme disease

Lyme disease was first identified in Connecticut in 1976, but it is now found in almost every state. Carried by ticks that feed on infected white-footed mice, Lyme disease affects humans as well as horses and dogs. Symptoms include fever, fatigue, and depression. The common rash and swelling at the site of the tick bite will be hard to notice on a horse, but many horses suffering from chronic lameness, laminitis, or arthritis-like symptoms are eventually diagnosed with the disease. Early treatment with antibiotics can be effective.

➢ *See also Parasites; Pest control; Ticks*

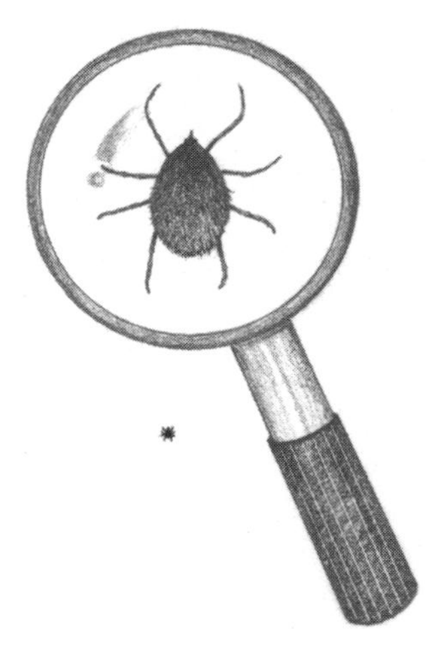

The tiny deer tick can create a number of health problems if it transmits Lyme disease to your horse.

m

Maiden class

Classes open to riders and horses who have not yet won blue ribbons in certain events; numbered as specified by show rules.

➢ *See also English competition; Novice; Western competition*

Maiden mare

A mare that has never been bred.

➢ *See also Broodmare*

Malocclusion

Abnormality of the jaw and teeth.

➢ *See also Dental care; Overbite (parrot mouth); Sow mouth; Teeth*

Malpresentations, foaling

➢ *See Foaling presentations [box]*

Management plans

➢ *See Cleaning stalls; Deworming; Feeding and nutrition; Grooming; Manure, handling; Picking out feet; Seasonal care; Signs of illness*

Mane brush

Many horsekeepers use a human hair brush on their horse's mane. Be sure to use the kind with plastic bristles that are widely spaced.

Mane, care of

Ideally, your horse already has a neatly groomed mane that combs out easily every time you groom him. However, after a lengthy pasture turnout, especially in bad weather, you might be faced with a knotted, snarled mess of burrs. Don't cut it all off and don't attack it with a comb. Start by thoroughly coating the entire mane with a detangling spray or gel. Work the solution into the mane with your fingers. Wearing a pair of old leather gloves (which will build up a slick layer of detangler), begin to gently untangle the worst knots. Horsehair tends to twist together, so look for the direction of the twist and work the hairs free.

Once the biggest snarls are loosened, comb through the mane with your fingers, trying not to yank too hard. Most horses don't mind having their manes tugged, but pulling at major knots can be irritating and can also lead to loss of hair. The next step is to work with a human hairbrush, the kind with widely spaced plastic bristles. Start at the ends and brush gently as you work your way up to the crest, checking for burrs and other debris that you might have missed.

Don't use a comb until the brush goes the length of the mane without catching. A long mane should be treated regularly with detangler to keep it from snarling. It can be braided loosely into sections before a lengthy pasture turnout to prevent tangles and to keep burrs at bay. Most owners prefer the mane to be about four inches long, though several breeds are shown with long, flowing manes.

➢ *See also Bathing horses; Grooming; Western banding*

Mane Styles

GIVING A HORSE A HAIRCUT

To cut your horse's mane, follow these steps:

1. Make sure the mane is clean and well brushed.
2. With a tape measure, estimate from the crest the length you want the mane to be.
3. With thinning shears, a thinning comb, or electric clippers, begin to cut the mane, starting at the middle, not at the ends. *Note:* You want the edges to be ragged at this point — don't try to make them even.
4. The next step is **pulling the mane** to even it up. Applying veterinary liniment to sections of the crest as you work may make the process easier.
5. Select a few (six to eight) hairs that are somewhat longer than the others and hold them in your left hand. Comb against those hairs to push the rest of the mane away from them.
6. Wrap the hairs around the comb a couple of times.
7. Holding the wrapped hair with your thumb, yank out the hairs with a quick pull.

At first the mane may seem *too* neat and tidy, but in a few days, the hairs will look more natural.

Mane comb

Your grooming kit should contain a wide-toothed comb (usually plastic) and a metal mane-pulling comb with a handle.

Mange

A condition caused by several different kinds of mites, which burrow into and bite the skin, producing intense itching. Horses will rub affected areas and may lose patches of hair and develop crusty, weeping areas and thick, wrinkled skin. **Sarcoptic mange** most commonly appears on the neck, where thick, brown scabs form. **Psoroptic mange** generally is found where there is a lot of hair, such as the base of the tail and the crest of the mane, though it also appears in hairless areas, such as the udder. **Chiroptic mange** (or leg mange) is most active in cold weather and causes terrible itching. Horses can damage their legs by stomping in an attempt to alleviate the itching.

All itchy rashes should be seen by a veterinarian, who can determine the cause by examining a skin scraping under a microscope. There are several effective sprays and ointments to treat mange. Affected horses should be quarantined until they are well, as the condition can spread from horse to horse by direct contact, as well as through shared equipment.

➢ *See also Parasites; Skin problems*

"Manger tie"

➢ *See Knots*

SHOW BRAIDING

For most hunter and dressage classes, the horse is expected to have his mane neatly braided in a particular style. You'll need a mane comb, strong thread or yarn, and scissors. A step stool is also useful — your arms will get pretty tired.

1. Make a regular three-strand braid, starting with the right section over the center one, and the left over the middle.

2. At the end of the braid, bring a folded piece of string behind the braid and loop the ends through the folded middle.

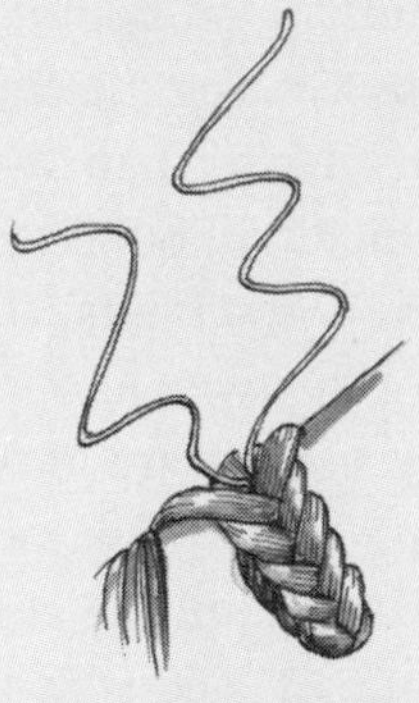

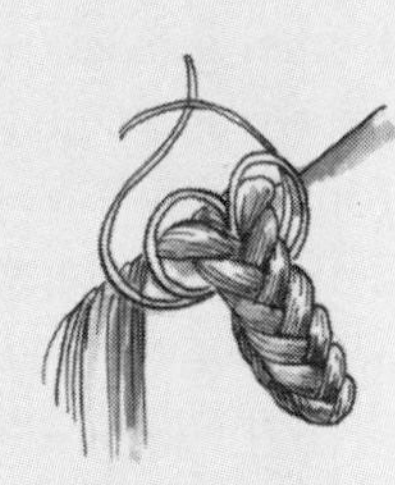

3. Poke the ends of the string through the top of the braid and pull snugly, folding the braid in half.
4. Separate the strings, then loop them around the top of the braid and tie off with a square knot.

Manners, of the horse

➢ *See Handling horses; Leading; Safety*

Manty

A piece of canvas used to protect the load on a pack animal.

Manure, as indicator of health

Mucking out your horse's stall every day lets you keep an eye on how his intestines are functioning. Generally, healthy manure forms into balls that are moist and stay in a heap when passed. There are usually signs of some undigested hay and grain. Very hard and dry manure balls might indicate that your horse is not drinking enough water (this sometimes happens in the winter when horses tend to drink less). Loose manure or diarrhea can be an indication of illness, of a change of feed, or of a sudden increase in the intake of salt and, subsequently, water. Some horses have loose stools when they are nervous, upset, or in heat. If the manure is scant, check to see that the horse is passing what he eats and colic is not beginning. If you suspect sand colic, immerse some manure in water. If sand settles out to the bottom, take the horse off sandy ground and consult your vet about dealing with the sand already in his intestines.

➢ *See also Signs of illness*

Manure, handling

Most farms or stables have a manure pile, and every manure pile has parasite larvae. The management of the manure pile is an important part of running any operation that involves animals. With manure, you've basically got three options: having it hauled away, spreading it directly on pasture or crop land, and composting it. Fresh manure can be put on grass pasture or fallow fields but should not be applied to vegetable or flower gardens until it has composted for 6 to 8 weeks. A compost pile must be properly aerated and kept at the correct moisture level to produce good compost. Ideally, you should have three piles going: one that's ready to use as fertilizer, one that's well on its way, and one that is currently being added to.

A key to handling all this waste is daily stall cleaning. Piles of manure and spots of urine-soaked bedding in a stall provide ideal breeding grounds for parasites, and they also emit strong vapors that can irritate the eyes and lungs.

➢ *See also Mucking out stalls*

Manure spreader

A piece of equipment that is pulled behind a tractor or, less commonly, by a team of horses. As it moves, it broadcasts manure over a field.

Mare

A female horse of more than four years of age.

➢ *See also Breeding; Broodmare; Filly*

A CELEBRATED HORSE

By the time Man o' War, also known as Big Red, died in 1947, he was the most famous racehorse of all time. He won 20 out of 21 races during his life.

This is that very Mab
That plats the manes of
horses in the night.

—William Shakespeare,
Romeo and Juliet

TONS OF BENEFITS

Horse manure is valued for its high nitrogen content and ability to break down rapidly. Fortunately, there is plenty of it to go around. The average horse produces about 50 pounds of manure a day. That's ten tons each year!

Markings

➢ *See Face markings; Leg markings*

Martingale

A strap that prevents the horse from tossing his head too high, called a **tie-down** in Western parlance. There are two basic types, the **standing martingale** (more restrictive) and the **running martingale** (less restrictive). The standing martingale consists of two straps: one strap that circles the base of the horse's neck, and another that runs from the back of a cavesson noseband, through a slit in the neck strap, to buckle around the saddle girth. The running martingale attaches to the girth and the neck strap in the same way. Two straps come up from the neck strap, one attaching to each rein with a metal ring. Both types of martingale should be carefully fitted.

➢ *See also Running martingale; Standing martingale; Tacking up; Tie-down*

Mash

A warm bran mash is a time-honored recipe to perk up a cold or sick horse. However, any grain can be made into a mash (which is basically a porridge) and bran has been found to be harmful to horses if fed in large quantities. To make a mash, add enough boiling water (or as hot as you can get it from the tap) to make the normal feed ration wet, but not dripping. Start with one part water to two parts grain and experiment for best results; flaked grains will absorb more water than whole grains, while pellets will quickly turn to mush, so the amount of water depends on what grain you're feeding.

Stir the mixture well and cover the bucket for 5 or 10 minutes to let it steam. Feed while it is still warm but not too hot for you to handle with bare hands. After use, make sure to clean the bucket and the feeder well, as wet feed spoils quite quickly.

➢ *See Bran; Feeding and nutrition; Grains*

Massage therapy

People love massages, and so do horses. Their bodies are composed of more than 60 percent muscle, so there are lots of opportunities for aches and pains. A good equine massage therapist can manipulate tissue to increase circulation, relax tight spots, relieve spasms, and just generally make your horse feel better. Massage can be part of a treatment program for an injury and can actually help prevent injuries, especially in horses that are worked hard.

Mastitis

A serious infection of the udder. Although mares have two teats, there are four quarters to the udder and each teat has two openings for milk to flow through. Mastitis is an infection in one or more quarters of the udder; it is generally found on one side or the other, not both at the same time. While usually found in lactating mares, especially around weaning time when the udder is full and painful to begin with, the condition can also

MARWARI HORSES

The Marwari horse is the product of centuries of careful breeding in northwest India, particularly in the desert state of Rajasthan. The Marwari, descended from Mongol ponies, is the ideal horse for the Rajput warrior, wiry, muscular, and renowned for its courage and loyalty. Two characteristics of the breed are the curved, almost crescent-shaped ears and the natural pacing gait, called the *revaal.* Marwaris' hooves are so strong and hard that they rarely need shoeing.

affect mares that are not pregnant or lactating. Symptoms are varied, depending on the severity of the infection, but the mare will definitely show signs of discomfort as she tries to avoid bumping her sore udder with her leg. She may stop eating; run a fever; have swelling, heat, or edema in the affected area; and she may kick at a nursing foal.

Mastitis must be treated with antibiotics, both oral and topical. Untreated mastitis can destroy mammary tissue in one or more quarters and may make it impossible for the mare to produce milk for future foals. In rare cases, the infection can completely destroy the udder and even kill the mare.

➢ *See also Udder problems*

Mats

Rubber mats that provide traction in stalls and trailers, on ramps, in aisles, and anywhere that surfaces might become slick. They also prevent horses from digging holes in dirt floors, and they make excellent cushions for preventing calluses and easing arthritic horses' joints.

➢ *See also Facilities; Stall*

Maturity classes

Classes for older horses that are no longer competing at the highest levels but still have plenty to offer.

McClellan saddle

The official saddle of the United States cavalry. The McClellan is still used by many long-distance riders who swear by its comfort for both them and the horse. Others find it comfortable for the horse, but not so comfortable for the rider.

➢ *See also Saddle; Tack*

Measuring a horse

➢ *See Hand, as unit of measurement; Weight, of horse*

Mecate

Reins of braided horsehair, used with a bosal (noseband) instead of a bit in Western tack.

➢ *See also Hackamore*

Melanoma

A skin tumor to which all dark-skinned horses are susceptible and which most gray horses eventually develop. Frequently, there is little cause for concern, though your vet should check any suspicious growth. Lumps generally appear around the genital area and under the tail and may at first be benign, but they can become cancerous and spread internally where they might interfere with breeding, foaling, or even digestion. Internal melanomas are more dangerous. The external ones are often very slow-growing, and your veterinarian may decide to leave them alone unless they

Different horses fall to one's lot at different times, and the same horse serves you one way at one time and another at another.

—Xenophon, *The Art of Horsemanship*, 430 B.C.

m

are interfering with bowel function or use of a saddle. All lumps should be checked regularly for changes in appearance and size.

Surgical removal of the tumor and freezing or burning the remaining area can be done if necessary, but there is a risk that disturbing the growth may stimulate the proliferation of cancer cells.

Memory, horse's

There are many variations of the saying that it takes years to make a good horse, but only a moment to make a bad one. Their excellent memories make horses highly trainable, but it also makes them hard to handle in situations where they've previously been frightened or mistreated.

➢ *See also Body language, of horses; Handling horses; Pecking order, in herds*

Merychippus

A primitive ancestor of today's *Equus,* which lived in the Miocene period. *Merychippus* was about the size of a Shetland pony.

➢ *See also Evolution of the horse*

Mesh wire fences

➢ *See Fencing materials*

Mesohippus

A deerlike, three-toed ancestor of the modern horse. Its fossils from the Oligocene period have been found in South Dakota.

➢ *See also Evolution of the horse*

Mexican-style sidesaddle riding

A style of sidesaddle riding in which the reins are held in one hand, while the other hand holds the riding whip, which rests across the lower arm of the rein hand.

➢ *See also Sidesaddle, riding*

Mice, controlling

➢ *See Pest control*

Middle-ear infection

Ear infections can be hard to diagnose in horses. Signs of infection might include fever, odd head carriage, trouble eating, and reluctance to have the ears handled. If left untreated, a middle-ear infection can lead to inflammation of the bone at the base of the skull and the development of bony growths that are constantly being broken by the action of the tongue. This can actually cause a fatal skull fracture. Early treatment with antibiotics will clear up most ear infections, so have your veterinarian help you figure out what's bothering your horse if he shows any sign of discomfort in his ears.

➢ *See also Ears*

MESSENGER

Messenger was an 18th-century English Thoroughbred stallion related to all three Arabian foundation sires. He in turn was the progenitor, through Hambletonian, of the American Standardbred breed, prized as the world's best harness racers and trotters.

MEXICAN HORSES

The Spanish *Conquistadores* brought 16 horses to Mexico in 1519 as part of their brutal conquest. Hernan Cortez later claimed that the horses were responsible for the victory, because they were so terrifying-looking, clad in armor and carrying armored warriors. The native Aztecs had never seen horses before.

Midges ("no-see-ums")

Tiny flies that attack at nighttime and may swarm in the thousands. Their itchy bites cause the horse to rub and scratch; he may injure his skin or develop signs of hypersensitivity. Keep horses in at night if midges are a problem, or apply fly spray in the late afternoon.

➢ *See also Parasites; Pest control*

Military and horses

Over the thousands of years that humans developed their partnership with the horse, a major concern was how to engage an enemy force. Consequently, many of the conformation traits that we still deem desirable in horses arose from the need for a quick, strong, hardy animal that could go long distances without breaking down. As the horse spread across the world, indigenous people soon realized the benefits of mobility and power that the horse offered. The horse played a vital role in military history for several centuries before the era of the tank and high-powered artillery.

➢ *See also Cavalry*

Milk replacer

Sometimes a foal needs to be raised on milk replacer. If you find yourself with an orphaned or rejected foal and no nursemare, consult your veterinarian about an appropriate product. There are a number of products on the market, each with different formulas for the various needs of newborn foals. There are several home recipes based on cow's milk that are also effective (see box). Straight cow's milk is higher in fat and protein but lower in carbohydrates than mare's milk. Most foals will also do well if you substitute goat's milk.

A young foal needs about 2 gallons of milk every 24 hours. Many small meals are better than trying to stuff him a few times a day. Try to accustom the foal to drinking from a bucket right away; buckets are easy to clean and you can leave enough milk replacer to last the foal several hours. In fact, he should be fed a half-pint every half-hour for the first five days, increasing to a full pint for the next five days. After 10 days, the foal can handle two-hour intervals at night if he is still fed hourly during the day. Gradually, by the time the foal is a month old, the amount fed can increase as the time between feedings increases to 4 or 5 hours.

If your orphan foal is not doing well on a particular milk replacer, find a different one. If he develops diarrhea, substitute electrolyte solution for the cow's milk for a few days to give the gut a chance to heal. Your foal should gain 1 to 3 pounds a day, and should be introduced to alfalfa hay, green grass, and/or pellets as soon as he shows interest. Weaning can take place at around 6 to 8 weeks.

➢ *See also Foal care; Nursemare; Nursing; Weaning foals*

Milkvetch (crazyweed, locoweed, poisonvetch) (*Astragalus* and *Ozytropis* spp.)

➢ *See Locoweed; Poisonous plants*

HOMEMADE MILK REPLACER

Mix up the following to get an orphan foal off to a good start:

- 1 can (12 ounces) reconstituted evaporated milk, 4 tablespoons limewater or tap water, 1 tablespoon sugar or corn syrup
- 1 pint milk, 4 ounces limewater or tap water, 1.5 teaspoons sugar
- 1 pint milk, 2.5 ounces water, 2.5 ounces limewater or tap water, 2 teaspoons lactose
- For several feedings, use 3 pints milk, 1 pint water, ¼ package Sure-Jell pectin or 20 mL of 50% medical glucose

Note: For milk, use 2% cow's milk. White sugar is hard to digest so use another type. Limewater, which used to be sold in drugstores, has extra calcium, but tap water is fine unless it's particularly soft (lacking in minerals) or high in fluoride. You can use distilled water if necessary.

Milo (sorghum)

➢ *See Grains; Sorghum grain (milo)*

Minerals

➢ *See Feeding and nutrition; Supplements; Vitamins; individual minerals*

Miniature horse

These tiny horses are the favorite companions of many people who either can't handle a large horse or just don't want to. They are popular driving animals, and a good choice for people with physical limitations. The largest "minis" can stand no more than 38 inches tall, while another division exists for those 34 inches and under. A mini is properly measured to the last hair of his mane, rather than the highest point of his withers. He should look like a small pony, not a misshapen horse. Large heads and short legs are considered a defect. Minis tend to have problems with their teeth and mouth conformation, so check carefully if you plan to use a bridle on your pony. Minis can be any solid color, Pinto, or Appaloosa.

➢ *See also Falabella*

From right to left, here are the comparative sizes of a horse, a pony, and a miniature horse.

Missouri Fox Trotter

Found primarily in the rural south-central states, these smooth-gaited horses amble without overly exaggerating their movement. They make great trail and pleasure horses and are used in some western parks, where they are crossed with donkeys to produce comfortable riding mules.

➢ *See also Amble; Gaited horse*

Mites

Tiny insects that cause dermatitis and are most active in the spring and winter. **Trombidiform mites** infest harvested grain (the larvae develop on small rodents, who then infiltrate the grain) and can transmit disease to animals. Affected animals will develop itchy, scaly areas around the lips and face.

Mange mites live in the hair follicles and cause chronic inflammation and intense itching, as well as secondary infections. Some mange mites burrow under the skin and can be felt as small bumps, again typically found around the face.

➢ *See also Mange; Parasites; Skin problems*

Mitt

A sheepskin finishing mitt is used at the end of a thorough grooming to "set" the coat and bring out a shine. It can also be used to groom the horse's face.

➢ *See also Grooming*

Molars

➢ *See Dental care; Teeth*

MITBAH

Arabic term for throatlatch, or attachment of head and neck. The word translates literally to "the place where the throat is cut" because it is the same for camels, sheep, and goats. A fine, long mitbah is much desired in an Arabian horse.

Molasses

Molasses provides carbohydrates and is often added to grain to produce sweet feed. Too much molasses will cause loose stool, so the proportion of molasses to grain should be less than 10 percent. Molasses has around 54 percent digestible nutrients.

➢ *See also Feeding and nutrition; Sweet feed*

Mold (fungi) in feed

Mold microorganisms thrive in moisture and multiply rapidly in warm, damp weather. Many molds, which are included in the fungi family, are very dangerous to horses. Grain and hay must be checked regularly for quality. Grass hay is less likely to be moldy than legume hay, which is richer in nutrients and has a higher moisture content. Molds to look out for are black patch disease, found on clovers and other legumes; ergot, a fungus that sometimes infects the seed heads of grasses; and the deadly *Fusarium moniliforme*, which causes moldy corn poisoning. *Always* look at and smell hay before feeding it, and discard any that contains mold.

➢ *See also Ergot (fungus); Fungal infections;* Fusarium moniliforme; *Hay; Toxic substances*

Monday morning disease

➢ *See Rhabdomyolosis*

Monensin (Rumensin) poisoning

A drug often added to cattle feed that is fatal to horses. Just a few grams produce symptoms of weak and irregular pulse, high heart rate, abnormal lung sounds, and blue mucous membranes. The poison attacks the muscles, and the horse dies of heart failure.

➢ *See also Poisonous plants; Toxic substances*

Moon blindness (iridocyclitis; periodic ophthalmia)

➢ *See Periodic ophthalmia*

Mor-Ab

A half-Morgan, half-Arab cross. The Mor-Ab can be registered as a Mor-Ab, Half Arab, or Half Morgan.

➢ *See also appendix*

Morgan

A true American horse, the Morgan was developed in Colonial New England. Justin Morgan was a singing master who, according to legend, acquired a small, sturdy stallion (just 14 hands high and weighing 800 or 900 pounds) as payment for a debt. Called "Figure," the little horse made a name for himself as an incredibly hard worker, able to pull a plow all day and beat all comers in races in the evenings and on weekends. One eyewitness reported seeing him shift a log that much heavier horses were unable to budge. Apparently, he was never beaten in a race, either under saddle or in harness, and he did everything he was asked willingly and with gusto.

SWEETENING THE BIT

If you are training a young horse to be bridled and he doesn't want to open his mouth for the bit, smear a little molasses on it and he will get into the habit of willingly opening his mouth for it.

THE MONGOL HERD

The Mongolian Pony is a direct descendant of the Asian Wild Horse *(Equus caballus przewalski poliakov)* and exhibits the "primitive" qualities of incredible toughness and hardiness. Genghis Khan's Mongol warriors conquered most of Asia because of the speed and stamina of their horses.

Figure seems to have died in obscurity after a very long life (at least 28 years), but horse lovers in the part of Vermont where he lived soon noticed the little stallion's remarkable prepotency, as the numerous foals he sired grew up and became just like him. Today's Morgan is still a small horse (14.2 to 15.2 hands high), but strong and well-built, with muscular hindquarters and dished face. Morgans are dark horses with few white markings, and are known for their versatility and good nature.

A typical Morgan horse

Mosquito

Mosquitoes carry deadly diseases such as West Nile virus and equine encephalomyelitis and, of course, leave itchy bites that can drive your horse to distraction. They breed in standing water, so every effort should be made to provide excellent drainage and thus reduce the opportunities for them to lay eggs. Insect repellents can also work; check your horse around dusk to see how big a problem mosquitoes are in your area.

➢ *See also Pest control*

Mounted police

Police departments all over the world use mounted officers for crowd control, street patrols, and directing traffic. The oldest known mounted police unit is the London Bow Street Horse patrol, which was established in 1738. Mounted police have existed in the United States since the mid-1800s. Even as cars and motorcycles became the backbone of most police departments, many urban areas kept their mounted officers.

Being on horseback makes an officer six times more visible and gives him/her far greater range of vision, both of which increase effectiveness. Horses can get into narrow, twisting alleys and down sidewalks where patrol cars can't, and horses can move through a crowd of people — even a rioting crowd — far more easily than an officer on foot.

➢ *See also Police horses; Royal Canadian Mounted Police (RCMP)*

Mounting, technique

Tips to remember when mounting a horse:

- The object is to hop up as lightly as possible, without slamming down on the horse's back or grabbing at his mouth. The quicker you get up, the less opportunity for the horse to fidget or move away from you.
- Check your girth and stirrup length before getting up. Check the girth again after you've been riding for a few minutes.
- Keep both reins firmly in your grasp.
- Do not mount a tied horse — he might try to move and panic when he finds he can't.
- If your horse won't stand still, have someone hold him or face him into a corner while you mount. All horses should stand quietly while being mounted and wait for the signal to walk on.

LEARNING FROM THE GROUND UP

All riders should know how to mount a horse from the ground, even if they usually use a mounting block. A mounting block not only makes it easier for the rider to get on, but also it puts less strain on the girth (and, consequently, the horse) and allows the rider to settle more lightly into the saddle. However, if you rely only on a mounting block, you may be in trouble if you have to dismount on the trail and get back in the saddle. So learn the hard way first!

HOW TO MOUNT A HORSE

1. With the reins gathered in your left hand and the horse's head turned toward you, stand at the horse's left shoulder, facing his hindquarters. If you're facing forward and he starts to walk away once you've got your foot in the stirrup, you're at a considerable disadvantage.

2. Place your left foot in the stirrup. (You're allowed to grab it with your right hand and lean as far back as necessary to get your foot up that high.)

3. Once your foot is securely in the stirrup, turn to face the saddle, with your left knee pointed toward the horse's nose. Grab the cantle with your right hand and rest your left hand on the withers.

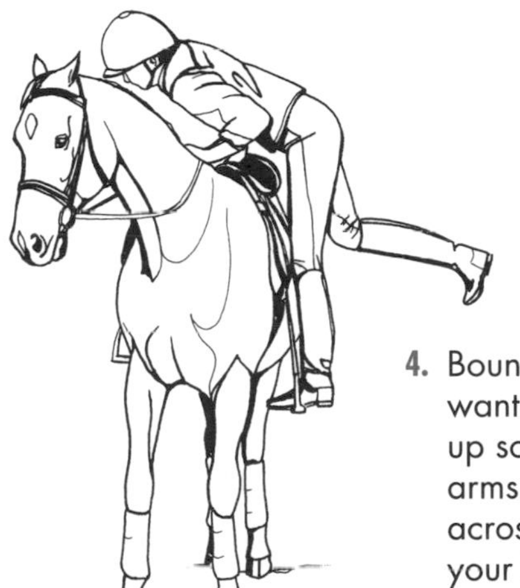

4. Bounce up from your right leg (you might want to bounce two or three times to get up some steam) while you pull up with your arms. Your momentum should carry you across the horse's back, not straight up. As your left leg straightens, swing your right leg over the rump and sit gently in the saddle.

- If the saddle slips as you go up, check your girth again; some horses can really blow up their bellies while the girth is being tightened. If the girth is fine, make sure you're not going straight up instead of over the horse's back, and make sure you're not pulling the saddle down rather than pushing off from it with your arms.

Mounting block

A stepstool that gives you some additional height when getting on a horse. The common type is made of molded plastic with two steps and a carrying groove. You also might build a set of wooden steps with a platform or use a kitchen stepstool, but be sure it's one that won't tip over too easily or damage your horse's legs if he steps on it.

Mouth, hard or soft

A term to describe a horse's responsiveness to the reins. A horse is said to have a soft mouth if he responds readily to cues from the reins. A hard-mouthed horse will ignore the rein aids, fight the bit, or even clamp it between his teeth. Once a horse like this gets going fast, he can be very hard to stop, as normal aids to halt aren't effective. A hard mouth usually results from poor handling, either by a brutal rider or a series of inexperienced ones. It is possible to retrain a hard-mouthed horse with patient and tactful handling, with great reliance on aids from the seat and leg, and very light rein contact. The best way to avoid a hard mouth is to develop a balanced seat with good leg contact. Do not yank on the reins excessively or exert constant pressure on the horse's mouth.

➢ *See also Aids; Hands, rider's; Legs, use of when riding; Seat*

Mouth problems

➢ *See Dental care; Teeth*

Movement, conformation and

The way a horse moves is directly connected to his conformation. This is known as form-to-function. For example, a horse with a long, 45-degree sloped shoulder and long hip will move with a free-flowing stride, whereas a horse with a short, straight (55- to 65-degree) shoulder and short, steep hip will move with short, choppy strides.

➢ *See also Conformation; Leg, of horse; Gait; Gaited horses; Way of going*

Moxidectin (Quest)

A dewormer that kills parasites in the larval stage.

➢ *See also Deworming; Parasites*

MOVING A TIED HORSE

Often when your horse is tied, you find yourself wanting him to be standing a couple of feet either closer to or farther away from where he is. Or perhaps you want him to face in a different direction. Before getting him to move while he is tied, make sure he knows the basics of ground handling, such as moving away from your hand when you press against him and responding to verbal cues such as "walk on" or "back up." Teaching him to turn on the forehand with you at his halter is also a useful maneuver. Be sure as you cue him to move forward or backward that you are aware of the limits of the rope to which he is tied.

See also Handling horses; Tying

Mucking out stalls

Part of the pleasure of owning a horse is taking care of him and keeping him safe, clean, and healthy. Mucking out stalls is a vital component of horse care. Stalls should be thoroughly cleaned every day, with additional manure and wet spots picked up as frequently as possible. It is important to remove all soiled bedding to prevent moisture and fumes from building up and to cut down on potential parasite problems.

People have individual techniques for cleaning stalls and you will find a technique that suits your needs. The basic procedure is to pick out the intact manure and messy bedding with a pitchfork or manure fork, dumping it in a wheelbarrow as you go. Scoop out wet spots with a shovel, which can also be used to get any bedding and waste that falls through the tines of the fork. Use old bedding to soak up any really wet places and remove it. The idea is to get rid of all bedding that is wet or dirty, while reusing bedding that isn't soiled. If possible, allow the floor to air-dry before adding fresh bedding.

However, if the stall's occupant is coming right back in, you can sprinkle lime or another odor neutralizer on the wet spots and rake the remaining old bedding over them. Fresh bedding should be heaviest where the horse tends to lie down, and should be banked up along the walls somewhat in case the horse gets cast against a wall and can't get up. You don't need to put bedding under the feed bin and water bucket or right in front of the stall door.

➢ *See also Bedding; Facilities; Manure, handling; Stall*

Mud and health concerns

Mud is not healthy for horses. Walking in it is hazardous, both when it's wet and sloppy and can practically suck a shoe right off a hoof, and when it's rutted and frozen and just waiting to trip up a fragile leg. Rolling in wet weather can coat your horse in damp, chilly mud, matting his coat and sapping his body of warmth. Standing in mud gives rise to a host of foot problems. In general, you want to avoid muddy conditions as much as possible by spreading gravel or sand in damp areas, grooming frequently in bad weather, and making sure that outdoor turnout areas have dry places for horses to stand.

➢ *See also Facilities; Seasonal care*

Mule

The sterile offspring of a male donkey (a **jack**) and a mare. A stallion and a female donkey (a **jenny**) produce a **hinny,** which is generally smaller than a mule. Depending on the size of the parents, a mule can stand anywhere from 12 to 17.5 hands high and weigh between 600 and 1,500 pounds. They come in all colors. Mules have a reputation for being stubborn and mean, but mule lovers describe them as smart, curious, and sensible, with the ability to forge close relationships with people and other animals. Mules are surefooted, agile, generally sound, and not known to put up with foolish or abusive handling. Mules can do anything horses can, and, according to their fans, do it all better. In addition to their well-known role as superb pack animals, they are used in harness, under saddle, for jumping, and even for hunting.

➢ *See also appendix*

Muscling and conformation

Horses that are used for long-distance or extended types of work (racing, dressage, jumping,and so on) need long, lean muscles to move smoothly through their paces and keep up a steady pace. Heavily muscled horses can achieve sudden bursts of speed and quick acceleration but usually cannot sustain the effort.

➢ *See also Conformation; Rhabdomyolosis; Types of horses*

STALL-CLEANING TOOLS

If you're cleaning your own stalls, you'll need the following equipment:

- ❑ A wheelbarrow
- ❑ A four- or five-tined pitchfork or manure fork (plastic ones are lightweight, but the tines break easily) to sift clumps of manure from the bedding and spread new bedding
- ❑ A shovel to scoop out very wet bedding; also useful in loading up your wheelbarrow with new bedding
- ❑ A broom and dust pan for sweeping the aisles
- ❑ An odor neutralizer of some sort; a number of commercial products are very effective, and baking soda and hydrated lime are other options

JUMPING JACKS

In some competitions, mules jump without riders; apparently, they love to jump and can clear poles taller than themselves from a standstill.

Mustang

A legend of the American West, the mustang has been both romanticized and reviled. Chased for sport, tamed as cow ponies, and slaughtered for dog food, the famous bucking broncos of the rodeo have had an up-and-down relationship with humankind. Technically, mustangs are not wild but feral, meaning that their forebears escaped from domesticity and survived to breed. Recaptured (or stolen) by Native Americans, these horses soon spread across the West and became a vital part of Native American life. Those remaining in the wild bred so successfully that fierce arguments have raged between ranchers, wild horse advocates, and the United States Bureau of Land Management (BLM; the agency responsible for dealing with mustangs and wild burros) as to how best to control the population. The United States Congress passed acts in 1959 and 1971 prohibiting the hunting and harassment of the animals on federal land. These acts also provided for the monitoring and management of wild horse and burro herds.

A mustang is a feral horse, not a true wild horse.

Today, the BLM runs adoption programs in an effort to maintain wild herds without resorting to slaughtering excess animals or letting them overrun their environment. Wild horses have very few natural predators anymore and are fairly successful at reproduction; there are between 40,000 and 45,000 of them ranging across 10 Western states. Approximately 6,000 to 7,000 mustangs and burros are adopted each year.

Descended from the tough little Spanish horses brought into Mexico by Cortez in the early 16th century, mustangs are known for their wiry build, agility, and stamina. They come in all colors and a variety of conformations, from attractive horses with good conformation to scrubby, scrawny creatures sometimes called "broomtails." A newly captured mustang should be handled only by an experienced rider, but a mustang with proper training could make a very good mount.

➢ *See also Spanish Barb; appendix*

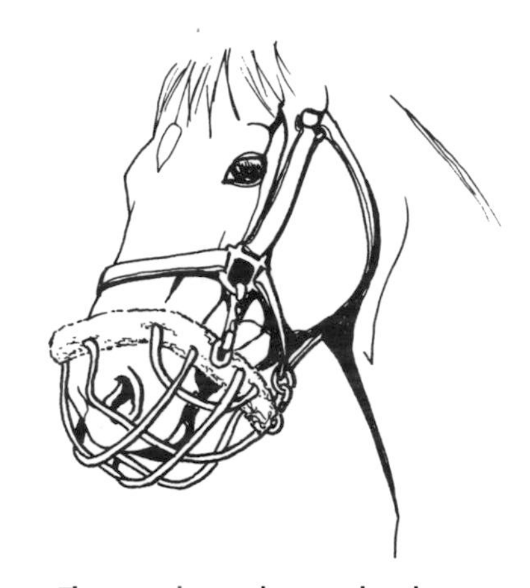

This muzzle attaches to a breakaway halter and should be checked regularly for proper fit.

Mutton-withered

Having poorly defined and well-padded withers. This conformation is common in ponies, and it usually requires a fairly tight girth or cinch to keep the saddle from slipping forward or sideways. A **crupper strap,** which attaches to the saddle and runs under the tail, may also be helpful, as may a breastplate, which attaches to each side of the saddle and runs across the horse's chest.

Muzzle (device)

A device that prevents a horse from opening his mouth. There are several reasons why you might find it necessary to muzzle your horse. If he is tearing at a bandage or chewing at a wound, a muzzle will give the injury a chance to heal. A muzzle will also keep a chronic wood chewer off your fencing. In some cases, broodmares are muzzled while turned out with their foals. The babies often need the rich pasture more than their mothers, who might put on too much weight if allowed to graze all day.

Muzzle (horse's)

The part of a horse's face comprising his lips and nose. A horse's muzzle is as much a part of his senses as his keen sight and acute hearing.

➢ *See also Anatomy of the horse; Muzzle hair; individual senses*

Muzzle hair

Horses have whiskers for a reason. Like other animals, horses need these stiff facial hairs to help them find their way through narrow spaces and in dimly lit areas. Horses have extremely sensitive lips and muzzles and can use their lips to find very small items on the ground that they can't see. Be aware of this when trimming muzzle hair and make sure it's something that is absolutely necessary before undertaking it.

Mycotoxins

Fungal toxins that can contaminate grain and other food sources. They damage the liver and kidneys and destroy red blood cells.

➢ *See also Feeding and nutrition; Grain; Mold (fungi) in feed*

Mythology, the horse in

The horse figures in the myths and legends of many different cultures. Perhaps the most familiar is Pegasus, the winged horse of Greek mythology who assisted the hero Bellerophon and later became a constellation.

Gods associated with the sun often drive chariots drawn by fiery horses. Although Apollo is associated with the sun in Greek mythology, Helios is the actual charioteer who steers the chariot of the sun across the sky. Surya, a Hindu deity, drives seven horses or, in later traditions, one seven-headed horse. In his incarnation as the sun stallion, Vivasat, he sired twin sons called the Ashvins, horseman who were benefactors of mankind. Poseidon, Greek god of the sea, also drove a chariot, pulled by horses with golden manes.

Not surprisingly, the patron gods of warfare are often associated with horses. Epona, the Celtic "Great Mare," was worshiped by Roman cavalry and probably reflects the important role of the horse in early Celtic society. Romans also revered Castor, a horse tamer, and his twin Pollux, the sons of Jupiter who were believed to protect horse soldiers in battle.

The Hindu god Vishnu has many incarnations, the final one being Kalki, a warrior on a white horse (sometimes portrayed as the horse itself), who drives evildoers from India and saves the true believers.

Less heroic and noble are the ferocious flesh-eating mares of Diomedes, which become tame when Hercules feeds their owner to them. Bringing the mares to King Eurystheus is one of his twelve labors.

Instead of having man-eating horses, ancient Scandinavians evidently sacrificed sacred horses to the god Frey, as well as holding horse races and stallion fights in his honor.

There was no idling-time for a pony-rider on duty. He rode fifty miles without stopping, by daylight, moonlight, starlight, or through the blackness of darkness — just as it happened. He rode a splendid horse that was born for a racer and fed and lodged like a gentleman; kept him at his utmost speed for ten miles, and then, as he came crashing up to the station where stood two men holding fast a fresh, impatient steed, the transfer of rider and mail-bag was made in the twinkling of an eye, and away flew the eager pair and were out of sight before the spectator could hardly get the ghost of a look.

—Mark Twain, *Roughing It*

n

Nails, for horseshoes

Nails are used to securehorse shoes onto hooves. It doesn't hurt the horse to have his shoes nailed on, because the nails go into the tough outer wall, not into the sensitive sole or frog.

➢ *See also Foot care; Shoes, horse*

Nasal bone

Like the human nose, the equine muzzle is made of cartilage and soft tissue. The nasal bone runs along the front of the face to just above the nostrils. The tip of this bone is fragile and sensitive, so care must be taken when fitting a halter and handling the face.

➢ *See also Anatomy of the horse; Halter*

National Associations

➢ *See appendix*

National Show Horse

Established as a breed in 1981, the National Show Horse (NSH) crosses the Arabian with the American Saddlebred to produce a flashy, exciting horse for the show ring. These beautiful horses are valued for their elevated movement, athletic ability, and brilliant presence. They are shown under saddle and in driving classes.

➢ *See also American Saddlebred; Arabian*

Navel (Joint) Ill

Sometimes a diseased uterus in the mare will transmit infection to the bloodstream of the foal before birth, which appears as inflammation in the foal's joints. This is caused by **septicemia,** or infection of the bloodstream. Infection might also be transmitted shortly after birth through the navel, before it is sealed off.

The main symptom of this disease is high fever, which makes the foal weak, dull, and too listless to nurse. Unless treated immediately, the condition can be fatal. Even infection halted early can affect the joints. Tenderness or lameness is the first sign of trouble, and the veterinarian will need to administer large doses of antibiotics, as well as possibly flushing the affected joints to remove destructive enzymes.

Prevention includes attending all births; disinfecting the navel stump as soon as the umbilical cord breaks; making sure the mare is not suffering a uterine infection when she is bred or pregnant, and treating it if she has an infection; keeping the birthing area as clean as possible, free of feces and with new bedding at and right after birth; and making sure the foal gets colostrum or an adequate replacer colostrum serum as soon as possible after birth to introduce needed antibodies.

➢ *See also Joint ill (navel ill)*

Navel stump, of foal

➢ *See Foaling; Joint ill (navel ill); Umbilical cord, breaking*

For want of a nail
the shoe is lost,
for want of a shoe
the horse is lost,
for want of a horse
the rider is lost.

—George Herbert, *Jacula Prudentum,* 1651

NATURAL HAZARDS

There are a number of "natural" hazards that you should be alert for — whether you are trail riding or in your familiar barn. These include:

- ❑ Sudden storms
- ❑ Dogs and other farm animals
- ❑ Snakes and other wildlife
- ❑ Unfamiliar horses in roadside pastures
- ❑ Wasps, bees, and other stinging creatures
- ❑ Hunters, bikers, all-terrain vehicles

See also Poisonous plants; Safety; Trail riding

Navicular disease (navicular syndrome)

Disease of the **navicular bone,** a tiny bone located well inside the hoof, where the deep flexor tendon passes over it. Wild horses rarely suffer from navicular disease, but it is common in domestic horses whose feet are too small in proportion to their weight, whose pastern angles are too steep, who are used for athletic activities not suitable for them, or who are kept unnaturally confined.

The main cause of the degeneration of the navicular bone is not fully agreed upon, even by vets. Once the bone begins to break down, pain and lameness follow. Initially, the horse may seem occasionally tenderfooted in one or both front feet. As the disease progresses, he may stumble frequently and move with a short, choppy stride as he seeks to avoid putting his heels down. There is no cure for navicular disease, but early treatment can relieve symptoms and prevent further damage. Changes in exercise, special shoes, and sometimes arthritis medication can be effective. Surgery, which severs nerves to the hoof, is possible for advanced cases, but only as a last resort.

➢ *See also Lameness; Leg, of horse*

A TINY BONE WITH A BIG PROBLEM

Although small, the navicular bone plays a crucial role in soundness.

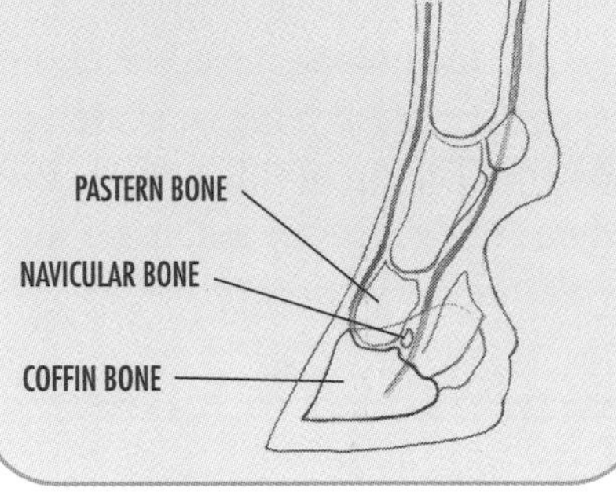

Near side

The horse's left side, even when it's farther away from you.

➢ *See also Off side*

Neat's-foot oil

A pale yellow fatty oil made from the bones of cattle ("neat" is a centuries-old word for the domestic bovine.) Neat's-foot softens stiff, new leather and helps prevent dryness and cracking. It will darken light-colored leather and can be used to stain tack, if you want a darker color. Apply with clean rags until the desired shade is obtained.

Neck

The horse's neck is an important balancing mechanism and should be well proportioned to his body, without a pronounced crest. He should be able to flex his head easily from the poll and move his neck up and down and side to side freely. Proper neck conformation allows him to collect and extend himself as he moves.

➢ *See also Anatomy of the horse; Conformation*

The top horse has correct neck conformation, while the middle horse is cresty-necked (too thick and short). The bottom horse is ewe-necked, or too thin and weak.

Neck-reining

A Western reining style in which the rider holds the reins in one hand (usually the left, if she's right-handed) and instead of pulling against the bit, places the rein on the horse's neck to signal a turn. The horse moves away from the rein, so if you want to turn left, you apply the right rein against the neck and press your right leg against the girth. Western horses are trained to respond primarily to leg and weight cues, with a very light rein.

➢ *See also Aids; One-handed riding; Western riding*

Needs, of horse

➢ *See Facilities; Feeding and nutrition; Shelter; Temperament, of horses*

Neonatal isoerythrolysis (Rh foal)

A condition that arises when a foal is born with a different blood type than his dam. The antibodies that he ingests while nursing attack his red blood cells, causing weakness, anemia, jaundice, and bloody urine. Fatal unless treated, the problem appears a day or two after birth. An alternative source of colostrum must be found and the mare should be hand-milked until her regular milk comes in. After three days of life, the foal's intestines will no longer absorb the antibodies, so he can resume nursing.

Once a mare has given birth to an Rh foal, she should be bred only to stallions with compatible blood types.

➢ *See also Foal care*

Neoprene

A stretchy, nonbreathing fabric used in a variety of horse-related products. It is flexible, waterproof, and airtight and can be molded to various parts of the horse's anatomy. It is often used to wrap tails, as it grips well and doesn't slip off, and for protective boots and neck sweats. It's important to use moderate tension when wrapping neoprene, however, as it can easily be put on too tightly, thus cutting off circulation.

➢ *See also Tail wrapping*

Net or mesh wire fencing

Safer than barbed wire, net or mesh wire makes a good horse fence if the openings are smaller than hoof-sized. It can last a long time if properly installed and reinforced. A pole across the top prevents horses from leaning over the top.

➢ *See also Fencing materials*

Neutraceuticals

Nutritional supplements that have pharmacological effects and are sometimes useful in calming a horse. Consult with your veterinarian about what to use and at what dosage. Also, make sure you test the substance on your horse several times so that you know what the effects are before using it "for real." For example, you might test it during trailering or some other nerve-wracking procedure. Some horses panic over many routine chores — teeth floating, shoeing, injections, a full-fledged bath — and can use a little help settling down.

➢ *See also Handling horses; Tranquilizers, use of; Twitch (restraint)*

Nevada neck-rein

➢ *See Neck-reining*

WHAT YOUR HORSE NEEDS TO BE CONTENT AND HEALTHY

Food. Good hay or pasture (lots of bulk to keep digestive system functioning properly) and some grain (a little goes a long way, especially for horses that aren't working hard).
Water. Fresh, clean, and always available, no matter what the weather (4 to 20 gallons a day).
Shelter. Basic protection from wet, cold, or windy weather or from heat in the summer.
Exercise. Regular riding or other work is good; pasture turnout counts too. A horse that is stabled all the time will develop all manner of health problems (mental and physical).
Companionship. Horses are herd animals and are happiest with another equine companion. Even if you visit every day, a single horse will spend a great deal of time alone. Donkeys, goats, and even cows are better than no company at all.

New Forest Pony

A breed of ponies from the New Forest in Southern England, where they roam in semiwild herds. They are friendly and docile, making them a favorite mount for children. Standing between 13.3 and 14.2 hands, they have expressive faces, long, strong backs, and bushy manes. They can be any solid color.

Nicking

A fortuitous combination of genes that results in offspring of a higher quality than might be expected of either parent. Nicking is sheer breeding luck; the two individuals or family lines involved produce better foals with each other than with other gene pools.

➢ *See also Breeding; Genetics; Pedigree*

Night blindness

An inherited disorder predominantly affecting Appaloosas. Both parents must be carriers in order for the trait to show up, which it does fairly early on. Individuals with night blindness have difficulty seeing at night, and they may also have some impairment of day vision as well. The condition does not worsen over time, unlike moon blindness, which is a different disorder. The eyes of an affected horse look normal, and electronic tests are necessary for diagnosis. There is no treatment for night blindness, and mares or stallions that have produced offspring with this trait should not be bred again.

➢ *See also Periodic ophthalmia*

Nightshade (*Solanum* spp.)

The nightshade family includes garden cultivars such as eggplant, tomato, and potato. The fruits are not toxic, but other parts of the plant are.

➢ *See also Poisonous plants*

Night vision

Horses have better night vision than humans do, but their eyes take longer to adjust when moving from shade to light.

➢ *See also Vision, of horse*

Noninfectious diseases

➢ *See Diseases*

Norman cob

➢ *See Cob*

Noseband

The part of a halter or bridle that fits around the horse's nose, above the bit.

➢ *See also Bridle; Halter*

POISONOUS PLANT

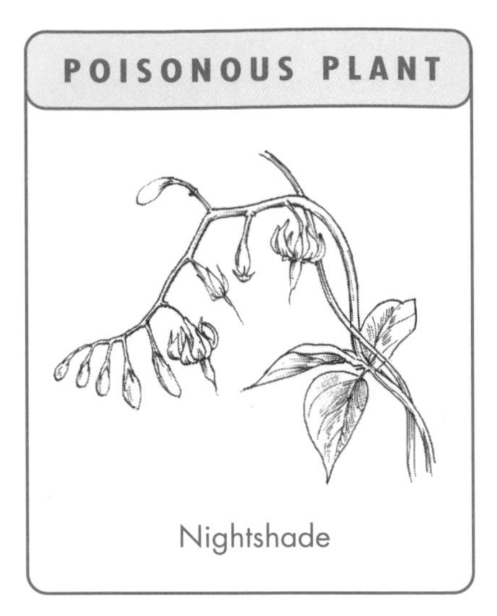

Nightshade

"NO JERKDOWN" RULE

In calf roping competition, rodeos and some associations have a rule to prevent the horse and rider from roping a calf and then stopping so suddenly that the calf flips over. The idea is for the horse to hold the calf at the end of the taut rope while the rider approaches. It's then up to the rider to throw the calf.

See also Calf roping; Cattle roping; Team roping

Nose chain

A chain that runs through the cheek rings and over or under a horse's nose so that it corrects him if he throws his head around. Nose chains can be useful in handling a horse that is nervy or badly behaved, and they are routinely used when leading stallions. You can buy a chain separately that can be attached to any lead rope, or purchase a lead rope with a chain at one end. The chain should be applied intermittently, not with constant pressure, and should be released the instant the horse behaves appropriately. *Caution:* Never tie a horse with a nose chain.

➢ *See also Handling horses; Physical force, use of; Punishment*

NOSTRILS

A horse's nostrils should be large so he can draw deep breaths. Horses do not breathe through their mouths.

See also Respiration; Respiratory system

Novice

An inexperienced rider. In most shows, the novice division is for riders who have not competed very often. After winning several first-place ribbons or earning a certain number of points in specified shows, the rider graduates to higher levels of competition.

➢ *See also English competition; Maiden class; Western competition*

Nursemare

A mare that nurses a foal not her own. It is often possible to persuade a new mom to accept an orphan, but it is a tricky process. The mare is most likely to bond with a second foal if her own has died or if you introduce the orphan right after her own foal is born.

➢ *See also Foaling*

A new mother or one with a sore udder may kick or bite at her foal when it nurses.

Nursing

A healthy foal will be on his feet and nursing within two hours of his birth. It is crucial that he nurse within several hours, as that is when he is most able to absorb the antibodies provided by the mare's colostrum. Most foals do not need assistance finding the teat, although it may look as though they're awfully confused at first. The mare might squeal and kick a little if her udder is sore but rarely enough to prevent her baby from nursing. If she is serious about keeping him away from her, you'll have to restrain her to help the baby get his first few feedings.

➢ *See also Foaling; Milk replacer; Weaning foals*

Nursing hobbles

A device applied to a nursing mare's hind legs when there is reason to believe she may kick the foal — for example, when she is being induced to take on an orphan. Hobbles are generally lined with fleece so they won't cut into the mare's legs, and they are used only until she accepts the foal she is expected to nurse. They have enough give to allow her to walk around the stall.

➢ *See also Foal care; Nursemare*

Nutrition

➢ *See Feeding and nutrition; Grains; Hay; Supplements; Vitamins*

O
36

Oat hay, cautions

Oat hay can be a suitable component of a horse's diet if it is cut before the seed heads mature. The mature straw is hard to digest and offers little nutritional value. It often causes gas colic, which can become life threatening. Under certain growing conditions, oat hay may contain nitrates which are harmful to horses.

➢ *See also Feeding and nutrition; Hay*

Oats are a favorite horse food and offer a good balance of fiber and protein.

Oats

A staple food that horses love, oats are nutritious and offer a good balance of fiber (roughly 11 percent) and protein (9.4 percent). A bushel of oats weighs between 32 and 40 pounds. Oats can be safely fed in quantity; they are about 30 percent hull, which gives bulk and keeps the feed moving along the digestive tract.

➢ *See also Feeding and nutrition; Grains*

Obstacles, negotiating

It is important for a horse to be comfortable going over, around, and through a variety of obstacles. There are a number of ways to train a horse at home so that you can safely ride on trails or in fields or compete in trail classes.

Obstacles can be constructed of poles on the ground, railroad ties forming a box or platform (covered with rubber mats), barrels, and many other objects. Gradually accustoming your horse to going through narrow spaces, through water, on and off platforms, and over poles will increase your confidence in each other. This training will also help when teaching your horse to load onto and unload from a trailer.

➢ *See also Trail class; Trail riding*

OCD

➢ *See Osteochondritis dessicans (OCD)*

Offset knees

When viewed from the front, the bones of the upper and lower leg should create a straight line through the knee. Offset, or crooked, knees are a serious fault that often leads to downtime from lameness, not to mention expensive vet bills.

➢ *See also Conformation; Leg, of horse*

Off side

Refers to the right side of a horse, no matter which side is closer to you. The left side is the "near" side. Although horses are usually led from the near side, it makes sense to accustom your horse to being led from the off side as well. Off and near side should not be confused with inside and outside, which require references to the direction you are riding around a ring or circle.

Off the bit

A fully trained or "finished" Western horse is said to work "off" the bit, because the rider exerts very little pressure on the mouth, relying primarily on the seat and legs as aids.

While a horse in a bosal is working completely "off the bit," that is not what the term means.

Oilseed meals

A very concentrated protein extracted from seeds (for example, linseed, cottonseed, and soybean), which should be fed in limited amounts as a supplement to grains and hay. A horse should never be fed more than one pound of oilseed meal a day.

➢ *See also Feeding and nutrition; Supplements*

Oldenburg

Originally bred as a powerful and elegant carriage horse, this German warmblood has been crossbred with Thoroughbreds and other lighter horses to produce an all-around sporthorse with light, springy action and good jumping ability. Oldenburgs are big, muscular horses, standing 16.2 to 17.2 hands high. They are generally dark in color, with chestnuts becoming more common. They are known for their energetic yet equitable temperament.

➢ *See also Breed; Warmblood; appendix*

The Skin Horse had lived longer in the nursery than any of the others. He was so old that his brown coat was bald in patches and showed the seams underneath, and most of the hairs in his tail had been pulled out to string bead necklaces.

—Marjorie Williams Bianco, *The Velveteen Rabbit*

Older horses

A horse is considered in his prime at around eight to ten years old, but many horses can be ridden into their twenties, though not as rigorously. Horses commonly live into their thirties, though life span varies greatly depending on a number of factors, such as breed, individual bloodlines, life experience (humane vs. inhumane treatment, athletic vs. sedentary lifestyle), level of care, and the immeasurable variable known as "heart" or will to live.

The bottom line for owners or caretakers is learning how to adapt the care of an older horse to its changing needs. Some older horses are easy to manage and have relatively few special needs. Others require a level of care that could be compared to the geriatric care given to humans in nursing homes. The best way to learn how to care for your aging friend is to work closely with your veterinarian and another horseman who has cared for an aging horse in a manner you respect.

Older horses can provide years of enjoyment and productive use if cared for properly. A beginning rider will generally do better on an older horse with plenty of training and handling, while younger horses need more experienced riders. So beware not to miss an excellent investment in a solid teacher or outstanding breeding animal by avoiding a horse just because it is beyond middle age.

➢ *See also Age; Feeding and nutrition*

Oleander *(Nerium oleander)*

An ornamental shrub common in warm areas that is fatal to horses if even small amounts are ingested.

➢ *See also Poisonous plants*

Olympic equestrian events

For many years the only disciplines represented in international competition were Dressage, Show Jumping, and Eventing. However, Combined Driving was added to the Olympic Games in 1979, Endurance Riding in 1992, and Reining in 1998.

Dressage dates back to the Renaissance, when it gained recognition as a great training method for cavalries. The sport was introduced into the Games in 1912 and was open only to military riders. By 1952, the rules opened the event to everyone.

Show jumping has been popular since the early 1800s. It remains the most popular Olympic equestrian event. Grand Prix show jumping was first established in 1866 in Paris. The first civilian won the gold in the 1952 Helsinki Games. Prior to that time, this event was also only open to military riders.

The triathlon of equestrian competition is the Three-Day Event. It combines dressage, cross-country, and show jumping. With its debut at the 1912 Stockholm Games, it remains the ultimate test of horse/rider teamwork.

➢ *See also Show associations*

In one-handed riding, the reins are generally held in the left hand, leaving the right hand free.

On deck

A term describing the next rider in a competition. Typically, an announcement is made over the public address system at a show to signal that the next competitor will take her turn. For example, "Number 33 is on deck" means that rider number 33 is next to compete.

One-handed riding

While English riders usually keep two hands on the reins, Western riders generally ride with the reins in one hand. Also called **neck reining,** one-handed riding leaves the other hand free for swinging a rope, opening a gate, carrying a tired calf, or firing a weapon.

➢ *See also Neck-reining*

One-on-one (penning)

A Western event in which a single rider separates one cow from a small herd and pens it in a portable pen within the arena.

➢ *See also Team penning*

On the bit

When a horse is responding to the rider's aids and moving forward in a balanced and willing manner. The term describes a "feel" that advanced riders have when they are effectively communicating with their horse. A horse is said to be "on the bit" when he is moving forward freely, yet yielding to the pressures of the bit. This requires an advanced level of skill that reflects a rider's ability to coordinate the natural aids. The rider's legs and back drive the horse forward to the bit (keeping him "on the bit"), while the rider's hands keep a sensitive contact with the horse's

mouth. Heavy mouthed horses often are the result of riders who lack this coordination of aids, riders who rely too heavily on their hands.
➢ *See also Collection*

On the forehand

A horse moving on the forehand has too much weight on his front legs and looks unbalanced. Because the hindquarters provide the primary propulsion, a horse needs to distribute his weight evenly to be engaged and moving well. Correct use of the aids helps the horse shift his weight off the forehand and become balanced.
➢ *See also Aids; Collection; Turn on the forehand*

> **ONE-EAR BRIDLE**
>
> Most Western bridles do not have a noseband. Some don't have a brow band, either, but instead have a loop that goes around one ear to keep the bridle in place.

Open-front blanket

A blanket that has buckles across the chest.
➢ *See also Blankets and blanketing; Sheet*

Open shows

A show in which any breed can compete in the events.
➢ *See also Show associations*

Ophthalmoscope

A device used by veterinarians to examine a horse's eye.

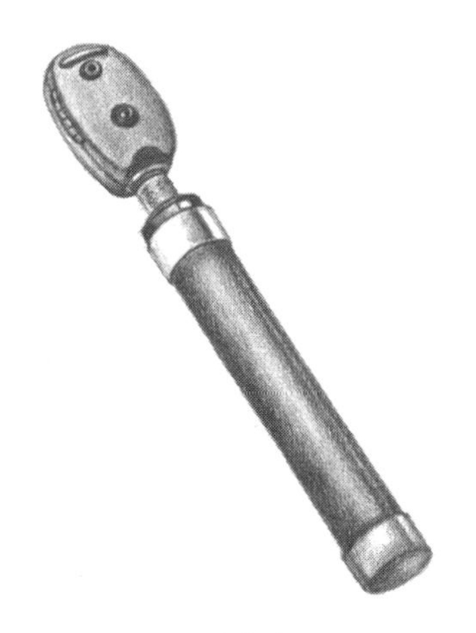

An ophthalmoscope

Orphan foal

While raising an orphaned or abandoned foal isn't easy, it can be done. The preferred method is to introduce the orphan to a nursemare as soon as possible after the birth of her own foal. Many mares will raise a second foal as a twin or can be persuaded to accept an orphan after losing their own foal. Using a camphor rub on the mare's nose to confuse her sense of smell and rubbing the afterbirth of her own foal on the orphan helps the bonding process. An adoptive mare should be carefully watched for several days until she has fully accepted the strange foal. She may need to be hobbled at first to prevent her kicking the newcomer, or she may need to be separated from him by a see-through barrier in the stall between feedings.

If a nursemare isn't available, it is possible to raise an orphan by hand, but it is just as much work as caring for a human baby. Newborn foals nurse every half hour for the first few days of life and hourly for several weeks after that. They begin nibbling at their mother's hay and grain at just a few weeks old but need to nurse for at least two months; they are generally not fully weaned until they are five or six months old. Once he's strong enough, a hand-raised orphaned foal should live with a gentle older horse that can help him learn about proper equine behavior.
➢ *See also Foal care; Foaling*

Orthocide

➢ *See Captan*

Osteochondritis dessicans (OCD)

A developmental disorder in growing foals caused by overfeeding of high-protein and high-energy foods. Foals need a carefully balanced diet and plenty of exercise to develop strong bones. Too much protein can cause inflammation and interference with the normal development of cartilage into bone. A high-protein diet can also create weak spots in the legs of young horses, which can result in further inflammation, tearing, breaking, or twisting. Stress on vulnerable joints and growth plates can produce deformities, abnormal growth of the long bones, and lameness.

Outcrossing (outbreeding)

The opposite of inbreeding, outcrossing is the mating of unrelated animals.

➢ *See also Breeding; Genetics; Inbreeding*

"Out in the country" leg conformation

Refers to the hind legs being set too far back, rather than lining up straight under the haunches. Also called **camped out.**

➢ *See also Conformation; Leg, of horse*

Outside (leg or rein)

The outside aids are those on the outside of the bend or direction of movement. In other words, when circling, the outside aids are the ones on the outside of the turn. If you are making a clockwise circle, your outside aids are your left leg and rein. On a counterclockwise turn, the outside aids are your right leg and rein.

➢ *See also Aids; Hands, rider's; Legs, use of while riding*

The outside aids change sides depending on which direction you are riding. Here, the right leg and hand are "outside" aids.

Ovaries and oviducts

The ovaries produce eggs, which travel down the oviducts to the uterus where the egg is fertilized and the embryo is formed and nurtured.

➢ *See also Anatomy of the horse; Breeding; Reproductive tract*

"Over and under" quirt

A long, ropelike whip that is sometimes used in timed Western events. It can be held against the leg, where even a slight swing can be effective in urging a horse forward.

Over at the knee (buck knees)

A conformation flaw in which the knees look as though they are slightly bent forward or even buckled. Buck knees, while unattractive, are not as serious as the opposite fault, called **"back at the knees."** Many buck-kneed horses perform quite well.

➢ *See also Back at the knee; Conformation; Leg, of horse*

Overbite (parrot mouth)

Just like people, horses can have buck teeth, although they don't get braces to fix them. A mild overbite, also called parrot mouth, is nothing to worry about, as long as your horse can comfortably eat and wear a bit. Serious overbite can interfere with eating and wearing a bit, and may cause long-term problems.

➢ *See also Dental care; Teeth*

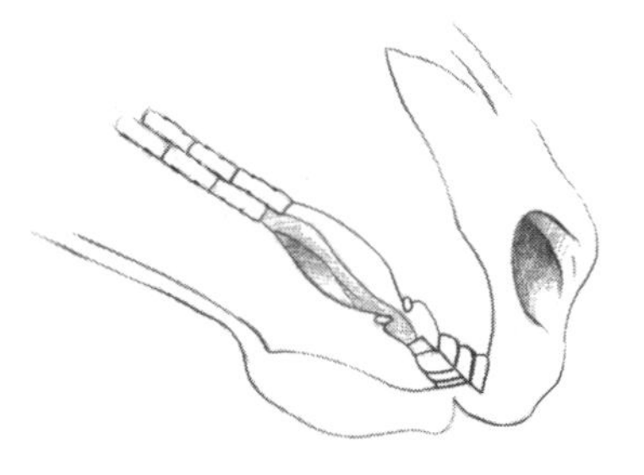

A horse with an overbite should not be bred, as the condition is heritable.

Overfeeding

Feeding a horse more than he needs to maintain adequate body weight and energy output can lead to serious problems. A fat horse cannot perform well and is prone to unsoundness and other health problems. An overfed foal may develop permanent deformities of the leg.

Overfeeding of grain, apples, or other treats (even hay or grass if a horse is unaccustomed to it) can lead to colic, laminitis, and other illnesses.

➢ *See also Feeding and nutrition*

Overheating

While many horses are blanketed both indoors and out during the winter (or during cold, windy, or rainy weather), it is important to judge their needs. A heavy, warm blanket may be necessary for cold nights in a stable but could easily be too hot for a sunny, still day in the pasture, even if the temperature is low. An active horse can work up a sweat in a too-heavy blanket and then become chilled, which saps his energy and could lead to respiratory problems.

➢ *See also Blankets and blanketing; Sheet*

Riding a horse that is too much for you to handle can get you into all kinds of trouble.

Overmounting

Getting on a horse that is too strong or too temperamental for you or trying to maneuver a horse through an exercise that you or he is not ready for. Being well-matched to your mount increases your chances of riding safely, enjoying yourself, and continuing to learn.

Overo

According to the American Paint Horse Association, an overo meets these specifications: An overo may be either predominantly dark or white; generally, the white is irregular and is rather scattered or splashy; the white usually will not cross the back of the horse between his withers and tail; generally, at least one and often all four legs are dark; head markings are distinctive, often bald-faced, apron-faced, or bonnet-faced; the tail is usually one color.

➢ *See also Paint and Pinto; Tobiano; Tovero*

Overreaching

A fault in moving during which the hind foot catches the heel of the front foot on the same side before the forefoot leaves the ground. Also called **grabbing,** overreaching often results in lost shoes. Bell boots offer protection against lost shoes and injury to the heel.

Overstepping

Placing the hind foot beyond the hoofprint of the same side's forefoot.

Ovulation

Although they are in heat for anywhere from 1 to 10 days, mares generally ovulate 24 to 48 hours before their heat cycle ends. It is common practice to breed a mare on the third day of her cycle and every 36 hours after that until she goes out of heat. It is also possible to palpate a mare to determine the time of ovulation and decrease the number of breedings, but this may disrupt her cycle and create enough stress to impede fertilization.

➢ *See also Breeding; Estrus*

Oxer

A type of jump with two rails that maybe even or uneven.

➢ *See also Jumps, types of; Spread*

Oxibendazole

A worming agent effective against small strongyles.

➢ *See also Strongyles*

Oxygen deprivation

A condition that affects foals during the birthing process. Foals can suffer birth defects similar to humans as a result of oxygen deprivation during foaling. Although this condition is rare, it supports the need for human observation during the foaling process, especially of mares with unknown foaling histories. The primary risk comes during an abnormal birth presentation that could result in the umbilical cord wrapping around the foal's neck, or the cord breaking long before the foal is able to take its first breath.

➢ *See Foaling; Pulse and respiration*

Oxyuris equi (pinworms)

Pinworms are not as harmful as other types of worms, but they can cause a great deal of itching around the anal area, where females lay their eggs. A horse with pinworms may break much of the hair off his tail or develop sores and secondary infections from rubbing, and he may lose weight if the itching is severe enough to interfere with grazing.

Pinworms are more common in horses kept in small areas than those pastured in roomy fields. The parasites live in the large intestine and infection occurs when horses bite or lick themselves where eggs have been laid or eat feed contaminated by infected manure. Pinworms can be treated with a good deworming medication. The stall and stable area also must be thoroughly cleaned to eliminate eggs that may stick to walls, floors, and bedding.

➢ *See also Parasites*

Take the life of cities!
Here's the life for me.
'T were a thousand pities
Not to gallop free.

—Anonymous, *Riding Song*

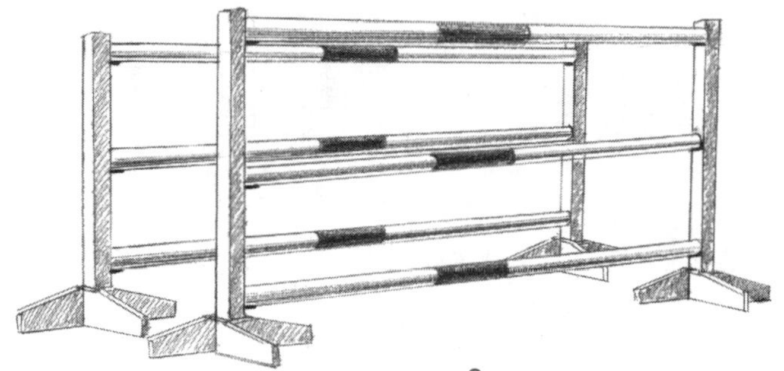

Oxer

p

Pace

A gait in which the front and back leg of the same side advance simultaneously, so that the two left legs strike the ground together and then the two right ones. Some Standardbreds exhibit this gait naturally, though other Standardbred bloodlines produce trotters. A few individuals can go at both gaits, but generally horses perform better at one or the other. Sometimes horses that are slow at the trot can be successfully switched to pacing in mid-career, but changing from the pace to the trot is uncommon. The pace is generally a little faster than the trot, and although pacers were ridden in Colonial times, nowadays the gait is seen almost exclusively in harness racing.
➢ *See also Gait; Standardbred*

In the pace, the legs on the same side advance simultaneously.

Pacer

A Standardbred that naturally paces rather than trots.
➢ *See Pace; Standardbred*

Pacing, habit of

Some horses, like other confined animals, pace along the perimeter of their pasture fence or back and forth in their stalls. This is often a sign of boredom or distress.

My foot in the stirrup,
my pony won't stand,
Good-bye, Old Paint,
I'm a-leavin' Cheyenne.

—Traditional cowboy song

Paddling

A characteristic of horses with turned-in toes, who often swing their feet outward in an unattractive "paddling" motion when trotting.
➢ *See also Anatomy of the horse; Conformation; Leg, of horse*

Paddock

A fenced area near a barn or stable that is used for exercise. Paddocks can range from quite small to an acre in size; larger areas are referred to as pastures. **Corral** is a Western term for paddock. Smaller paddocks tend to be drylots without grass.
➢ *See also Facilities; Pasture*

Paint and Pinto

Both terms refer to horses with large splotches of white and another color (usually black, bay, or chestnut), but a **Paint** is specifically a Western-type horse with a stocky build. In order to be registered as a Paint, a horse must have two registered Paint parents or one registered Paint parent and one Quarter Horse or Thoroughbred parent. Recent rule changes designed to allow more Paints to race permit the registration of a horse with two Thoroughbred parents, but only if his coloring is correct.

The term **Pinto,** on the other hand, relates strictly to coloring, so a horse of any breed (for example, Saddlebred or Shetland pony) can be registered as a Pinto if his coloring is correct. It is not uncommon for Paints to also be registered as Pintos, increasing the number of shows they can enter.

More specific terms for the coloring of these horses include **tobiano** and **overo.** A tobiano has big blocks of color across his body, and his face is usually, but not always, colored with a star or blaze. His legs tend to be all or mostly white, and his tail is often two colors. An overo has spots patterning his belly and spine, with more white on his sides. His head may have a lot of white, even going over both eyes — this pattern is called bald-faced, and it is often accompanied by blue eyes. The Paint registry calls a horse showing characteristics of both tobiano and overo coloring a **tovero**.

Other terms for Paints and Pintos that you may hear are **piebald** and **skewbald.** A piebald horse is black and white, while a skewbald horse is white and any other color (usually red or brown).

While many people love Paints because of their wonderful colors, be aware that they are sometimes bred to have small, tidy feet that can create problems as the horse matures and gains weight. Paints can be susceptible to navicular disease, a degenerative bone disease of the hoof, and to hyperkalemic periodic paralysis (HYPP), a genetic illness that causes tremors, mild to severe seizures, and sometimes even death.

➢ *See also Coloring*

Every Paint is a Pinto, but not every Pinto is a Paint. A Paint is a breed, while a Pinto is a coloring.

Palisade worms

➢ *See Bloodworms*

Palomino

A bright golden color ranging from pale honey to deep copper. The mane must be at least 85 percent white or silver, and white markings on legs and face are permitted. Because Palomino is a color, not a breed, any breed of horse that is the correct color can be registered with the Palomino Horse Breeders of America (PHBA).

➢ *See also Coloring*

PALFREY

Archaic term for a light riding horse, intended particularly for a lady's use.

Palpation

Examining or exploring by touch. Palpating can refer to either an external examination (checking legs for warmth or swelling) or an internal examination (checking for uterine problems). Internal palpation is a highly risky procedure that can lead to fatal complications, even when performed by highly experienced veterinarians. So be sure to discuss the necessity of the procedure with your veterinarian, and weigh the risks against the expected benefits. Despite the popularity of rectal pregnancy exams and the new ultrasound technology, these are not always the best choice for all mares.

P&R

➢ *See Pulse and respiration*

Panic snap

A clip for a lead rope, cross tie, or trailer tie that can be opened with one hand in an emergency. Some panic snaps are designed to release if a horse falls or gets hung up; excessive weight or force triggers release.

➢ *See also Bull snap*

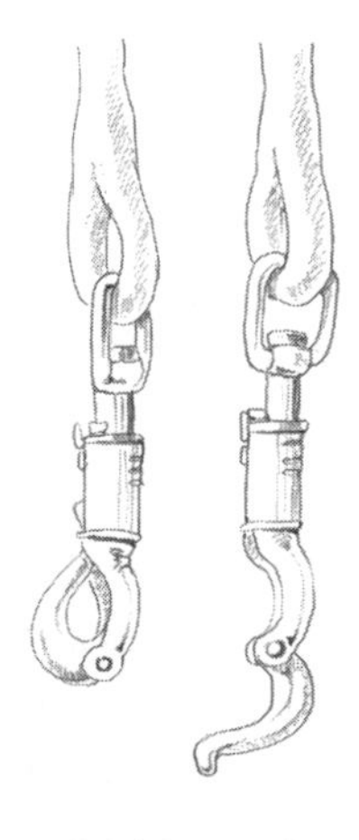

When using a lead rope with a panic snap, be careful not to inadvertently release the catch while leading your horse.

Pants for riding

While many people happily ride in jeans, and, in fact, nearly all Western riders prefer jeans, most English riders favor breeches (worn with tall boots), jodhpurs (worn with ankle boots), or some type of leggings or tights. Riding pants should be comfortable, fairly close-fitting, and somewhat stretchy. They should not be made of shiny or slippery fabric or you'll have trouble staying in the saddle. There are many different pant styles and colors available, including jeans with some stretch in the fabric. Pants designed especially for riding often have reinforced inner knees and seats to protect against wear.

Color is strictly a matter of preference, except in competitions where rules determine attire. For many riders, the color of their pants depends on the color of their horse. Darker colors don't show dirt and may be more practical — but perhaps not if you ride a gray horse.

If you do ride in jeans, make sure that they are not so tight that they hamper your getting on the horse, or so baggy that they bunch up and chafe your legs. Also be aware that thick inseams can be very uncomfortable and can rub your skin raw quite quickly.

➢ *See also Attire; Breeches; Chaps; Jodhpurs*

Riding pants should be comfortable, close-fitting, stretchy, and nonslippery.

Parascaris equorum

➢ *See Ascarid; Deworming; Parasites*

Parasites

All horses are vulnerable to a variety of parasites whose presence ranges from merely annoying to dangerous or even fatal. Common internal parasites include bloodworms (also called palisade worms or large strongyles), bot flies, lungworms, pinworms, roundworms, small strongyles, tapeworms, and warbles. Signs of intestinal parasites may include general debility or unthriftiness; a harsh, dull coat; unexplained bouts of colic; a potbelly; and loss of appetite.

Most internal parasites conduct their life cycle through the digestive system; that is, the larvae hatch in manure, are ingested by the horse as he grazes near infected manure, and mature into worms, which lay eggs in the intestines and start the cycle all over again. The worms themselves can cause serious problems by burrowing into organs and blood vessels, depriving the horse of nutrients and causing internal bleeding, which can lead to anemia and ulcers. Scar tissue from repeated infestations narrows the arteries and decreases the flow of blood to the legs and intestines. Large masses of worms can cause intestinal blockage and torsion of the gut.

The bad news is that all horses have some of these parasites all the time. The good news is that a conscientious deworming program and good stable hygiene can keep the infestation under control and keep your horse healthy. Adult horses should be dewormed every couple of months with one of several dewormers available from horse-supply stores or your veterinarian. Although foals will eventually build up some immunity to parasites, young horses should be dewormed more frequently, as they can become quite ill from an overload of worms.

PESKY PARASITES

Parasites affect all horses and can only be controlled, not eradicated. Discuss with your veterinarian the best way to manage parasites found in your area. If you do not administer your own worming medication, make sure your barn manager has all the horses in the barn on a regular worming schedule.

External parasites to be aware of are flies, gnats, ticks, lice, mosquitoes, and a variety of mange mites. The buzzing, biting type of pest is just as annoying to your horse as they are to you. Large swarms of gnats, face flies, and stable flies can drive a horse to distraction, preventing him from feeding and causing injuries from kicking and biting at the pests. Individual insect bites can be painful; hundreds of them can cause severe blood loss and open the door to skin infections and lesions. Many external pests also carry diseases.

Good stable hygiene can reduce pest problems considerably. Manure management, judicious use of insecticides, both in the stable and on the horses, pest strips, and the cleaning up of damp or swampy areas are ways to cut down on the parasite population. If there are cows on the property, rotating grazing areas between cows and horses can also help, as many parasites are host-specific.

Flies can be controlled somewhat by introducing predator wasps, which prey on the pupae of flies that lay eggs in manure. These tiny creatures are nocturnal and are not harmful to people or animals. However, they are just as vulnerable to insecticides as the pests you are aiming for, so use care when spraying. Another nonchemical fly control is diatomaceous earth, an inert feed additive that passes into the manure and inhibits pest reproduction. Read the product label carefully for amounts and proper usage. In addition, apple cider vinegar acts as a fly repellent when added to feed or water — mix ½ to 1 cup with the horse's daily grain, starting with a few drops to accustom the horse to the taste and determine the minimum effective amount for each horse. Flies will still land but won't bite, perhaps because the horse smells unpleasant to them. Vinegar can also be used externally on the coat to repel flies.

➢ *See also Deworming; individual parasites and medications*

Most parasites are ingested as larvae during grazing. They mature in the digestive tract and lay eggs in manure to start the cycle again.

There's nothing so good for the inside of a man as the outside of a horse.

—Attributed to Lord Palmerston

Park horse

An English performance class. Park horses have a flashier, more elevated action than do English pleasure horses, yet are not as brilliant as animated show horses.

The criteria used for evaluating park horses include: outstanding refinement and elegance; suitability of horse to rider; manners; expression; willingness; quality; gaits performed with brilliance on command.

➢ *See also English competition; Pleasure horse*

Parrot mouth

➢ *See Overbite (parrot mouth)*

Parturition

The process of delivering a foal.

➢ *See also Breeding; Estrus; Foaling; Ovulation*

Pasi-trote

A term used by Peruvian Paso riders to refer to an undesirable, uneven diagonal gait with four beats.

➢ *See also Paso llano; Peruvian Paso; Sobreandando*

Paso Fino

Paso Fino horses were brought to the New World by the Spanish, who transported them in slings on the top deck of ships and pushed them overboard to swim ashore when the months of sailing were finally over. From those tough beginnings came a sturdy, compact horse with a powerful neck and flowing mane and tail. Bred in Peru, Puerto Rico, the Caribbean, and Central America, Pasos are amblers, developed for stamina and smoothness of gait. Many of them have a distinctive gait in which the front feet swing forward and to the side, an adaptation of moving through heavy sand. A good Paso is extremely comfortable to ride, can be quite fast, and will go all day.

Popular in California but less well known elsewhere in the United States, Pasos are common in Puerto Rico and Central America, where they are used for ranch work. They can be any color and tend to be fairly small (14 to 15 hands high).

➢ *See also Amble; Gait; Gaited horse; Peruvian Paso; Running walk*

Paso Llano

The most desirable gait demonstrated by the Peruvian Paso. This four-beat gait is very evenly timed and falls between a pace and a trot. The footfall pattern is: RH, RF, LH, LF.

➢ *See also Pasi-trote; Peruvian Paso; Sobreandando*

Passage

A slow, measured trot with the knees and fetlocks lifted very high, performed by high-level dressage horses.

➢ *See also Dressage; Lipizzaner*

Passing, on horseback

When two or more riders are sharing a ring, there is some potential for confusion and collision. Just as you follow rules of the road when driving a car, you need to know the rules of the ring when riding a horse. Enter the arena only when the gate is clear and move quickly to the center to avoid getting in the way of riders on the rail. More experienced riders are expected to keep an eye out for beginners. Keep your horse at least 8 feet — one horse length — away from other horses. If your horse is crowding the horse in front of you, make a small circle or cut across the ring without getting in thc way of other riders.

Paste dewormers

A premeasured dose of deworming medication that comes in a ready-to-administer syringe.

➢ *See also Deworming*

MORE PASO TERMS

Here are more terms used to describe Paso gaits:

Paso corto. A moderate gait demonstrated by the Paso Fino horse, executed with medium length of stride and extension.

See also Paso Fino

Paso largo. The fastest speed of the smooth gait demonstrated by the Paso Fino horse. Actual speed will vary with individual horses.

See also Paso Fino

PASSING ETIQUETTE IN THE RING

- ❑ Try to avoid passing on the outside (between the horse ahead of you and the rail); doing so places you and your horse in danger.
- ❑ If you must pass on the outside, let the rider of the horse you are passing know you are passing by saying "Rail" loudly. (This is a good idea even when passing on the inside.)
- ❑ When passing a horse that is coming toward you, always remember the rule, "Left shoulder to left shoulder," and stay to the right of oncoming traffic.

Pastern

The angled portion of the horse's lower leg, located just above the hoof and below the fetlock. The pasterns act as the primary shock absorber for the legs and should be well-proportioned and set at approximately a 50-degree angle. Short, upright pasterns can mean a jolting ride and eventual damage to the horse's feet and legs, while long or sloping ones are weaker and prone to strain and injury.
➢ *See also Anatomy of the horse; Conformation; Leg, of horse*

Pastern Angles

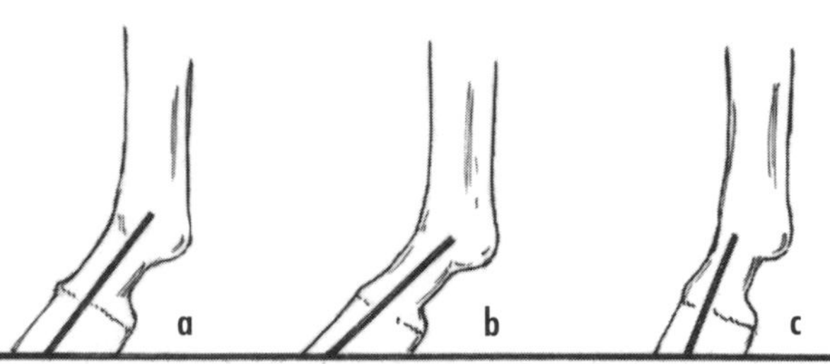

Normal (a); sloping, prone to weakness (b); upright, causing harsh gait (c).

Pasture

As grazers, horses are usually happiest when turned out on grass. In fact, living out in a pasture with a couple of equine buddies and sufficient shelter against the wind and rain may seem like the ideal life for a horse. In many ways, it is. Pastured horses are less likely to suffer from respiratory or gastrointestinal ailments than those confined to stalls all day. They generally get more exercise, strengthening their legs and keeping fitter. It's a lot less daily work for the owner, who can skip the hours of mucking out stalls and sweeping floors that are required in a barn. There is no reason why a healthy horse can't spend all winter, even in a harsh climate, outdoors, as long as he has adequate shelter and food.

However, there are certain disadvantages to keeping a horse on pasture all the time. First of all, just because the horse is carefree doesn't mean the owner can relax. Pastured horses must be checked at least twice daily; partly just to make sure they're still inside the fence! Injuries are more common, as horses can hurt each other, often inadvertently, while playing or establishing dominance. Too much grass can make a horse fat and lazy, and too much grass too soon after a winter without it can cause colic or founder. There is also the problem of catching a horse that isn't interested in being ridden.

Pastures need safe, sturdy fencing and must be free of poisonous plants and other hazards such as old farm equipment or hidden debris. Horses are very hard on grass — they're heavy, have sharp hooves, and eat all the time. Pastures must be large enough, or rested often enough, to support the number of horses kept on them. Overgrazing compacts the soil, encourages the growth of nonpalatable weeds, and can lead to erosion and long-term damage.
➢ *See also Facilities; Fencing materials; Grass toxicity; Poisonous plants*

Pattern

A predetermined series of movements used to educate and evaluate the skills of a horse, handler, or rider. From an educational perspective, patterns are used by instructors of both showmanship and riding students to teach a variety of skills. For example, a showmanship instructor will run students through patterns similar to those used in the showring, in order to teach the skills necessary to excel in that division. Riding instructors often combine movements like circles, figure eights, serpentines, changes of gaits or leads, and work without stirrups into patterns that teach students how to handle their horses safely in a variety of situations. Inherent to such use of patterns is the fact that skills tested are vital to safe handling of horses from the ground (such as stopping, walking, trotting, backing and turning a horse in hand) and while riding.

Patterns used in competition range from those used for showmanship to the highly complex patterns used in reining competition. Judges use patterns to separate the truly skilled handlers and riders from those trained only to have good personal form or equitation. It takes significant skill and coordination of aids to maneuver a horse, whether mounted or from the ground, through most patterns used in competition.

➢ *See also English competition; Equitation; Showmanship classes; Western competition*

Pavement, riding on

When riding along a road, do your best to avoid the pavement by staying on the shoulder. Horses have much better traction on grass and dirt — even dry pavement is a tricky surface for your horse to negotiate. If you can't avoid riding on pavement, do so at a walk. If it's something you plan to do with some frequency, consult your farrier about special shoes or pads that can improve traction.

Pawing (habit)

For a horse that ordinarily keeps his feet to himself, pawing can be a sign of colic or other discomfort. Any unusual pawing should be evaluated in connection with the horse's general demeanor. However, pawing is often a sign of restlessness, nervousness, or boredom, and in those cases it should be discouraged.

Horses paw at each other in play and when establishing dominance, so be careful when working around a group of horses. Horses that are confined for long periods of time may begin to paw out of boredom. Habitual pawers can create dangerous holes in stall floors and paddocks. They can also wear down bare hooves to the point of lameness and cause shoes to wear out sooner. Place rubber mats in stalls and bury them in trouble spots in paddocks (along fences and gates, for example) to help prevent damage. A habitual pawer might respond to the self-correction of a length of chain fastened to a leather strap around the girth; the chain will knock against him whenever he paws. A small wooden block can also be attached to the end of the chain. This measure may seem harsh, but the horse is more likely to hurt himself by pawing than by connecting with the chain a few times.

If your horse paws while tied, make sure there are no insects bothering him, that he's not too crowded (especially in a trailer), and that he's not trying to get at food he can't reach.

Peanut pusher, peanut roller

A term used to describe show horses who work with their heads lower than their withers, giving the appearance of rolling peanuts across the arena with their noses. This frame became popular in the 1980s when

Western Riding Pattern

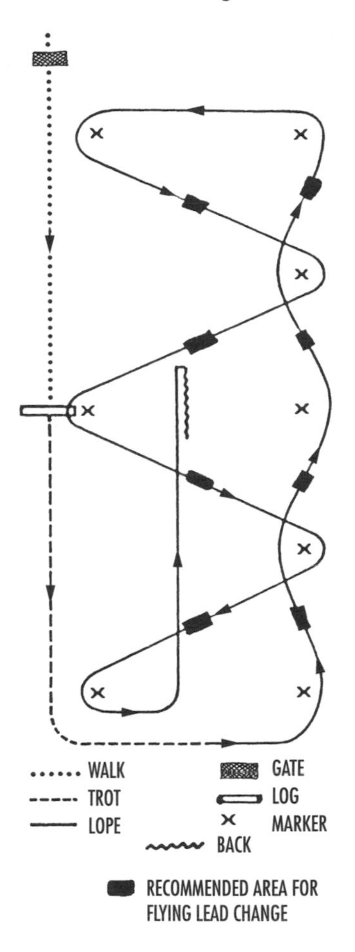

National Reining Horse Association Pattern

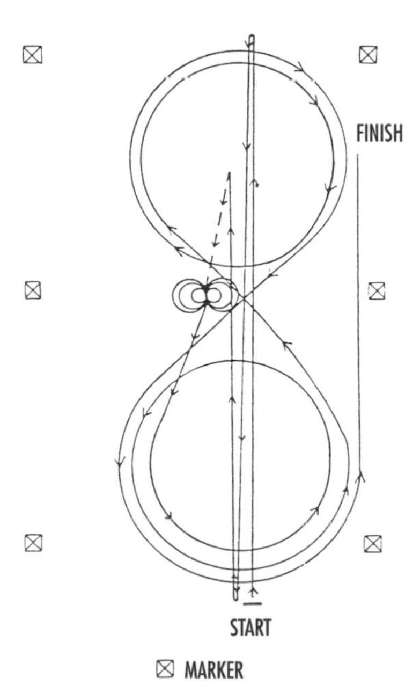

winners at high levels of competition worked in this manner. It was predominant in Western pleasure and horsemanship classes; however, it also was seen among some breeds in hunter classes. Throughout the 1990s, organizations that set rules for the showring defined acceptable standards for the frame of performance horses. The most commonly accepted standard is maintaining a frame in which the poll is carried no lower than the withers.

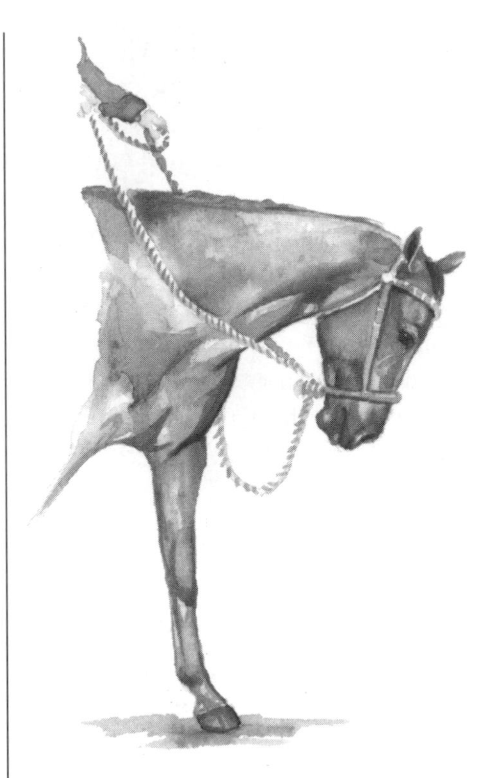

A "peanut roller" carries his head lower than his withers.

Pecking order, in herds

Horses are herd animals whose social order is arranged in a hierarchy or "pecking order" of dominant and passive animals. A herd of horses establishes its pecking order through a series of interactions between individual members of the herd. Horses establish dominance through behaviors ranging from laying back ears to biting and kicking. Each time animals are removed from or introduced into a herd, the rituals to reestablish pecking order are repeated.

Understanding this phenomenon can greatly improve a human's ability to interact with and care for horses. It's important to establish yourself as the dominant partner in your relationship with horses. Otherwise, you'll be on the receiving end of dominating behaviors, such as crowding, biting, and kicking, from the horses you work with.

Sound management of a herd of horses relies on understanding this phenomenon. The top horse eats and drinks first and will often drive lower ranked individuals away, so when feeding a group of horses, it's important to spread out the hay so that all have access. It is also a good idea to introduce a newcomer slowly, preferably from an adjoining pasture at first. If the new horse accepts the current pecking order, his transition to the herd will probably be smooth. But if he challenges the leader, there could be some serious kicking and biting as they sort things out.

➢ See also Body language, of horses; Punishment

Pedigree

A horse's family tree. If you're buying a pleasure horse, it's unlikely that you'll need to know a lot about his pedigree, but if you're in the market for a serious performance horse (on the racetrack or in the ring) you'll want to investigate his ancestors pretty thoroughly. Breeders of horses study pedigrees very carefully in order to pass along desirable traits and weed out weaknesses. They will go back at least three generations (parents, grandparents, and great-grandparents on both sides) to see which traits a new foal is likely to inherit.

➢ See also Breeding; Genetics

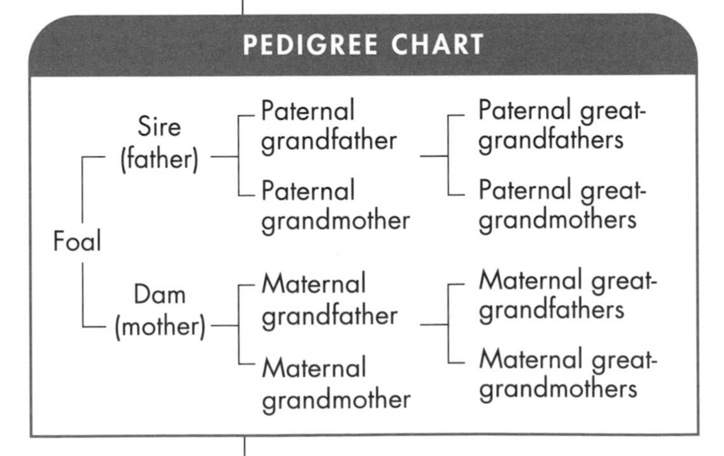

Peel out

A term used in cutting classes to indicate separating one or more cows from a herd.

➢ See also Cutting; Team penning

Pegasus

The famous winged horse of Greek mythology sprang from the womb of Medusa when she was slain by Perseus. Tamed by Athena, he lived with the Muses on Mount Helicon. When the hero Bellerophon needed help slaying the monstrous Chimera, he sought out Pegasus and was able to overcome the fire-breathing beast by flying around him and shooting him with arrows.

After many other heroic adventures, Bellerophon thought he deserved to visit the gods on Mount Olympus and set off on Pegasus. Annoyed at this presumption, Zeus sent a gadfly to sting Pegasus. The resulting rough ride sent Bellerophon headlong into a thorny bush, leaving him lamed and blinded because of his arrogance. Pegasus, however, went on to carry thunderbolts for Zeus and ultimately became a constellation.

To turn and wind a fiery Pegasus And witch the world with noble horsemanship.

—William Shakespeare, *Henry IV, Part One*

Pelham bit

A combination bit requiring two sets of reins. The Pelham can be used like a snaffle if just the upper reins are employed. The lower reins control the curb bit, which uses the leverage action of the shanks to provide more control. They come in a variety of styles.

➢ *See also Bit; Bridle*

Pelham bit

Pelleted feed

Pelleted feeds come in three types: grains, hay, and complete pellets with both hay and grain as well as vitamins and minerals. Pellets are ideal for horses with bad teeth, older horses, or individuals that have developed respiratory problems from eating hay. This feed is convenient to store and measure, and there is little waste. Horses on pellets don't get hay bellies and don't produce much manure, due to the concentrated nature of the feed.

The disadvantage to pellets is the lack of bulk, which the horse needs to feel satisfied and to maintain bowel regularity. Many pellet-fed horses chew wood in an effort to increase their roughage. This habit usually tapers off as the horse adjusts to the smaller portions, but you should be very careful when turning a pellet-fed horse out on grass or giving supplemental hay. His stomach will have shrunk somewhat and won't be able to handle a sudden intake of unaccustomed roughage.

➢ *See also Feeding and nutrition*

Even a small pen provides an opportunity for fresh air.

Pen

A turnout area for a horse, usually measuring no more than 50 feet by 50 feet, that provides a horse with outdoor living space. Because pens are small, they usually have no grass and do not allow the horse sufficient room to exercise. Pens can be surrounded by permanent fencing or created by putting together portable, metal-barred panels. In the case of the latter, extreme care should be taken when selecting the panels to ensure that horses cannot get their legs caught between the bars. The panels also need to be high enough to prevent a horse from jumping or climbing over them.

➢ *See also Round pen*

Penis

Male horses have retractable penises, which are housed in a sheath of skin. When urinating or mating, the male extends the penis.

Both geldings and stallions should be made accustomed to having their sheaths and penises handled for routine cleaning. Once or twice a year, you should clean the sheath carefully to remove any buildup of smegma, a smelly, waxy substance made up of fatty secretions, dirt, and dead skin cells that accumulates around the penis. Smegma can create a hardened bean in the diverticulum, a small pocket near the urethral opening. If large, the bean can partially or entirely block the flow of urine.

➢ *See also Anatomy of the horse; Grooming; Sheath; Sheath, cleaning; Smegma*

Despite their size and strength, Percherons are admired for their elegant movement.

Penning

➢ *See Team penning*

Percheron

A large draft horse from the La Perche district of France. Percherons are generally black, gray, or white and lack the feathered legs characteristic of Clydesdales. They were esteemed as war-horses because of their great strength and courage. With their long, wide backs, they are often used by bareback riders in the circus, where they are called Rosinbacks because of the rosin used on their backs for traction. This breed is noted for its free, elegant movement.

➢ *See also Breed; Draft horse*

Performance class

In the showring, events are divided into two basic types: halter and performance. **Halter classes** judge the conformation, quality, and movement of the horse, while **performance classes** focus on the horse and variables such as how he handles himself, how he is turned out, or how fast or well he gets through a specified task (for example, jumping or barrel racing).

A number of performance organizations sponsor shows in which horses of all breeds compete in specific activities. There are many different kinds of performance classes in both English and Western riding (see box). Riders might focus on one type of event or compete in several different ones. Some horses are more suited to one kind of event or another, as well.

➢ *See also English competition; Equitation, classes; Western competition*

PERFORMANCE CLASSES

English Pleasure

Combined training or three-day eventing
Dressage
Hunter/jumper classes
Saddle seat or hunter pleasure classes
Park horse
Show horse

Western Performance

Cutting
Working cow horse
Trail
Western pleasure
Western riding or reining

Periodic ophthalmia (moon blindness)

Originally thought to be affected by the phases of the moon, this disease is actually the result of a local or systemic infection that causes inflammation of the inside of the eye. Leptospirosis, a bacterial illness spread by rodents, is a common starting point.

A horse with periodic ophthalmia will be quite sensitive to light, and treatment and recovery should take place in a darkened stall. Other early symptoms might include a watery eye and constricted pupil, with apparent redness and swelling. Although symptoms may seem to improve, the eye will not look normal between attacks; look for a bluish ring around the cornea. In advanced cases, the pupil will be cloudy and widely dilated. Early treatment of the eye is important, as the infection will recur and may eventually cause permanent blindness.

Perlino coloring

Almost white with rust-colored mane and tail.

Peruvian Paso

Similar in looks and action to the Paso Fino, the Peruvian Paso is slightly larger (14 to 15.2 hands high). A wide range of solid colors is accepted; a lot of white markings are undesirable. The Peruvian Paso's gaits are the **flat walk,** the **paso llano,** and **sobreandando.** The action is marked by the characteristic of termino, a swinging motion of the front leg from the shoulder. This breed offers an exceptionally comfortable ride.
➢ *See also Amble; Gait; Gaited horse; Paso Fino; Sobreandando*

Pest control

Barns and stables are very attractive to a variety of pests; flying insects and rodents are the most common. Virtually all equine parasites depend on manure to complete their life cycle, so scrupulously cleaning stalls and pens and composting manure can really cut down on those pests. Flies are often a terrible problem around horses, and can be controlled somewhat with sticky paper traps or baited water traps. Pesticides can be useful if applied with caution, but they also kill beneficial insects that prey on flies and larvae, as well as harming people, horses, and other animals.

Rodents are another common barn pest that can cause a lot of damage. They spread disease, spoil feed, create fire hazards by chewing wires, and destroy tack. Cats are the best way to handle a rodent problem, but there are a variety of traps on the market as well. Feed bins and all storage areas must be tightly sealed around lids and doors to keep mice and rats away.
➢ *See also Parasites*

Pets around horses

Pets can pose hazards to horses, although in many cases they live in complete harmony. Your average barn cat doesn't pose much of a problem, although a skittish horse may react violently to one streaking across the ring or slinking along a wall. Many barns have resident dogs that are used to horses and know to stay well away from their hooves and teeth. Most horses will ignore the "locals," but might be fearful of or aggressive toward strange dogs. One well-aimed hoof can seriously injure the average dog, so it's better for both species if they are kept apart.

Do *not* to let your dog, cat, or even small children come with you into a pasture with a group of horses. Horses chase smaller predators, and they may not know the difference between Fido and a fox. Also, a dog or a child that knows how to behave in a barn may get excited and unpredictable when faced with loose horses, which can be dangerous for everyone.

PESTICIDES, USE OF

While it is sometimes necessary to use pesticides around your barn to control insects, they should be used sparingly and applied very carefully. Never spray them near food and water buckets or in areas where a horse might be fed hay. Store all pesticides away from feed areas and where a curious horse can't knock them over or even sniff them.

See also Parasites; Pest control

Most barns have cats, though most cats aren't this chummy with the horses.

p

Phenylbutazone (bute)

A common horse medication used to reduce swelling and pain. Bute, which is also known as Butazolidin, is relatively safe for horses and can be very useful in managing injuries and lameness. Like any drug, it should not be overused and must be prescribed by a veterinarian. Because it alleviates pain, horses are often given a dose before racing or showing, and different organizations have different rules about how much bute can be present in a horse's bloodstream during an event.

Caution: Bute should never be taken by people as a substitute for ibuprofen or aspirin; it can cause a fatal condition called aplastic anemia.

➢ *See also Drugs and horses [box]*

PHF

➢ *See Potomac horse fever (PHF)*

Phosphorus

The most important mineral in your horse's diet, along with calcium. Together they build strong bones and teeth. Phosphorus is found in grain, while calcium is obtained from hay and other forage. It's important to balance the two minerals so that they are absorbed by the body and used effectively. Too much grain can impair calcium absorption and hinder proper skeletal development in growing foals. A severe calcium-phosphorus imbalance can deform the bones of even mature horses.

➢ *See also Feeding and nutrition; Supplements; Vitamins*

Photosensitization

More serious than sunburn, photosensitization occurs when a horse has eaten certain plants (including alsike clover, dried buckwheat, St.-John's-wort, and perennial ryegrass) that damage unpigmented skin cells, usually around the eyes and muzzle or where the horse has white markings. Liver damage can also trigger photosensitization. The damaged skin tissue reacts to ultraviolet rays by releasing histamines, which lead to swelling and itching under the skin. Outer layers of skin will peel and then begin to blister and scab over. If untreated, a horse with this condition can go into shock.

The first step to take is get the horse out of the sunlight. He can be turned out at night, but he needs to be protected during the day. Your veterinarian may prescribe antibiotics and antihistamines; corticosteroids can also help.

Physical force, use of

Because horses are so large, it may seem that great strength is needed to handle or train them. In fact, brute strength is far less useful than patience, a gentle voice, an even temper, and the ability to remain calm. Most horses resent harsh treatment and will either resist your commands or become fearful and mistrustful. A few days of rough treatment can take months to correct. While many horses that have been physically abused can learn to trust a humane handler, some become intractably vicious.

➢ *See also Body language, of horse; Handling; Pecking order, in herds; Punishment*

They say Princes learn no art truly, but the art of horsemanship. The reason is, the brave beast is no flatterer. He will throw a Prince as soon as his groom.

—Ben Jonson, *Explorata*

Piaffe

A high, collected trot in place, characterized by an extended period of suspension. It is a movement made famous by the Lippizaner horses and is the ultimate example of collection, in which a simple release of hand allows the horse to smoothly move forward into the passage and eventually into a working trot.

➢ *See also Lippizaner*

The piaffe is the ultimate example of collection.

Picket line

A rope strung between two poles (or trees or trailers) to which you can tie your horse.

Picking out feet

Cleaning dirt from the feet. When doing this, be sure to scrape the hoof pick in the direction away from your face so that you don't accidentally flick dirt or manure into your eye.

➢ *See also Foot care; Grooming; Picking up feet*

Picking up feet

It's important that your horse stand quietly while you examine, clean, or treat his feet. Your farrier will also appreciate being able to work on a well-behaved horse.

➢ *See also Farrier; Foot care*

Many horsemen prefer a horse with large eyes to a pig-eyed horse with small eyes.

Piebald

A black and white Pinto. Piebald is an uncommon color.

➢ *See also Coloring; Paint and Pinto*

Pigeon-toed

Having hooves that turn in toward each other. Like many leg conformation faults, this predisposes a horse to problems because the joints are more subject to strain.

➢ *See also Leg, of horse; Paddling*

Pig-eyed

A horse with small eyes. A pig-eyed horse is generally considered less attractive or desirable than a horse with larger eyes. A pig-eyed horse might have somewhat limited peripheral vision and therefore be more prone to shying as objects appear to loom from the side or rear. While not enough of a fault to eliminate him from active use, a pig-eyed horse should be ridden, handled, and bred with his visual limitation in mind.

Piggin string

A small rope for hog-tying calves and steers in roping events. Both hind legs and one foreleg must be tied. The rider carries the rope coiled in his teeth, with one end tucked into his belt as he ropes the calf with the lariat.

Use a piggin string to tie a calf.

Pinch test

A way to measure a horse's hydration level. To do a pinch test, grasp a fold of skin in the neck or shoulder area and pull it gently away from the muscle. When you release it, the fold of skin should flatten out almost immediately. Don't worry if the fold is still visible for a couple of seconds, but be aware that a marked peak after that point indicates some loss of body fluid. If the fold is still visible after 5 or 10 seconds, your horse is probably severely dehydrated and possibly in need of veterinary care.
➢ *See also Dehydration*

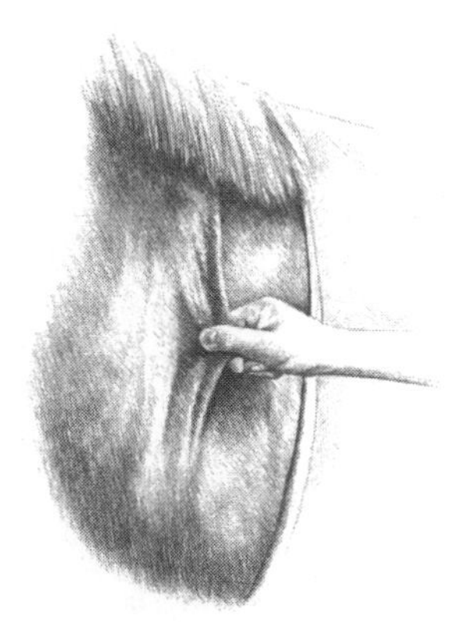

Performing a pinch test

Pinto

➢ *See Paint and Pinto*

Pinto Warmblood

Though most Warmbloods are dark in color with minimal body markings, you will occasionally see one with Pinto coloring.
➢ *See also Breed; Paint and Pinto; Warmblood*

Pinworms *(Oxyuris equi)*

Pinworms are annoying to horses but not as dangerous as roundworms or strongyles. They inhabit the large intestine and cause intense itching around the rectum, where the female migrates to lay her eggs. Pinworms are gray, yellow, or white and can be as long as 6 inches. An infected horse may rub much of the hair off his tail trying to relieve the discomfort. A good dewormer will take care of pinworms, but stalls and pens should be carefully cleaned to get rid of eggs that may have stuck to walls, floors, and bedding.
➢ *See also Deworming; Parasites*

Piroplasmosis

An often fatal disease carried by ticks and other bloodsuckers. Death results from severe anemia. If the horse recovers, he can be a carrier for a year. Symptoms include intermittent fever, depression and weakness, loss of appetite, and rapid pulse and respiration. At an advanced stage, signs of anemia and jaundice may appear, along with dark-colored urine. Treatment with medication can help, but there is no vaccine, and care must be taken to quarantine infected horses.

Pirouette

A very controlled circle at the walk, trot, or canter, with the forehand moving around the hindquarters. A pirouette is similar to a **spin** in Western riding. The radius of the circle equals the length of the horse, so he appears to be spinning in place while maintaining the rhythm and foot placement of the gait being performed.

Pitch

In Western riding, to drop the reins completely, or to toss a rope.

Placental insufficiency

A common cause of abortion, especially in older mares whose uterine lining has become less efficient with age. The placenta cannot attach adequately to the uterine wall, depriving the developing embryo of oxygen and nutrients.

Placenta previa

When the placenta detaches from the uterine wall during the early stages of labor and begins to emerge before the foal does. This is a life-threatening situation to both mare and foal. Call your veterinarian immediately. Be sure to ask for guidance as to what to do until he arrives.

Placenta shedding

After a normal birth, the mare will generally rest, lying down for perhaps 10 or 20 minutes, giving the placenta time to finish detaching from the uterine wall and pumping its blood supply to the foal through the umbilical cord. When she stands up, the umbilical cord breaks naturally and she should expel the placenta with a few more contractions.

Pleasure class

Western or English competitions designed to evaluate a horse's manners, performance, quality, conformation, and presence as he travels in both directions around the showring.

Pleasure driving class

Various pleasure driving classes are regulated by the American Driving Society; others are offered by different breed associations. In general, the pleasure driving horse or team must demonstrate their overall ability to provide a pleasant drive. They are judged in both directions before lining up in the center of the ring. They must be able to stand quietly and back willingly.

➢ *See also Combined driving; Driving*

Pleasure horse

Rather than indicating a particular breed, this term refers to a type of animal that can be found in nearly any breed. Pleasure horses, part of both English and Western traditions, are distinguished by a smooth, comfortable ride at all three gaits; an even, dependable temperament; and attractive, well-balanced conformation.

➢ *See also English competition; individual breeds; Western competition*

Plussed

Having scored above the average of 70 points in a National Reining Horse Association routine.

➢ *See also Reining class*

A gigantic beauty of a stallion, fresh and responsive to my caresses, Head high in the forehead and wide between the ears, Limbs glossy and supple, tail dusting the ground, Eyes well apart and full of sparkling wickedness. . . ears finely cut and flexibly moving.

—Walt Whitman,
The Stallion

Pneumonia

Horses can develop two types of pneumonia. Aspiration pneumonia is often fatal, especially in foals. Food particles or medication that gets into the windpipe and into the lungs causes irritation and infection. Whenever a horse is unable to put his head down to cough (for instance, if he's tied in a trailer or straight stall), he is in danger of developing aspiration pneumonia.

More commonly, horses develop pneumonia from airborne pathogens, including bacteria, viruses, parasites, and fungi. If a horse is already ill or very stressed, he is more vulnerable to pneumonia. Although often a secondary infection, pneumonia can easily overwhelm a weakened system and kill the horse. Symptoms include increased respiration and a cough, listlessness, and loss of appetite. You will probably be able to hear wheezing and rattling in his lungs if you use a stethoscope. Prompt medical attention is required, but only a bacterial infection will respond to drugs. Viral pneumonia is much harder to treat and may not be curable. The sick horse should be kept warm and quiet, out of drafts, with light feed (moistened to reduce dust) and complete rest.

A barrel racer about to turn in the pocket.

Pocket

The area around a pole or barrel in which you turn your horse during timed events.

Points

The mane, tail, lower legs, and nose.

Buckskin horses have black points.

Poison hemlock *(Conium maculatum)*

A member of the carrot family, poison hemlock is common in pastures and along roadsides. Although horses generally avoid it, if ingested it causes death from respiratory failure.

➢ *See also Poisonous plants*

Poisonous plants

Because they cannot vomit, horses are particularly susceptible to poisoning. While most horses will avoid poisonous plants if more palatable fare is available, any horse might eat them if the pasture is overgrazed, if he is bored, or if he is just plain greedy. Make sure that any area where your horse is turned out is free of toxic plants, including branches hanging over fences. Your county extension agent can tell you which plants to look out for in your area and how to get rid of them.

It is crucial that your horse not be allowed to graze in any yards or gardens or near any ornamental plants. Many common garden plants and wildflowers are poisonous to horses, including rhubarb, buttercup, daffodil, lily of the valley, and delphinium (or larkspur). Do not let your horse graze or snatch at leaves and plants while you are out on the trail — not only is it bad manners, but also it could be fatal. The chart below lists only a few of the many hundreds of plants that can kill a horse, some within minutes of being ingested.

➢ *See also Toxic substances; individual plants*

POINT-TO-POINT

This amateur sport involves jumping obstacles on a course, though originally the course went cross-country from one "point" to another.

See also Hunter Pace; Steeplechasing

COMMON POISONOUS PLANTS			
PLANT	RANGE AND HABITAT	TOXIN AND SYMPTOMS	TREATMENT & PREVENTION
Black locust *Robinia* spp.	Native deciduous tree of eastern North America often forming pure stands, naturalized in much of West. Tolerates wide range of environmental conditions from wet to dry areas.	Robin, a glycoprotein similar to ricin (see castor bean), found in all parts of the plant. Highest concentrations in new growth, seeds, and bark. Ingesting the leaves, twigs, or bark causes colic, diarrhea, dilated pupils, weak and irregular heartbeat, depression, and death.	No known antidote; supply supportive therapy. Remove *Robinia* species from areas frequented by horses.
Black walnut *Juglans nigra*	Large deciduous tree native from New England and southern Canada west to Great Lakes and south to Gulf and Atlantic Coasts. Naturalized in parts of the West.	Toxin unidentified, possibly juglone. Nibbling on bark and branches or bedding containing black walnut shavings leads to colic and laminitis.	Treat for laminitis. Remove black walnuts from pastures and other areas around horses, avoid using black walnut shavings for bedding.
Brackenfern (pasture brake) *Pteridium aquilinum*	Tall fern native to much of the United States and southern Canada except the Southwest. Tolerates many environments including waste ground, wetlands, forest, and burned areas.	Thiaminase, an enzyme that can induce thiamin (vitamin B_1) deficiency. Symptoms of thiamin deficiency include blindness, weakness, and depression. Ingestion can also depress bone marrow, dangerously lower blood clotting factors, produce massive hemorrhaging and possibly cancer. Horse may stand with back arched and legs apart.	Poisoning most often occurs from ingestion over extended period of time. Remove plant from hay, hayfields, and pastures; provide prompt injections of thiamin hydrochlorate when symptoms of thiamin deficiency appear.
Castor bean *Ricinus communis*	Shrubby tender perennial naturalized in southern regions of North America. Commonly grown as an annual as far north as southern Canada.	Ricin, an extremely toxic protein found throughout the plant but most concentrated in the seeds. A few ounces of ingested seeds can be lethal to horses. Symptoms appear a few days following exposure and include colic, diarrhea, sweating, greatly increased heart rate, and death.	There is no known antidote for ricin poisoning. Administer activated charcoal and intravenous fluids that include vitamin C. Avoid planting castor bean near areas frequented by horses.
Cherry *Prunus* spp.	Deciduous shrubs and trees native to southern Canada from Atlantic provinces west to Rocky Mountains and south to Texas and Florida.	Cyanogenic glycosides are present in all plant parts except the flesh of ripe berries. Damaged or wilted leaves more toxic than fresh. Ingestion results in release of cyanide that inhibits blood hemoglobin from releasing oxygen to tissues, resulting in death from anoxia (lack of oxygen). Symptoms include difficulty breathing, cardiac arrest, and sudden death. Ruminant animals are more susceptible than horses.	Animal should be treated with intravenous therapy using sodium thiosulfate and sodium nitrite. Remove plants from areas frequented by horses.
Cocklebur *Xanthium strumarium*	An annual found in most areas of North America. Tolerates a wide range of environments: tilled and fallow fields, waste ground, pastures, and near lakes, rivers, and roadways.	Carboxyactractgloside, a toxin that inhibits cellular metabolism, is present in seedlings and burrs. Symptoms include convulsions; vomiting; liver damage; rapid, weak pulse; labored breathing; leg and neck spasms.	Supportive therapy is only treatment. Cultivate areas where seedlings may appear in spring. Remove plants from pastures and other areas used by horses.
Crotalaria (rattlebox) *Crotalaria* spp.	A large group of annuals and perennials native to many warm climates worldwide. Most poisonings result from species grown in southeastern United States from Atlantic Coast to Texas. Listed as noxious weed in some Southern states.	Seeds and leaves contain toxic alkaloids that lower blood pressure and heart rate.	Provide supportive therapy. Avoid planting *Crotalaria* species as cover crops near pastures and other areas frequented by horses. Do not use hay from fields containing *Crotalaria* plants.

COMMON POISONOUS PLANTS (continued)

PLANT	RANGE AND HABITAT	TOXIN AND SYMPTOMS	TREATMENT & PREVENTION
Death camas *Zigadenus* spp.	Perennials native to the southeastern United States west to Great Plains and Rocky Mountains and north to southern Canada. Some species also occur in Far West.	Plants contain toxic alkaloids that lower blood pressure. Bulbs and mature leaves contain highest alkaloid content. Symptoms include difficulty breathing, excessive salivation, weakness, and coma.	No known antidote. Less than 10 pounds will kill a horse. Administration of atropine sulphate may be helpful.
Equisetum (horsetail, scouring rush, jointfir) *Equisetum* spp.	A primitive perennial with an erect leafless stem native to most areas of North America. Tolerates a wide range of environments including waste areas, wetlands, pastures, and roadsides. Listed as noxious weed in parts of Pacific Northwest.	Thiaminase, an enzyme that can induce thiamin (vitamin B_1) deficiency, is the suspected toxin. Most often ingested from consuming baled hay. Cumulative effects include weakness, trembling, staggering, excitement, diarrhea, and loss of condition.	May be present in hay. Remove source of equisetum exposure and administer thiamin injections to restore proper levels of vitamin B_1. Avoid using hay from fields where equisetum grows.
Fiddleneck (fireweed, tarweed, yellow burr weed) *Amsinckia* spp.	Annuals native to western North America from British Columbia south into Mexico. Occasionally naturalized east of the Rocky Mountains. Common weeds in fields of wheat and grain.	All parts of the plant contain pyrrolizidine alkaloids that inhibit proper cell division and act as a liver poison. Symptoms include photosensitization, depression, weakness, odd behavior, diarrhea, and weight loss.	May be present in hay. Even small amounts can be lethal if ingested. Administer supportive therapy. Avoid using hay from fields where fiddleneck grows. Remove the plant from pastures and other areas frequented by horses.
Lantana *Lantana* spp.	Mostly shrubby perennials naturalized from southeastern United States west to California. Prefer sandy, dry environments but tolerate a wide range of conditions including coastal areas and alkaline soils.	All parts of the plant contain triterpene acids that produce symptoms including photosensitization, reddening of the eyes, difficulty breathing, increased heart rate, and bloody diarrhea.	Horses will eat lantana if grazing is poor; fatal dose is 20 to 30 pounds. Provide supportive therapy, especially fluids. Remove the plant from all pastures, hay fields, and areas frequented by horses.
Locoweed (crazyweed, milkvetch, poisonvetch) *Astragalus* & *Ozytropis* spp.	Most common cause of livestock poisoning in Western North America especially Great Plains where many species are native. Listed as noxious weed in Hawaii.	Different species of locoweed contain a variety of toxic substances including the alkaloid swainsonine, the glycoside miserotoxin, and selenium in concentrations high enough to produce poisoning. Swainsonine produces heart failure, reproductive disorders, locoism, edema, and birth defects. Miserotoxin produces locoism, difficulty breathing, and incoordination of rear legs. Selenium can produce both a syndrome called the blind staggers where animals walk in circles and have impaired vision and poor appetite, and alkali disease, indicative of chronic poisoning with symptoms including malformed hooves and hair loss.	No effective treatment known. Avoid pastures, range, and hay that contain locoweed. Animals not killed by locoweed poisoning are often permanently impaired. Some animals become habituated to eating locoweed.
Nightshade *Solanum* spp.	Large group of mostly Eurasian plants frequently naturalized over wide areas of North America. Close relatives include potatoes, tomatoes, peppers, and jimsonweed, all of which are similarly toxic. Common in waste areas, along roadsides and stream banks, and in cultivated fields.	Plants contain alkaloids including solanine that mimic atropine and impair normal nervous system function by inhibiting the enzyme acetylcholinestrase, causing muscle tremors and weakness. Some species also contain saponins that can produce digestive problems including excessive salivation and diarrhea.	Provide supportive therapy as indicated including the administration of intravenous fluids. Remove plants from pasture and range areas.

COMMON POISONOUS PLANTS (continued)

PLANT	RANGE AND HABITAT	TOXIN AND SYMPTOMS	TREATMENT & PREVENTION
Oleander *Nerium oleander*	Large shrub naturalized and used as a landscape plant in warmer regions of North America. Tolerates a wide range of conditions including drought.	All parts of the plant contain the extremely toxic cardiac glycosides oleandrin and neriine. Plants with dark-colored flowers may be most toxic. The glycosides inhibit normal cellular electrical conductivity resulting in erratic heart rate, rapid breathing, dilated pupils, diarrhea, and heart failure. Less than a quarter of a pound of this common ornamental will kill a horse.	Administer activated charcoal as soon as ingestion is suspected. Treat heart irregularities as they manifest with anti-arrhythmic drugs. Provide supportive therapy as needed. Avoid giving the animal calcium or potassium as these elements exacerbate the effects of the toxins. Remove plants from areas frequented by horses.
Poison hemlock *Conium maculatum*	Large biennial herb naturalized across much of North America. Found in pastures, along roadsides, and in wet areas. Listed as noxious weed in parts of Midwest and West. Not to be confused with water hemlock, which is actually more poisonous but less frequently eaten.	All parts of the plant contain toxins including the potent alkaloid gamma-coniceine. Symptoms include slow heart rate, drooling, staggering, uncoordinated movement, and dark urine.	Administer activated charcoal when ingestion is suspected. Treat presenting symptoms and provide supportive therapy. Remove plants from areas frequented by horses.
Rayless goldenrod (jimmyweed) *Isocoma* spp.	Woody perennials common to arid regions of Western North America.	Plant contains tremetol, which produces muscle tremors, weakness, stiffness, and depression when ingested over several weeks. The toxin is present in the milk of lactating animals and can poison nursing foals.	Contains the same toxic agent as white snakeroot. Remove plant from pastures where possible. Place animal on healthy diet.
Red maple (swamp maple) *Acer rubrum*	Large tree native from Maritime Provinces to Minnesota and south to Texas and Florida. Common in a wide range of environments from forests to fields, pastures, and a variety of wetlands.	Leaves contain unknown toxin that damages hemoglobin and is most concentrated in wilted leaves, autumn foliage, and bark. Symptoms usually appear about 3 days after ingesting leaves and include rapid breathing and heart rate, weakness, depression, brown urine, and brownish blood. Small amounts can be quite harmful.	Supply supportive therapy including administration of intravenous fluids as needed. Blood transfusions may be necessary. Remove red maple trees from areas frequented by horses. Dried, wilted, or damaged leaves are more toxic than fresh foliage.
Russian knapweed (Russian thistle) *Acroptilon repens* syn. *Centaurea repens*	A close relative of yellow star thistle and more toxic. Naturalized in waste areas, pastures, and along roadsides through most of from southern Canada and Midwest United States south to Mexico and west to Pacific Coast. Reported less frequently in Eastern North America. Listed as noxious weed in most of the United States.	Unknown toxin produces "chewing disease" with symptoms including uncoordination, restlessness, and repetitive chewing motions but inability to eat. Poison accumulates over many days of ingesting the plants. More toxic than close relative yellow star thistle. Disease is most common in Western North America in early summer and midfall.	Once symptoms appear condition is untreatable. Supportive therapy is sometimes administered in mild cases. Euthanasia often indicated to prevent animal from slowly starving to death. Avoid pastures, range, and hay contaminated with Russian knapweed or yellow star thistle.
St.-John's-wort (Klamath weed, goat-weed) *Hypericum perforatum*	A naturalized perennial of meadows, roadsides, waste places, and prairies throughout much of the United States. Listed as noxious weed in Hawaii and parts of the West.	Toxic agent is hypericin that produces photosensitivity and in severe cases convulsions and coma. Unpigmented areas become sore, swollen, and itchy, with peeling skin.	Remove animal from any contact with direct sunlight. Mild cases can be treated with anti-inflammatory medication. Fresh leaves are much more toxic than dried foliage. Remove plants from pastures.

COMMON POISONOUS PLANTS (continued)

PLANT	RANGE AND HABITAT	TOXIN AND SYMPTOMS	TREATMENT & PREVENTION
Senecio (includes groundsel, stinking willie, ragwort, and tansy ragwort) *Senecio* spp.	At least 70 species that thrive in a wide variety of habitats throughout much of North America. Listed as noxious weed in Hawaii and parts of the West.	All parts of the plant contain pyrrolizidine alkaloids including jacobine and seneciphylline, which when ingested produce a syndrome called "Pictou disease." Symptoms include cirrhosis of the liver, jaundice, abdominal swelling, lung congestion, and agitation.	Liver disease initiated by pyrrolizidine poisoning has no effective treatment. Keep animal away from direct sunlight. Eradicate the plant from all pasture and range land.
Sorghum grass (includes Sudan grass and Johnson grass) *Sorghum* spp.	Group of drought-tolerant perennial and annual grasses most common in pastures throughout the South and Southwest. Listed as noxious weed in many states across the country.	All parts of the plant contain dhurrin, a cyanogenic glycoside. Ingestion results in release of cyanide that inhibits blood hemoglobin from releasing oxygen to tissues, resulting in death from anoxia (lack of oxygen). Symptoms include difficulty breathing, cardiac arrest, and sudden death. Sorghum can also accumulate toxic quantities of nitrates.	Keep animal calm and administer sodium thiosulfate and sodium nitrite intravenously. Removal of plants from pastures can be difficult. Use of herbicides followed by reseeding with select forage is indicated in severe infestations.
Tall fescue grass *Festuca elatior* syn. *Festuca arundinacea*	A common grass that is not harmful in itself but is often infested with a toxic endophytic fungus *(Acremonium coenophialum)*. Listed as noxious weed in many Mid-Atlantic states.	A problem only for pregnant mares; may cause abortion, retained placenta, and failure to produce milk.	Test field for endophytic content of grasses if poisoning is suspected. In pastures and fields containing high levels of endophytes use herbicides to remove infected grass, and reseed with select forage.
White snakeroot *Ageratina altissima* syn. *Eupatorium rugosum*	Found in sandy areas in the Midwest. Often stays green in early fall, when horses might be tempted to eat it if other plants have dried up. Listed as noxious weed in Hawaii.	Contains the alcohol tremetol, which damages the heart muscle and liver; symptoms of poisoning include weakness, lethargy, loss of coordination, and stiffness. A foal can be poisoned through nursing if the mare eats the plant.	Contains the same toxic agent as rayless goldenrod. Remove plant from pastures where possible. Place animal on healthy diet.
Yellow star thistle *Centaurea solstitialis*	An annual closely related to Russian knapweed and similarly toxic. Naturalized in pastures, fields, roadsides, and disturbed ground throughout much of North America west of the Great Plains, less common in eastern areas. Listed as noxious weed in Hawaii and parts of the West.	Unknown toxin produces "chewing disease" with symptoms including uncoordination, restlessness, and repetitive chewing motions but inability to eat. Poison accumulates over many days of ingesting the plants. Disease is most common in Western North America in early summer and midfall.	Once symptoms appear condition is untreatable. Euthanasia often indicated to prevent animal from slowly starving to death. Supportive therapy is indicated in mild cases. Avoid pastures, range, and hay contaminated with Russian knapweed and yellow star thistle.
Yew *Taxus* spp.	Native and naturalized needled evergreens; some species commonly planted as ornamentals. Native species prefer moist forests along Pacific Coast and throughout Eastern North America. Introduced species most commonly planted from Eastern North America west to Midwest and south to Gulf and Atlantic Coasts.	Needles, twigs, and seeds contain the alkaloid complex taxine, which depresses the action of the heart resulting in heart failure. The toxin is most concentrated in the needles during the coldest months of the year.	No effective treatment is known. Supportive therapy should be provided as symptoms indicate. Atropine sulfate can sometimes increase slowed heart rate. Remove ornamental plantings of yew in areas where horses have access. Native species should be removed from pastures.

Poisonvetch (crazyweed, milkvetch) (*Astragalus* and *Ozytropis* spp.)

➢ *See Locoweed*

Pole bending

An event in which the rider and her horse must run a course of eleven turns around six poles as fast as possible without touching the poles. The 6-foot high poles, usually made of plastic pipe, are set 20 to 21 feet apart. From the starting line, the rider gallops to the farthest pole, bends around it, weaves through the remaining five, doing a flying lead change at each pole. At the last pole, horse and rider circle and repeat the weaving pattern back through the poles in the opposite direction, then go at a dead run for the finish line. An average run takes 20 to 25 seconds.

A pole bending horse must be very fast and light on his feet, with great athletic ability and the quick reflexes of a good reining horse. With about two strides between poles, the horse must change leads in an instant and turn on a dime around the end poles, while the rider sits quietly and straight and guides her mount with verbal cues and a lot of leg.

➢ *See also Western competition*

Police horses

Although police horses must be large and strong, they are generally chosen for their temperament rather than for their breed. A police horse must be intelligent, calm, and kind, in order to handle crowds, noise, smoke, dogs, traffic, and all the other distractions and hazards that the mounted police face routinely. The horse has to tolerate frequent trailering and hordes of strangers wanting to pet him, and he must be able to work long hours in all kinds of weather.

Many police departments accept donated horses, but the animals must pass a series of tests and an intensive training period. The horses must learn to stand still in the face of frightening noises, go forward into strange situations and over scary surfaces (sometimes at considerable speed), respond immediately to signals from the rider, and stay calm in large crowds of people.

Mounted police have the advantage of height and visibility over foot and car patrols, and they are effective in controlling and dispersing crowds. They have the disadvantage of presenting a larger target, but many police departments find that horses actually seem to diffuse tension in potentially ugly situations.

➢ *See also Royal Canadian Mounted Police (RCMP)*

Poll

The spot between the ears where the skull meets the vertebrae. The poll is quite sensitive, and striking it hard enough can kill a horse or cause permanent brain damage. This is one good reason to discourage a horse from rearing, especially indoors.

➢ *See also Anatomy of the horse*

Polo

Possibly originating as a training exercise for mounted warriors, this ancient game, played on horseback with long mallets to drive a ball into a goal, has been played for thousands of years across the plains of Asia. In the mid-1800s, polo was discovered in India by the British, who brought the game back to England and formalized rules of play. By the end of the 19th century, polo was very popular in the United States as well.

Played on a field measuring 160 by 300 yards, an average polo match takes about an hour and a half, divided into six seven-minute chukkars. Each team has four players, and since they may cover as much as 3 miles in a single chukkar each player needs several ponies. With such a huge playing area, the game is very fluid. While players take offensive or defensive positions, those often change as play progresses. Rather than setting up specific plays or maneuvers, the teammates rely on their ability to follow the rapidly changing course of the ball and to anticipate where the next shot will go.

A typical score might be 10–7, with teams switching direction after every goal. Games are played on the flat or the handicap. Each player has a handicap based on skill, ranging from C (or –2 goals, a beginner's handicap) to 10 (the highest). A handicap of 6 is considered very impressive. In handicap play, the total handicap of each team is added up and the difference between the two is awarded as goals to the lower rated team before the game begins.

Now a polo pony is like a poet. If he is born with a love for the game, he can be made.

—Rudyard Kipling, *The Maltese Cat*

Polo pony

Though not technically ponies anymore, these horses are still referred to as such. The original polo ponies were quite small, but as the game spread to Europe and America, larger animals became the norm. Today's polo ponies are generally between 14.5 and 16 hands tall and weigh around 1,000 pounds. The ideal mount combines the speed and endurance of a Thoroughbred with the quickness and maneuverability of a good cow pony. A well-trained and experienced polo pony is a valuable addition to the team.

Polyethylene-coated wood fences

A fencing material that offers the benefits of wood fencing with the added advantages of being easier to maintain, more resistant to rot, and less attractive to horses that chew wood.

➢ *See also Fencing materials*

Polyvinyl chloride (PVC) fences

Although initially more expensive than wood, PVC tubing is sturdy, rot-resistant, and doesn't need to be painted. Horses won't chew it, it doesn't splinter or crack, and it lasts for years.

➢ *See also Fencing materials*

Pommel

The part of the saddle that fits over the horse's withers.

➢ *See also Saddle; Tack*

Pony

Essentially, a small horse. Exact size limits depend on the breed and/or rules of the competition, but in general, a pony is a horse that measures 14.2 hands (58 inches or 145 centimeters) or less. His body is generally longer than it is tall, while horses tend to be equal in height and body length.

Like horses, ponies come in a number of distinct breeds and in every color imaginable. Some common breeds are the well-known Shetland, the Welsh pony, the Connemara, the Hackney, the Haflinger, and the Pony of the Americas (POA). There are pony divisions of nearly every kind of class and event, and ponies have their own shows and associations, as well. There is a wide variety of registries available for ponies, just as there is for horses. Most register ponies based on ancestry; however, some register based on size alone.

Ponies have a reputation, often well-deserved, for being clever (even devious), stubborn (possibly intractable), and cranky (downright mean). However, many ponies have carried generations of children safely and handily through shows, lessons, and trail rides and given those children a solid base of horsemanship for life. Ponies can be excellent teachers for children and small adults of all levels of experience. The golden rule of horses applies equally to ponies: The younger or less experienced the human, the older and more experienced the pony. Green ponies are not good choices for novice horsemen of any age.

➢ *See also individual breeds; Miniature horse*

Pony Club

There are more than 400 pony clubs in the United States with over 9,000 members. The clubs are designed to introduce young riders, many of whom participate with horses instead of ponies, to the basics of horse care and the art of English riding. Participants can progress through as many as nine levels of proficiency, the last of which qualifies you to teach. Pony clubs are run by volunteer instructors, and some groups own donated horses for people without mounts of their own to use.

➢ *See also appendix*

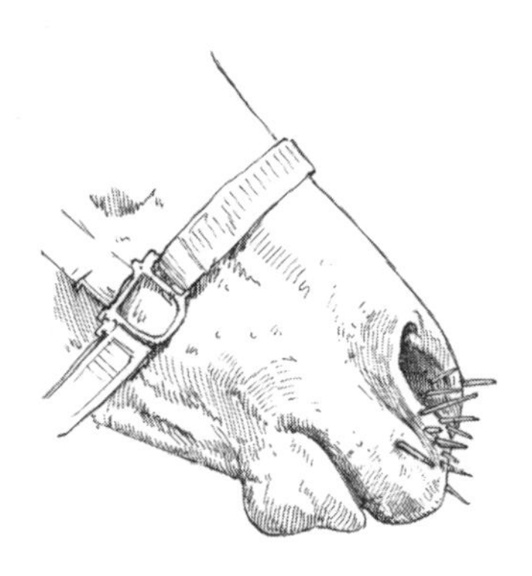

Porcupine quills **are quite short and can twist all the way into the lips and nose.**

Ponying

Leading one horse while riding another. Ponying allows you to condition or exercise two horses at once.

Pony of the Americas (POA)

Developed by crossing Shetland ponies with Appaloosa horses as well as some Quarter Horses and Arabians, the POA shows the same vivid coloring as the Appaloosa but is generally no taller than 13.2 hands high.

Porcupine quills, first aid for

It's not uncommon for a curious horse to poke his nose where it doesn't belong, and a muzzle full of porcupine quills may be the end result. The quills should be removed as soon as possible so they don't work further into the tissue. It's a two-person job: One person should hold and distract the patient while the other yanks the quills out with pliers. Pull straight out with a quick, smooth motion.

After the quills are all out, check the lips and mouth for broken-off pieces or quills that have worked deeply into the tissue. If the horse is too agitated, you may need to put him in stocks or call your vet to give him a tranquilizer.

Position, in saddle

There are a lot of things to remember when you get on a horse. It is very important that you sit in the saddle in a balanced and comfortable position. If you are balanced, you will be able to help, not hinder, your horse's balance. If you are relaxed, you will be able to follow his rhythm, give effective aids, and respond easily to unexpected movements. If you are stiff and unbalanced, you will bounce around the saddle, your aids will be ineffective or even contradictory, and you could fall off if the horse shies.

Remember to keep your:

- Eyes forward and chin up. Don't forget to keep your tongue in your mouth!
- Shoulders relaxed, back, and down, which opens your chest for effective breathing.
- Back straight, not stiff or hollowed out.
- Hands in a straight line with elbows and the bit.
- Hips relaxed and supple, with your weight on your seat bones.
- Thighs, knees, and calves resting lightly on the horse, not gripping tightly.
- Heels down, toes up and pointing forward.

➢ *See also Aids; Seat*

This rider is sitting correctly with her ear, shoulder, hip, and heel lined up.

Posting trot

Rising out of the saddle at every other stride of the trot. Posting allows you to move comfortably and rhythmically with your horse as he trots and to stay in control at a more active trot. You can also sit the trot, but it is usually easier to learn to post first.

To get the feel of posting, practice at a standstill for a few minutes. Your movement should be in the knees and hips, while your feet should remain steady. Move your hips forward and back, rather than straight up and down, and touch the saddle lightly rather than plopping back down in it. Keep your weight in your heels and your hands low and flat. Don't balance yourself by pulling on the reins; if you need to hang on to something at first, grab some mane.

Once you feel comfortable and balanced posting at the halt, you're ready to try it at the trot. Rather than trying to stand up and down in the saddle, let your horse's hindquarters push you forward as you rise and fall to the rhythm of his gait. Your motion should be subtle and supple, not stiff and bouncy. Keep your legs long and relaxed, with your knees pointing straight ahead.

POSTING AS AN AID

Once you've become proficient at posting, you can use the motion to communicate with your horse. If you slow your up-and-down movements, your horse should slow down his trot. Conversely, a more active post with more leg will make him extend his gait.

To complicate matters, you need to remember to post on the diagonal, meaning that as the outside front leg comes forward, you rise. As the inside leg comes forward, you sit. With enough practice, you'll know when you're on the correct diagonal, but at first you can check by glancing down (with your eyes only!) to see which leg is doing what. If you find you're on the wrong diagonal (rising to the inside front leg), just sit for two beats and pick up the correct one. Don't forget to change diagonals when you change direction.

Most riders don't understand the horse's role in the posting trot. A posting trot requires sufficient impulsion for the horse's hindquarters to essentially push you out of the saddle. One of the most common mistakes while learning this skill is seen when riders try desperately to "lift" themselves out of the saddle, rather than allowing the horse to push them out. This usually occurs because the rider is so focused on posting that she doesn't notice that the horse has lost impulsion.

➢ *See also Aids; Collection; Position, in saddle; Seat; Sitting trot*

When posting, it's important to move with the horse's motion, not against it.

Post-legged

Having hind legs that are *too* straight, without a proper angle at the hock and stifle.

➢ *See also Conformation; Leg, of horse*

Posture as sign of illness

Because your horse can't tell you if he's sick or in pain, you must be alert to his body language to determine his condition. Be familiar with his usual stance and attitude so that you notice any changes. Any time your horse seems dull or listless, check his vital signs. A horse with mild abdominal pain may get up and down more often than usual, may kick or bite at his belly, and may stand with his legs stretched to the front and back in an effort to relieve gas pains. However, if he's standing with his back humped up and stomach muscles tensed, he's in severe stress and needs medical attention.

➢ *See also Signs of illness*

One's . . . liver . . . goes up and down like the dasher of a churn in the midst of the other vital arrangements, at every step of a trotting horse. The brains are also shaken up like coppers in a money-box.

—Oliver Wendell Holmes, *The Autocrat of the Breakfast Table*

Potomac horse fever (PHF)

A serious illness first diagnosed in Maryland in 1979 and now found in nearly all states. Primarily a summertime disease, it's most commonly found in horses that graze in irrigated pastures and near streams. PHF is spread by the larvae of parasitic flatworms that live in freshwater snails.

If your horse is off his feed and water, running a fever, and then develops severe diarrhea, he probably has PHF. Immediate treatment with IV fluids and tetracycline is usually effective, though acute cases can be fatal or can cause founder. There is a PHF vaccine; it is given in the first year, with an annual booster after that. Consult with your veterinarian about the necessity of vaccinating horses in your area.

Povidone iodine, as a wound wash

A dilute iodine solution (Betadine). It can be diluted further in distilled water with a small amount of salt to make an effective disinfectant for wounds.

➢ *See also First aid*

WOUND WASH

To make 1 gallon:
- 1 gallon distilled water
- 2½ tablespoons table salt
- 1½ teaspoons povidone iodine

To make 1 quart:
- 1 quart distilled water
- 2 teaspoons table salt
- ½ teaspoon povidone iodine

Praziquantel (Droncit)

A canine dewormer that is effective against tapeworms in horses.

➢ *See also Deworming; Parasites; Tapeworms*

Predator wasps

A nocturnal insect that lays its eggs in the pupae of flies that lay eggs in manure. The hatching wasps eat the pupae, effectively interrupting the breeding cycle of many flies that pester horses. The wasps do not bother people, horses, or other animals.

➢ *See also Deworming; Parasites; Pest control*

Pregnancy

Pregnancy should be treated as a natural condition for a mare. Mares have been ensuring the survival of their species for a lot longer than humans have been around to help them. Most mares conceive, carry, and birth a foal without much need for complex involvement from humans. Yet humans' mastery of technology can easily make it seem that we need to intervene for the best interest of mare and foal.

The most important guideline for caretakers of a pregnant mare to remember is that it takes a healthy mare to produce a healthy foal. Our primary responsibility to a pregnant mare is essentially the same as it is to all domesticated horses:

A healthy mare showing signs of her pregnancy.

- To provide adequate nutrition based on a ration, including free-choice water, developed for the individual's needs, which gradually increases as the fetus develops
- To provide supplements, if needed, to meet the high demands that a fetus places on a mare's system
- To provide adequate exercise, since horses were intended to cover a lot of ground each day while grazing
- To provide adequate health care (including hoof care) that takes into consideration the individual's special needs. A pregnant mare requires special vaccinations against diseases like rhinopneumonitis that threaten the health of both her and her foal. It is even more vital that her parasite control program is carefully planned and strictly adhered to, since any depletion of her system can lead to serious health problems for her and her foal

Mares should not be bred before the age of three or four, depending on when they mature. A healthy broodmare can continue producing foals into her twenties. The approximate length of an equine pregnancy is eleven months and one week.

➢ *See also Breeding; Broodmare; Estrus; Foaling; Gestation; Ovulation*

Premature foal

Though they are very weak and vulnerable to infection, premature foals can be saved if they are kept warm and isolated while they develop strength and resistance to germs. Some might have to go into intensive care at a clinic. At first the mare might have to be kept behind a barrier where she can see and smell her foal but can't knock him over or step on him. Twins may need supplemental bottles if the dam isn't producing enough milk for two.
➢ *See also Orphan foal; Twin foals*

Premolars

The teeth used for grinding and chewing, located between the incisors and the molars.
➢ *See also Dental care; Teeth*

Prepotency

The ability of a stallion to produce foals that look like him, no matter what type of mare he breeds. The stallion Justin Morgan, named for his owner, was so prepotent that he is the only horse to have an entire breed named after him.
➢ *See also Breeding; Genetics; Pedigree*

Prepurchase exam

Before buying a horse, have him checked out by a veterinarian. If he passes this exam, he is said to have "vetted out." In a routine prepurchase exam, the vet will examine the eyes, heart, lungs, legs, and hooves, and he may draw blood for a Coggins test and possibly a screening for drugs or genetic tests. The Coggins test looks for equine infectious anemia and is required in many areas. You will also want the vet to confirm the age of the horse (it should be within a year of what you've been told). You may want to have x-rays taken to rule out navicular disease.

This is your opportunity to ask lots of questions and to receive an unbiased opinion. Some buyers prefer to use a vet that doesn't work for the seller. In any case, it's a good idea to have some time to discuss the horse without the seller present. No vet can tell you for sure that a horse is or will remain problem-free, but the prepurchase exam can reassure you that you know what you're buying.

ASK BEFORE YOU BUY

Find some time to discuss the horse with the veterinarian when the seller isn't around. Ask:

- ❑ Is this horse the age he's claimed to be?
- ❑ Are there any obvious problems that I should be aware of?
- ❑ Is his vision okay?
- ❑ Do you think he's physically capable of performing as I want him to?
- ❑ Does he seem suited to my needs and experience?
- ❑ How do the x-rays, if any, look?
- ❑ Is it possible the horse has been drugged?
- ❑ What other tests would you recommend?

Pressure and reward

Horses do not react to strong pressure by giving way to it, but by resisting or fighting against it. The average rider can't force a horse to do something by using physical strength, because most horses are stronger than most people. For example, if you pull back steadily on the reins to get your horse to come onto the bit, he is probably going to lean against your hands and push his nose into the air to get away from the pressure. If you use a series of light squeezes on the reins, he is more likely to move away from the intermittent pressure on his mouth by dropping his head and flexing his poll. Once he has done what you ask, reward him by avoiding steady pressure and maintaining smooth contact.

Similarly, if you lean against your horse to get him to move over in his stall, he is likely to lean toward you. But if you tap him lightly a couple of times, he will move away from your hand. His reward in that case might be a scratch on the neck or a "good boy!"

➢ *See also Aids*

Pressure bandage

For wounds that bleed profusely and haven't stopped after a few minutes, apply the cleanest materials you can find firmly against the wound. Keeping sanitary napkins or disposable diapers in your first-aid kit is a good idea. A roll of bandages or something hard like a rock or piece of wood (well wrapped in clean cloth) can be pressed against the wound and bound in place with bandages or strips of cloth. Wrap firmly until the bleeding stops — but do *not* apply a tourniquet. Call your veterinarian and leave the bandage in place until he or she arrives.

➢ *See also First aid*

Preventing injury

➢ *See Accidents, prevention of [box]; Safety*

Preventive medicine

The basics of keeping your horse healthy are providing a balanced ration, sufficient exercise, regular deworming, good dental care, and routine immunizations against common diseases. Consult your veterinarian about the usual vaccination and deworming schedule in your area; timing the medication to the life cycle of the common kinds of parasites and diseases makes it more effective. Teeth should be checked at least once a year and whenever there seems to be a problem (if your horse is avoiding the bit or dribbling grain while eating, for example). All horses should be vaccinated against tetanus and encephalomyelitis. Your veterinarian might also recommend immunizing against equine influenza, rhinopneumonitis, strangles, Potomac horse fever, and more, depending on the conditions and climate in your area.

If you show your horse or transport him along public highways, most states have laws requiring a regular Coggins test, which checks for equine infectious anemia, a devastating disease for which there is no effective vaccine.

➢ *See also Accidents, prevention of [box]; Deworming; First aid; Parasites; Safety; individual diseases*

PREVENTIVE CARE FOR OLDER HORSES

The most important aspect of preventive care for horses older than 20 is regular dental care. The veterinarian should examine the teeth of an older horse every three months, to file points and check for loose teeth that should be removed. In addition, a farrier should trim the horse's feet every six to eight weeks to keep them in top condition.

Prices of horses

People say not to look a gift horse in the mouth, but you should, in fact, closely examine any horse you're thinking of owning, especially a free one. There are many different ways to become a horse owner, from adopting a rescue horse or retired racer to paying top dollar for a fancy mover with papers. Only you can determine how much is too much, based on your budget, needs, and interests. In general, you'll pay more for a purebred than for a grade horse, more for a younger horse than for an older one (unless he's so young he needs extensive training), and more for an experienced show horse than for a pleasure mount.

It's useful to spend a lot of time looking at ads for horses in newspapers and horse magazines and to go to a few auctions before you consider buying. You'll find an amazing array of prices, from a few hundred dollars to tens of thousands. Do some research on the Internet, too, especially if you're interested in a particular breed; many breed organizations have Web sites. Another possibility is to let it be known around local barns that you're looking. A retiring school horse might be just what you need.

In figuring out the cost, don't forget the monthly stabling fees, veterinary bills, farrier bills, and the cost of riding lessons and equipment for both you and the horse. Once you've paid for the horse, the expense is just beginning!

➢ *See also Prepurchase exam*

Prizes

In addition to or instead of awarding ribbons, some show sponsors give trophies or cash prizes to winners. Most trophies become the property of the winner, but some remain the property of the show, with a replica being given to the winner. Challenge trophies are donated by an individual or business and are returned by the winner after a year. In addition to traditional trophies, prizes might include blankets, coolers, halters, saddles, and other useful items.

Cash prizes are usually based on the amount of the entry fees for that class, plus an added purse. The total amount is distributed to the winners on a previously determined scale.

➢ *See also Ribbons*

Professional horsemen

The term is defined differently by various associations, but if you're over 18, you're generally considered a professional if you are paid to ride, drive, or show at halter; train or board other people's horses; give lessons, seminars, or clinics; or endorse products. The term also sometimes applies if you accept prize money in competitions.

Profiling a class

When a judge looks at the side view of a group of horses in a halter class. This allows the judge to make comparisons between individual horses.

➢ *See also Judging*

Progesterone

A hormone that stops a pregnant mare's heat cycles and maintains the pregnancy. Insufficient levels of progesterone can cause an abortion. A mare that habitually aborts early in her pregnancy can be treated with supplemental progesterone to help sustain future pregnancies.

Progression of the aids

Also known as an "increasing of aids," this technique is used while training a horse or to reinforce previous training he's chosen to ignore. A perfect example of progression of aids is how a rider is taught to increase leg pressure if a horse does not respond to the first cue given. If ignored by the horse, this increased pressure might be followed by a light tap of the crop or perhaps a light spur. The aids progress in severity until the horse responds. Riders must consistently ask first with the lowest aid.

➢ *See also Aids; Pressure and reward; Progressive halt; Seat*

Progressive halt

When your horse is moving forward at a good clip, you can't expect to just haul back on the reins and get an instant stop. You need to use a series, or progression, of aids to convey the message that you want to slow down and come to a halt. The description in the following box may seem counterintuitive: Why would you push your horse forward when you want him to stop? But, in fact, your legs and back move him onto the bit, giving the reins more leverage than if you just pull on them.

➢ *See also Aids; Pressure and reward; Seat*

FOUR AIDS TO THE HALT

1. Sit down. This is the cue for you to stop following the horse's movement with your seat and to tuck your tailbone under you without letting your legs come forward.
2. Set your hands by pinching the reins between your thumb and forefinger and resisting any pressure from the bit.
3. Squeeze your legs, using the upper part of your calf without pinching or gripping.
4. Push further into your hands by rotating your tailbone even more underneath you, again without bringing your legs forward or collapsing your spine completely.

Prop

A pole or barrel used in Western timed events, such as barrel racing or pole bending.

➢ *See also Western competition*

Proprioception

A sense that uses cues from your inner ear and from receptors in your muscles, tendons, and joints to convey to your brain how your body is moving and where it is located in space. It allows you to reach down and scratch your ankle without looking where your hand is going. In riding, you rely on proprioception constantly to adjust your legs, seat, and hands while you look ahead to see where you're going.

Protein

Regardless of the mammal being fed, protein is required for cell growth, maintenance, and repair, in muscles, bones, blood, skin, hair, and hooves. The building blocks of proteins are amino acids. A horse's diet must provide approximately half of the required amino acids; his body manufactures the other half.

Insufficient protein can lead to poor growth, poor performance, insufficient muscle development, and a lack of appetite. Mares need extra protein during the last trimester to ensure proper development of the fetus and healthy nursing. If a horse is slow to shed out his winter coat, a little extra protein can help the new hair grow in more quickly.

On the other hand, too much protein can be dangerous. It can lead to kidney damage and eventually toxic levels of the by-products of protein digestion. If too much protein is fed to young horses, their growth rate may lead to contracted tendons, bone malformation, and inflammation of the cartilage and bone. Too much protein can make a horse "hot" or overly energetic and hard to handle.

Because protein produces heat when digested, it also literally warms a horse and can help maintain body temperature in the winter. For the same reason, protein can be a poor choice of feed in the summertime.

All of the common feeds used for horses provide some protein. Soybeans and legume hay provide the highest concentration of protein; they should be fed only to horses with high-level needs for protein.

➢ *See also Feeding and nutrition; Grains; Hay*

You might see Przewalski's horses in a zoo, but not in a stable.

Proud flesh

Puffy flesh protruding from a wound that won't heal; scar tissue. Deep cuts below the knee or hock are susceptible to proud flesh. It can be prevented by the use of caustic material provided by a veterinarian, but applying a healing ointment for a few days, and then an application of readily available alum (found in many grocery stores and pharmacies), usually keeps proud flesh at bay.

Provided horse

A horse provided by the organization hosting a show. For example, at an Intercollegiate Horse Show Association (IHSA) show, you ride a horse provided by the host college, rather than a horse with which you are familiar.

Horse sense is the thing a horse has which keeps it from betting on people.

—W. C. Fields

Przewalski's horse

These stocky, primitive animals resemble donkeys and are the only truly wild horses in existence. (The mustangs of Western legend are actually feral horses that escaped from captivity hundreds of years ago.) Przewalski's horse evolved on the Mongolian steppes and has never been domesticated. Named for Nikolai Przewalski, the Russian explorer who first wrote about them, these horses can survive great extremes of temperature. Their narrow nostrils limit the amount of freezing air they take in per breath, and their large internal air passages warm the air before it reaches the lungs. Their shaggy coats are dun-colored, with a dorsal stripe and a dark, bushy mane and tail. They resemble the horses in paintings from the Stone Age.

Psoroptic mange

➢ *See Mange*

Pulley rein

A rein used only in extreme circumstances, when your horse is running and out of control. This is a very strong check on the bit, in which you shorten your reins, brace one hand against the horse's neck and pull up and back on the other rein. It's a good idea to practice the motions of a pulley rein, without the final strong pull, to get a feel for it.

➢ *See also Aids; Progression of the aids; Progressive halt*

Pulling back (vice)

A horse that resists being tied by pulling back on his lead rope is a hazard to himself, to you, and to everyone around him. He might fall while still tied or break his halter or rope and hurt himself, a nearby horse, or you. If he pulls back suddenly enough, your fingers could get caught in the rope and badly hurt or even severed. All horses should be trained to stand patiently while tied.

PREVENTING PULLING PROBLEMS

Many handlers accustom young horses to being tied by tying the lead rope to a rubber inner tube that is firmly attached to a solid wall or a post at least 8 feet high. When the youngster pulls against the inner tube, it gives with him and teaches him that he can't fight against a yielding object. Eventually, he learns to stand quietly.

Pulling the mane

To even up the mane by pulling longer hairs out a few at a time.

➢ *See also Mane, care of*

Pull-off

A long-handled tool designed to grip nails and pull them from hooves. Also used to pull shoes.

➢ *See also Shoes, horse*

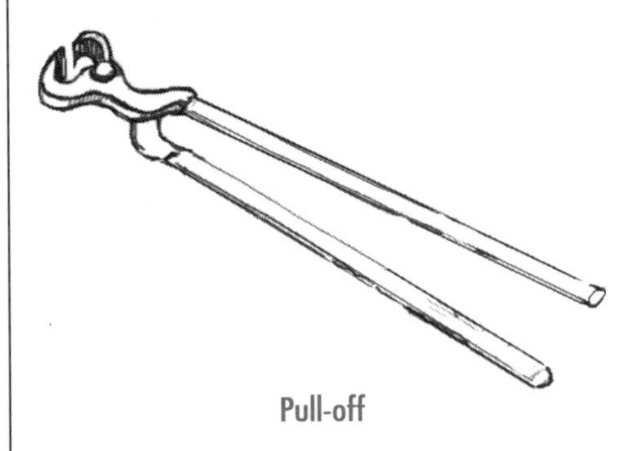

Pull-off

Pulse and respiration

A horse's normal pulse is between 36 and 40 beats per minute, and his normal respiration is 8 to 10 breaths per minute. These are averages, and you should establish what your own horse's normal pulse and respiration are so you know what might indicate illness. Check for pulse and respiration when your horse is at rest and has not been exercising. To count respiration, just watch his nostrils or flanks as he breathes. Count either exhalations or inhalation, but not both. To get his pulse, you can use a stethoscope to count the heartbeat or put your fingers on the large artery that runs across the lower jaw (under the bone). The pulse can also be felt at the fetlock joint.

➢ *See also Signs of illness; Temperature, of horse*

CHECKING YOUR HORSE'S PULSE

Place a stethoscope just behind the left elbow.

Place your hand on the left girth area.

Place your fingers against the large artery under the jaw.

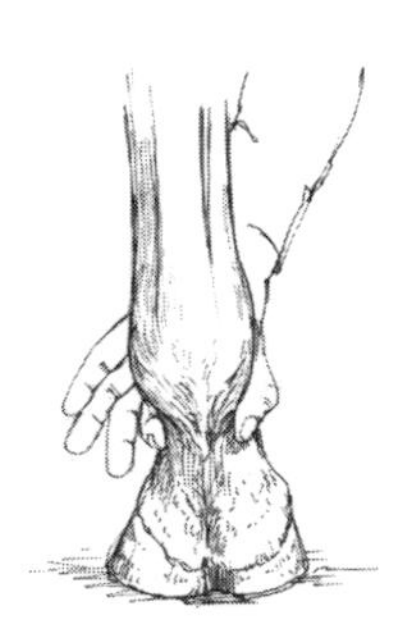

Or place your fingers against the artery in the fetlock joint.

Puncture wounds

Although they often don't bleed profusely, puncture wounds can be very serious if not treated immediately. Flesh punctures can abscess or lead to blood poisoning or tetanus if the horse is not vaccinated. Punctures of the sole of the foot can cause permanent damage to the hoof. All puncture wounds should be thoroughly cleaned with a mild disinfectant (not tincture of iodine or other strong medications, which can burn tissue) and then kept covered. Change dressings daily until the wound heals. Hoof punctures should be seen by your veterinarian.

➢ *See also First aid; Tetanus (lockjaw)*

A rider's hands turned downward from the wrists are sometimes described as "puppy paws."

Punishment

Excessive punishment is never effective with horses. Instead of learning what you want them to do, they become frightened, headshy, resentful, or aggressive, none of which improves their behavior. Punishment for unacceptable behavior must be prompt and brief. Often a sharp vocal reprimand or one smack will do, as it mimics the swift correction applied by other herd members.

➢ *See also Body language, of horses; Pecking order, in herds; Physical force, use of; Pressure and reward*

Puppy paws

When a rider holds the reins with hands turned down from wrists like a begging dog, some instructors will call attention to her "puppy paws."

➢ *See also Aids; Hands, rider's*

Purebred

Refers to a horse (or any other animal) with proven lineage of the same breed. A purebred should not be confused with a Thoroughbred, which is a specific breed. Each breed registry establishes requirements for the number of generations of proven ancestry. You can have a purebred Thoroughbred but "thoroughbred Quarter Horse" is a meaningless term.

Putting down a horse

➢ *See Euthanasia*

Pyrantel tartrate (Strongid C)

An effective drug against ascarids (roundworms) that can be given in feed daily to reduce infestation.

➢ *See also Deworming; Parasites; Roundworms*

Quagga

An extinct relative of the horse. The quagga was a subspecies of the plains zebra that ranged throughout South Africa but was wiped out by Boer settlers in the 1880s. It was yellowish brown in color with stripes on its head, neck, and shoulders only.

Quality

A term that describes a horse's overall conformation, action, and refinement. Quality in horses is determined by the ideals of the breed and gender you are evaluating. One breed may consider refinement of bone and joints and smoothness of skin and coat to be indicators of high quality. Another breed may consider these same points to be signs of poor quality.

Quarter (of hoof)

The outer sides of the hoof, separated by the toe.

➢ *See also Foot care; Hoof*

Quarter crack

A crack that develops in the quarter of the hoof. Cracks usually occur when a horse is barefoot and the hoof wall becomes too long or the hoof is brittle. There are several ways to heal a crack, but time is the main factor. As with a broken fingernail, the hoof just has to grow until the crack is completely trimmed off. In the meantime, your farrier might trim the area surrounding the crack slightly shorter than the rest of the hoof to relieve pressure on the crack. He might also rasp a small notch at the top of the crack, which spreads out the pressure when weight is put on the foot. In some cases, special clips hold the crack together while it grows out.

➢ *See also Foot care*

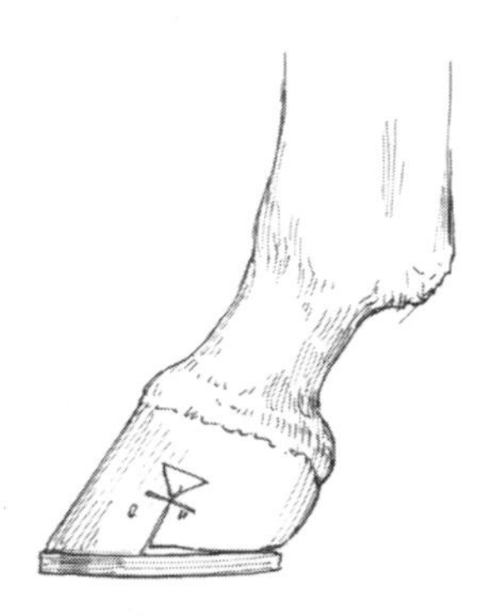

It may be necessary to groove the top of a quarter crack to speed the healing process.

Quarter Horse

Although the Quarter Horse is thoroughly associated with the American West, the breed originated in Virginia, where colonial farmers bred heavy English horses with quick Indian ponies to create a muscular but speedy work horse. Faster than the Thoroughbred at distances of a quarter mile or less, the Quarter Horse was the first American racehorse, referred to in early documents as the *Celebrated American Quarter Running Horse*, or CAQRH. But as longer distance Thoroughbred racing prevailed over sprinting, the Quarter Horse moved westward with the settlers, where his toughness, reasonableness, and quick reflexes made him a natural cow pony.

Today, Quarter Horses outnumber most other North American breeds by a factor of ten to one. Cowboys still ride them on the range, and they make wonderful pleasure horses. Sturdy and muscular, a typical Quarter Horse has short, strong legs and a sloping croup or rump. They can be

any color, with some white markings. (Too much white is not allowed by the American Quarter Horse Association.) Along with their many good qualities comes a disadvantage: a genetic tendency toward navicular disease, a degenerative disease of the hoof.
➢ *See also Appendix Horse; Western competition*

Quarter method

A term used to describe a handler's method of movement during the inspection phase of a showmanship class. The handler should always be in a position that prevents him from obstructing the view of the judge, thereby allowing the judge an optimum view for evaluating conformation and balance. The judge's primary objective is to evaluate a handler's mastery of skills necessary for showing a horse at halter. *Note:* Some regulating organizations have adopted rules for showmanship requiring that handlers *not* use the quarter method of showmanship. Learn from a coach who is knowledgeable about the breed regulations in your area.

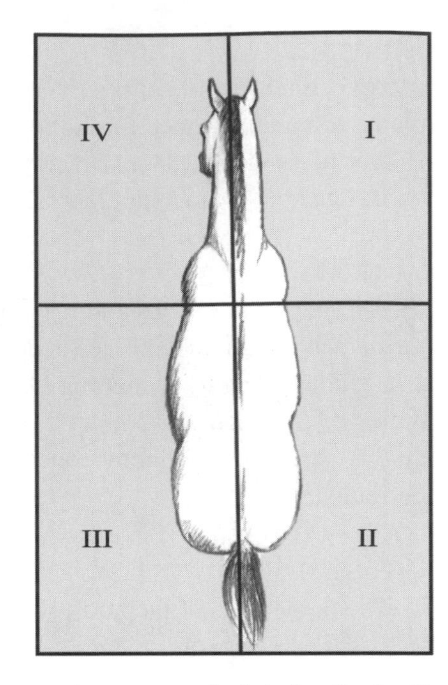

In the quarter method, the handler should always be one quadrant away from the judge. For example, when the judge is in quadrant I, the handler should be in quadrant IV. When the judge moves to quadrant II, the handler should move to quadrant I. As the judge moves behind the horse to quadrant III, the handler should move back to quadrant IV. When the judge moves up to quadrant IV, the handler should move back to quadrant I.

Note: The numbering of quadrants may vary with coaches and regulating organizations; however, the underlying principle remains the same.

Quarter Pony

Quarter Ponies start out as Quarter Horses but don't get tall enough, growing to be less than 14.2 hands high 58 inches or 145 centimeters. They can be shown in International Quarter Pony Association shows.
➢ *See also appendix*

Quarter-turn

A 90-degree turn in either direction.

"Quicked" hoof

This term most commonly refers to injury caused by a farrier who has either trimmed the hoof too deeply and exposed sensitive sole or driven a nail into the sensitive laminae. In either case, the horse is prone to serious infection and permanent damage to the coffin bone if not treated properly. The first situation is usually best treated by placing pads and shoes on the hoof, possibly with an iodine-based hoof packing material intended to toughen the sole. The second should be treated like any other puncture wound in the hoof: Clean it well, soak it in Epsom salts, and have it seen within a day by a veterinarian.
➢ *See also Foot care; Hoof*

Quick-release knot

For obvious reasons, you should always tie a horse with a knot that won't get tighter if the horse pulls against it. It is also important to be able to free your horse in a hurry if he is panicking or if an emergency arises. See the following illustrations for two variations of quick-release knots.

Note: When tying a quick-release knot, make sure to keep your fingers out of loops to avoid getting them caught if your horse pulls back suddenly. If your horse learns to pull the free end of the rope (thereby releasing himself), run that end through the loop; this is referred to as a "safety"

TWO QUICK-RELEASE KNOTS

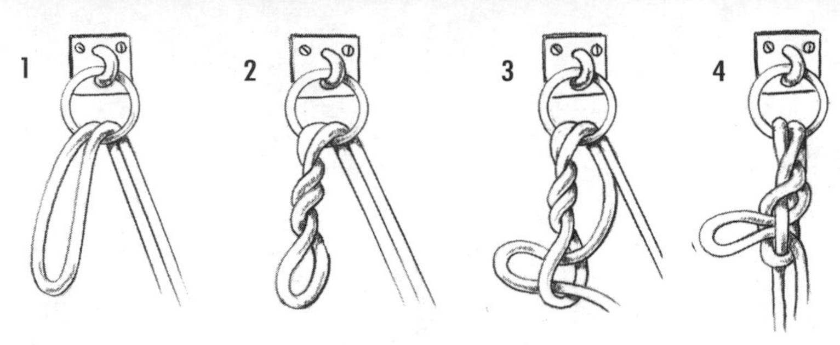

(1) Pass a loop of the rope through the ring. (2) Twist the rope several times, leaving a loop. (3) Take the free end of the rope, make a fold (bight) and slide the bight through the loop. (4) Slide your hand up the rope to tighten the knot.

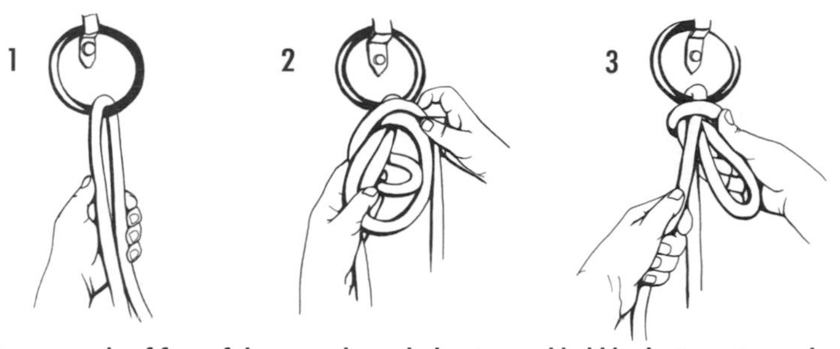

(1) Run a couple of feet of the rope through the ring and hold both pieces in one hand. (2) Make a fold (bight) in the tail end of the rope, cross it over the two pieces in your hand and through the resulting loop. (3) Pull the bight through the loop until it is about six inches long and the knot is secure, then slide the knot up to the ring.

Give a man a horse
he can ride,
Give a man a boat
he can sail.

—James Thomson,
Sunday Up the River

knot. He won't be able to untie himself, but be aware that you won't really have a quick-release knot any more.
➢ *See also Knots; Tying*

Quidding

Partially chewing hay and then spitting out pieces of it.

A quirt

Quirt

A Western-style crop, also called a **romal,** with a short handle and rawhide lash. Also, in a closed or "California" type rein, the quirt connects the side reins and provides a built-in whip.

Quitting

➢ *See Stopping in front of a jump (quitting)*

Quittor

An infection in the lateral cartilage of the hoof.

r

Rabies

Although generally considered a disease of small animals, all mammals are vulnerable to this fatal virus, which affects the central nervous system. It is carried in the saliva of infected animals, who transmit it by biting or by licking wounds or lesions on the skin of another animal. Rabies can have a long incubation period, because the virus travels through nerve cells rather than the bloodstream. Once it reaches the brain it is always fatal. People can be treated successfully after exposure, but there is controversy over whether this will ever work for animals.

Symptoms vary widely and can be mistaken for many other illnesses, but some common signs are difficulty swallowing and drooling. There is no test for this disease; a positive diagnosis can only be made from an autopsy of the brain. Rabies takes two forms, the dumb or paralytic form and the furious form. The latter is more commonly associated with the disease, manifesting itself in aggressive and even vicious behavior. However, the horse might also be lethargic and uncoordinated, with evident weakness in the hindquarters.

If rabies is common in your area, ask your veterinarian about vaccinating your horse. If he is exposed to a rabid animal, a booster will be necessary and he should be quarantined for 90 days. An unvaccinated horse that is exposed may have to be put down or put into strict quarantine for 6 to 9 months, with as little human contact as possible.

RABICANO

A horse with white hairs in the base of the tail is called a rabicano.

An American Saddlebred doing the rack.

Racehorse

➢ *See Horse racing; Thoroughbred*

Race sulky

➢ *See Sulky, race*

Racing

➢ *See Flat racing; Gaits; Harness racing; Pace; Speed, of horse; Standardbred; Steeplechasing; Thoroughbred; Triple Crown*

Rack

A four-beat gait in which each foot meets the ground at equal, separate intervals. The rack is smooth and highly animated, performed with great action and speed, in a slightly unrestrained manner. Desired speed and collection are determined by the maximum rate at which a horse can rack in form. Racking in form should include the horse remaining with a good set head. It should be achieved by the horse in an effortless manner from the slow gait, at which point all strides become equally rapid and regular.

Although the American Saddlebred is the breed most commonly associated with this gait, it is performed by horses of other breeds. In fact, there is a registry known as The Racking Horse Breeders Association that specializes in horses that rack.

➢ *See also Amble; American Saddlebred; Missouri Fox Trotter; Pace; Racking horse; Rocky Mountain Horse; Tennessee Walker; Tolt*

Racking horse

Breeds other than the Saddlebred that exhibit the racking gait. These include the Missouri Fox Trotter and the Rocky Mountain Horse. The Racking Horse Association has established a registry and requires pedigrees, rather than registering any horse that can rack. These horses do not generally snap their hocks in the flashy showring style but can move quite quickly and smoothly at the rack.

➢ *See also Amble; American Saddlebred; Missouri Fox Trotter; Pace; Rocky Mountain Horse; Tennessee Walker*

Rail

The fence line of an arena or ring.

Rail class

Term for a flat class (without jumps or obstacles, as in trail).

Rail work

Working a horse primarily on the outer edge of an arena or ring. Rail work forms the basis for most basic riding education of horse and rider because the rail provides a form of security. The term also refers to the portion of a competitive class that's judged "on the rail"; this is often combined with performance of a pattern with or without obstacles. An example of such a class would be Hunter Hack, in which horses are judged equally on rail work and their performance over two fences.

I took my horse
to market,
And lost her on the way,
For dappled were
the moors with rain,
And she was
dapple gray.

—Traditional nursery rhyme

Rain, horses and

Most people don't like to stand in the rain, but it doesn't seem to bother the majority of horses. They are well protected against the average summer precipitation and it doesn't hurt them to get wet, but they should always have shelter in case of a thunderstorm or if it's cold and windy as well as rainy. In areas where it rains a lot, be aware of **rainrot,** a skin disease caused by a fungus that thrives in dirty, wet conditions.

➢ *See also Rainrot (rain scald)*

Rain, riding in

If you find yourself riding in the rain, make sure you watch the footing very carefully. Grass can be quite slippery with just a little rain on it, and slick mud can lurk in grassy areas that look dry enough to canter through. Many instructors recommend not riding in the rain. Not only is it a potential hazard to the horse, but also it can damage leather tack.

You can buy raingear especially designed for riding. It is important to protect your saddle as much as possible. Do not wear a poncho or anything that will flap, rustle, or otherwise startle your horse. And don't leave your helmet at home because you don't want it to get wet; there are plastic helmet covers for that very reason.

Rainrot (rain scald)

A skin condition caused by a funguslike organism that horses pick up when they roll in the dirt. When the dusty hair gets wet, the organism can multiply and cause the skin to form scabs, while the hair mats into clumps and falls out. Before rainrot becomes visible, it can be felt as bumps under the skin that may be sensitive when groomed.

Treatment depends on the climate and standards of your region. Usually horses are bathed with a medicated shampoo, which is followed by liberal application of antifungal cream or spray. Obviously, this treatment must be modified in cold, northern climates. Always seek advice from your veterinarian.

Ratcatcher

A collarless shirt worn with formal hunt or show attire. To complete the look, the rider wears either a bow tie of the same fabric as the shirt or a white **stock tie,** which wraps around the throat like a scarf.

➢ *See also Attire; Stock (as item of clothing)*

Rating a cow

Getting in the right position to keep a cow under control without crowding it or backing off too far. Each cow has its own control zone (similar to "personal space"), and both horse and rider need to gauge the optimal distance for maneuvering an individual cow most effectively.

Rating the horse to the jump

Positioning a horse and adjusting his stride so that he takes off in the correct spot at a jump. This requires an advanced understanding of a horse's gaits and striding, as well as a high level of coordination of aids. Improper rating can lead to unsafe and dangerous jumping.

Rations

The overall combination of feeds provided to a horse. A ration should provide a well-balanced diet that meets the needs of each individual.

➢ *See also Feeding and nutrition*

Rayless goldenrod (jimmyweed, rosea) (*Isocoma* spp.)

Also called rosea or jimmyweed, a bushy shrub with yellow flowers that is most common in the Southwest. It is toxic to horses.

➢ *See also Poisonous plants*

Rearing

When a horse rears, or stands up on his hind legs, it is usually a defensive response that if left unchecked can develop into a dangerous habit. Horses that rear not only threaten the humans that handle, ride, or drive them, but they pose a danger to themselves as well. Horses can rear so far up that they hit their heads or, worse, flip over and break their necks. A horse that rears habitually should be handled by a professional trainer.

RANUNCULUS ALERT

When fresh, buttercups (also known as *Ranunculus*) are slightly poisonous to horses. When dried (as found in hay), they are harmless.

Rearing is dangerous for both horse and rider.

Here are a few safety tips to remember if your horse rears.

- If your horse rears while being led or on the longe line, give him some slack in the line. If you put pressure on his halter or jerk on his lead while he's rearing, you may inadvertently cause him to resist the pressure and flip over backwards.
- If your horse rears up and seems to strike out at you or comes at you in any manner, he needs to be evaluated by a professional trainer who specializes in rehabilitating problem horses. This is not work to be done by amateurs.
- If your horse rears while being ridden, keep the horse's head down by leaning forward, lowering your hands down toward your knees, and exerting steady but not hard pressure on the reins; then urge your horse forward. It is impossible for a horse to continue rearing if his head is down and he's moving forward.

Reata, Riata

A braided, leather rope used in Western roping events and seen coiled and hanging from the saddle when not in use.

Recessive gene

Each gene has the potential to transmit a trait or characteristic. A recessive gene carries information for certain traits but is always overridden by a dominant gene. Only when a foal receives the same recessive gene from each parent will the recessive trait (for example, chestnut coloring) appear.

➢ *See also Dominant gene; Genetics; Pedigree*

Recessive traits

Traits such as coloring, height, and conformation are determined by genes. While dominant traits are always expressed if a dominant gene is present, recessive traits can also be passed along to future generations as part of the genetic code. However, two recessive genes must be combined for the trait to show.

➢ *See also Dominant trait; Genetics; Pedigree*

Reconditioning tack

➢ *See Tack, cleaning*

Rectal palpation

➢ *See Palpation*

Red dun coloring

A light red and tan body color, with reddish, white, flaxen, or mixed mane and tail. Like a regular dun, a red dun has a dorsal stripe and often has striping on the legs and withers.

➢ *See also Coloring; Dun coloring*

RED HORSES

Red horses include:

Bay. Medium red with black mane and tail; can come in a range of shades from sandy bay to mahogany and blood bays.
Blood bay. A clear, even red, with black mane and tail. The most common shade of bay.
Chestnut. Medium red with red or flaxen mane and tail.
Liver chestnut. Darkest red.
Red corn. Distinct red spots that resemble kernels of corn.
Red dun. Very light red, usually with darker mane and tail.
Red points. Red mane, tail, and lower legs.
Red roan. Red and white hairs evenly mixed.
Sorrel. Light, clear red, lighter than chestnut; or for draft horses, three or more shades of red, with light points.
Strawberry roan. Light red evenly mixed with white, often with red points.

Red mange (itching dermatitis)
➢ *See Mange; Parasites*

Red maple (swamp maple) *(Acer rubrum)*
A tree, also known as swamp maple, found in the eastern United States whose wilting leaves are extremely poisonous to horses.
➢ *See also Poisonous plants*

Red roan coloring
Coat color in which auburn and white hairs are equally mixed, giving a reddish appearance.
➢ *See also Coloring; Roan coloring*

Red worms (bloodworms)
➢ *See Bloodworms; Deworming; Parasites*

Refusal, at a jump
➢ *See Stopping in front of a jump (quitting)*

Register of Merit (ROM)
An award given in specific events by the American Quarter Horse Association at its breed shows. The ROM counts toward lifetime performance awards.

Rein back
A Western term for backing up your horse.
➢ *See also Backing up, backing while riding*

Reined cow horse
A Western competition that combines a "dry" performance of reining patterns with actual cow work. Each contestant performs basic reining maneuvers and then must work a single cow, keeping it at one end of the ring, turning it back after it is allowed to start for the other end of the ring, and circling it without letting it get away.
➢ *See also individual events; Western competition*

Reining class
In Western competition, reining classes, sometimes referred to as "cowboy dressage," are a popular choice for people who have worked closely with their horses to develop skill and precision. The horse must perform changes of pace and direction, large and small circles, spins, and the showy sliding stop, in which the horse halts with his hindquarters set deep underneath him, sliding his hind feet. This type of stop should be solid and square, without a great shift of the rider's balance, which tends to cause the horse to hollow his back and brace his front end.

A good reining horse listens to his rider without anticipating the pattern and responds immediately to the aids. He performs calmly, but has quick

POISONOUS PLANT

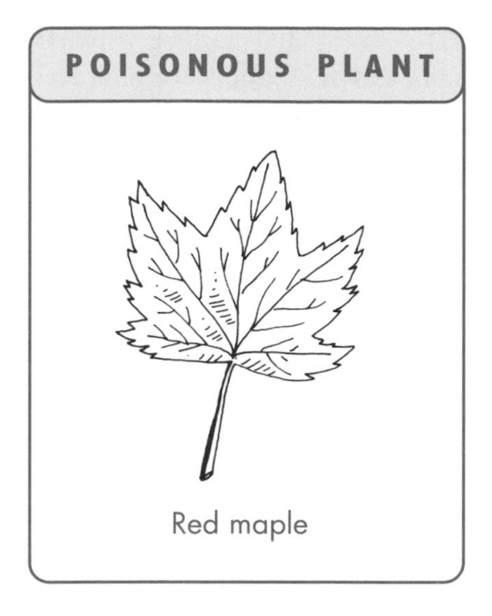

Red maple

REMUDA

A remuda is a group of trained stock horses used on a ranch to work cattle.

reflexes and moves fast while staying controlled. For more information, contact the National Reining Horse Association .
➢ *See also Aids; Western competition*

Reins

The reins attach to the bit and connect the horse to the rider, who should hold them lightly and with respect for the horse's mouth. There are many kinds of reins, generally made of flat or braided leather, though reins can be rubber, nylon, or braided mohair rope.

In English riding, the reins are connected in a single loop from one side of the bit to the other, usually buckled in the middle, and are almost always leather. In Western riding, the reins (or roping reins) might be connected, as in English riding, but are more commonly open or split. Open reins are 6 to 8 feet long, made of a variety of materials, and not connected at the ends. Closed reins, also called California reins, have three parts. The two side reins are connected by a long **quirt** or **romal,** which hangs down by the horse's shoulder.
➢ *See also Bridle; Tack*

A reining horse comes to a sliding stop from a full gallop.

Renvers

Also referred to as haunches out, a suppling exercise in which the horse carries his hindquarters toward the outside of the circle being ridden. The maneuver can be executed along the rail or in a schooling figure, such as a circle or figure eight.
➢ *See also Shoulder-in; Travers; Two-track (half-pass)*

Reproduction

➢ *See Breeding; Foaling; Gestation; Ovulation; Pregnancy*

Reproductive tract

The mare's reproductive tract, like that of other female mammals, consists of ovaries to produce eggs, fallopian tubes to conduct the eggs to the uterus for fertilization, and a uterus to nurture the fetus. The cervix between the vagina and the uterus remains tightly closed except during estrus and foaling. The vagina is the birth canal through which the foal passes, while the vulva is the external opening located just below the rectum.
➢ *See also Breeding; Estrus; Foaling; Gestation; Ovulation; Pregnancy*

Reproductive Tract of the Mare

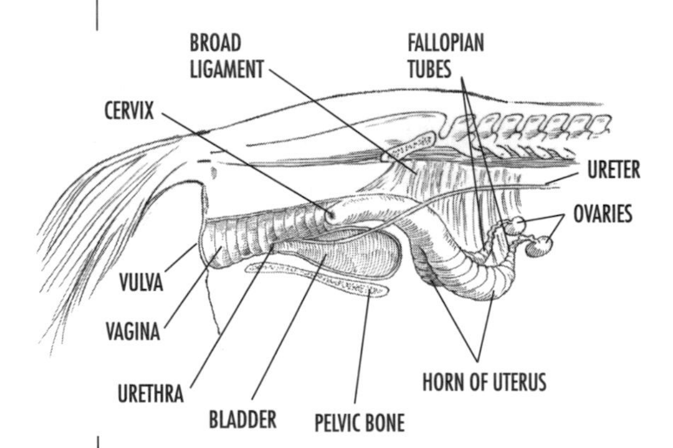

Reserve champion

The champion is the judge's first choice; reserve champion is the second choice. Championship classes are held in a variety of disciplines at all levels of competition. A reserve champion is essentially the second best horse in a discipline, at a particular level, on a given day of a specific competition.
➢ *See also Prizes; Ribbons*

Respiration

The number of breaths per minute. One breath equals one inhalation and one exhalation. A horse normally takes between 8 and 10 breaths per minute.

➢ *See also Pulse and Respiration*

Respiratory system

Horses are unable to breathe through their mouths, which is why they have such large nostrils. When they are working hard, their nostrils flare to bring in more oxygen. Horses are vulnerable to respiratory ailments, especially when stressed by travel, showing, or poor condition.

When buying a horse, make sure his wind is sound. Some horses have obstructions in their windpipes that restrict their breathing during exercise. Because of the labored sound such horses make, they are sometimes referred to as "roarers." The condition can sometimes be fixed with surgery.

Restraints

Humans have devised methods for restraining horses for thousands of years. Some form of restraint is usually required for veterinary care or special situations that might trigger the horse's defensive responses. The type of restraint used depends on many factors, including but not limited to the situation, the personality and experience of the horse and handler(s), and the environment or facilities available.

The most common forms of physical restraint available are:

- **Halter and lead shank with chain.** The chain can be run either over or under the horse's nose to add a higher degree of control.
- **Creative physics.** That is, using the horse's body to force the desired behavior. For example, holding up the left foreleg encourages a horse to keep its right foreleg on the ground so the veterinarian can examine it.
- **Twitch.** This device was once thought to only distract the horse; however, it is now believed that it actually stimulates mechanoreceptors in the skin that activate a release of endorphins in the brain similar to those released during acupuncture. The most common types of twitches are the humane twitch, the rope twitch, and the chain twitch.
- **Stocks.** Built on many larger farms as a form of restraint for physical examination, treatment, grooming, and breeding. Stocks restrict the movement of the horse by enclosing him on all sides. Most stocks are built out of tubular pipe, which reduces the risk of injury.
- **Chemical restraint.** Many drugs are available for the restraint of horses. The drugs have a tranquilizing or sedative effect that makes many procedures much less stressful for the horse. Only a veterinarian should administer chemical restraints.

➢ *See also Cross-tying/cross ties; Handling horses; Picking up feet; Twitch (restraint); Tying*

TRACKING RESPIRATION

Observe a horse's respiration by standing in front of him and counting how many times per minute his ribs rise and fall as he breathes.

AVOIDING RESPIRATORY PROBLEMS

Always feed a horse who has breathing problems with his head down, to let mucus drain out, not into his lungs. Feed good-quality, dry hay to eliminate fungal spores.

Retained caps

When a foal begins to grow adult teeth, his baby teeth become hollow and are pushed out by the new ones. In some cases, these hollow caps do not detach from the gum and can cause painful inflammation and even sinus problems if the tooth is in the upper jaw. Signs of a retained cap include bad breath, drooling, and trouble eating. Your veterinarian should check out any tooth problems, as infections in the gums can get into the bloodstream and cause serious problems.

➢ *See also Dental care; Teeth*

> **REVAAL**
>
> A revaal is a natural lateral pacing gait performed by a number of Asian horses, including the Kathiawari and Marwari horses of northwest India.

Retained placenta

In a normal birth, the mare will expel the placenta within two hours of delivering the foal. In some cases, the placenta will not detach properly from the uterine wall and may hang down her legs. If she does not completely shed the placenta after giving birth, she can develop a serious, even life-threatening, infection of the uterus, which can lead to laminitis or peritonitis. Call your veterinarian for assistance.

➢ *See also Foaling*

Return-in-season

A clause in a breeding contract that means that the owner of the mare must pay the stud fee at the end of the breeding season, no matter what the results of the pregnancy.

➢ *See also Live foal guarantee*

Rhabdomyolosis

A severe cramping of the muscles, particularly of the hindquarters. Also called **azoturia** or **tying up,** this condition can affect any horse that exerts himself excessively, although it is most likely to strike a horse that is not accustomed to exercise. Tying up often happens after a long ride but can also occur after a strenuous romp with another horse, a long trailer ride, or a struggle with the veterinarian or farrier. There are two types of this disorder, both of which generally occur in out-of-condition horses that overexert themselves.

Type A is often called **Monday morning disease** because it typically affected draft horses that worked hard during the week, ate rations of grain, and would be unable to move after a day of rest. It usually occurs in horses that are fed grain regularly but not exercised every day. The condition is caused by lactic acid and waste products that build up during strenuous exercise and can't be cleared away fast enough by the bloodstream. Symptoms come up very suddenly, sometimes 15 minutes to an hour after starting to exercise. The horse will halt and be reluctant to move. His muscles will feel rigid and he may sweat and want to lie down. He should not be forced to move, as this makes the condition worse. Your veterinarian can administer pain relievers and muscle relaxants to help the muscles return to normal.

Type B rhabdomyolosis appears at the end of a long ride, usually in horses not accustomed to heavy work. It can also happen if a hot, sweaty

horse is hosed off with cold water before being cooled out. The condition is caused by lack of oxygen and nutrients to the muscles as well as trapped waste products. While the symptoms of type B are similar to type A, treatment is different. In type B, slow walking can improve circulation and help flush waste from the muscles. It is very important to replace fluids and electrolytes. Do not feed grain, although hay and green grass are all right. You should call your veterinarian any time you suspect tying up.
➢ *See also Azoturia; Rhabdomyolysis*

Rh foals

➢ *See Neonatal isoerythrolysis (Rh foal)*

Rhinopneumonitis

A highly contagious disease caused by equine herpes virus (EHV). "Rhino" causes respiratory illness and can also lead to abortion and, rarely, nervous system disorders. Young horses are particularly vulnerable to the disease because they haven't built up immunity. Symptoms, which can occur 2 to 10 days after exposure, generally include a runny nose, elevated temperature, and cough. Keeping your horse warm, unstressed, and properly fed will help him recover in a couple of weeks. A high fever may indicate a secondary infection requiring antibiotics.

Pregnant mares and horses that are exposed to many strange horses can be vaccinated against rhinopneumonitis, but in order to protect against abortion the vaccine has to be repeated several times during a pregnancy. It may not be necessary to vaccinate your horse, depending on your circumstances. Ask your veterinarian if your horse is at risk.

Rhizoctonia leguminicola

A toxic fungus that causes black patch disease on clovers and other legumes. If a horse eats infected plants, he can get "slobbers" or irritated salivary glands. More serious effects are fetal malformation or abortion, digestive upset and weight loss, or even death if the fungus is eaten in sufficient quantities.
➢ *See also Poisonous plants*

Ribbons

Ribbons are given to the top riders in each class. The number of ribbons awarded depends on the type and level of competition.
➢ *See also Prizes*

Ribs, as an indicator of condition

When you run your hand over your horse's barrel, you should be able to feel his ribs but they should not be easily seen. A layer of flesh over them indicates good condition. A layer of flesh that pads the ribs so that they can't be felt at all means that your horse is too fat. If your horse grows a heavy coat during the winter, it is wise to check his ribs periodically by hand to make sure he is maintaining his condition, especially if he's older. Winter coats can be deceiving.

UNITED STATES RIBBON COLORS

RANK	COLOR
First place	Blue
Second place	Red
Third place	Yellow
Fourth place	White
Fifth place	Pink
Sixth place	Green
Seventh place	Purple
Eighth place	Brown
Ninth place	Dark gray
Tenth place	Light blue
Grand	Blue, red, yellow, and white or purple
Reserve grand	Red, yellow, white, and pink; or purple and white
Champion	Blue, red, and yellow or purple
Reserve champion	Red, yellow, and white; or purple and white

Note: In Canada, red is first and blue is second.

Ride a buck

➢ *See Dollar bill class*

Ride and tie

This unusual athletic event combines endurance riding with marathon running. Each team consists of two humans who switch off between riding a single horse and running on foot. Most ride and tie events cover 10 to 20 miles, but can be as short as 6 or as long as the World Championship contest at 40 miles.

All teams start out on horseback at the same time. After the rider has covered as much distance as she thinks the runner can handle, she dismounts, ties the horse, and takes off on foot. The runner finds the horse, climbs on, and goes after her partner. During the race, the humans must switch off a minimum number of times, and the horse must go through vet checks along the way.

Teams can be any age or gender. A "century" team has three members whose ages add up to 100 or more. In addition to human partners who can run long distances under rugged conditions, the ride and tie takes a particular type of horse that is able to handle the stop-and-start activity, not to mention being left behind by his rider and then passed by other horses.

On the wings of the morning they gather and fly,
In the hush of the night time I hear them go by,
The horses of memory thundering through
With flashing white fetlocks all wet with the dew.

—Will H. Ogilvie,
The Hooves of Horses

Rider, levels of experience

There are various systems for classifying the experience levels of horsemen. They begin with the well-known "beginner/intermediate/advanced" naming system. This oversimplified system can be confusing as well as frustrating for both students and instructors. But it is common because of its simplicity and ease of use. Therefore, it's important for both student and instructor to clarify the definition of terms being used. One possible set of definitions could include:

A **pre-beginner** is interested but knows very little about horses or riding. She needs to learn about handling horses on the ground as well as being introduced to riding. She may be timid or fearless, but has little direct experience.

A **beginner** has some experience actually riding and can handle a quiet horse at the walk and trot. She can ask the horse to turn and halt but is not yet comfortable enough to lope or canter.

The **advanced beginner** is confident on a well-schooled horse at the lope or canter. She should be comfortable grooming and tacking up her own horse.

An **intermediate** rider is probably more serious about riding regularly and may want to compete. She knows how to control the horse at all gaits, knows her diagonals and leads, and can maneuver the horse through circles and serpentines. She may begin learning to jump at this point.

An **advanced intermediate** rider can perform simple and flying changes of lead, can vary the speed of each gait, can turn on the forehand and the haunches, and is familiar with lateral movements. She is confident on a variety of horses, is developing the skills to train a horse, and can take the horse over a course of jumps.

The **advanced** rider knows how to train a horse. She has a strong sense of balance and timing and a deep understanding of horse behavior. She performs advanced maneuvers in her style of riding without obvious aids to the horse.

Perhaps the most complete system of classification is that used by Pony Club members and instructors. This system is fairly objective because there are international standards that every student must meet in order to achieve advancement to higher levels.

Rider up

At a show, this term signals that the rider is mounted and ready for her class.

Riding fence

Riding along the fence line on large ranches and farms, riding fence is an important maintenance chore. All pastures and other fenced areas should be checked regularly for breaks, protrusions, holes, or sagging posts.

Riding ring

➢ *See Arena events; Arena work*

Riding style

➢ *See Attire; English competition; English riding; Tack; Western competition; Western riding*

Rigging

The straps that hold the cinch and the saddle tree together on a Western saddle.

➢ *See also Tack*

Right-of-way

➢ *See Passing, on horseback*

Ringbone

A bony growth most often found on the pastern. It can also affect the coffin bone within the hoof. It is classified as *high* when it affects the proximal phalange (pastern bone closest to the fetlock joint), and *low* when it affects the medial (pastern bone just above hoof) or distal (coffin bone) phalange. It usually only causes lameness if there is growth within the joint. Many cases of ringbone go virtually unnoticed.

Ringbone that affects the joint does not have a good prognosis for treatment, so prevention is the best medicine. It is believed that ringbone is a heritable trait; essentially the predisposition for a weakness in bone tissue has been passed on by a parent. (This can be researched by studying incidences of ringbone among siblings and other related horses.) It can best be prevented by choosing not to breed horses affected by it.

WHAT'S A RIGGING DEE?

The D-ring that holds the tie strap or latigo of a Western saddle. The latigo runs through a ring on the cinch and loops several times between the two rings to hold the cinch in place.

See also Saddles; Tacking up

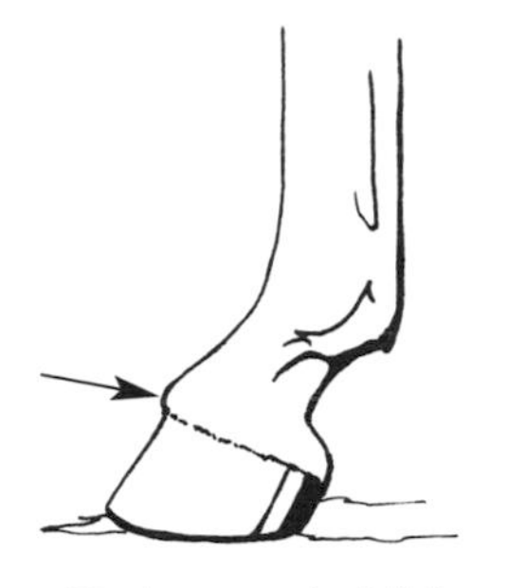

Ringbone on the left foot

Ring sour

Reluctant to perform. A horse that has been worked too hard or too repetitively may become ring sour. He'll look for ways to leave the ring or evade your commands, and will generally exhibit a negative attitude. Variety and change of pace while training and exercising keep both horse and rider fresh and interested.

Ringworm

A fungal infection of the skin that causes hair to fall out in distinct, round spots. Lesions form a week to a month after exposure. Ringworm spreads between horses that share tack and grooming tools. Some families or breeds seem to be more susceptible than others. Young horses, older horses, and ones with compromised immune systems are more vulnerable.

Ringworm should be treated promptly by isolating the infected horse and applying a topical medication. All grooming tools and other equipment must be disinfected. Make sure your horse has enough vitamin A, zinc, and selenium in his diet, as a deficiency of these can encourage the development of ringworm.

➢ See also Parasites

Rising trot

➢ See Posting trot

Roach

To cut or shave a mane or tail very short. This practice is common in cutting and other fast action events to prevent the reins from tangling in the mane. Roaching is traditional in some breeds, such as the sturdy Cob-type horses of Great Britain. Primitive breeds tend to have naturally upright manes, often with little forelock.

Road founder

Road founder can result from riding a horse hard and fast on a solid surface, like pavement. It can also happen if a horse is worked hard and fast after a radical change in shoeing. For example, a horse that has been barefoot but is shod for a show and works hard at the show can develop road founder. The symptoms and physiology are similar to founder.

➢ See also Laminitis

Roads, riding on

➢ See Pavement, riding on

Roan coloring

Roan is a mix of dark and light hairs, giving the overall coat a mottled or speckled appearance. A mix of black and white hair is called blue roan, while brown or auburn and white is called red or strawberry roan.

➢ See also Blue roan coloring; Coloring; Red roan coloring; Strawberry roan coloring

RISING THREE

When a horse is described as "rising three" it means he is almost three years old.

A **roached mane**

ROARER

A horse that makes a loud, roaring sound when breathing heavily. This unsoundness can interfere with breathing and limit use of the horse.

See also Respiratory system

Rocky Mountain Horse

A rare breed of amblers that are wonderful pleasure horses because of their sturdy conformation, calm dispositions, and smooth gait. These horses are not actually common in the Rocky Mountains, but that is where the founding sire lived. Generally found in Kentucky and southern Ohio, these horses aren't your best bet for jumping or competition. If you're interested in learning more about them, contact the Rocky Mountain Horse Association.
➢ *See also Amble; Racking horse; appendix*

Rocky Mountain Horse

Rodents

➢ *See Pest control*

Rodeo

A demonstration of Western riding skills that seems to have originated in the mid-1800s on large Texas cattle ranches where groups of cowboys would gather to socialize, swap stories, and show off their roping and riding skills. These early cowboys learned a great deal from the Mexican cattlemen who ran large herds of cattle on vast tracts of land. This influence is seen in much of the language of the rodeo (from the Spanish "rodear" — to round up or encircle). The words "lariat," "lasso," "mustang," and "bronco" all come from the Spanish language.

The first recorded rodeo involved teams from different ranches and took place in Colorado in 1869. Before becoming regulated contests with arenas, loading chutes, and grandstands, rodeos were held in town squares or on main streets, no doubt with some consequent property damage to the town buildings. The events that make up the typical rodeo are rooted in the daily work of the ranch, such as branding cattle and breaking horses. Events also include bareback and saddle bronc riding, bull riding, bulldogging or steer wrestling, and calf roping. Many modern rodeos also include barrel racing and other speed contests, as well as youth divisions.

If wishes were horses, beggars might ride.

—John Ray, *English Proverb*

Bareback and Saddle Bronc Riding

In bareback riding, the horse wears a halter without reins, a wide strap around his girth with a sturdy handle attached just behind the withers, and a bucking strap or cinch that is pulled tight around his loins. The rider mounts in a chute and spurs the horse into the ring, holding onto the handle with one hand. In order to have a successful ride, the other hand must stay clear of the horse for an entire 8 seconds, and the rider can't change hands. If he's still on board after the 8-second whistle sounds, the rider looks for the pickup men who gallop alongside to pull him clear of the bucking horse.

Staying on a frantically bucking horse might seem easier with a saddle but can, in fact, be more complicated. For one thing, while the stirrups in the saddle bronc event offer added balance, if the rider drops one during the 10-second ride, he's disqualified, and if his foot slips through and he's thrown, he could be dragged and badly hurt. The reins give little control since they are attached to a halter, not a bridle and bit. Riders have the option to remove the saddle horn, as grabbing it ("pulling leather") will also disqualify them, and it could hurt them if the horse decides to drop and roll.

In both bareback and saddle bronc riding, points are given to both horse and rider by two judges. A maximum of 100 points is theoretically possible, but scores in the 70s are considered remarkable, while many very good riders are happy with a score in the 60s.

Calf Roping and Bulldogging

Unlike bronc riding, calf roping and bulldogging involve teamwork between rider and horse. In calf roping, the rider ropes a galloping calf and immediately leaps off his horse, running to the calf to flip it over and tie three of its legs together. While he moves to the calf, the horse must keep the rope tight enough to prevent the calf from escaping, but not so tight that it pulls over the calf before the rider can get to it. The tie must hold for five seconds to count. A good run takes about 12 or 13 seconds.

In bulldogging, or steer wrestling, two riders gallop alongside a steer. One of them acts as the hazer, keeping the steer from veering away. The other rider must leap off his horse and grab the 700- or 800-pound steer and yank its horns around to bring it to the ground. The slightest miscalculation will leave the bulldogger sitting in the dust looking foolish while the steer careens around the arena. Experienced bulldoggers with well-trained horses can bring down a steer in about five seconds.

Bull Riding

Although no rodeo event is without risk, surely the most dangerous is bull riding. Bulls are not only a lot bigger than the average bucking horse, they are generally quite aggressive. Once the bucking strap is loosened, most horses calm down, but bulls will often try to gore the rider or the skilled rodeo clowns, whose job it is to distract the animals while the rider gets out of the arena. Bull riders have to complete an 8-second ride holding with one hand to a rope loosely tied around the bull's middle. The other hand cannot touch the bull.

➢ *See also Western competition*

Rollback

A maneuver common to reining and Western stock horses that the horse performs from a sliding stop by rolling back over his hocks (180 degrees) to go in the opposite direction. He brings his front feet down into his original tracks at a lope on the opposite lead.

Rolled grains

Rolled or slightly crushs grain that are easier for the horse to chew and digest.

➢ *See also Feeding and nutrition; Grains*

Rolling

Most horses roll in the dirt to relieve itchy skin and stretch weary muscles after a workout. Like dogs, they will also roll right after a bath,

Rodeo Events

A bareback rider in action

Roping a calf at full speed takes lots of practice and a good horse.

Bull riding is probably the most dangerous rodeo event.

preferably in a very muddy area. To get onto the ground, a horse lowers his front end first and then collapses his hindquarters. Some horses can roll completely over onto their other sides, but many have to get up and start over again. When standing up, horses push up with their front legs, sitting up for moment like big dogs before scrambling up on their hind legs. For some reason, cows, which go down the same way as horses, push up with their hindquarters first and then their front legs when getting up.

After rolling, your horse will usually give himself a big shake and wander off looking for something to eat. Sometimes rolling or getting up and down frequently can be a sign of colic, so be aware of your horse's normal behavior. A horse that rolls because of colic can twist a gut; he should be kept moving, preferably by hand-walking.

➢ *See also Colic*

Roll top

A solid, curved jump with a rail at the top.

➢ *See also Jumps, types of*

Romal

A piece of leather that connects the ends of Western closed reins and forms a Y. The tail of the Y hangs down the horse's neck and can be used as a quirt.

Roman nose

A term used when the profile of a horse's head is somewhat convex from just above the eyes down through the nose. It was derived from the original name, "ram's head," which perhaps more appropriately describes this shape of head.

➢ *See also Conformation; Dished face*

Rope, types of

➢ *See Lead rope*

Rope halter

Halters made of rope have some advantages over web halters. They are stronger, and the narrower rope gives more effective control with some horses. However, they can be harder for an inexperienced rider to fasten and, because the lead rope is fastened with a knot, they can be hard to remove quickly. To put on a rope halter, slip the noseband over the horse's nose like a web halter, but tie the throatlatch instead of buckling or snapping.

➢ *See also Halter*

Roping

➢ *See Calf roping; Cattle roping; Rodeo; Team roping; Western competition*

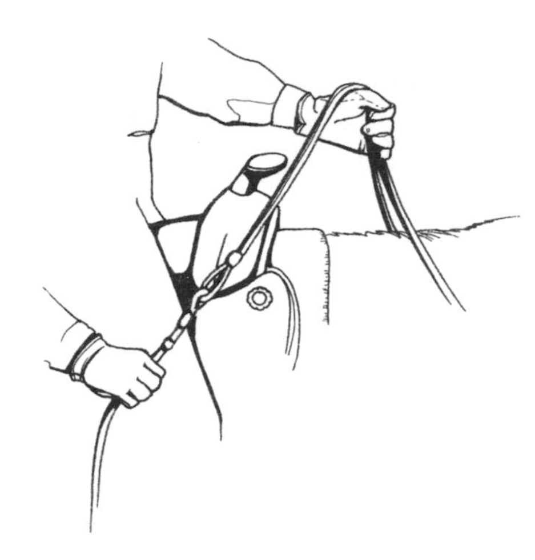

A romal connects the two ends of Western closed reins.

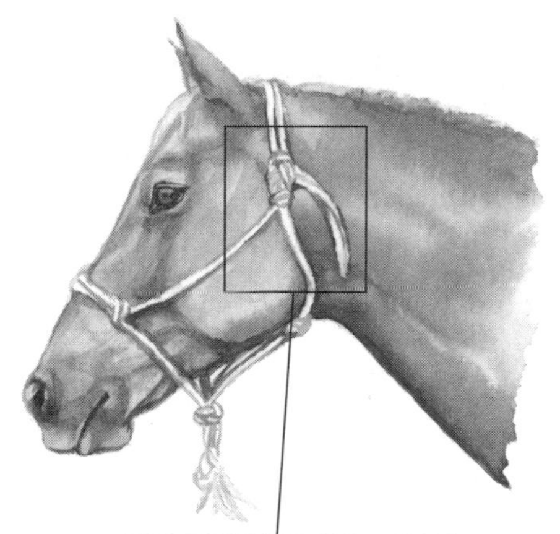

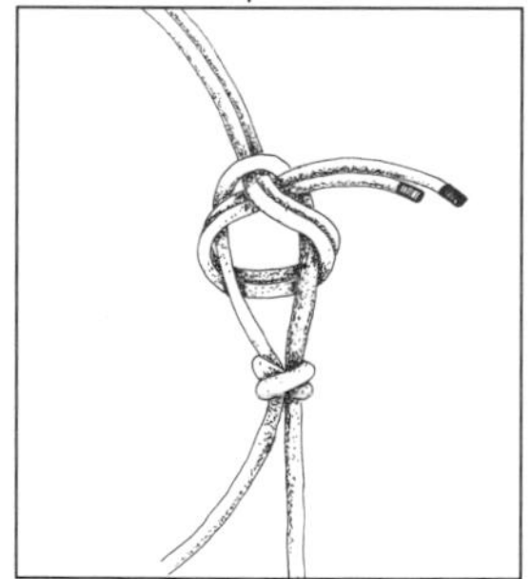

A rope halter should be fastened with a sheet-bend knot.

Rosea (jimmyweed, rayless goldenrod) (*Isocoma* spp.)

➢ *See Poisonous plants; Rayless goldenrod*

Roughage

➢ *See Hay*

Round bone

When viewed from the side, the cannon bone (between the fetlock and the knee or hock) should appear flat, with the tendon set away from the bone. When the tendon and the bone are too close together, the horse is said to have round bone, which indicates instability.

Round pen

A round pen is a useful training space if you have the room for one. Usually measuring 66 feet in diameter, the pen serves as a safe place to turn out young horses and provides a smaller space to work a horse from the ground as well as in the saddle. For training young or very excitable horses, a breaking pen with a 35-foot diameter and solid, 7-foot high walls provides a place to work with no distractions. A fractious horse can't climb over the walls, but there's no quick way out for the trainer either.

➢ *See also Facilities*

Roundworms

➢ *See Ascarid; Deworming; Parasites*

Routine care

Horses like to know what to expect and are happiest when kept on a regular schedule of feeding, grooming, turnout, and exercise. Of course, each day won't go in exactly the same way, but an overall plan for daily care will serve your horse best. If you board your horse, become familiar with the barn routine and notify the barn manager if you need to change that routine. For example, if the horses are usually turned out in the mornings but you plan to ride then, you might request that your horse stay in his stall so you don't have to catch him. It's important that you respect the barn's daily schedule as much as possible.

If you take care of the horse yourself, you'll need to feed him at least twice a day and clean his stall every day. When you feed him, give hay first to take the edge off his appetite so that he's less likely to gobble his grain. Having hay in his digestive tract will also slow the passage of grain, thus increasing the efficiency of the feed.

You should groom your horse once a day or at least every other day, making sure you pick out his feet. After he's turned out, clean the stall and let it air out as much as possible. In addition to a daily routine, your annual routine should include regular visits from your veterinarian for deworming and dental care, and from your farrier for trimming and shoeing.

➢ *See also Cleaning stalls; Deworming; Feeding and nutrition; Grooming; Manure, handling; Picking out feet; Seasonal care; Signs of illness*

ROUTINE CHECK

When you first see your horse in the morning, spend a few moments noticing his general appearance and attitude. Observe the following:

- ❑ Does he greet you in his usual way?
- ❑ Does he have any cuts, scrapes, or bumps, especially on his legs?
- ❑ Has he left any feed from the previous night?
- ❑ Has he drunk a reasonable amount of water?
- ❑ Does his manure look normal?
- ❑ Does his attitude seem normal?

Rowel

A small, pointed wheel on the shank of the spur.

➢ *See also Spurs*

Royal Canadian Mounted Police (RCMP)

Established as the Northwest Mounted Police in 1873 as a temporary body to protect settlers moving into the western territories, the RCMP has evolved into one of the world's most visible law enforcement agencies. The word "Royal" was added by King Edward VII in 1904, and in 1920 the unit was officially declared the national police force. Today, the RCMP deploys more than 20,000 officers (not all of them on horseback) in a variety of duties, from municipal policing to international peacekeeping. Until 1966, all recruits went through equestrian training, but now officers must apply to become part of that unit.

The famous red tunics and broad-brimmed hats have been part of the uniform from the beginning, and the vivid image of the Mountie "always getting his man" has inspired hundreds of stories, novels, and films. The famous musical rides evolved from riding drills in the 1870s that were open to the public and have since become popular internationally.

The RCMP established a breeding facility in 1939 (the current farm is located in Packenham, Ontario) to produce the large, athletic, black horses that are as famous as their colorful riders. A cross of primarily Thoroughbred and Hanoverian bloodlines, with Trakehner and Anglo-Arab genes mixed in, the standards for an RMCP horse are: 16 to 17 hands high and 1,200 to 1,400 pounds; an elegant look; strong, clean legs; and a pleasant disposition. They must be able to jump and do dressage movements, as well as tolerate miles of travel and hours of performing (more than 125 shows a year). About half of the approximately 20 foals born each year make it to the intensive training program, that begins when they turn three.

➢ *See also Mounted police; Police horses; appendix*

Rug

A heavy horse blanket.

Rumensin

➢ *See Monensin (Rumensin) poisoning*

Run (enclosure)

A turnout area designed specifically for exercise rather than grazing. An enclosure measuring 20 by 100 feet is sufficient for your horse to trot, but if you want to encourage him to gallop, the pen should be at least 200 feet long, with enough room at the ends for him to turn around safely while moving fast.

➢ *See also Arena; Corral; Facilities; Paddock; Pasture; Round pen*

Run (gait)

The gallop. This term is used in Western riding.

➢ *See also Gait*

REASONS FOR RUBBING

Horses rub against things for a variety of reasons, but mostly to relieve itching. Excessive rubbing off the tail area or legs often indicates parasites or skin disorders. If not treated, the horse might rub off patches of hair or break off much of his tail.

When a horse rubs his head against you, it may seem like an endearing gesture of affection, but it is actually a dominance behavior that can easily get out of hand. Discourage your horse from rubbing his head against you and try showing your affection for him by patting or scratching his neck and withers instead of his nose and forehead.

Run-in shed

Horses need to have some shelter, such as a run-in shed, if they are turned out for long periods of time.

Although horses can live perfectly well outdoors without manmade shelter, they do need some place to get out of the wind and rain. If there isn't a natural windbreak in your pasture, a simple three-sided shed provides relief from the elements. Flies prefer hot, sunny areas, so having a shady spot can make a big difference in your horse's comfort level.

A run-in shed must be large enough to comfortably hold all the horses in the pasture (allow 142 square feet for each horse), with one side completely open to avoid injury if all the animals should decide to enter or leave the shed at once. The shed should be built on the highest ground available and be faced away from the prevailing winds (usually coming from the north). The walls need to be sturdy, with no sharp protrusions or gaps, and the roof should pitch slightly away from the front of the structure to keep moisture from accumulating.

➢ *See also Barn; Facilities*

Running away

A truly panicked, out-of-control, runaway horse is an unusual occurrence. If your horse is frightened and bolts, he will probably only run a little way before you get him calmed down. However, it is not uncommon for a horse to get going faster and faster, perhaps on the way back to the barn or on a group ride, and suddenly that canter turns into a gallop without the rider's permission. The best way to deal with a runaway is to prevent him from getting too fast in the first place. It is the rare beast that doesn't move more quickly going home, so never (absolutely *never!*) allow your horse to go any faster than a walk when returning the last mile or so to the barn. Use half-halts, ride in a zigzag pattern, turn him in circles, but keep him walking. Also be aware that horses tend to race each other when cantering in a group, so keep your horse well in control in that situation.

If you do find yourself on a galloping horse that doesn't respond to your initial request to slow down, use a pulley rein. Use it immediately to get your point across.

➢ *See also Aids; Progression of the aids; Progressive halt; Pulley rein*

Just a-ridin', a-ridin, —
Desert ripplin' in the sun,
Mountains blue among
the skyline —
I don't envy anyone
When I'm ridin'.

—Badger Clark, *Ridin'*, 1915

Running martingale

A piece of equipment used by riders in various disciplines to aid in the control of a horse that carries his head too high. It attaches to the girth/cinch, runs between the front legs, and divides into two straps, each of which has a ring attached. Each rein runs through a ring, thus providing a point of leverage to aid in teaching the horse to yield to pressure from the rider's hands. If adjusted properly, a running martingale does not put pressure on the horse's mouth, except when the horse raises his head too high.

➢ *See also Bridle; Martingale; Standing martingale; Tack*

A running martingale

Running up stirrup leathers

When you dismount after a ride, make it a habit to immediately loosen your horse's girth and run up your stirrups by sliding the iron to the top of the back stirrup leather (the one closest to the saddle) and pulling the whole loop forward through the iron. This keeps the irons from flying around and striking your horse, and also makes it easier to put the saddle on and take it off.

Running up the stirrup leathers works only on English saddles. With a Western saddle, the stirrups hang free. When you tack up, loop the off stirrup over the saddle to get it out of the way.

➢ *See also Saddle; Stirrups; Tack*

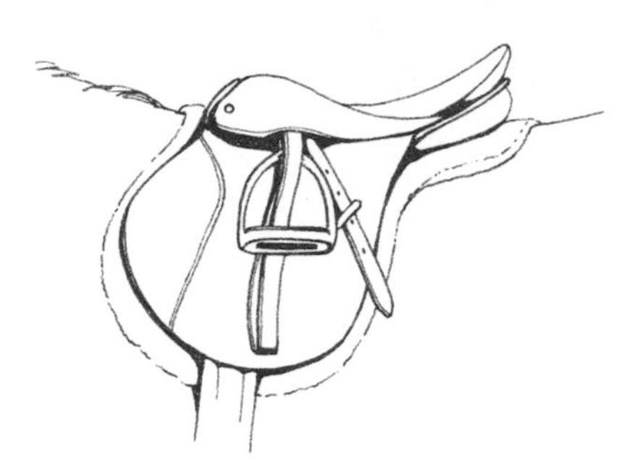

Stirrup leathers properly run up

Running walk

A smooth, fast gait natural to Tennessee Walking Horses and some other breeds.

➢ *See also Gait; Gaited horse; Tennessee Walker*

Rushing fences

Coming up to a fence too fast. This usually means the horse jumps flat or inverted rather than rounding his back and popping his hocks over the obstacle. This isn't a problem in steeplechasing, where speed is of the essence and the jumps (usually hedges) are forgiving, but in most kinds of jumping, you don't want your horse to drag his feet. Taking a shorter approach to the jump gives your horse less time to get up speed.

When a horse rushes a fence he may drag his hind feet and strike the bar.

Russian knapweed (Russian thistle) *(Acroptilon repens* syn. *Centaurea repens)*

A member of the same family as yellow star thistle, causing progressive, irreversible brain damage.

➢ *See also Poisonous plants; Yellow star thistle*

Rust

A variety of fungus that infects grain and grass, rusts appear as reddish brown patches. Ingesting contaminated feed can cause colic, while breathing in rusts can create respiratory problems.

S

Sacking out

Introducing a young or nervous horse to potentially frightening or startling situations in an effort to desensitize it. The term comes from a common practice of taking an empty feed sack and flapping it near the horse, laying it across his back, sliding it across his legs, having him step on it, and so forth, to accustom him to strange situations. A round pen is a good place to sack out a horse.

➢ *See also Round pen*

Saddle

Shortly after taming the horse, humans began to devise easier ways to ride them, ranging from the rough pads used by Native Americans to heavy wooden cavalry saddles. Today, there are two basic types of saddles: English and Western, with a great deal of variety in each style. The purpose is essentially the same: to make the horse and rider comfortable by helping the rider maintain proper position by distributing the rider's weight evenly across the horse's back.

To achieve this goal, however, the saddle must fit both horse and rider properly. Poorly fitting saddles can cause problems ranging from saddle sores to training difficulties. A good quality, well-fitted saddle, if appropriately cared for, can last a lifetime. Ask a knowledgeable horseman whom you trust to help you evaluate saddle and fit before you buy. Be sure to research the types of saddles available, and choose a saddle appropriate for the type or riding you'll be doing. Many saddles might look the same to novices, but competition rules can exclude and disqualify riders who use the wrong equipment.

➢ *See also English riding; Sidesaddle, riding; Tack; Western riding*

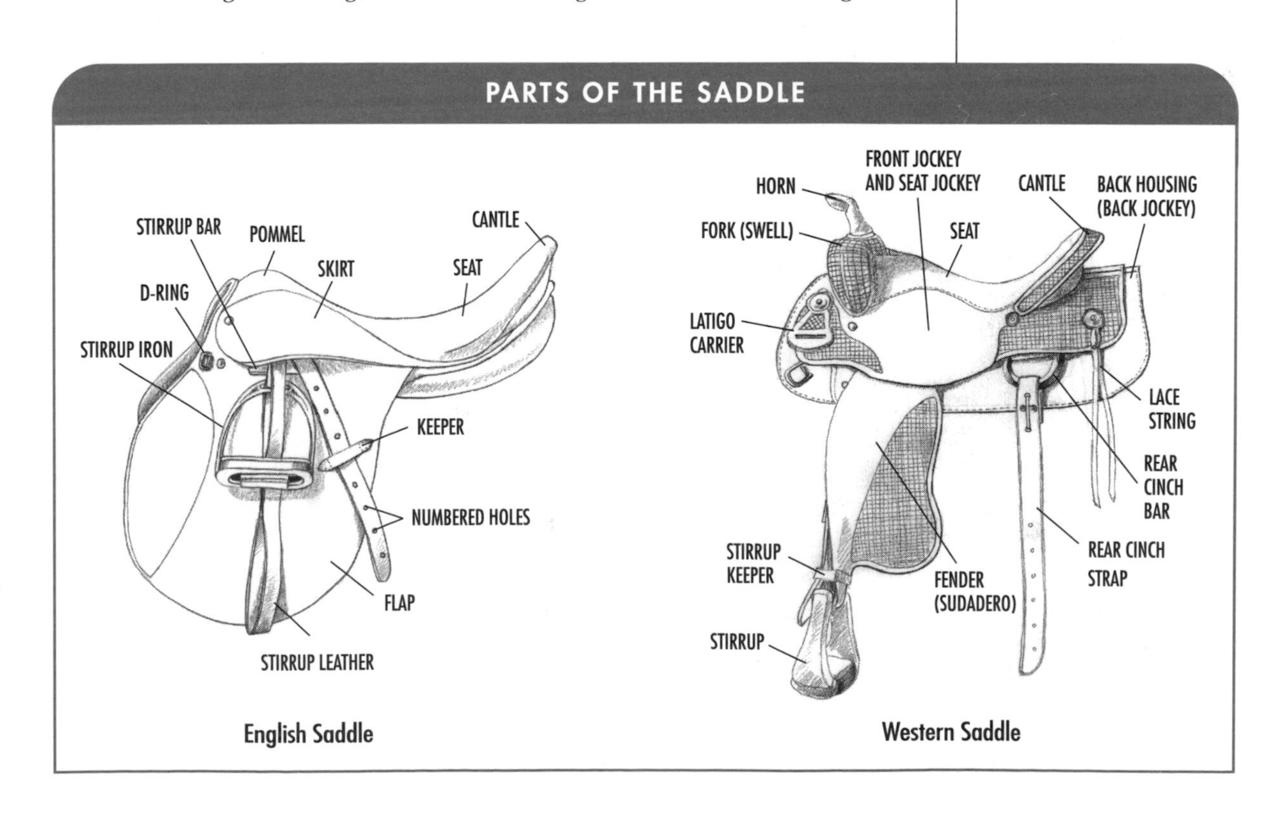

DIFFERENT TYPES OF SADDLES

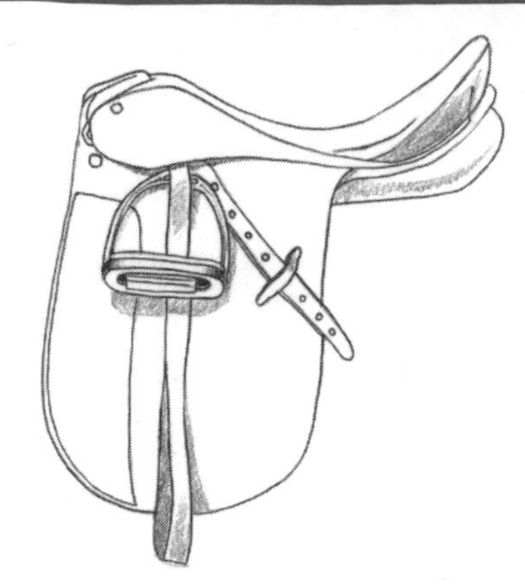

Dressage saddles have deep seats; long, straight flaps; and stirrups that are set farther back.

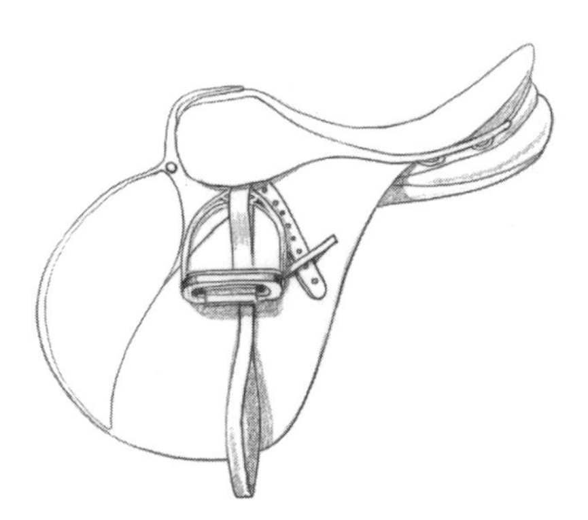

Jumping saddles have a more forward seat and flaps with deep knee pads for a secure grip.

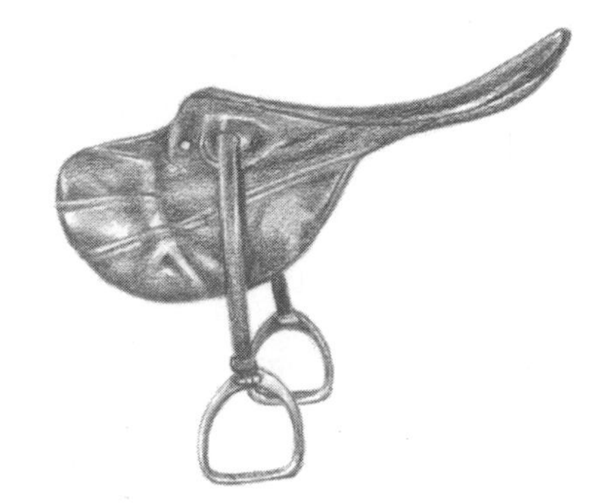

Because jockeys spend most of their ride crouched on the horse's back, racing saddles are tiny and flat, with very short stirrups.

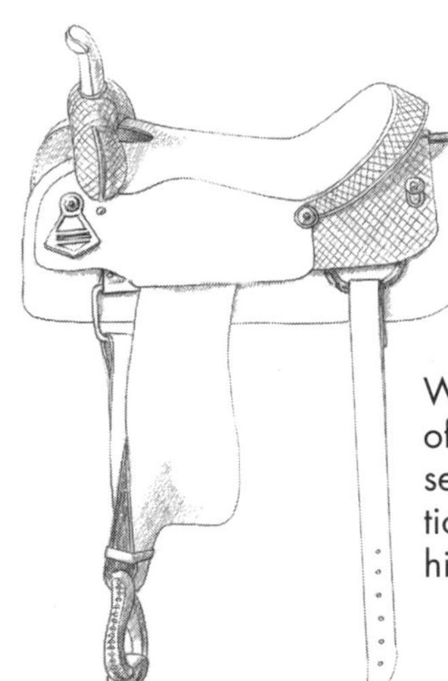

Western saddles offer a very secure seat with the traditional horn and a high cantle.

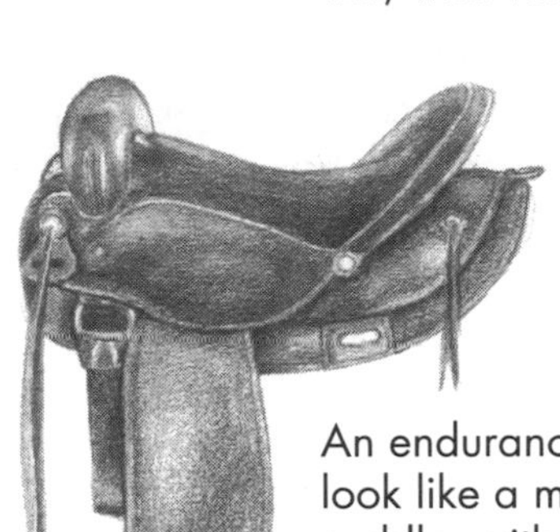

An endurance saddle may look like a modified Western saddle, with a high pommel and cantle, but it also has heavy knee pads for comfort and safety.

Saddlebags

Bags designed to balance across the back of a saddle to carry supplies like a first-aid kit, water, and snacks.

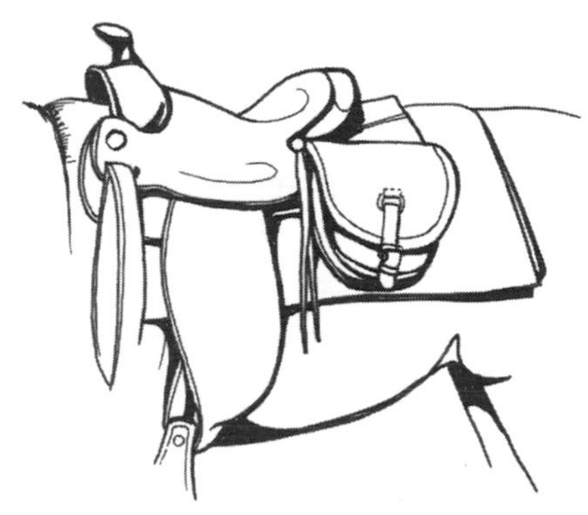

Saddlebags allow you to carry lunch, a first-aid kit, and other important items on a trail ride.

Saddle blankets and pads

After you purchase a saddle, you'll need to protect your horse's back with a pad or blanket under the saddle. Pads are designed for use under different kinds of English saddles (dressage versus jumping, for example) and come in a variety of fabrics and thicknesses. Blankets used under Western saddles are usually rectangular, folded in half, and brightly colored.

The choices can be overwhelming, but keep in mind that the primary purpose of this piece of equipment is utilitarian. It is meant to protect your horse's back by absorbing sweat and cushioning pressure created by the weight of rider and saddle. It is more important to choose a pad or blanket that will do its job effectively than one that looks good. Be careful to buy

a pad or blanket that fits your saddle and horse well. Excessive padding can lead to dangerous slippage of saddles.

Saddlebred

➢ *See American Saddlebred*

Saddle galls

Spots of white hair in the saddle area, where the skin has been rubbed raw by improperly fitting tack. This is a sign of previous neglectful care.

Saddle rack

A structure upon which a saddle is hung. The rack must give good support to the saddletree and provide sufficient ventilation to dry the underside of the saddle. Saddles should not be stored on the floor, and it's best for each saddle to have a separate rack. Collapsible versions are handy in aisles or grooming areas to hold tack while getting ready to ride.

➢ *See also Tack, cleaning; Tack room*

Saddle seat equitation

Equitation is the art of riding a horse intelligently, gracefully, and with the greatest degree of comfort and enjoyment to both horse and rider. Saddle seat equitation is nothing more, or less, than performing this art in the saddle seat style; that is, with a flat saddle and a full bridle.

Many, though not all, riders of saddle seat equitation compete in classes of the same name. These classes are judged on the rider's skill more than on the horse's performance. Judges look for riders who can show their horse in an animated and somewhat flashy fashion while maintaining the standards of equitation previously mentioned.

➢ *See also English competition; Equitation*

Saddle seat showmanship

Showmanship in saddle seat attire, sometimes with a full bridle rather than a halter.

➢ *See also Showmanship classes*

Saddling a horse

➢ *See Tacking up*

Safety

All horse-related activities should begin with a solid understanding of basic safety rules. Horses are domesticated *wild* animals that frequently weigh ten times our body weight and always have a mind of their own. Safety rules must be learned and adhered to in order to minimize risk.

Safety rules protect both horses and humans. Although rules may range from essentially nonexistent in some barns to overbearing in others, there are some commonsense rules that all horsemen should consider golden.

SPECIAL PADS

Sometimes a horse develops back soreness even with a well-fitted saddle and pad. A horse with very high withers, for example, might need extra padding or a cutback pad. There are therapeutic pads for horses with back problems; consult your veterinarian or riding instructor.

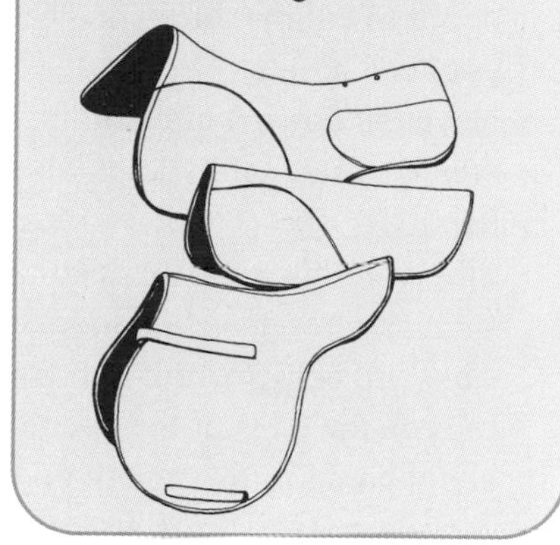

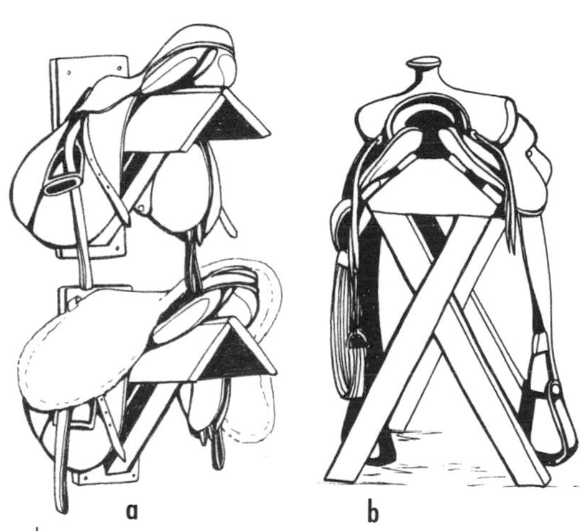

Wall-mounted saddle racks (a) keep saddles away from rodents and free up valuable floor space. Floor stands (b) take up more room, but are often used for heavy Western saddles. They are portable and can provide a convenient place to clean and repair tack.

Rules for Approaching and Handling Horses

- Never ride alone! To be truly safe, never handle horses while alone. Consider what may happen if you get injured and have internal bleeding. How long might it be before someone finds you if you are alone?
- Be sure the horse is aware of your presence as you approach.
- Approach the horse from the side at a three-quarter angle to his shoulder and speak to him. Never approach from or stand directly in front or behind because these are blind spots in the horse's vision.
- Feed treats from buckets, not your hand. How is a horse supposed to tell the difference between your fingers and the treats you feed him?
- Never sit down while holding a horse.
- Always use a safety knot when tying. (See *Tying* for more details.)
- Stop your horse before leading him through narrow openings such as gates and stall doors; teach him to let you enter first. This will help prevent him from running you over or hitting his hips on the edge of the opening, which could lead to anxiety about going through such openings.
- Always move calmly around horses. No running or horseplay.
- Never ride in the barn aisle.
- Bend from the waist, rather than kneeling down, to work on the horse's lower legs and hooves.

Guidelines for Safe Dress

- Always wear a helmet! Eighty percent of all deaths from horse-related accidents when mounted result from head injuries. (See *Helmet, safety* for more details.)
- Wear long pants at all times when around horses since serious lacerations can occur if shorts are worn and an accident occurs.
- Wear safe footwear: boots that are at or above the ankle, have hard soles, and at least a ½-inch heel. This helps prevent life-threatening injuries that occur when a rider's foot slips through the stirrup during a fall.
- Avoid loose clothing. Loose garments can catch on the saddle during a fall, and cause a rider to be dragged. (This situation often results in life-threatening injuries.)

➢ *See also Accidents, prevention of; Body language, of horses; Facilities; Handling horses; Passing, on horseback; Progressive halt; Pulley rein; Tack, cleaning; Tacking up; Vices, how to handle*

St.-John's-wort (Klamath weed, goatweed) *(Hypericum perforatum)*

A plant causing photosensitivity in horses that eat it.

➢ *See also Photosensitivity; Poisonous plants*

HORSETALK

Horses have become part and parcel of the American idiom. How often do you use or hear horsetalk?

- ❑ Horse sense
- ❑ Charley horse
- ❑ Horse of a different color
- ❑ One-horse town
- ❑ Riding the high horse
- ❑ Strong as a horse
- ❑ Eats like a horse
- ❑ Putting the cart before the horse
- ❑ Beating a dead horse
- ❑ From the horse's mouth
- ❑ Shutting the barn door after the horse is gone
- ❑ Horsing around
- ❑ Hold your horses
- ❑ Wild horses couldn't drag me away

See also Deworming; Grooming; Mucking out stalls; Parasites; Pasture; Pest control

POISONOUS PLANT

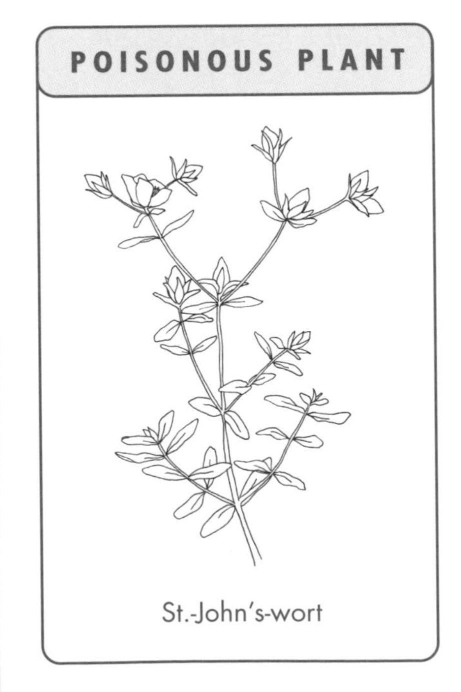

St.-John's-wort

Salmonellosis

A disease caused by a common bacterium that makes many animals, including humans, quite ill. Salmonella may be present in a horse's gut without causing problems, but it can flare up if the horse is ill or otherwise stressed. Infected individuals may have a high fever with watery diarrhea and they may be very thirsty. Pulse and respiration are rapid. Because the disease develops so quickly, some horses die within a few days of extreme dehydration; others may linger for weeks before either dying or slowly recovering. Still others may have chronic diarrhea for months.

Treatment consists of giving fluids and electrolytes, both orally and intravenously. Some veterinarians give antibiotics, but there is concern about strains of the bacteria that have developed resistance to certain drugs. Ill horses must be completely quarantined, to the extent that your boots and clothing should be changed before you work with healthy horses. The bacteria spread through manure, so all tools and equipment used around the sick horse should be kept separate and disinfected regularly. There is no effective vaccine at this time.

Salt (sodium chloride)

All animals need salt to maintain body fluids, and horses are no exception. Salt should be available to your horse at all times, and it is usually offered in blocks that can be left in stalls and pastures. Different horses have different salt needs, so putting it in his food won't always ensure that he's getting the right amount and may provide an overdose.

Salt block

Solid blocks offered freely for horses to lick and chew for proper salt intake. These are an important part of a balanced ration. Be careful to situate them in places that have good draining, to minimize waste.

Sand colic

Horses can develop colic from grazing in sandy areas and ingesting sand and grit along with the grass. The bulk of the sand settles in the intestines instead of passing through and can cause a fatal bout of colic. The only way to diagnose sand colic is to dilute the horse's manure with water and see if any sand settles out. Treatment consists of mineral oil administered through a stomach tube. Then the horse is given a product containing psyllium powder, which is fed to the horse and swells when wet, trapping the sand and moving it out of the gut.

➢ *See also Colic*

Sarcocystis neurona

A protozoan that lives in opossums and can cause a neurological disease, called equine protozoal myeloencephalitis (EPM), in horses.

➢ *See also Equine protozoal myeloencephalitis (EPM)*

SANITATION, IMPORTANCE OF

A clean barn is a healthy barn, or at least a healthier one. Many parasites and infectious diseases thrive in dirt, breed in manure, and are transmitted by contact. Help your horse stay healthy by making sure that his stall is cleaned out daily, the manure piles at your barn are properly managed, and the pasture isn't overused. Maintain a regular schedule of deworming and keep your horse clean and well-groomed. Avoid sharing brushes, saddle pads, and other equipment, and clean all of your equipment often.

See also Deworming; Grooming; Mucking out stalls; Parasites; Pasture; Pest control

Sarcoid tumors

Like warts, some sarcoid tumors are caused by a virus and may appear for no apparent reason. Other sarcoids develop after an injury; in fact, wounds that are not properly cleaned or that are subject to continued irritation (on the joint, for example) are most susceptible to sarcoids. Initially the tumor looks like proud flesh (excessive granulation tissue that sometimes protrudes from a healing wound), but it will not respond to medications used to heal wounds. While sarcoid tumors will not spread through the bloodstream, they do spread and grow locally. Regrowth is often stimulated by removal, but sometimes surgery is successful, as is burning or freezing the tumor. An effective technique has been to create a vaccine from the tumors to stimulate the horse's own immune system, essentially by feeding the tissue to the horse after removal.

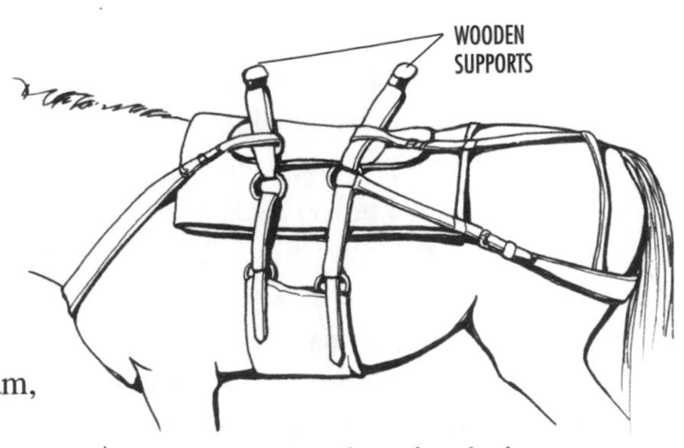

A traditional sawbuck

Sarcoptic mange

➢ *See Mange*

Sawbuck

A traditional pack saddle formed by pairs of wooden supports onto which you tie your load.

School figures

Standard maneuvers that all beginning riders should master: circles, figure eights, changes of direction, half-turns, cross-the-school, cross-the-diagonal, up-the-centerline, and serpentine. These figures provide a standard language that can be used by student and instructor. They also provide the student with an opportunity to develop coordination of aids because such figures require skill to perform, especially at gaits above a walk.

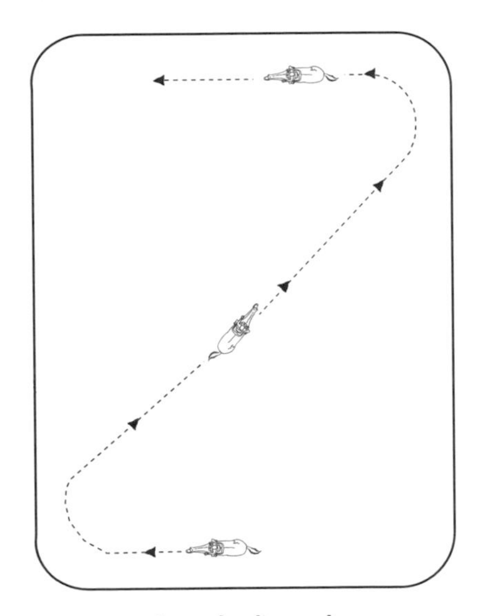

Cross-the diagonal

School horses

Horses that are used for teaching lessons. If you are looking for a place to take lessons, ask about the age and experience of the horses. You will learn more if you have several school horses to ride.

➢ *See also Lessons, riding*

Schooling shows (practice shows)

Shows that are designed as practice shows for both novice and experienced riders. Often these shows have no official affiliation with any equine organization, but they are a good way to gain experience for both rider and horse. Some schooling shows are just for fun, with innovative classes, games on horseback, and no formal dress code. Others offer the opportunity for you to receive feedback from the judge after your performance. Some are designed to more closely resemble regular competitions. Make sure that you understand what the class titles designate — sometimes the terms used at schooling shows ("First Time Ever" or "Baby," for example) are not used in regular shows.

➢ *See also Showmanship classes*

Scoop for grain

Feeding your horse a scoop of grain in the morning and evening may seem pretty reasonable. But because scoops come in so many sizes and shapes, you should actually weigh the feed you are giving rather than rely on volume. Also, one crop of oats or other grain may weigh more than another due to higher moisture content.

Once you've figured out how much your particular scoop holds, you still should weigh the feed regularly, especially if you change suppliers or brands.

➢ *See also Feeding and nutrition; Grains; Weighing feed*

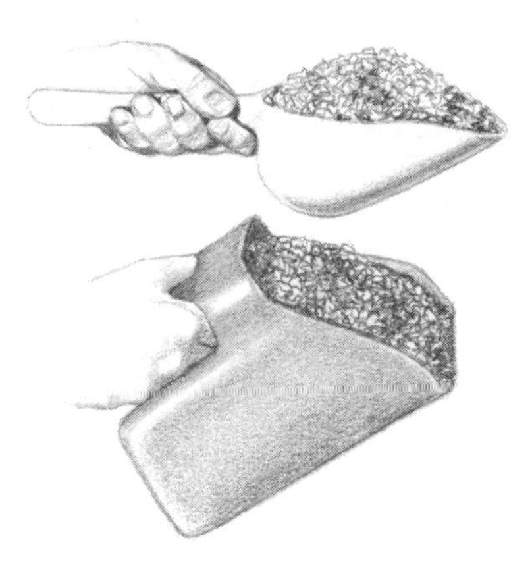

Make sure you know how much grain you are feeding. Weigh the feed rather than relying on volume. The bottom scoop holds nearly twice as much as the one on top, but how much does the feed weigh?

Scoop shovel

A handy piece of equipment to have around the stable. It can be used to handle anything from wet bedding and manure to bulk feed to snow. Just be sure to have separate shovels for stall cleaning and feed shoveling.

➢ *See also Cleaning stalls*

Score

This term has different meanings to different horsepeople. Working Western riders define a score in two ways: (1) to break quickly from the gate to overtake cattle in a working cow class; (2) points received on a "run" or portion of competition. The second definition is most common among competitors in a wide range of disciplines.

SCRATCH

To withdraw an entry from a class or race before a competition begins.

Scotch

In reining, a horse that appears to have half-halted while in the midst of a run-down prior to the sliding stop. It appears most often when a rider misreads a pattern or is unaware of the exact location of the marker. It does not result in a penalty but lowers the score for the maneuver.

Scouring rush (equisetum, horsetail, jointfir) (*Equisetum* spp.)

➢ *See Equisetum; Poisonous plants*

Scratches (grease heel)

A fungal (sometimes bacterial) skin problem picked up when horses walk through wet areas. Scratches affects the lower legs, entering through a cut or scrape and causing the area to become crusty, scabby, and swollen. Light-colored horses seem to be more susceptible to the disease, which can be prevented by keeping the horse out of wet areas and treating small lesions promptly, before the infection spreads.

Treatment is to clean the area thoroughly and apply a mix of one part nitrofurazone ointment (an antibiotic), one part dewormer paste (a paste containing a benzimidazole), and one part dimethyl sulfoxide (DMSO). Do not apply bandages; you want the area to stay open to the air to promote healing.

➢ *See also Skin problems*

Scribe

Person who assists the judge by recording scores during events that require detailed scoring. This allows the judge to remain focused on the horse's performance while recording scores for each maneuver (in reining), obstacle (in trail), fence (in hunter), or element of a test (dressage) or pattern (showmanship). The use of scribes is required in many events because effective judging is virtually impossible without them.

Seasonal care

While your horse's basic needs for food, shelter, and health care are the same year-round, each season brings with it particular routines. Throughout the year, your horse should be on a regular schedule of deworming and vaccinations established with your veterinarian, which will vary depending on where you live. The horse should also be seen regularly by a good farrier, who will trim his feet and check his shoes every 6 to 8 weeks, even in the winter. Get the horse's teeth checked at least once a year. Even when not ridden regularly, horses need to be fed at least twice a day and checked for overall health and safety.

The sturdy steed now goes to grass and up they hang his saddle.

—Francis Beaumont

Seat

The position in which a rider sits on a horse. There are numerous types of seats that correspond to different disciplines. The most common and versatile style is called either "basic" or "balanced" seat, which can be adapted easily as a rider progresses to different disciplines.

The basic seat or balanced seat positions the rider squarely in the saddle, with weight distributed evenly on both sitting bones down through both stirrups. Your head is up and your eyes should be looking ahead. Shoulders should be square and relaxed with arms hanging comfortably. You should be able to draw an imaginary line from your elbow through your wrists and reins to the horse's mouth. The height of the hands, therefore, depends on the conformation of horse and rider.

Your upper body is held in a position comparable to proper posture for sitting in a chair, with suppleness and softness in your abdomen that allows for absorption of the horse's motion at the trot and canter. Your hips should be level, with even weight on each buttock. Your thighs should hug your horse with knees pointing straight ahead, not pinching inward or flopping outward. Your calves should lightly contact the horse's sides (exact amount of your leg that contacts horse depends on the size of horse's barrel since smaller horses' barrels tend to slope inward just below the rider's knees).

The lower leg is carried in such a manner that an imaginary line can be drawn perpendicular to the ground from the rider's knee to her toe. The heel is down but not pushed down in a manner that locks the ankle. The ankles should be softly held in place. Rigidity creates a stiff and ineffective leg. The stirrup is usually set so that it reaches to just below the ankle when the feet are disengaged. Variations of leg position are normal because physical differences do not enable all riders to conform to this standard.

➢ *See also Aids; Position, in saddle; Posting trot; Sitting trot; Two-point seat (hunt seat, light seat); Weight, as an aid*

SEASONAL CARE

Winter

- Outdoor horses grow thick, protective coats but still need shelter from very cold temperatures, strong wind, and icy rain. Soaking wet or muddy hair lies flat and loses its insulating properties, so keep your horse as clean and dry as possible.
- If he's out all winter and is not being ridden, pull his shoes so that he'll have better traction. Sand any icy patches on walkways, paths, and near water troughs.
- If your horse is blanketed and stabled all winter, adjust his blankets according to the weather. A heavy overnight rug can be too warm for a sunny day, even if it's very cold. If your horse sweats and then becomes chilled, he could get sick. If you ride him, make sure he's cool and dry before you blanket him for the night or turn him out.
- Horses need extra feed in the winter to generate body heat. Hay and other roughage provide more heat during digestion than does grain, and in extremely cold weather most horses should have all the hay they can eat.
- Keep an eye on your horse's water consumption (6 to 10 gallons a day in the winter) and make sure he has access to water that doesn't freeze over. Inadequate water intake affects digestion and can lead to impaction and inability to maintain body temperature. Horses may or may not eat snow for moisture, but they waste energy melting and warming it in their digestive tracts.

Spring

- During mud season, pay particular attention to skin and feet. Wet conditions promote mildew, fungi, and parasites so keep your horse as clean as possible. Check his feet daily for signs of thrush and other problems. Limit turnout rather than letting him stand for hours on soggy ground.
- Mud also is a hazard for walking, as it forms bumps and trenches and then freezes solid. Horses can easily strain a ligament, pull a tendon, or even break a bone slipping or falling in mud, so keep trouble spots well sanded to minimize the potential for harm.
- Longer days will trigger shedding. Proper nutrition, frequent grooming, and massage to promote circulation and stimulate oil glands will help your horse shed out faster and develop a shiny, healthy summer coat.
- As lush spring grasses fill the pasture, make sure your horse becomes accustomed to them slowly. Too much grass after months of only hay can cause colic. Start with an hour or two of turnout a day and gradually increase it. Also, feed your horse a full hay ration prior to pasturing for the first week, as he'll be less likely to overeat.

Summer

- While overly wet hooves are a concern in the spring, summer heat often dries out hooves to the point of cracking or splitting. If there is no natural moisture, you can create a wet spot in the pasture or near the water trough to help maintain a proper moisture balance. Also, dry hooves can be treated with a hoof dressing — consult your veterinarian on usage.
- Strong sunlight can burn the skin of light-colored horses or the skin under white markings on darker horses. Make sure your horse has a shady spot to retreat to when the sun is high. You can also apply human sunscreen or zinc oxide to thin-skinned areas.
- Use a fly mask to cut down on the irritation of flies in the eyes.
- Make sure your horse has some shade in his pasture to escape the heat and get away from flies (most flies avoid shady places).
- Very hot weather can be dangerous, even if your horse is not being worked. Be alert for signs of heat stroke: erratic behavior, disorientation, rapid breathing, dry skin, high temperature, and lack of appetite. Prevent overheating by providing shade, plenty of water, and appropriate feed (with not too much protein). Become familiar with the heat index if high humidity is common in your area.

Fall

- Fall is a good time for extra maintenance around the barn. As the weather cools, wash and put away flysheets and check over winter blankets for any needed repairs.
- Keep checking for bot eggs and give a boticide after the first hard frost.
- Pull shoes from horses that will go barefoot for the winter or put snow pads on horses that will stay shod. Snow pads prevent ice from balling up in the frog of the hoof.
- Keep in mind that rainy days can chill horses quickly in cool weather.

VARIATIONS ON THE BASIC SEAT

The **classical seat** is primarily used for dressage. It is essentially the same as the basic seat with perhaps a bit more style added by individual riders. The stirrups are a bit longer and the foot is more level than in basic seat, but otherwise the position is the same.

The **forward seat** in its most extreme is used by jockeys. It allows the rider to move her center of gravity forward to remain above the horse's center of gravity. The stirrup leathers are usually noticeably shorter and the rider remains in a more forwardly inclined position at each gait.

Hunt seat riders demonstrate a position between the forward seat and basic seat. At the walk and sitting trot, they tend to be less forwardly inclined. But at the posting trot, canter, and gallop, they move their center of gravity forward to remain balanced with their horse. Both the forward and hunt seats are used by riders who jump obstacles with their horses.

The **saddle seat** is standard on high-trotters, which tend to be highly elevated in their forequarters. It is the seat used on three-gaited, five-gaited, and park horses. It is also used for a variety of other disciplines that require elevated motion. The horse's high headset and elevated forequarters place the center of gravity further back, so the cutback saddle allows the rider to move his position closer to the loin. The rider's hands are often held much higher than in basic seat since the horse's head and neck are more elevated; this allows the same imaginary line to be drawn from the elbow through the wrist and hand to the bit. The lower legs also often do not maintain the same type of contact as in basic seat because of their position on the horse's barrel. Otherwise, this seat is very similar to the basic seat.

Seat of corn (of hoof)

A part of the sole of the foot, located near the heel.

➢ *See also Foot care; Hoof*

Secondary shock

Horses can go into shock from the trauma of colic, serious burns or injury to tissue, or toxic infection. Although there is no actual blood loss, pain and fear can cause small blood vessels and capillaries to shut down. Horses with any kind of shock must be treated immediately with intravenous fluids, adrenaline, and steroids.

➢ *See also First aid; Shock*

Seeds, embedded

Seeds can cause problems for your horse if they become caught in his mouth or eye. Grasses like cheatgrass and foxtail have sharp seeds that can get wedged in the gum or cheek and create an abscess. If the horse has trouble eating or develops a swelling anywhere in his mouth, have your veterinarian check for an embedded seed, among other possibilities. Similarly, bits of grass or slivers of seeds, like burdock burrs, can get into a horse's eye and cause serious problems. Suspect a foreign object in the eye if your horse becomes sensitive to light, holds his eye closed, has any discharge or weeping from the eye, or has a constricted pupil. If your

Then will I sell my
goodly steed,
My saddle and my bow;
I will into some far
countrey,
Where no man doth
me know.

—Traditional ballad

veterinarian is unable to flush out the particle, you may have to take your horse to a clinic that is equipped to handle eye problems.
➢ *See also Dental care; Eyes*

Selenium

Although needed only in trace amounts, selenium is vital for proper muscle development. A deficiency is a suspected cause of azoturia. This mineral is also important in maintaining immune function. However, it is toxic if eaten in large quantities — certain plants such as locoweed, most vetches, and some varieties of aster store high levels from the soil.
➢ *See also Azoturia; Poisonous plants; Rhabdomyolosis*

Semen

➢ *See Artificial insemination (AI)*

Senecio (groundsel, ragwort, stinking willie, tansy ragwort) (*Senecio* spp.)

A species of plant (including groundsel, stinking willie, ragwort, and tansy ragwort) that causes serious liver damage in horses.
➢ *See also Poisonous plants*

POISONOUS PLANT

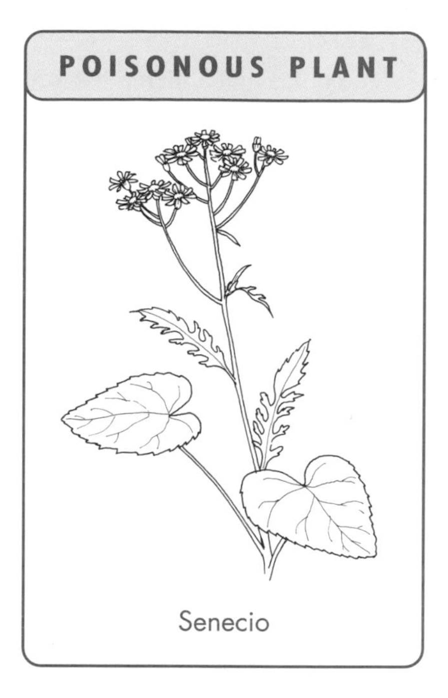

Senecio

Septicemia

A serious bacterial infection of the blood.

Serpentine

A series of half circles performed in an S-like pattern on a center line that is used to teach or test riders and horses; excellent for developing coordination of aids in riders and suppleness in horses. Riders can be asked to ride a serpentine at any gait, showing changes of lead or diagonal.
➢ *See also School figures*

Serum hepatitis (Theiler's disease)

➢ *See Hepatitis*

Set

To hold your hands steady in one position and keep even pressure on the bit without pulling back on the reins. This term can also refer to a group of competitors in cutting or team penning that uses the same bunch of cattle. After the cattle have been used a certain number of times, a new bunch is brought in for the next set.

Settle

This term can be used in several contexts: (1) to gather a bunch of cattle into a calm, compact herd; (2) when a mare becomes pregnant, it might be said that she is "settled"; and (3) some patterns require that a horse be "settled" before proceeding with the next maneuver; that is, the horse must be calm and standing quietly for a specific time period.

SENSES

Horses are acutely aware of their environments and may react violently to loud noises, sudden movements, or objects that appear out of their blind spots. These animals have keen hearing and eyesight, as well as sensitive skin — especially around the muzzle. Smell is particularly important in identifying familiar people, animals, and places.

Horses respond to the emotions of the people around them. They sense fear or anger as well as confidence or calmness, and will usually modify their own behavior as a result.

See also Vision, of horse

Shafts

The poles that attach a wagon or other horse-drawn vehicle to the horse that is pulling it. The horse stands between the shafts, which are connected to the harness with traces and sometimes belly-bands and tugs.

➢ *See also Driving; Harness*

A horse in draft harness, with shafts clearly visible

Shaker foal syndrome

The result of botulism poisoning, which is nearly always fatal. Affected foals quiver and get very weak before collapsing.

➢ *See also Botulism*

Shampooing

➢ *See Bathing horses; Grooming*

Shank

A lead shank is one to which a chain is attached. The shank of a curb bit is the part that extends below the mouth. Reins are attached at the bottom of the shank. There are rules that limit how long a shank can be in competition. The longer the shank, the more pressure it exerts on the bars and chin groove when a rider pulls back on the reins.

To "shank" a horse means to jerk on the lead attached to the chain that runs either over its nose or under its chin.

➢ *See also Bit; Leading; Lead rope; Lead shank*

SHALES HORSE

The Shales Horse is a British trotting horse descended from the Norfolk Trotter, considered exemplary both under saddle and in harness. Typically it stands 15 hands high and is esteemed for its speed and its even temper.

Shear mouth

A condition in which the lower jaw is abnormally narrow, so the top molars can eventually wear the bottom ones down at such an angle that the teeth begin to slide past one another. If the abnormality is noted in time, proper dental care can keep the grinding surfaces normal and prevent a serious problem.

➢ *See also Dental care*

Sheath

The pocket of skin that protects the penis of a male horse.

➢ *See also Anatomy of the horse; Penis; Sheath, cleaning; Smegma; Swelling*

Sheath, cleaning

Male horses should have their sheaths cleaned periodically to remove smegma, a combination of dirt, dead skin cells, and fatty secretions that build up in the folds of the sheath. Smegma is somewhat waxy and smells nasty. Some horses produce more of this substance than others, so when to clean the sheath is an individual determination; it could be once a month or even just once a year. Most horses will become used to having their penises handled if it's done regularly, and many will let the penis down completely, which makes the job easier. It's best to have a veterinarian teach you how to do this procedure safely.

To clean the sheath, cover your hand with a rubber glove and then a clean tube sock. Wet the sock with warm water and a little bit of mild soap and gently wash the sheath inside and out. You should be able to remove chunks of smegma this way.

To rinse the sheath, use a hose with warm water and low pressure. The hose can be inserted 2 to 3 inches in the sheath. Most horses will quickly get used to the sensation of the water, but be alert for an initial impulse to kick.

Finally, check the penis for a "bean" or mass of smegma that sometimes forms in the diverticulum near the urethral opening. A large enough bean can interfere with urination. Move back the skin at the end of the penis until you find a small pouch of skin. If there is a bean, it should roll out quite easily. Again, your horse might be inclined to kick at this point if he's not used to the procedure.

If the horse will release his penis while you are cleaning the sheath, you can also wash it off with the wet sock, using just warm water (no soap).

➢ *See also Penis*

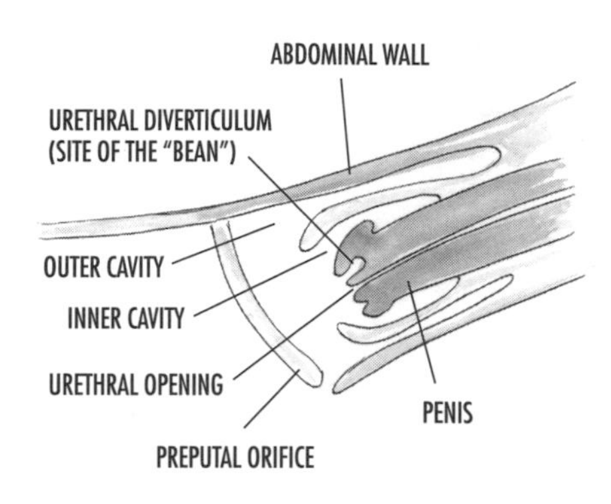

Anatomy of the sheath of the male horse

Sheath swellings

A gelding's sheath may become sore and swollen because of a buildup of debris (dirt and urine) inside the sheath. Glands in the sheath lining produce a secretion called **smegma,** which sometimes forms into soft, dark, waxy deposits or flakes; these, too, can irritate the sheath and penis and cause swelling. To prevent swelling from buildup, cleanse the sheath as often as needed.

Other causes of swelling to be considered if regular cleansing doesn't alleviate the problem are recent castration, local injury, local eruption of skin conditions, and bladder stones.

Shedding blade

A curved metal blade with serrated teeth for pulling off loose hair. The serrated edge of a shedding blade can be used after the rubber curry comb to remove loose hair. Lay it against the horse's coat and sweep in the direction of the hair. The smooth edge can be used to scrape off water after a bath.

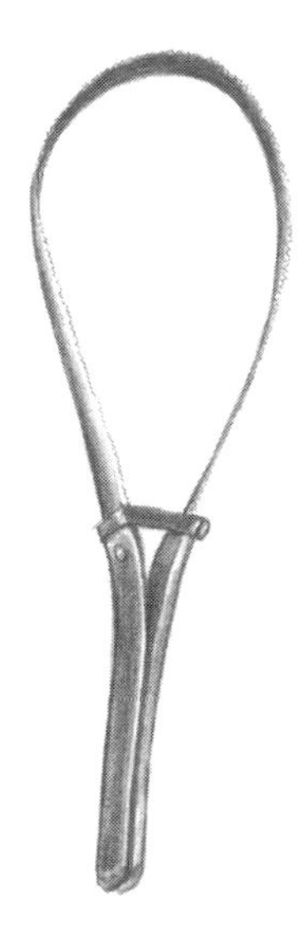

A shedding blade

Shedding in spring

As the days increase in length, horses begin to shed their heavy winter coats. Even horses that have been blanketed all winter will lose some hair. It's important to groom your horse often in the spring to encourage the shedding of old hair and the growth of a sleek summer coat. A thick mat of winter hair is a perfect environment for organisms that cause skin problems, especially in wet conditions.

If your horse is slow to lose his winter coat, check that he is eating a balanced diet with enough protein and sufficient mineral intake. Parasites, illness, and stress also delay shedding, as does hormonal imbalance. Older horses may have Cushing's disease, which is a tumor in the pituitary gland that interferes with hormone levels.

➢ *See also Cushing's disease/Cushing's syndrome; Grooming; Seasonal care*

Sheet

A lightweight covering available for a variety of purposes: to protect against flies, to keep a newly washed horse clean, to reduce sweating on very hot days, to cool out a sweaty horse on a cold day, or to keep rain and dirt off a horse at pasture. Some sheets are designed for turnout, while others are only appropriate for use in stalls or while the horse is tied or being walked.

➢ *See also Blankets and blanketing*

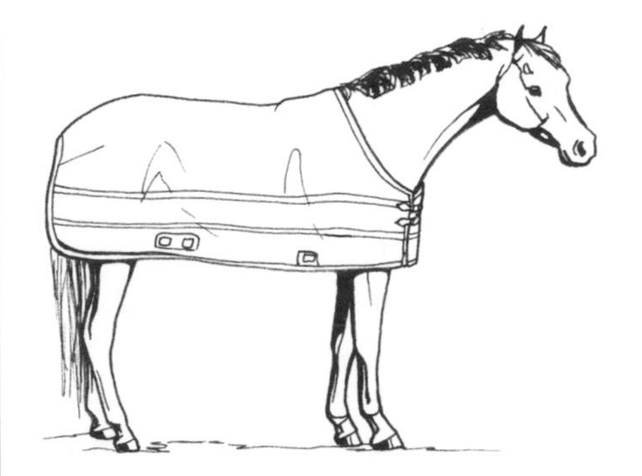

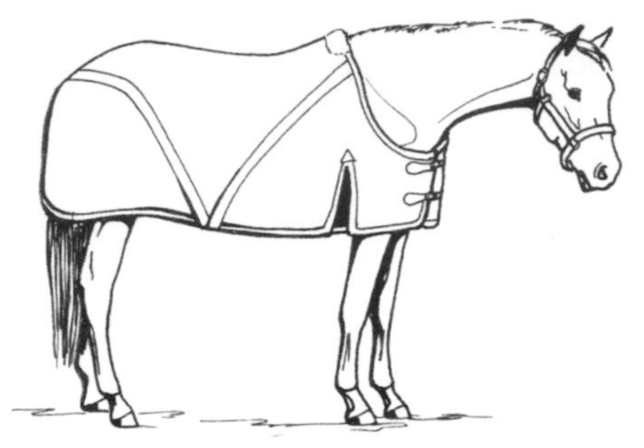

Two styles of sheets; the sheet at top has a belly band and leg straps, which work best to keep a blanket in place.

TYPES OF SHEETS

SHEET	DESCRIPTION
Antisweat sheet	Made of mesh fabric, often cotton, to wick away sweat and keep the horse cool. For use indoors in very hot weather. Some can also work as flysheets.
Cooler	Designed to wick moisture from wet hair; often used after a bath or hard workout on a cool day. Allows the horse to cool off slowly without becoming chilled. Made of wool, wool blend, or polyester fleece, a cooler covers the horse loosely from poll to tail.
Fly sheet	Usually made of a synthetic mesh that keeps flies from landing on horse. Some models have leg straps and can be used for turnout. A good alternative to fly spray during heavy fly seasons. Fly hoods are also available to protect the head.
Stable sheet	A light sheet designed to protect the coat from dirt and dust. Used in stall after a bath or before a show to keep the horse clean.
Turnout sheet	Heavier than other sheets and designed for wear in pasture or paddock. It offers some protection against wind and rain.

An old-style Shetland

Shelter

Wild horses can withstand a fair amount of bad weather as long as they can find adequate protection on very cold, wet, or windy days. But most horse owners provide some sort of structure for their horses, ranging from a simple three-sided shed to an elaborate stable with box stalls, warm water, and an attached riding arena. In some states, Humane Societies have fought to make at least minimal shelter a requirement by law.

➢ *See also Barn; Facilities*

A modern or show Shetland

Shetland pony

First bred as sturdy work ponies in the Shetland Islands north of Scotland, Shetlands have become the beloved playmate and first mount for generations of children. The original Shetland pony was a miniature draft horse, capable of carrying heavy loads (even adult riders), pulling small plows, and working long hours in the mines hauling coal cars. He had a shaggy coat and thick mane for protection against the severe weather of his homeland.

After Shetland ponies were imported to the United States, breeders developed a slimmer, sleeker version, more like a Hackney pony, standing between 10.3 and 11.2 hands high at the withers, These show Shetlands are lively and flashy, while the fuzzy, rotund, old-fashioned Shetland (a maximum of 10.2 hands high) tends to be gentler. Well-known for their ability to work bolts and latches with their lips, Shetlands are smart, curious, and known to outsmart an inexperienced or absent-minded rider.

➢ *See also Breed; Pony*

Some horses need time to get used to feeling the pressure of leg wraps or shipping boots. Make sure you stay well to the side if your horse is making exaggerated movements like this. Let him walk around his stall with his legs wrapped if he seems bothered.

Shipping boots and shipping wraps

Whenever your horse travels in a trailer, his legs should be protected from bumps and cuts with specially designed boots or heavily padded wraps. Travel or shipping boots come in many different styles, but generally have self-fastening straps and a durable outer layer of heavy nylon or even Kevlar (the same material used in bulletproof vests). Shipping wraps are similar to other leg wraps but use taller, thicker quilts and wider bandages. The most important part in both cases is to protect the coronary band and the bulb of the heel and to avoid damage caused by poorly fitting wraps or boots.

As part of training your horse to load and unload from a trailer, accustom him to the sensation of the boots or wraps on his legs. Right before a show isn't the time to discover that he won't load because the shipping boots make him crazy.

➢ *See also Leg wrap; Trailering a horse*

PUTTING ON SHIPPING WRAPS

1. Wrap the quilt around the leg, starting on the inside of the cannon and keeping it low on the leg and fairly snug (without getting too tight). Close the self-fasteners on the edge of the quilt to keep it in place while you wrap the bandages.

2. A 6″ x 12′ bandage is best. Beginning in the middle, wrap the bandage in the same direction that you did the quilt, moving down to the coronary band. Leave some of the quilt showing.

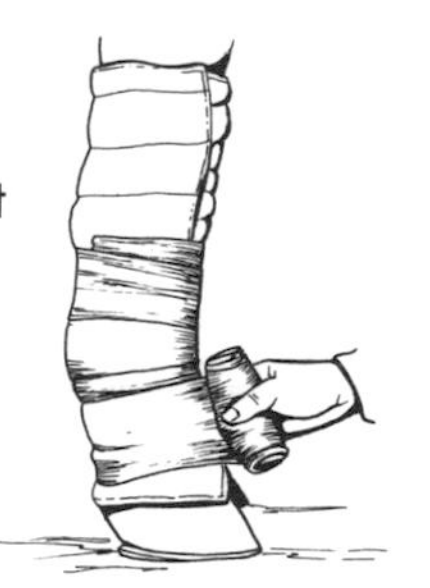

3. As you wrap back upward, apply a fair amount of pressure to keep the bandages from sliding off in transit. The quilts are thick enough that you won't cut off circulation. Finish wrapping at the top and tie off the bandage, making sure to tuck in the ends of the tie. (Self-fastening tape is considered safer by some horsepeople.)

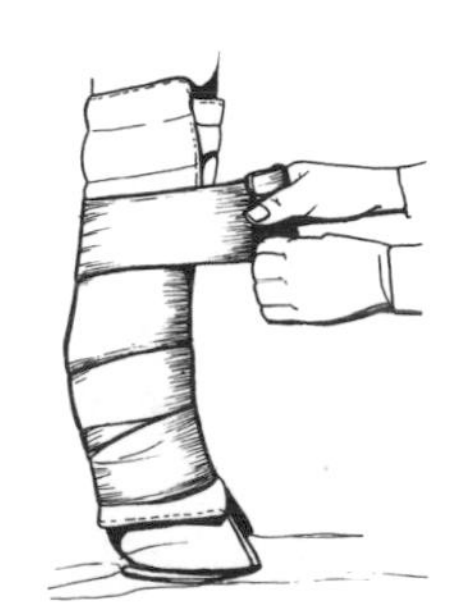

4. For further protection, you can wrap both quilt and bandage quite low on the leg, as shown here. Alternatively, you can use a large bell boot over the bottom of the wrap, which not only increases protection of the heel but also keeps the bottom of the wrap from getting stepped on and torn off.

Shire

Shire

Among the largest of the draft horses, an average Shire stands 17 hands high and weighs over a ton. Generally black, brown, or gray, Shires have long, thick hair on their lower legs. They were developed in England (*shire* means "county") from the heavy warhorses of the Middle Ages. Like other draft animals, the Shire was used for centuries for farming and hauling. In spite of its size, this is a gentle and good-natured animal.

➢ *See Breed; Draft horse*

Shock

If your horse is badly injured, be alert for signs of shock. Sometimes the smaller blood vessels and capillaries shut down, and the resulting loss of blood circulation can be more serious (and even fatal) than the initial trauma. Signs of shock include pale mucous membranes, a lower temperature than normal, falling blood pressure, and shallow breathing. The horse is sweaty but his skin feels cold. He may be anxious and upset or seem dull and depressed.

If shock is due to heavy bleeding, keep the horse warm while you try to stop the bleeding. He will need intravenous fluids or even blood transfusions to replace lost blood volume. Secondary shock can occur with severe burns, colic, or traumatic injury that causes great pain and tissue damage. In this case, the horse needs immediate treatment with adrenaline, steroids, and intravenous fluids to prevent permanent damage from lack of oxygen.

➢ *See also First aid*

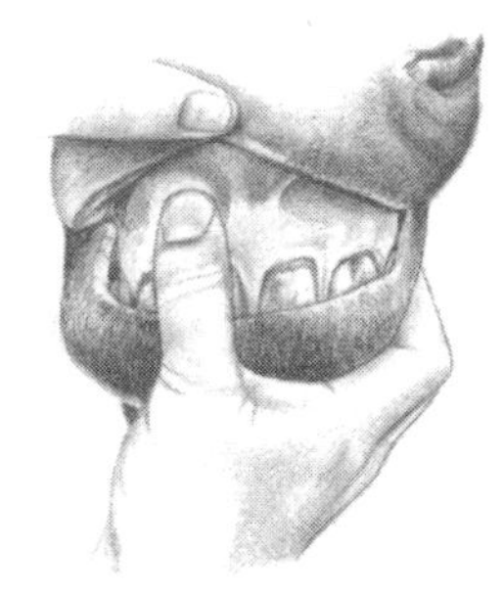

If you suspect that your horse is in shock, check his capillary refill time. Press your thumb against his upper gum for about 2 seconds, squeezing the blood away from that spot. When you remove your thumb, the resulting white spot should disappear in about a second. If it takes 5 to 10 seconds to go away, your horse is having circulatory problems, a possible sign of shock.

Shoes, horse

Like our fingernails, horse's hooves grow continuously and need to be trimmed regularly, whether or not shoes are worn. Over the course of a year, the entire hoof is replaced from the coronary band to the toe. Many ponies and some horses have particularly dense hooves and do not need shoes (Arabians and Appaloosas are known for having strong feet), and horses that are not being ridden or worked are generally not shod. However, most horses are shod with iron shoes to protect their feet from cracking and bruising. Horses that are ridden on gravel or pavement should always have shoes, as should jumpers.

Shoes can also correct a number of problems of the feet and legs. A good farrier can help an unbalanced horse move more evenly, can reduce "forging" (when a horse kicks his own ankles with his rear feet), and can keep a split in a hoof from leading to more serious problems.

When putting on new shoes or resetting old ones, the farrier first pulls off the old shoes. He then trims the sole and toe, carving away the excess with a curved blade. (Dogs seem to be quite fond of these trimmings.) Before being nailed on, new shoes are precisely fitted or old ones remeasured to make sure the fit is the same.

Horses often lose a shoe when turned out or while working. Sometimes the hind foot catches against the front shoe, loosening the nails, which is why you see many horses wearing bell boots. Frequent stamping of the feet against flies is another way nails come loose. Whatever the cause, once a shoe is loose, it needs to come off immediately and be reset by your farrier. Loose shoes can come part of the way off and cause quite a lot of damage to either the hoof itself or to the leg.

➢ *See also Foot care; Hoof*

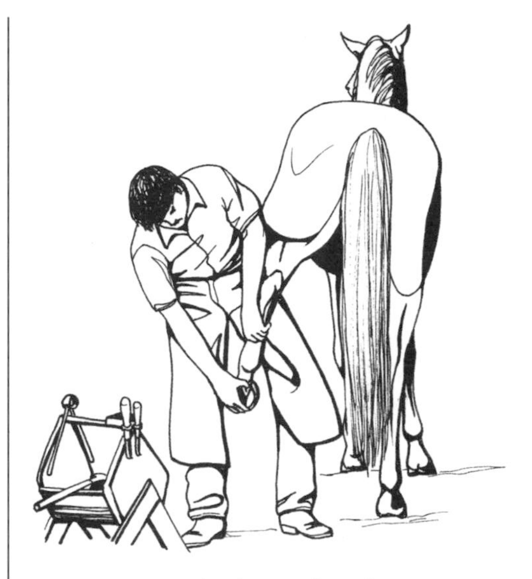

Horses need to be seen by a farrier every 6 to 8 weeks.

PULLING SHOES

To pull off a loose shoe, you'll need some basic tools: a hammer, a clinch cutter (a flat screwdriver also works), pull-offs, and a crease nail puller (ordinary pliers or vice grips can work but not as effectively).

Hold the hoof in the normal shoeing position (front foot between your knees; hind foot across your thigh). Starting at the heel, loosen the clinches with the chisel end of the clinch cutter. (The clinch is the little hook formed by the nail on the outside of the hoof.) Use the pull-offs or pliers to pull each nail in turn, alternating sides as you move toward the toe. You may need to gently tap the shoe back toward the hoof to get the nails to pop out a little.

If the nails aren't loose enough to pull, you can use the pull-offs to work the shoe off. Start at the side of the heel where the shoe is the tightest and pry *toward the frog*. Do the other side of the heel and then the toe, again prying toward the frog. Once the nails are loose enough to pull out, use the pull-offs to remove them.

Caution: Don't pry the shoe *away* from the hoof as pieces of the hoof wall could break off with the nails. Pulling the nails individually straightens them so they come out more easily.

While waiting for the farrier, you can wrap duct tape around the hoof or use a boot to keep it from chipping or splitting.

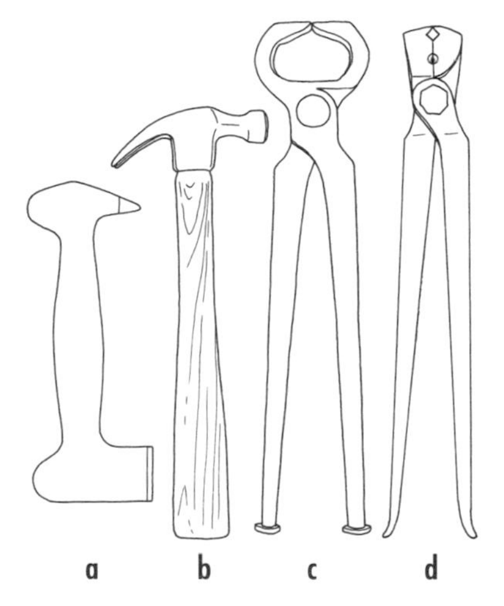

It's a good idea to have these tools around in case you need to pull a loose shoe: (a) clinch cutter, (b) hammer, (c) pull-offs, (d) crease nail puller.

Shooting horses

The decision to humanely end a horse's life is never easy. Most of the time we are lucky enough to be in a location where a veterinarian can handle the situation for us. However, it is actually quite important that horsemen become familiar with this method of ending their horse's life. For example, if a horse is severely injured as a result of a trailering accident, the most humane decision might be to end its life as quickly as possible. Even if you don't carry a firearm in your vehicle or if you are uncomfortable considering handling one, you can save your equine friend a lot of pain and suffering *if* you know how to talk a policeman or rescue person through the process of ending your horse's life. (Without such guidance, it may take several attempts for an inexperienced person to successfully end a horse's life.) Talk with your veterinarian about the steps for ending a horse's life so you are prepared if ever faced with this difficult decision.

Shortening the mane

➢ *See Mane, care of; Pulling the mane*

Shotgun start

When all competitors in a race leave the starting line at the same time (often at the signal of a starter pistol) rather than in staggered heats.

The horse on the left has good shoulder conformation. The one on the right has upright shoulders, which are less desirable.

Shoulder, conformation of

Horses with long, well-sloped shoulders generally have good withers and proper pastern angle, giving them more flexibility and length of stride. A straight or upright shoulder usually accompanies more upright pasterns and a choppier, shorter stride.

Shoulder-in

A lateral movement that many horsemen believe is one of the most important suppling movements available to riders. This movement can be done anywhere, but it's easiest to understand if you imagine it being done on the rail in an arena. The horse's forehand comes off the rail about thirty degrees, which puts the horse's forehand and hindquarters on separate but parallel tracks. This movement is distinguished from the leg yield because the horse is bent toward the direction he is moving, whereas in a leg yield his body is bent away.

➢ *See also Travers; Two-track (half-pass)*

Shoulder, popping

A horse is said to pop his shoulder when you ask him to turn and he resists by bending his head toward where you want to go but continues to move his front legs and shoulders in the original direction. This is a dangerous habit. He needs to be straightened out and then asked firmly, with strong outside aids, to do as he's told.

➢ *See also Aids*

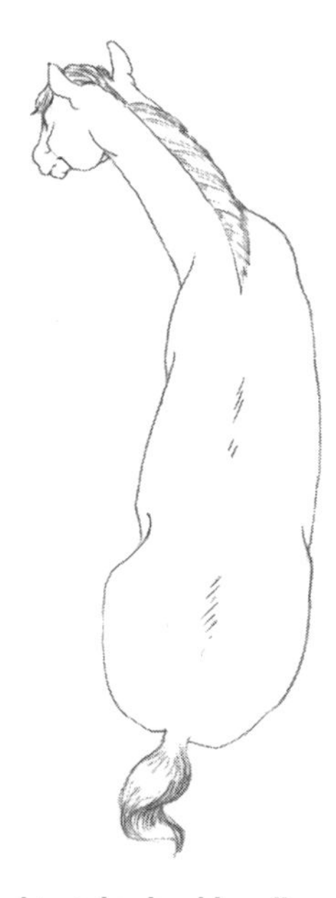

Popping his right shoulder allows a horse to ignore your desire to turn left.

Show associations

There are thousands of horse shows held in the United States each year. The vast majority are regulated by organizations that dictate how the shows are run as well as the rules judges use to evaluate the competition and determine awards. See the appendix for a listing of some of the most common horse show associations.

➢ *See also Events; English competition; Western competition; appendix; individual classes*

Showmanship classes

Showmanship classes are competitive events in which handlers are judged on their mastery of the skills required to show a horse at halter. The horse's quality and conformation are not considered, except as a reflection of the handler's ability to show the horse to its best advantage. Judges of higher level classes expect handlers to be aware of their horse's shortcomings so they can condition, groom, and present him in a manner that accentuates his strengths while minimizing his weaknesses.

The criteria used to evaluate participants vary with the breed and rules of the regulating organization. Most judges use an objective scoring system that minimizes the influence of personal preference and opinion. Participants are typically asked to perform a pattern that demonstrates mastery of basic handling skills, such as walk and trot in hand, back, turns of various degrees (90 degrees, 180 degrees, 360 degrees), and setup (standing horse in a position set by breed standards). Each segment of a pattern is scored. The final score for each participant often also includes a score for grooming and conditioning of the horse; poise, confidence, and alertness of the handler; cleanliness and proper fit of tack and appointments for horse and handler; and a general impression score.

Showmanship classes are an excellent place for novice showmen to get started, but at the same time offer incentives for seasoned competitors to hone their skills. Learning showmanship can help beginners gain confidence in their ability to control an animal that is often ten times their size. It also lays a foundation in ground handling skills that carries forward into work done in the saddle.

Shrubs, toxic

➢ *See Poisonous plants*

Shying, spooking

A horse shies when he leaps suddenly away from an object that startles him. Horses can shy while being led or ridden, at any gait, and from a complete standstill. Some calm, well-schooled horses rarely shy, while others will startle at any unusual sight, including things that he has seen before. While shying is a genuine fear response, some horses make a habit of it and may just be playing around with you.

You can try to prevent shying by being alert for situations that may startle your horse: flapping plastic, a van or truck parked in a new place, a pile of logs by the side of the trail — anything, in short, that might make him feel threatened. When leading your horse by a potentially scary spot, get between the object and the horse, so that he will leap away from you as well as the perceived threat. If he does shy, give him some slack on the line by dropping the hand nearest the halter and grabbing the rope farther down. There's no way you can hold a frightened horse by physical force, so follow a step or two before saying "whoa" and putting pressure on his head.

Most horsemen agree that it's best to try to work through this problem by using calmness and assurance. If your horse tends to shy often, perhaps it's best to get his vision checked before assuming it's strictly a behavior-based problem.

Sickle-hocked

A conformation fault of the hind legs, also known as "too much set to the hocks." It can be seen from the side view if you draw an imaginary line to the ground from the point of the buttock. An ideal set of leg allows such a line to touch the hock and travel down the back of the cannon bone and through the heel. On a sickle-hocked horse, the hock has too much angle,

When my feet is in the stirrups
And my hawse is on the bust,
With his hoofs a-flashin' lightnin'
From a cloud of golden dust,
And the bawlin' of the cattle
Is a-comin' down the wind
Then a finer life than ridin'
Would be mighty hard to find.

—Badger Clark, *Ridin'*, 1915

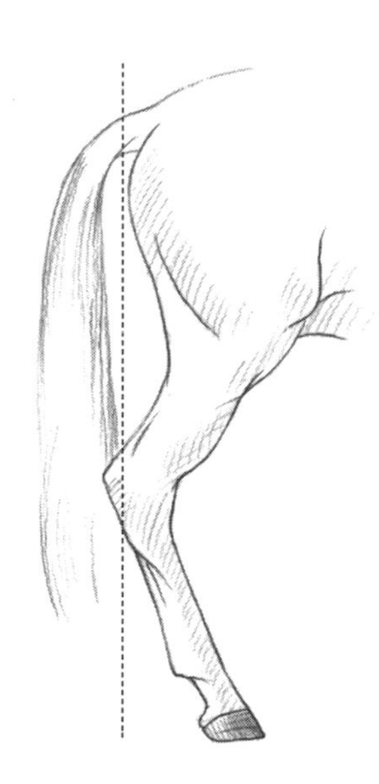

Sickle hock

which pushes the cannon bone forward and places the heel well ahead of the line that's drawn to the ground. This fault places enormous stress on the hock joint and can lead to serious lameness and problems with curbs.
➢ *See also Leg, of horse*

Sidebone

Sidebone is a hard, bony growth found at the hoof head or coronary band and back toward the heels. It is most often found in draft horses or horses with short pasterns. Lameness is uncommon except during the initial inflammation stage that indicates the beginning of a process that leads to ossification of the lateral cartilages.

Sidedness

While most people are right-handed, a good rider should develop strength and flexibility on both sides of her body. If the rider is equally strong when working to the right or to the left, the horse will become more supple and balanced as well.

Sidepass

➢ *See Side step (sidepass)*

Side reins

Training equipment used to help the horse improve flexion and develop a strong "circle" of muscle throughout his body. They are most often made of leather with a rubber ring that allows some elasticity and give during a longe-line workout. They attach to the bit and either a surcingle or a saddle ring.

It is vital that side reins be adjusted properly and readjusted as the horse develops in its training program. These are somewhat advanced tools that should only be used by educated horsemen. It's relatively easy to learn the adjustment process from someone who can explain both how to adjust them and why they are adjusting as they are.

Sidesaddle, riding

Sidesaddle riding, or riding "astride," is not just part of the history books. It is alive and well and growing in popularity each year. Nearly every major horse show–regulating organization has either adopted rules to allow sidesaddles in classes where traditional saddles were previously the norm, or their rules convention delegates have actively debated the merits of such rule changes.

Riding astride is not as difficult as most people imagine and in fact it is an excellent exercise for all riders to learn balance. The sidesaddle rider must be balanced, forward, and centered around her right thigh, with her right toe relaxed, pointing downward, or pressing against the horse's shoulder, but not pointing upward.

Investing in lessons with an instructor certified, or at least highly experienced, in riding astride is probably the best first step toward successfully adding a new dimension to your riding experience.

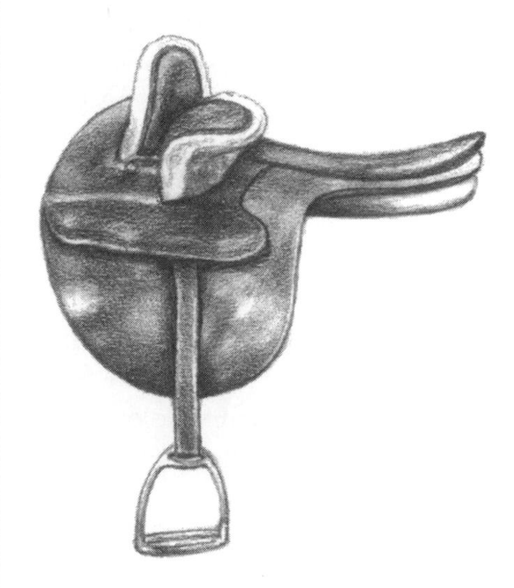

With a properly fitted sidesaddle, a rider can do almost anything on a horse, even jump and work cattle.

Position of the legs in a sidesaddle.

Side step (sidepass)

A horse side steps by crossing the right feet in front of the left (or vice versa) while moving sideways, rather than forward. This is also called a **sidepass** or **full pass.**

In a side step, the feet cross over as the horse moves away from the rider's leg.

Sight

➢ *See Vision, of horse*

Signs of illness

Other than providing routine medical care and proper nutrition and shelter, the most important way you can keep your horse healthy is by being completely familiar with his normal appearance and behavior. Check his resting vital signs (temperature, pulse, and respiration) every day for a week and take an average to establish what is normal for him. Individual horses will vary within a range.

Spend some time examining the horse's legs and getting to know how they look and feel. That way, if there is any heat or puffiness, you'll notice it right away. Become familiar with his mouth and teeth and check regularly for sharp edges or sore spots. Run your hands over his body to check for the bumps, scratches, and small wounds that all horses frequently get. In winter, check beneath the heavy coat for skin problems, especially if you notice that he's rubbing his tail or other parts of his body.

You should be aware of your horse's regular expression and temperament. If a normally placid gelding is pawing and pacing he may have colic, but a spirited youngster doing the same thing may just be bored. Watch your horse's ears, posture, and mood for clues as to whether he's feeling happy and healthy or whether he's got a problem that needs your attention. If you think he's just "acting funny," trust your instinct and try to figure out what's bothering him.

➢ *See also Body language, of horses; Colic; Diseases; First aid; Infectious diseases; Parasites; Poisonous plants; Pulse and respiration; Skin problems; Temperature, of horse; Vital signs*

SIGNS OF ILLNESS

- ❑ Unusual rolling or pawing
- ❑ Sweating without exertion (unless weather is very hot) or lack of sweat during or after exertion
- ❑ Lack of appetite or decrease in water intake
- ❑ Lying down more than normal or lying in an odd position
- ❑ Different posture (for example, legs bunched under body or stretched out; back humped)
- ❑ Keeping weight off a foot; reluctance to move
- ❑ Rapid respiration while resting
- ❑ Change in personality or usual behavior

Silage for feed

Silage is made from green plants that are allowed to ferment in storage. Although silage is commonly used for cattle feed, horses can get extremely sick from eating spoiled, moldy, or frozen silage, and they should never be allowed to eat silage that is intended for cows.

➢ *See also Feeding and nutrition; Hay*

Silent heat

Some mares do not display obvious signs of being in heat, such as mucous discharge, winking of the vulva, and a slightly raised tail. Mares with regular silent heats may need more social stimulation (that is, the presence of a stallion) to show signs of estrus. Occasionally, silent heats indicate some problem with the reproductive tract, although a mare with silent heats may also have normal, more obvious heats.

➢ *See Breeding; Estrus*

Simple lead change

To change the lead of the canter or lope from one side to the other by slowing to the trot or walk or halting in between.

➢ *See also Flying lead change*

Single-handed riding

Single-handed riding is an important part of every horseman's education as a rider. Riding with one hand on the reins can help a rider develop a seat independent of her hands. It can help a rider learn to not rely too heavily on her hands to guide the horse, because single-handed riding requires coordination of seat, legs, and shifts in body weight more than hand signals to maneuver safely.

Single tree

➢ *See Swingletree (single tree)*

Sire

The father of a horse. Some have a remarkable ability, called **prepotency,** to pass traits on to their foals. A founding sire is one to whom a whole breed of horses can be traced and upon whom a breeding program is built. A very famous sire is the small bay stallion owned by Justin Morgan.

Sitting trot

Sitting the trot can be quite difficult, especially on a horse with a bumpy or uneven gait, but it is a great way to develop a good seat and improve your sense of balance. Although it seems contradictory, the only way to be secure at the sitting trot is to relax. Gripping tightly with the knee and thigh makes you stiff and pushes you up and out of the saddle instead of allowing you to absorb the horse's motion with your body.

➢ *See also Posting trot*

Size, of horse

Horses are measured in hands, or 4-inch increments, from the withers to the ground. They range in size from the tiny American Miniature at 8½ hands (34 inches at the last hair of the mane, rather than the withers) to the gigantic Shire, which averages 17 hands high (5'8" at the withers). A horse measuring 14.0 to 14.2 hands or less is called a **pony.**

When measuring a horse to match a rider, a good general rule is that the rider's foot should come to about the bottom of the girth. Since your legs are such an important part of your communication with the horse, it's important that as much of your leg as possible be in contact with the horse's side. Most riding horses seem to be between 14.3 and 15.3, which is generally the right size for the average rider.

Another consideration is weight. If you're heavy, you need a strong horse, which doesn't necessarily mean a tall one. In fact, a sturdy 15-hand animal with a short back can probably carry more weight than a slender 17-hand horse with a higher center of gravity.

➢ *See also Aids; Hand, as unit of measurement; Leg, of horse*

SITTING (VICE)

Some horses will pull back so hard when tied that they actually sit down on their haunches. This is dangerous, as the horse could go right over and get hung up in the lead rope. The handler could get hit by a flying hoof or get a finger caught in a suddenly tightened rope.

You and your horse should be the right size for each other.

Skeid

A term for the rapid pacing gait of the Icelandic horse.

➢ *See also Icelandic horse; Tolt*

Skewbald

A coat color that has large patches of brown or any other color besides black on a white background.

➢ *See also Coloring; Paint and Pinto*

Skid boots

Skid boots protect the rear fetlock/ergot area from getting kicked or rubbed raw in the dirt when a horse stops hard and suddenly, as when reining, cutting, or roping.

A skid boot

Skin problems

Horses develop many of the same skin problems that humans do: hives, fungal infections, ringworm, and dermatitis, to name a few. Some horses are allergic to various foods or to insect bites. In rainy weather, they may be prone to rainrot or scratches, both of which cause scaly, scabby patches.

Check your horse often for skin problems. Run your hands all over his body to check for bumps, cuts, and scaly or flaky skin. Keep small wounds clean to prevent bacteria and fungi from creating further problems. Be alert to signs that your horse is rubbing his tail to relieve itchiness; he could damage his tail hairs and/or create raw patches on his skin.

➢ *See also Parasites; individual problems*

COMMON SKIN PROBLEMS

Allergies
Dermatitis
Fungal infections
Girth itch
Hives (wheals)
Melanoma
Photosensitization
Proud flesh
Rainrot (rain scald)
Ringworm
Sarcoid tumors
Scratches (grease heel)
Urticaria
Warts

Sleeping sickness

➢ *See Equine encephalomyelitis*

Slicker-broke

A horse that is slicker-broke will not shy, buck, or bolt when his rider puts on a slicker or other rain gear.

➢ *See also Sacking out; Shying, spooking*

SLEEP HABITS

As prey animals, horses have evolved to sleep lightly and take frequent naps. Although many horses will lie down for a nap, either in an upright position with legs tucked under or flat on the ground, most horses sleep standing up.

Slow amble or slow gait

This is one of the two additional gaits demonstrated by five-gaited horses. At full speed, this gait is called the **rack.** It is a four-beat gait that resembles the pace, except that each foot hits the ground individually.

➢ *See also Amble; American Saddlebred; Five-gaited horse; Gait; Gaited horse; Pace; Rack; Three-gaited horse*

Small intestine

The small intestine of an average-sized horse is approximately 70 feet long and 3 inches in diameter. Its contents are primarily in a fluid state, as the digestive process in this organ breaks down feed with bile and pancreatic juices. Enzymes break down the feed into usable nutrients that are absorbed through the intestinal wall.

➢ *See also Digestive system*

Small strongyles

Like large strongyles, these worms are bloodsuckers. They inhabit the intestine and cause anemia and debility, as well as ulcers that can lead to fatal hemorrhaging. All horses are infested with small strongyles to some degree, and a rigorous deworming program is essential in controlling them. Because the worms emerge seasonally from the intestinal wall, deworming must take place at certain times to be effective. In northern climates, this is in the late winter and early spring, and in the South in the fall and winter.

Ivermectin is effective, but many types of small strongyles are resistant to frequently used drugs. You should work with your vet to find a program that meets your needs.

➢ *See also Deworming; Parasites*

Smegma

A buildup of dirt, dead skin cells, and fatty secretions that accumulates in the sheath.

➢ *See also Penis; Sheath; Sheath, cleaning*

Smut

A fungus that appears as black powdery clumps on the seed heads of grasses. Grain contaminated with smut is toxic and causes convulsions, paralysis, and rapid death if eaten. Smut can also cause colic and bring on respiratory problems if inhaled.

➢ *See also Poisonous plants; Rust*

Snaffle bit

The snaffle is one of the most widely used bits because it is one of the gentlest available. Most trainers use a snaffle on the green horses they work. Most instructors put snaffles in the mouths of their lesson horses.

Most snaffles have jointed mouthpieces. Many horsemen believe it is this jointed mouthpiece that differentiates a snaffle from a curb, but some snaffles come with a straight, unjointed mouthpiece. Snaffles work through direct pressure on the corners of the mouth, the bars, and the lips. This differs from the leverage action of a curb bit. Snaffles come in a variety of sizes and styles, from the smooth, round-ring snaffle to the full-cheek Dr. Bristol snaffle.

The vast majority of snaffles have a single or double jointed mouth piece that lays across the horse's tongue. Double joints allow the bit to work more softly on horses with thick tongues or low palates. (Fitting bits is a science that requires years of experience, so ask for help from a professional if your horse has bitting problems.)

The thicker the mouthpiece, the milder its action on the horse's mouth. Thin mouthpieces can literally slice a horse's tongue, so they should only be used by experienced horsemen, educated about the proper use of such a bit.

➢ *See also Bit; Bridle; Tack*

Sing, riding's a joy!
For me, I ride.

—Robert Browning

Common Snaffle Bits

STRAIGHT SNAFFLE BIT

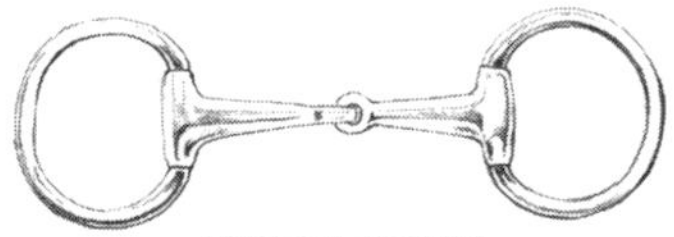

JOINTED SNAFFLE BIT

EGGBUT SNAFFLE WITH MULLEN (CURVED AND NONJOINTED) MOUTHPIECE

Snakebite, first aid for

If a venomous snake bites a horse on the leg, the bite is generally not fatal, as horses are large enough to absorb the poison without long-term ill effects. However, a bite on the nose or face can cause serious enough swelling to suffocate a horse. Any suspected snakebite should be reported to your veterinarian immediately, as an antivenin may be needed, as well as treatment to control swelling and tissue damage and to prevent infection.

Snatch

When either horse or rider grabs abruptly at the bit or reins. This is generally discouraged.

Sneakers, riding

Sneakers that are specifically designed for riding. These are the only kind of sneakers that you should wear while on a horse. Any shoe without a heel can slide through a stirrup and get caught. If you fall, you could be dragged. Regular sneakers aren't a good choice around horses even if you're not riding, as they offer very little protection against a sharp hoof.

➢ *See also Attire*

Snip

A patch of white hair between and below the nostrils, above the upper lip. It may be quite broad or just a few hairs.

➢ *See also Face markings*

Soaking feet

For many types of leg and foot injuries, your veterinarian will recommend that you soak the affected area in warm water, perhaps with Epsom salts added. The best choice is a special boot made for treating hoof rot in cattle, but a rubber bucket or tub will work also. Rubber won't clang like metal and will be softer if the horse bangs against it. Get him used to putting his foot in the tub or bucket (depending on how much water you'll need) before adding any water. Once he'll stand quietly, add water that is slightly warmer than his body temperature. As he accepts the water, you can add the Epsom salts and then gradually add hotter water (as hot as he'll tolerate, but not hot enough to burn).

It's easier to soak with two people — one to make sure the leg stays in the bucket and one to handle the water supply. Soaking for 20 to 30 minutes a day for 3 or 4 days clears up most infections. Usually, the veterinarian will recommend covering the wound between soakings to keep out dirt and wetness. There are also special boots to protect injured hoofs.

➢ *See also Boots, horse; First aid; Foot care*

Sobreandando

A Peruvian Paso lateral gait that is generally faster than the paso llano, with a longer stride and more overreach.

➢ *See also Paso llano; Pasi-trote; Peruvian Paso*

SNOW FOR WATER

While horses will eat snow in the winter if they are very thirsty, they use a lot of energy and body heat melting it and warming it. Always make sure your horse has access to unfrozen water in the winter, and keep an eye on his intake, as most horses drink less in the winter.

See also Seasonal care

S

Socks

White markings on the leg that extend above the fetlock but not above the knee or hock.

➢ *See also Leg markings*

Sodium chloride

➢ *See Salt*

Sole bruises

Corns and sole bruises happen when shoes are fitted improperly or left on too long or by stepping on a sharp rock. These bruises cause a horse to go lame as he shifts his weight forward or backward to avoid the sore spot. Corns are injuries to the heel, while sole bruises more often occur in the toe or quarter area. The bruised area appears red and will be especially apparent after trimming. The spot may be tender and hot to the touch. If the spot is not readily visible, use a hoof tester to locate the corn or sole bruise.

➢ *See also Foot care; Hoof; Shoes, horse*

Sole of hoof

The bottom of the hoof, in front of the frog. A healthy sole is slightly concave to absorb pressure from the weight it bears. It should be thick and hard to protect against bruises.

➢ *See also Foot care; Hoof*

Sores

Horses get sores from a variety of causes: nips and kicks from other horses; irritation from parasites, mites, or fungal infections; poorly fitting tack or blankets; or insect bites. Left untreated, small wounds can get infected and even become abscessed. A sore place on a leg can swell and cause lameness, while skin wounds invite parasites and fungal infections (if those aren't the problem in the first place).

You can treat most small sores by cleaning them thoroughly and applying a dab of antibiotic ointment. Some ointments come with fly repellent in them for further protection against infection.

➢ *See also First aid; Fistula of the withers; Saddle galls*

Sorghum grain (milo)

A good horse feed, though not available everywhere, sorghum grain should be rolled or crushed before feeding and mixed with a more bulky food such as oats. It is about 80 percent digestible nutrients with 8.5 percent protein, but only 2.3 percent crude fiber.

➢ *See also Feeding and nutrition; Grains*

Sorghum grass (Johnson grass, Sudan grass) (*Sorghum* spp.)

Common in pastures in the Southwest and Southeast, this grass can be quite toxic after a drought or hard frost, when it produces cyanide.

➢ *See also Poisonous plants; Sudan grass*

"SOFT IN THE FACE"

Guiding a horse with light reins and gentle but consistent contact produces a horse that is "soft in the face," that is, able to respond to the aids without stiffness or resistance in his head and neck.

BE VIGILANT!

Check your horse regularly for nicks, scratches, and bites. Take care of them early to prevent possible infection or lameness.

Soring

An inhumane practice, common in the 1950s and 1960s, used to accentuate a horse's gait and high-stepping action by irritating the forelegs with chemical or mechanical means. A sored horse quickly lifts its front legs to relieve the pain, which was thought to give some breeds a competitive advantage. The Horse Protection Act of 1970 and its amendment in 1976 ensure that responsible horse owners and trainers will not suffer unfair competition from those who sore their horses and that horses will not be subjected to the cruelty of soring. Tennessee Walking horses and other high-stepping breeds are the most frequent victims of soring, though the practice is widely prohibited.

Sorrel

A reddish brown or reddish gold coat color, often seen with a lighter mane and tail; synonymous with "chestnut."

➢ *See also Coloring*

Sound

A sound horse is free from any fault, injury, or illness that inhibits its ability to perform its designated purpose. Horses can be serviceably sound despite a minor unsoundness that only partly inhibits functioning. An example would be a horse that has a minor lameness. Such a horse might be serviceably sound as a trail horse or as an equitation mount or showmanship horse, but would not be sound for pleasure or halter competition. This same horse might be sound for breeding, though a detailed assessment of conformation should be conducted to be certain the lameness is not due to an inheritable conformation fault.

➢ *See also Conformation*

Soundness of wind

When buying a horse, make sure his wind is sound; that is, that he doesn't have any problems of the respiratory system. A horse with chronic breathing difficulties is said to be "broken-winded."

➢ *See also Respiratory system; Roarer*

Sow mouth

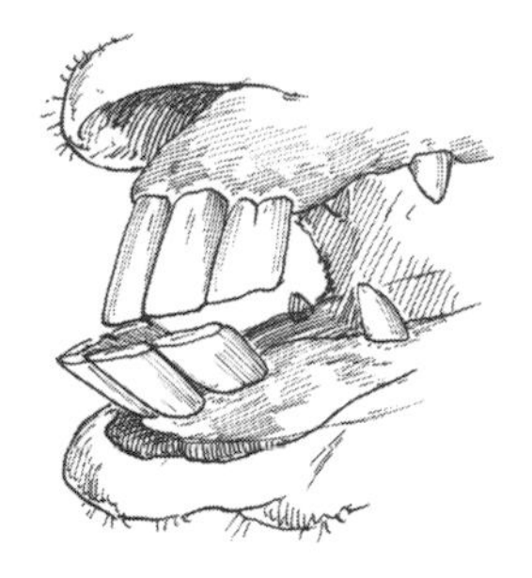

In sow mouth, the lower jaw juts farther than the top jaw, causing teeth to be misaligned.

Sow mouth, or "monkey mouth," describes an underbite; less common than its opposite, parrot mouth. The upper molars must be kept trimmed or they will form hooks that can puncture the lower gum.

Spanish Barb

An important influence on the history of many horse breeds, especially Western horses, the Spanish Barb today still shows the deep chest, strong neck, and close-coupled body that makes it a surefooted and quick mount. Averaging 14.2 hands high and weighing between 850 and 1,000 pounds, the Spanish Barb comes in all colors except gray and white. He is valued for his intelligence, agility, and stamina.

➢ *See also Barb; Breed*

Spanish mustang

A descendant of the Spanish horses that arrived in the New World in the 16th century, the Spanish mustang is wiry and solidly built, with long muscles and strong legs and feet. He has a Roman nose and an arched neck, with a flowing mane and tail. Ranging from 13.2 to 15 hands high, he weighs 1,000 pounds or less, and is known for his alert, even temperament, as well as stamina and energy. He comes in a variety of colors, both solid and spotted.

➢ *See also Breed*

Spanish Riding School

Perhaps the most prestigious school of riding in the world. The Spanish Riding School is best known for its performance of the great white stallions, the Lipizzanners.

➢ *See also Lipizzaner*

Spasmodic colic

➢ *See Colic*

Spavin of hock

There are two primary types of spavins, both affecting the hock joint: bone spavin and bog spavin.

Bone spavin, also known as "Jack" spavin, is the name given to a bony enlargement on the inside, lower part of the hock. The cause is not fully agreed upon by veterinarians; however, horses with weaknesses in their legs or hocks are more prone. Predisposing factors include, but are not limited to, hocks that are not "square" or tend to taper from top to bottom when viewed from the front; small hocks; tied-in hocks; sickle-hocks; cow hocks; as well as horses ridden before their hocks are fully developed.

Bog spavin is the name given to a spongy swelling of the lower inside of the hock joint. The most common cause is actually a predisposing factor — upright hocks, or hocks that lack angularity between the bones above and below the joint (tibia and metatarsal). It is most common in young horses that are placed in training before their hocks are fully developed and in breeding stallions. Symptoms other than the obvious swelling are rare, except when the spavin initially develops.

➢ *See also Leg, of horse; Sickle-hocked*

Speed, of horse

With the exception of cheetahs, several types of antelopes, and the brown hare, not many animals can outrun a horse. Horses can reach a top speed of about 43 mph, compared to a cheetah's 65 mph. A racing greyhound runs at about 42 mph.

➢ *See also Gait; Harness racing; Horse racing; Standardbred*

Splay-footed

A splay-footed horse has hooves that turn outward from the fetlock joint. This type of horse often causes himself harm, sometimes serious

SPANISH RIDING

The precursor of today's modern Western style, Spanish riding was developed in Europe before being brought to America with the Spanish conquerors. Developed for military purposes, Spanish riding (and similar styles used in France and Italy) depended on a very deep, balanced seat with long stirrups. Riders used harsh curb bits and spurs with large rowels to keep the horse collected. Saddles with horns and high cantles helped warriors stay on board during long marches and dangerous battles.

See also Lipizzaner; Western riding

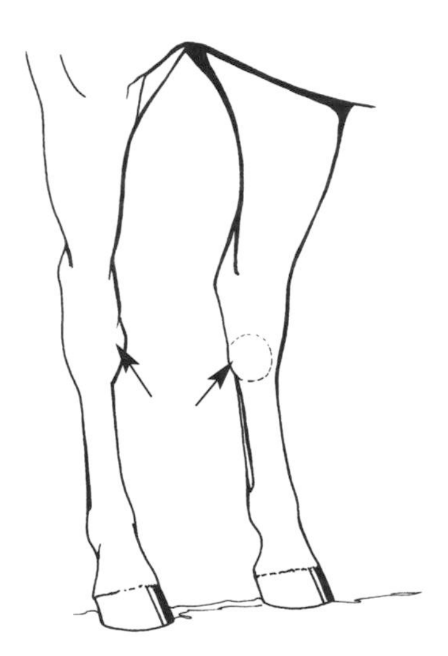

Bone spavin on right and left hock.

enough to result in unsoundness. He can't walk straight and true because his hooves are not centered under his legs, so his feet tend to "wing" inward and often hit the opposite leg. This interference of gait can cause serious damage, resulting in lameness

➢ *See also Leg, of horse*

Splint boots

➢ *See Boots, horse*

Spontaneous combustion

Freshly cut hay generates a startling amount of heat through bacterial activity, especially if high in clover or alfalfa content. Check stored hay frequently to make sure the internal temperature of the stack or bales is no higher than 150°F. At 175°F, the hay will begin to smolder from the center until reaching oxygen at the edge of the stack, whereupon it will burst into flames. If it gets to 185°F, call the fire department to supervise removal of the hay from the barn.

All hay should be stored in a building separate from the barn, if possible. If in the barn, have a buffer zone of at least 100 feet between the hay storage area and the stable and tack room. Make sure all hay is well-cured before storing it.

Spook; spooky

To startle easily; given to shying.

➢ *See also Shying, spooky*

Sport boots

➢ *See Boots, horse*

Sport horse

Any horse, purebred or not, that is suitable for a variety of activities, including dressage, jumping, eventing, or endurance riding.

Spread

A jump consisting of two to four fences set close together. Spreads can be from 2½ feet to 6½ feet wide.

Spring care

➢ *See Seasonal care*

Spurs

Spurs are one of the artificial aids used to reinforce the natural leg aid. Spurs should only be used by experienced riders who have been educated in their proper use, and only as reinforcement to cues given by the rider's leg when the horse does not respond to such cues. Riders must have full control of their lower legs and be able to ride with an independent seat,

Types of Spreads

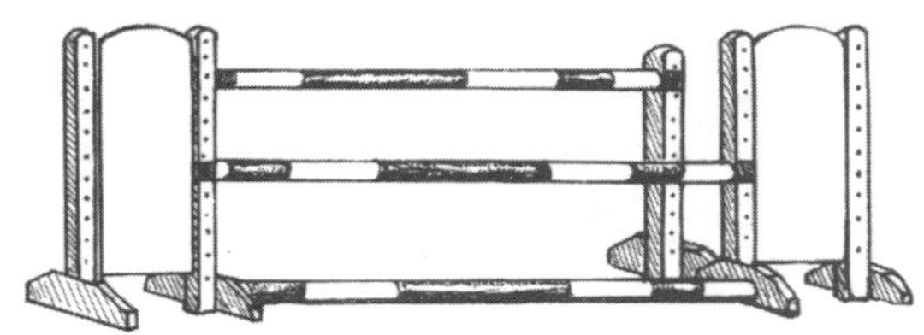

An uneven oxer, with the back bar higher than the front. Oxers can also have both top bars at the same height.

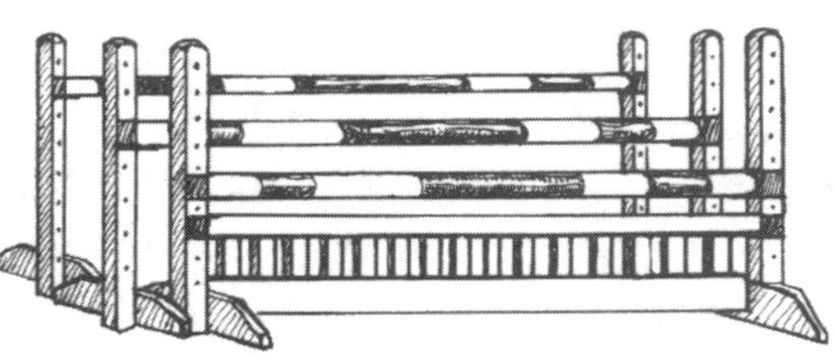

A triple bar, with three poles rising in height.

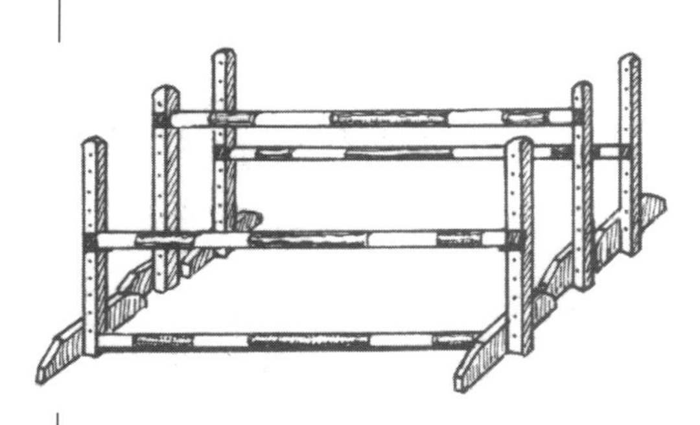

A hog's back, with the middle pole higher than the other two.

before advancing to wearing spurs. Riders who lack such skills often inadvertently jab their horse's sides with the spurs, sometimes with dangerous results.

There are many types of spurs, ranging from short blunt spurs commonly worn by dressage and hunt riders, to the large roweled spurs worn by rodeo cowboys. Regardless of style, spurs should always be worn pointing downward, with the longer side on the outside of the boot.

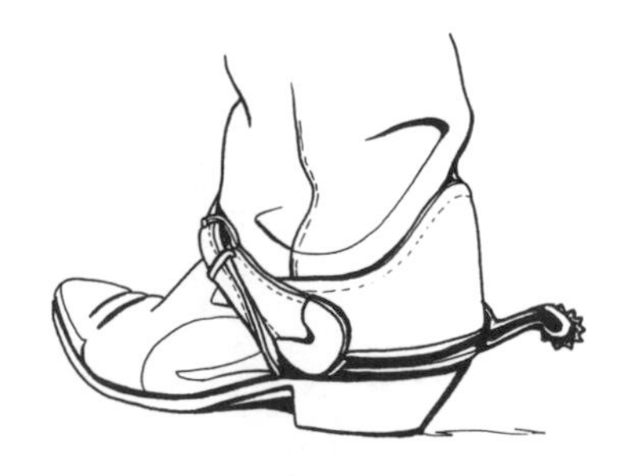

A Western rider with a rowel spur

Squamous cell carcinoma

Horses with little pigmentation around their eyes are vulnerable to squamous cell carcinomas. These growths may also appear around the vulva or sheath on light-colored horses. Horses with unpigmented eyelids should not spend a lot of time in the sun. Keeping them stabled on bright days or painting a dark, nontoxic substance around the eyes can help prevent carcinomas. This slow-growing cancer spreads more rapidly with exposure to strong sunlight. Any bump or sore around your horse's eyes that doesn't heal or go away should be examined by a veterinarian.

Carcinomas can be removed by radiation, freezing, or chemical burning, or by using immunotherapy (injecting a vaccine that stimulates anticancer cells). The location of the growth will determine the safest course of treatment. In some cases, removal of the eye is necessary.

Spur not an unbroken horse; put not your plowshare too deep into new land.

—Sir Walter Scott, *The Monastery* (1820)

Squinting

If your horse is squinting his eye and is reluctant to open it, suspect an eye injury and consult your veterinarian.

➢ *See also Embedded seeds; Eyes; Periodic ophthalmia; Photosensitization*

Stable, choosing

If, like most horse owners, you are unable to keep your horse at home, you need a suitable facility at which to board him. Like finding good daycare for an infant, the process of locating a stable should begin before you actually acquire the horse. To get an idea of what facility would work best for you, visit all the barns in your area and ask lots of questions of fellow horse owners. Check out the current boarders for healthy-looking coats and alert, interested attitudes. Ask about worming and farrier schedules, emergency procedures, hay and feed storage, and turnout policies. Some places charge extra for giving supplements and medicines or for putting on blankets and boots.

In addition to the obvious concerns of how well the horses are cared for and how safe the barn is for both horses and riders, you'll want to find out how the barn suits your riding style and needs. Is it convenient to your home or workplace? If you like to do trail riding, make sure you have access to well-maintained trails. If you do a lot of schooling, ask about availability of the arena for boarders versus students taking lessons.

➢ *See also Barn; Facilities; Fencing materials; Stall*

STABLE NEEDS

Your horse's stable needs

- ❑ A clean, safe barn with an experienced manager and conscientious staff
- ❑ A comfortable, well-ventilated stall with a level floor and good bedding
- ❑ High-quality hay and grain, fed on a regular schedule
- ❑ Regular turnout time with compatible horses; if there's grass, so much the better

Your stable needs

- ❑ Riding facilities that suit your needs (trails, jumps, an indoor arena) and schedule
- ❑ Trust in and access to the barn manager
- ❑ A safe place to keep your tack
- ❑ Knowledge that your horse is receiving the best of care in your absence

Unlike a twitch, the Stableizer allows the horse to move around and be led. Pressure can be set at any level with the quick-release button.

Stable flies

➢ *See Parasites*

Stableizer

A restraint device modeled on the Indian war bridle but more effective and humane. It fits over the head like a halter with a covered cord that goes under the upper lip. The application of pressure behind the ears releases endorphins, which block pain and relax the horse, while the pressure on the gums inhibits the release of adrenaline, keeping the horse even calmer. This device is made by Wheeler Enterprises, Inc.

➢ *See also Handling horses; Twitch (restraint)*

Stable wrap

➢ *See Standing bandage (stable wrap)*

Stadium jumping (show jumping)

Stadium jumping takes place in a ring as opposed to the outdoor course followed in cross-country jumping or steeplechasing.

➢ *See also Eventing; Jumper classes; Jumping; Jumps, types of; Olympic equestrian events; Steeplechasing*

Stake out (picket)

To tie an animal by a long rope or chain to a stake driven in the ground. Although some horses are sensible while tied on long lines, the potential for entanglement, rope burns, and other accidents is high.

Stall

Ideally, your horse should spend the better part of his day or night outdoors. Most horses, however, are kept in stalls either at night in the winter or during the day in the summer, with further confinement in very wet weather. Tie stalls, narrow chutes in which the horse stands with his head tied, are not satisfactory for long periods of time. The average horse needs a box stall measuring at least 10 feet by 12 feet (12 feet by 12 feet is even better), which allows him to move around, lie down, and, if he's so inclined, urinate and defecate away from his food. Ponies can live in smaller space, but no less than 10 feet by 10 feet.

Stall walls should be solidly built of wood, cement, or metal (possibly covered with rubber mats). They should be at a height of 4 to 5 feet, and can be up to 8 feet between stalls. The walls must be strong enough to withstand repeated kicking, both of the restless, "pay attention to me" type or the powerful "get me out of here" type. The top part of the stall wall should have wire grills or mesh with openings of 2 inches or less for light and ventilation. The ceiling must be at least 8 feet high. Openings need to be at least 4 feet wide to accommodate a horse, as well as a large wheelbarrow for cleaning.

There are a number of options for flooring including dirt, concrete, and wood (see box) and bedding. Rubber mats can protect against many of the disadvantages of each type of floor, offering further cushion for legs and keeping the floor itself safe from digging hooves and soaking urine.

➢ *See also Barn; Bedding; Facilities; Mucking out stalls*

HALTERS FOR BREEDING STALLIONS

To help the stallion separate his two tasks — breeding and riding or other work — keep two different halters, one for each purpose. The one for breeding might be used with a nose chain for greater control. In any case, always use one halter for the same purpose and do not confuse them. Each piece of information that is different for the stallion will help him remember and keep his mind on the task at hand.

FLOORING FOR STALLS

TYPE OF FLOORING	ADVANTAGES	DISADVANTAGES
Packed dirt or clay	Easy on the legs	Needs regular maintenance, especially if the horse paws and digs holes. Drains poorly (especially clay) and may hold odors if not cleaned and deodorized thoroughly. Uneven floors are hard on legs, so floors must be kept level.
Wood	Level (unless very old and warped); not hard on legs	Must be checked regularly for splinters, cracks, and other hazards. Must be well-bedded to protect against scrapes and splinters as horse lies down and gets up. Odors can be a problem, though chemical deodorizers help.
Concrete	Level (and will stay that way); easy to clean, especially after a contagious individual has been housed there	Can be very hard on legs, even with mats and thick bedding. Gets slippery when wet.

Stall chewing

➢ *See Cribbing*

Stallion

An intact male horse with at least one testes descended. Usually a male horse is referred to as a **colt** until it is mature, which ranges from three to five years, depending on the breed. A castrated stallion is a **gelding.**

Stallions require specialized care in nearly all areas from feeding and exercise to handling and training. It is a job best left for experienced horsemen.

Stall walking

Pacing back and forth in the stall.

➢ *See also Pacing, habit of; Vice, how to handle; Weaving and stall walking*

Standardbred

Named after the practice of requiring horses to meet a timed standard (originally a mile in 2½ minutes, though today's trotters are faster) before being allowed in the registry, Standardbreds either trot or pace (some do both, but most have a preference) at great speed without breaking stride. Most Standardbreds trace back to a foundation stallion named Hambletonian, whose foals were natural trotters. Although Standardbreds are the backbone of the harness racing industry, they also make fine endurance or trail mounts. However, they are not suited to jumping or galloping sports, such as barrel racing. Because retired racing Standardbreds are used to the commotion of the track, they are very safe mounts and are often retrained for use by police departments.

Measuring around 15 to 16 hands, these horses are generally dark in color and less finely built than Thoroughbreds, with plainer heads, heavier bodies, and stockier legs. They are valued for their toughness and stamina as well as their pleasant, calm personalities.

➢ *See also Breed; Harness racing; Pace*

Standing at stud

When a stallion is available for breeding to mares owned by persons other than the stallion's owner.

➢ *See also Breeding; Stallion*

Standing bandage (stable wrap)

This type of wrap, a piece of quilting held in place with a stretchy roll bandage, is often used with liniment on a lower leg to reduce swelling and fluid accumulation. Although it can stay on overnight (without liniment, which could burn the skin), a standing bandage should be removed for at least an hour out of every 12- to 16-hour period. It doesn't matter which direction you wrap the bandage, but it is important to use even tension without pulling on the flexor or extensor tendons. Make sure the quilt stays smooth, with no wrinkles or lumps.

Wrapping a Standing Bandage

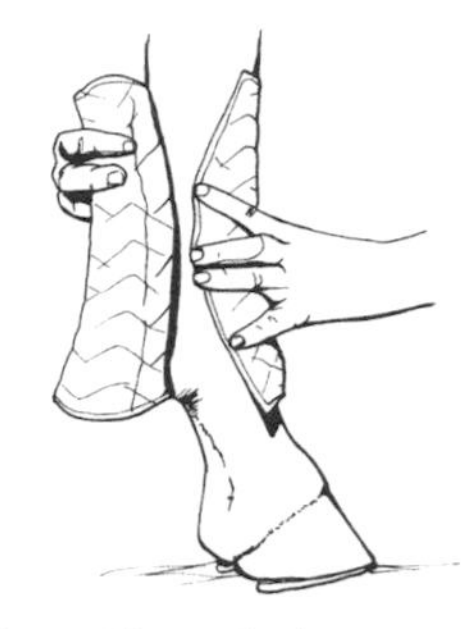

After rubbing the liniment into the leg (if needed), roll the quilt around the leg, starting on one side or the other, not the front or back of the leg. Make sure the quilt is below the knee to ensure a smooth fit.

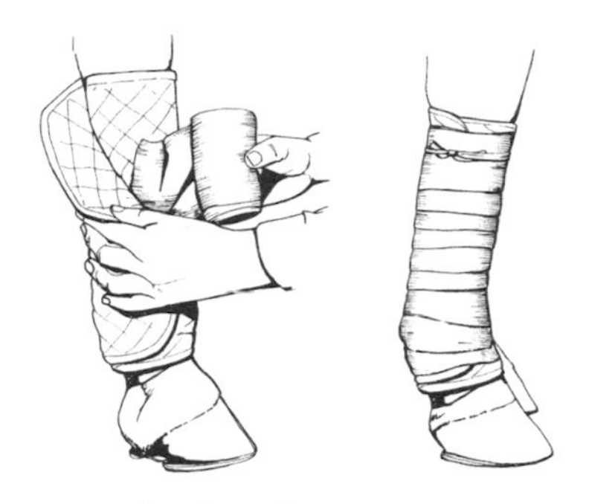

Your bandage should be 5 to 6 inches wide and at least 10 feet long. Starting midway on the cannon bone, tuck the end of the bandage under the end of the quilt and unroll it in the same direction, spiraling down to the hoof and back up to the top of the quilt. Once the bandage is wrapped securely, tie the ends and tuck them in.

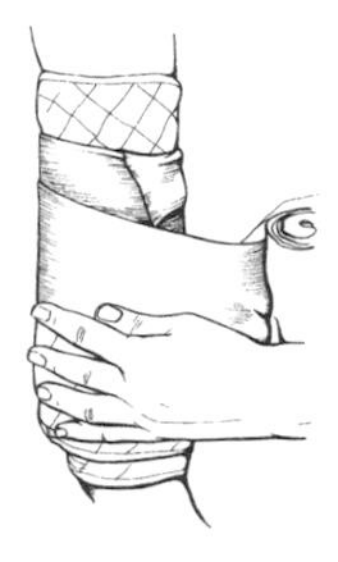

You can also start the bandage by wrapping it around once and folding a corner down to secure it.

Standing martingale

A standing martingale is designed to prevent the horse from tossing his head or holding it too high. It is more restrictive than a running martingale, and it consists of a leather strap buckled to the noseband and attached to the girth. The standing martingale is held in place by another strap that goes around the neck.

➢ *See also Bridle; Martingale; Running martingale; Tack*

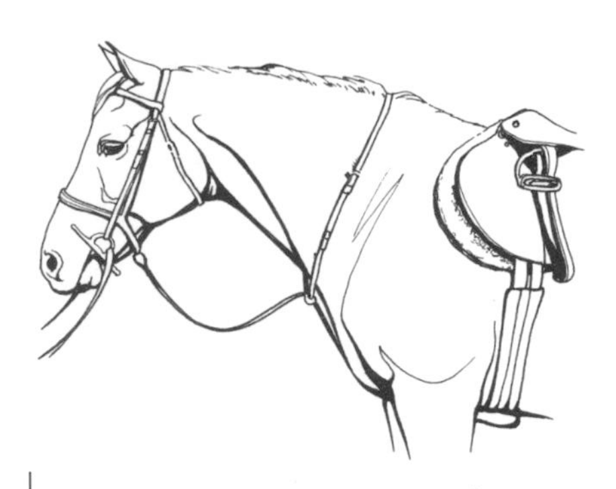

A standing martingale

Star

A white spot on a horse's forehead. A star may cover most of the upper part of the horse's forehead above and between the eyes, or it may be just a few white hairs.

➢ *See also Face markings*

Stare (standing hair)

A horse's coat is said to be "staring" when it is rough and harsh-looking and stands away from the skin. This can happen after a shampoo, but it can also be a sign of poor nutrition or another ailment.

Stargazer

A horse, often ewe-necked, that carries his head too high and can't see the ground right in front of him.

Steamed grains

Steaming grains makes them easier to digest and kills the bacteria that causes EPM (equine protozoal myeloencephalitis).

➢ *See also Feeding and nutrition; Grains*

Steeplechasing is not for the inexperienced rider or for the faint of heart.

Steel pipe fences

Pipe fencing is extremely durable, especially when painted with rust-proof paint. It's a good choice for containing stallions or for along a road where it is particularly important to have strong fencing. However, steel pipe can be unforgiving if a horse runs into it, and care must be taken to ensure that legs and heads can't get through the pipes.

➢ *See also Fencing materials*

Steeplechasing

Among the most grueling of equine events, steeplechasing involves galloping full speed over the countryside and jumping large obstacles. It is, in fact, similar to fox hunting, only without the fox and over a predetermined course. Beginning as "point to point" races, that is, from one jump or landmark to the next, the current name reflects the fact that tall church spires were often used as "points." The most famous steeplechase is probably England's Grand National, a 4½-mile run over 30 jumps.

Step-down dismount

➢ *See Dismounting*

Steroids

Because steroids (such as dexamethasone) and epinephrine (adrenaline) can be crucial in reversing the effects of anaphylactic shock due to severe allergic reaction, these drugs should always be kept in your first-aid kit. Steroids are also used to treat a variety of other conditions. They can be used as performance enhancers, but horse show rules restrict such use.

➢ *See also First aid; Shock*

Stethoscope, use of

A stethoscope can be useful in determining a horse's pulse or checking for the normal sounds of a healthy gut.

➢ *See also Colic; First aid; Pulse and respiration*

To check the pulse rate, place the stethoscope just behind the elbow on the left side (the heart girth). Listen for the rumbles and gurgles of digestion. An ominous silence could indicate life-threatening gastrointestinal problems.

Steward

The official at equestrian competitions who is responsible for interpreting the rules of the regulating organization. The American Horse Shows Association (AHSA) has the largest educational and licensing program for stewards in the United States and Canada. AHSA stewards are highly educated professionals who serve to protect the integrity of the sport. They assist judges in making sure there is a level playing field for all competitors by enforcing the rules. Stewards actually have a slightly higher level of influence than judges, in the sense that they can report inappropriate or illegal conduct by judges to the regulating organization.

Stifle

The stifle is located at the front of the hind leg, below the point of hip.

➢ *See also Anatomy of the horse; Leg, of horse*

Stirrups

Stirrups afford the rider greater stability and control of leg position. There are a number of styles available, depending on what kind of riding you do. English stirrups are made of steel and attach to the saddle with narrow leather straps, with holes to adjust the length. Western stirrups are thicker and wider than English stirrups, and are usually made of leather, but are sometimes made of wood.

The most important aspect of stirrups is that they fit properly, allowing your feet to slip out easily. Some stirrups have a modified design, while others have a rubber safety release. Rubber stirrup pads can improve your traction, but don't use them if you're wearing rubber-soled shoes, which may stick.

For most types of English riding, your stirrup leathers should be roughly the same length as your arm. To measure, place your fingertips at the top of the leather and stretch it out along your arm. The bottom of the stirrup should fit into your armpit. Jumping requires a somewhat shorter stirrup, while dressage riders use a longer one. Western stirrups are also longer, to allow for many hours in the saddle.

➢ *See also Running up stirrup leathers; Saddle; Tack*

He doth nothing but talk of his horse.

—William Shakespeare, *The Merchant of Venice*

STIRRUPS

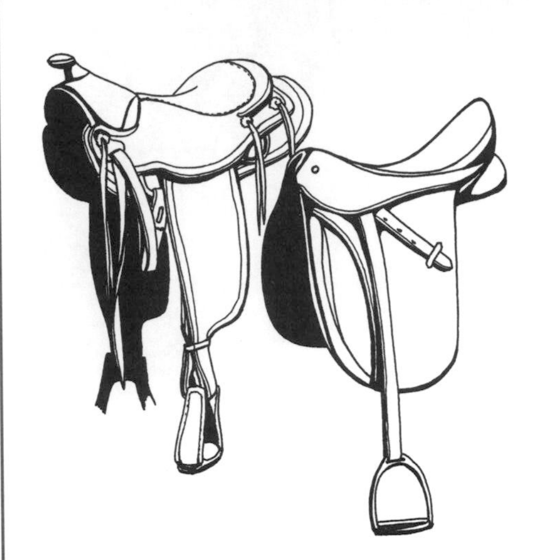

A Western stirrup hangs from a fender while an English stirrup hangs from a stirrup leather.

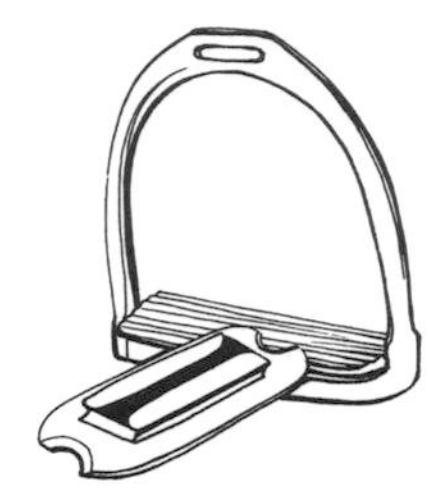

Stirrup pads can be removed if you are riding in rubber-soled shoes.

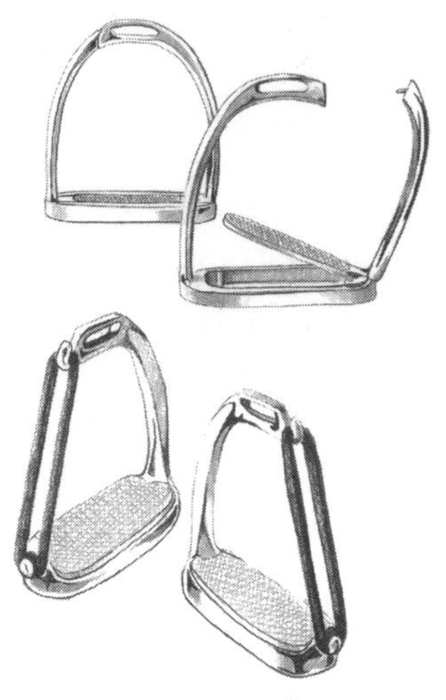

Many stirrups have safety catches that break in a fall.

Stitches

Not all serious wounds should be stitched. Some must be left open to allow drainage and prevent infection. Deep muscle wounds and injuries to joints or tendons are generally left to heal from the inside out. If you think a wound may need to be sutured, do not medicate the horse in any way until your veterinarian arrives. Do clean out any debris with running water and contain the bleeding with pressure.

Stitching must take place on a relatively fresh wound where there is a good supply of blood to the tissues. It should occur within 2 hours for a head wound, 6 hours for a leg wound, and 8 hours for a wound of the upper body.

➢ *See also First aid*

Stock horse

Traditionally refers to horses used to work livestock. The term can also refer to horses that compete in a division known as "stock horse classes." These classes are very similar to reining; however, they focus less on the smoothness and precision of maneuvers and more on the flashiness and style with which they are performed.

➢ *See also Western horse*

Stockings

Leg markings that go above the knee and/or hock.

➢ *See also Leg markings*

Stock saddle

A type of Western saddle traditionally used by cowboys and livestock ranch hands. It has a longer horn that allows for dallying ropes, when roping livestock, and a deeper seat that provides more stability and comfort for the long days of hard work in the saddle. The stirrups are usually much narrower than those found on the show styles of Western saddle, to enhance ease and speed of dismount.

➢ *See also Saddle*

Stock seat

The basic seat used for Western riding. Stock seat equitation refers to equitation of Western riders.

➢ *See also Seat*

Stock (tie)

Part of formal hunt or show attire, a stock is a narrow white scarf that wraps around the throat, rather like an ascot, and is often fastened with a small pin. Stocks were originally worn by hunters as a convenient way of carrying an extra bandage in case of injury.

➢ *See also Attire; Ratcatcher*

Stocks (as restraints)

Narrow stalls that prevent a horse from moving around while being bathed, shod, examined by a veterinarian, or undergoing any other procedure to which he might object. They come in several designs, allowing different ranges of movement and access to the animal. Some have solid panels at the front and rear to protect people from getting kicked.

➢ *See also Restraints*

Stock up

Become swollen. A horse's legs might "stock up" if he stands for a long time on a hard surface, if he kicks violently at something, or if he otherwise strains the legs.

Stomach

Horses have only one stomach, unlike cows, which have four. Horses chew their food only once. They are not only unable to regurgitate cud for chewing, but also they cannot vomit, making them vulnerable to digestive problems.

➢ *See also Colic; Digestion; Feeding and nutrition*

When the steed is stolen, shut the stable door.

—John Heywood, *Proverbs*

A stock tie worn with a ratcatcher shirt. A bow tie of the same fabric as the shirt may also be worn.

Stomach tube

A plastic tube about 8 to 10 feet long that is fed through the nostril to the back of the throat and down the esophagus to the stomach. The veterinarian will usually put a stomach tube in place so you can feed a foal that is not nursing. This way, you won't have to use a bottle, and you don't run the risk of his bonding with you as his mother. If the mare is alive and available, he can still nurse. If the situation goes back to normal, the tube can be removed.

Stomach tubes can also be used to introduce fluids in a horse that is dehydrated, or to administer mineral oil in a colicky horse.

➢ *See also Foal care*

Stopping or quitting at a fence can send the rider flying, although this one recovers her balance quickly.

Stop

Whatever cue is used, all horses should know how to stop on command while in-hand as well as while being ridden or driven. This simple concept could save lives, yet many horses and horsemen never learn full respect for its importance.

Stopping in front of a jump (quitting)

For many reasons, having a horse stop at a jump rather than go over it is undesirable. The most obvious reason is that the rider frequently continues in the original direction and winds up on the ground. In a show, a refusal is always penalized, and three in a row disqualifies the rider. Horses quit for a variety of reasons, including a real fear of jumping. But more frequently, your horse is either unenthusiastic or is sensing a lack of confidence or direction from you.

If the horse quits or runs out on a fence, it is imperative that you make him go over it immediately. A sharp smack right after a refusal can convey your displeasure. If you can, back him up a few strides rather than turning him away from the offending jump and trot him at it again, using plenty of leg to keep the pace. If he runs around a jump, turn him back toward it in the opposite direction and send him on again, paying particular attention to your leg and making sure your approach is straight.

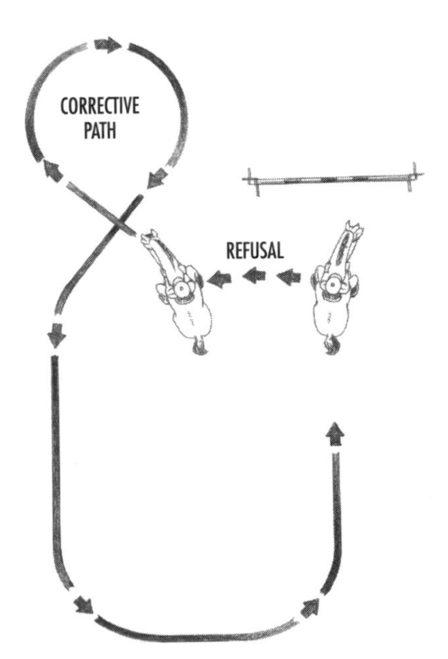

If your horse avoids the jump by swerving to the left of it, turn him back to the right and approach it again.

Storage (of bedding, equipment, feed, tack)

➢ *See Barn; Facilities; Hay; Tack room*

Straight bit

A bit with no joint or projections, just a bar. It is vital to fit this type of bit correctly because it is rather unforgiving. Work with a professional well-versed in proper fitting of bits if you are unfamiliar with the concepts of considering a horse's tongue size, palate depth, and bar width. (So-called professionals who are unfamiliar with these terms should be avoided since they won't likely know what is best for your horse's mouth.)

➢ *See also Bit; Bridle; Tack*

Straight-pull trailer

➢ *See Tagalong*

Straight stall (tie stall)

A narrow stall with two walls and a manger. Horses are tied while in straight stalls and cannot turn around or lie down.

➢ *See also Stall*

Strangles

An acute disease caused by bacteria spread through food or drinking water that has been contaminated by nasal discharge of an affected horse; it can also spread through direct contact with nasal discharge via nose-to-nose contact or through discharge left on surfaces. The most obvious signs of the disease are fever, abscess formation around lymph glands (most commonly under the jaw), and raspy breathing. Horses may appear dull, lethargic, lose their appetite, and have profuse nasal discharge before swellings appear.

Treatment is time-consuming but fairly simple. It is best to work with a veterinarian so the spread of the disease can be controlled. There is a vaccine available for prevention. However, be sure to weigh the risks of the vaccine versus the potential for exposure to the disease.

➢ *See also* Streptococcus equi

Straw bedding

Straw makes perfectly acceptable bedding but is becoming less popular because it causes colic in horses that eat it. Also, it is not as absorbent as other types of bedding, and you must use a lot of it to properly bed a stall.

➢ *See also Bedding; Mucking out stalls; Stall*

Strawberry roan

A mixture of red or brown and white hairs, giving an overall reddish color to the coat.

➢ *See also Coloring; Roan coloring*

Streptococcus equi

The bacterium that causes strangles, or distemper.

➢ *See also Strangles*

Stress

Stress affects horses in much the same way it affects humans; by depressing the immune system and leaving the animal open to infection and illness. Horses are particularly susceptible to respiratory ailments when stressed. Vaccinations should not be given during times of stress, such as weaning, training, or transporting, because the horse will not build up peak immunity.

Sound management and handling of horses requires that humans expand their awareness of the effects of stress on horses. This requires an understanding of how individuals vary because conditions that prompt

Go anywhere in England where there are natural, wholesome, contented, and really nice English people; and what do you always find? That the stables are the real center of the household.

—George Bernard Shaw, *Heartbreak House*

stress in one horse may actually serve as life-enhancing conditions for other horses. (Consider how some humans thrive on lives filled with complex demands, while others need simple, relaxed lives in order to thrive.)

> **STRESS BUILDERS**
>
> Horses can be stressed by many different factors, including sudden changes in their routine or surroundings, poor conditions, inadequate nutrition, lack of water, strange noise, pain, fear, toxic or allergic reaction, hard work, and exhaustion.

Stride

One measure of a gait; when each foot has hit the ground once.

Strike

A pawing motion in which the horse brings a foreleg up and out, usually with the intention of hitting something or someone with it. Though a horse usually strikes in response to fear or pain, some horses do so maliciously. A horse that strikes habitually needs to be handled by a professional trainer.

Stripe

A narrow, straight or slightly wavy line of white hairs running all or partway from forehead to muzzle. Some horses have a star, a stripe and a snip; when joined, they often look like a lightning bolt.

➢ *See also Face markings*

Strongid C

➢ *See Pyrantel tartrate (Strongid C)*

Strongyles

A type of blood-sucking worm found in either the bloodstream or the intestines.

➢ *See also Deworming; Large strongyles; Parasites; Small strongyles*

Strung out

A horse is said to be "strung out" if he is not moving in a collected fashion; that is, when his back legs are not squarely under him and look as though they're dragging behind somewhat.

It is fairly easy to determine whether a horse is strung out. Simply check to see if the horse is stepping its hind feet near, or into, the hoofprints made by the front hooves. When observed from the side, a horse that is strung out seems to be stepping its hind feet barely under its body; whereas a horse working "in frame" steps well up under its body.

➢ *See also Collection; Impulsion*

Stud

A stallion used for breeding purposes.

➢ *See also Breeding; Stallion; Standing at stud*

Studbook

Nearly every registry of horses keeps records known as studbooks that record vital information on each horse registered by that organization.

➢ *See also Pedigree*

Stud colt

An intact (un-neutered) male horse less than four years old.

Stud farm

A facility with a stallion or stallions standing at stud to which mares are sent to be bred. A mare can either stay at the farm until the birth of the foal or return to her own barn for the pregnancy and birth.

➢ *See also Breeding*

Stud fee (breeding fee)

The charge for breeding a mare to a particular stallion. An owner may charge a lower fee for a young stallion or even offer free breedings to get some foals on the ground. As the stallion's reputation becomes established, the fee generally goes up.

Stump-sucking

A regional term (from the southern and Appalachian regions of the United States) for cribbing or wind-sucking.

➢ *See also Cribbing*

Subcutaneous injections

➢ *See Injection*

Sucking wind

This terminology often leads to confusion because regional use is not consistent. In some parts of the United States, "sucking wind" refers to a mare who has a defect of the vulva that leads to air being sucked into her vaginal tract. Such mares are sometimes known as "wind suckers." In other parts of the country, the term refers to horses that put their upper teeth onto a solid object and suck wind into their stomachs, usually to combat boredom or possibly because of a nutritional deficiency. Because the term is used differently in some regions, it's best to ask what is meant when someone uses it.

Suckling

An unweaned foal.

Sudan grass (*Sorghum* spp.)

After a drought or frost, the glycosides in this plant change to cyanide, making it very dangerous to horses. The plant is common in the southwestern and southeastern United States.

➢ *See also Poisonous plants; Sorghum grass*

Sulky

A light, two-wheeled cart pulled by a single horse.

*Sometimes he trots,
as if he told the steps,
With gentle majesty and
modest pride;
Anon he rears upright,
curvets and leaps,
As who should say, "Lo,
thus my strength is tried;
And this I do to
captivate the eye
Of the fair breeder that is
standing by."*

—William Shakespeare,
Venus and Adonis

A pony pulling a sulky

Sull

To move forward reluctantly or slowly.

Summer care

➢ *See Seasonal care*

Summer sores (habronemiasis)

Although summer sores are an external problem, they are caused by the larvae of a stomach worm deposited by horse and stable flies that are feeding on skin wounds (often found on the lower legs) or moist areas, such as the corner of the eye or the sheath. The flies become infected by feeding on manure containing the stomach worm larvae, which create constant sores by trying to develop on the skin rather than the stomach.

Summer sores are marked by a refusal to heal; instead the lesion often grows bigger as time passes. Wounds, which look circular after a few weeks, will usually ooze fluid and be very itchy. If untreated, the wound may heal when cold weather arrives, but the larvae can live for up to two years in the skin and sores will reappear in the spring. Treatment with ivermectin will kill most of the larvae. Wounds should be kept clean and protected from flies to prevent summer sores.

A horse never knows who may buy him, or who may drive him; it is all a chance for us, but still I say, do your best wherever it is, and keep up your good name.

—Anna Sewall, *Black Beauty*

Sunburn

Although most horses are protected against the sun by their coats, light-colored horses or dark horses with white facial markings, especially around the mouth and eyes, are susceptible to burning in strong sunlight. Keeping the horse indoors on bright days or applying a human sunscreen lotion or thick ointment (like zinc oxide) around the eyes and muzzle can help prevent burning.

➢ *See also Melanoma; Photosensitization; Seasonal care; Skin problems*

Supplements

All that most horses need for a balanced diet is grass or hay, water, and salt. Technically, grain is a supplement, but most people use the term to refer to vitamins and minerals that are added to the grain. Be very careful when adding things to your horse's diet. Supplements are generally not necessary unless the horse is working extremely hard (like a race or show horse), has a health problem, is pregnant, or is not receiving sufficient nutrients from its basic diet of hay and/or grain.

Some supplements are toxic in large quantities, so read dosages carefully. Others are fattening, and fat horses are prone to performance and soundness problems. Overfeeding supplements to foals can create permanent skeletal malformation. Always determine the need for supplements on an individual basis. Ask your veterinarian for advice if you're unsure.

➢ *See also Feeding and nutrition*

Suppling

Improving flexibility, the foundation for balance. A stiff horse is usually resistant and unable to perform in a relaxed soft frame — a frame that

is as important to relaxed trail riding as it is to high level competition. Suppling, therefore, should become part of every ride. A variety of exercises can help your horse become more flexible and improve his ability to use his body effectively. Exercises can range from those done before mounting — for example, having your horse reach between his front legs to take a treat from your hand, which supples the neck and back muscles and spinal joints — to lateral work done under saddle.

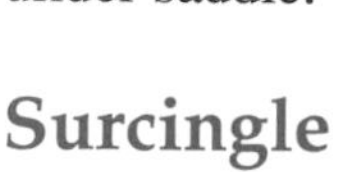

Surcingle

A term referring to two pieces of equipment that are similar but not identical. The first is a strap placed around the barrel of a horse to keep its blanket in place. It is commonly made of either leather or nylon. The second is a piece of training equipment that looks similar to the part of a harness that goes around the barrel of the horse. It is usually made of leather or nylon, has buckles similar to those used on English girths and metal rings in a variety of locations. This enables side reins to be attached to the surcingle and bit or driving lines to be run from the bit, through a ring on the surcingle to the driver's hands, to aid in ground driving training. Rings on the top of the surcingle can be used for attaching an overcheck and the crupper, which runs along the croup and under the tail. The crupper prevents the surcingle from being pulled up over the horse's withers.

Suspensory ligament

Ligaments connect bone to bone; tendons connect muscle to bone. The suspensory ligament is found in all four legs. It essentially connects the bones of the knee to the bones of the fetlock in the front leg, and the bones of the hock to the bones of the fetlock in the hind leg. Damage to this ligament can cause a lifetime of lameness problems.

➢ *See also Leg, of horse; Muscling and conformation*

Sutures

➢ *See Stitches*

Swamp fever

➢ *See Equine infectious anemia (EIA)*

Swamp maple (red maple) *(Acer rubrum)*

➢ *See Poisonous plants; Red maple*

Sweating

Horses sweat to dissipate heat and cool their bodies. They will sweat during exercise or on a very hot day, just as humans do. Sweating while at rest in cool weather can be a sign of illness. A lack of sweat after exertion or on an extremely hot day is a serious concern.

➢ *See also Signs of illness*

SWAY BACKED

Having a discernable sag to the backbone.

Swedish Warmblood

Known for brilliant performance in the dressage ring, the Swedish Warmblood also shines in jumping, driving, and other competition. This tall, elegant horse with excellent gaits and willingness to learn has won medals in nearly every Olympiad since 1912.

➢ *See also Warmblood*

Sweepstakes

A class or race that offers prize money to a certain number of winners, based on a previously determined scale. The money comes from entry fees as well as an additional "purse," which is sometimes donated by an individual or corporate sponsor.

Sweet feed

A mixture of grains and pellets held together with molasses, which cuts down on dust. Horses tend to love sweet feed because of the molasses, but it spoils quickly if not kept in a cool, dry place. Dry molasses still cuts dust but is less perishable. Different sweet feeds have different amounts of protein and fat in them, so check carefully to find a brand that offers the right nutrition for your horse.

➢ *See also Feeding and nutrition; Grains; Supplements*

Swelling

Swelling in horses is essentially the same physiological reaction as seen in humans and other mammals. It should be viewed as a cause for concern because it signals an inflammatory response that usually only occurs when tissues have been damaged. Although treatment of swelling is similar in horses and in humans, it is best to have a veterinarian examine your horse to determine the cause of the swelling. Treatment of the symptomatic swelling without treatment of its cause can lead to complications.

➢ *See also First aid; Lameness*

Swells

The front part of a Western saddle, under the horn, which is also called the fork.

Swingletree (single tree)

Also called a **whippletree** or **wiffletree,** this is the part of the harness that connects the traces to the cart or plow. There can be single or a double swingletree, depending on how many horses are being hitched.

➢ *See also Harness, draft; Harness, light-driving*

One white foot,
buy a horse;
Two white feet,
try a horse;
Three white feet, look
well about him;
Four white feet,
do without him.

—Anonymous

Synthetic saddle

Saddles made of manmade materials can be just as comfortable, sturdy, and well-constructed as saddles made of leather. They are easy to care for (no saddle soap needed!), lightweight, and can get wet without damage. When riding in a synthetic saddle, make sure to check the girth after mounting, as some types will compress with a rider on board.

➢ *See also Saddle; Tack*

t

Tack

The gear that people put on horses to control them — from halters to harnesses. Whole books have been written about tack. When mankind first tamed the horse thousands of years ago, the only equipment was a leather thong wrapped around the horse's jaw. Native Americans rode that way for centuries. The basic bridle as we know it was developed by the Egyptians more than 400 years ago. Today, a bewildering number of options are available to the rider.

The two basic styles of riding, English and Western, use quite different saddles and sometimes very different bridles. For either style, the average rider will use the following: a bridle with reins and a bit (or a bridle without a bit, called a hackamore); a saddle pad; and a saddle with a cinch or girth, stirrups, and stirrup leathers. Some horses also need a martingale, which offers further control of the head.

When shopping for tack, make sure it fits your horse and fits you. There are options for all budgets, from finding a good, used saddle in your local tack shop to custom-ordering a fancy dressage saddle. Well-made used tack is often a better bargain than new, cheap equipment that might not last as well. Don't be shy about seeking help when purchasing tack — it's a big investment and is worth doing right.

➢ *See also Bit; Bridle; Cinch; Hackamore; Halter; Harness, draft; Harness; Martingale; Saddle; Saddle blankets and pads; Synthetic saddle; Western tack*

Tack, cleaning

To get the longest life from your tack, you must clean it regularly. Even tack that is not used frequently should be cleaned once a week or the leather will dry out. It's a good idea to rinse off your bit after each use, as well. It's convenient to have a place, especially for cleaning tack, with a stand to hold the saddle and with hooks for bridles and harnesses while you work on them.

The first step is to wash off accumulated dirt and sweat with plain water and a clean sponge. You can add a little ammonia to the water to cut grime. Take your bridle apart and rub the sponge over all the pieces, rinsing the sponge often and changing the water as it gets dirty. The sponge should be damp, not dripping; excess water will ruin leather and weaken stitching. Do the same to your saddle, first removing the stirrup leathers and taking off the stirrups.

The next step is to protect the leather with saddle soap or leather conditioner. The standard products are Lexol or Murphy's Oil Soap and glycerin, which comes in bars. Dampen a clean, dry sponge with the product and rub it on the bar of glycerin to get a dry lather. Rub it into your tack to get a nice shine.

If your saddle gets soaked in the rain or going across a stream, let it dry naturally before using it again. It may take a day or two, but don't put it in the sun or near a heat source. After it dries, clean it normally and oil it lightly to restore suppleness.

Now shift the blanket pad before your saddle back you fling, And draw your cinch up tighter till the sweat drops from the ring. We've a dozen miles to cover ere we reach the next divide. Our limbs are stiffer now than when we first set out to ride. And worse, the horses know it, and feel the leg-grip tire, Since the days when, long ago, we sought the old camp-fire.

—Bret Harte, *The Old Campfire*

Very old, dried-out tack can be reconditioned by soaking it in neat's-foot oil (mink oil, baby oil, or olive oil also work) for an hour or so. But too much oil will just drip all over you and your horse and can rot the stitching. After the oil is absorbed, wipe off any excess and let the leather dry. Repeat the process the next day if necessary, or use a good saddle soap to finish the process.

To clean a synthetic saddle, simply wipe it off with a damp sponge. If it gets very dirty, use a scrub brush. A mild (20%) ammonia solution can be used to cut dirt or mildew, but as with a leather saddle, avoid soaking the material. Synthetic saddles should be stored under the same conditions as leather ones, that is, on a proper saddle tree in a room that isn't too damp or dusty.

This cleverly designed stand holds an English saddle in two different positions for convenient cleaning.

Tacking up

The process of getting equipment on a horse.

Saddling a Horse

After you've groomed your horse, the saddle goes on first. Place the saddle pad on the horse's back, well up on the withers, creating a channel along the spine to allow air to circulate. Check that the pad lies equally on either side before lifting the saddle gently into place. Slide the saddle and pad back and adjust the pad if it's wrinkled. Make sure the hair under the saddle pad is smooth.

For an English saddle, fasten the girth on one side and then bring it under the belly to attach it to the other side (usually right side first, then left). (If your horse needs a martingale, put that on first, so that you can run the girth through the loop and only fasten your girth once.) Don't tighten the girth too much at first; do it just enough to keep the saddle in place. Watch the horse's ears; some horses object to this part of the process.

For a Western saddle, adjust the blanket the same way as above and carefully swing the saddle into place. The stirrup and cinch should be looped over the horn to get them out of the way. If you tack up from the off side, you only have to lift the latigo strap out of the way along with the stirrup instead of the entire cinch. Western saddles are heavy, so be careful you don't slam the horse's back and hurt him. Slide the saddle back from the withers until it rests in a natural spot behind the shoulders.

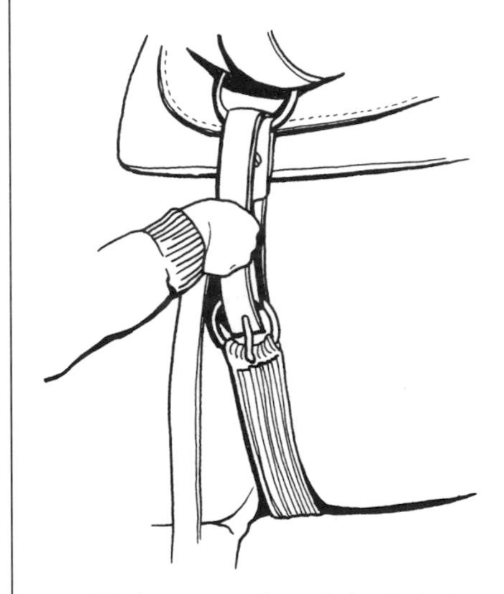

The latigo goes through the cinch ring twice before being buckled or tied.

Now bring the cinch forward and loop the latigo through the cinch ring, up to the rigging and back again before securing it in the manner you've been taught.

Bridling a Horse

Once you've got the saddle in place, put your bridle in a handy spot (hanging over your shoulder, for example). If your horse is in his stall, proceed with bridling him, but if he's tied, make sure he stays put by unbuckling his halter, slipping it off his nose, and rebuckling it around his neck so that you'll still have something to grab onto if he takes off.

To get a bridle on a horse, stand just behind his head on the left side. Holding the crownpiece in your right hand, pull it toward the ears, while you guide the bit into his mouth with your left hand. If he doesn't accept the bit easily, slide your thumb into his mouth at the corners (bars) where there aren't any teeth. Be gentle with the bit — how would you like to have a heavy piece of metal banging against your teeth? Once the bit is in place, your left hand is free to finish pulling the crownpiece over the ears. Fold the ears forward rather than pinning them back and then having to bend them under the crownpiece.

Make sure all the straps are laying flat and comfortably against his face. Now fasten the throatlatch, leaving room for four fingers to slide through, and the noseband, which should be fairly snug without pinching. If there's a standing martingale, don't forget to slip the loop onto the noseband before buckling. Unless you're leading your horse, put the reins over his head.

If you have to tie a saddled and bridled horse, put a halter on him and run the reins under the stirrup irons to keep them from falling over his head. *Never* tie a horse by the bridle!

Tack room

Most barns or stables have a separate room for storing tack. It's important to store tack off the floor away from nibbling rodents and to keep it out of damp and dusty conditions. A well-designed tack room has enough room for each rider to hang her saddle and bridle and store a trunk for grooming equipment, bandages, blankets, and other gear.

In the tack room it helps to have each saddle rack and bridle peg labeled with the horse's name.

TACKING UP: BRIDLING A HORSE

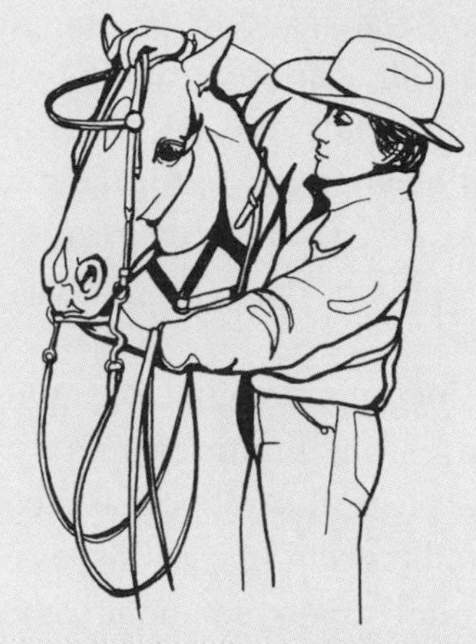

Stand next to your horse on his left side, not in front of him, where a tossing head could give you a bloody nose.

If your horse's head is too high for you to reach, you can hold the cheek pieces together across his nose while you guide the bit into place. Then both hands are free to finish the job.

Tagalong (straight-pull) trailer

This tagalong is correctly hitched to the truck. When the trailer is loaded, the trailer floor and the hitch should be level.

A trailer, usually designed to carry two horses, that attaches to a hitch that is welded to the frame of the towing vehicle. *Caution:* These trailers should never be attached to the bumper of a truck or SUV, even if they are referred to by the manufacturer as "bumper-pull" trailers. The truck's weight, towing capacity, and hitch must be appropriate for the trailer.

➢ *See also Gooseneck trailer; Trailering a horse*

Tail

Blocking, docking, wringing, and banging: there's a large vocabulary connected with this seemingly simple fly swisher. In addition to chasing away pests, the tail acts as a communication device. If your horse **wrings** his tail (swishing it in circles or moving it rapidly up and down), he's definitely annoyed about something. When mounted, keep your horse away from the heels of a tail-wringer or you could be inviting trouble. In the show ring, tail wringing can mean points off your total score in some classes.

Blocking is an unethical attempt to prevent tail wringing by partially paralyzing the nerves of the tail by injection. The effect lasts for several weeks, and although the practice has been banned by all major show associations, it still happens. Repeated nerve blocking can create permanent damage, preventing the horse from using his tail effectively. It can even interfere with normal defecation and urination. When purchasing a horse, make sure his tail hangs and moves normally; if you have any questions, ask your veterinarian to examine the tail during the prepurchase exam.

Another practice banned by some show organizations is **setting** a tail by cutting the muscles on either side of it and putting it in a tail set to

A simple plait started at the end of the tail bone and fastened with a child's double-ball ponytail holder will keep your horse tidy during turnout. The longer you make the braid, the shorter the tail gets, so during fly season, leave enough of a "switch" so that he can still use his tail.

TYPES OF TAILS

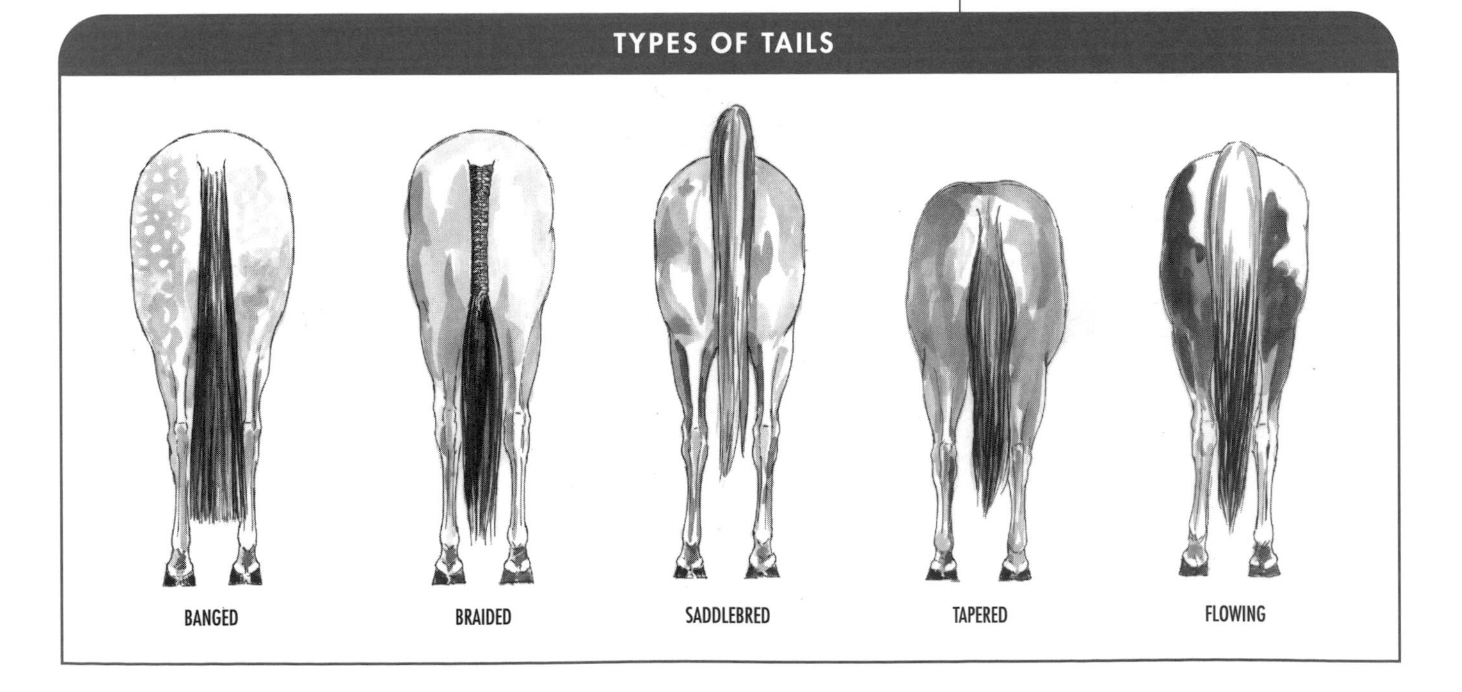

make a high arch. This used to be common on the Saddlebred circuit, but it is becoming less popular. Although artificial, the tail set doesn't permanently damage the tail in the way that blocking can. But if the tail isn't kept in the set, it will flop over.

Even more drastic than setting, the once common practice of **docking** the tails of carriage and draft horses has become less frequent, as well. This operation actually removes much of the tail, including a portion of the tailbone (or dock), and leaves the horse without any means of swishing flies. It may be more convenient to harness and drive a horse with a short tail and it's certainly easier to keep a docked tail clean, but it's pretty hard on the horse.

Except for brushing and the occasional shampoo, most horse's tails are just left alone. But show horses have all kinds of options. Dressage horses sport **banged** tails, neatly trimmed straight across at the fetlock, while hunters have a tidy braid down the length of the dock.

➢ *See also Anatomy of the horse; Tail, care of; Tail rubbing; Tail wrapping*

Tail, braiding

For competition purposes, horse and rider must present a certain appearance, depending on the class. For many classes, the mane and forelock are braided while the tail is left loose, but for many hunter competitions, the tail must also be braided. Never braid the tail without also braiding the mane and forelock.

HUNTER BRAID

Starting with damp hair, separate three narrow strands from the sides of the tail and begin to braid.

Continue braiding, bringing new strands in from the sides at each step. When you reach the end of the dock, stop adding new strands and just finish off the braid with the remaining hair.

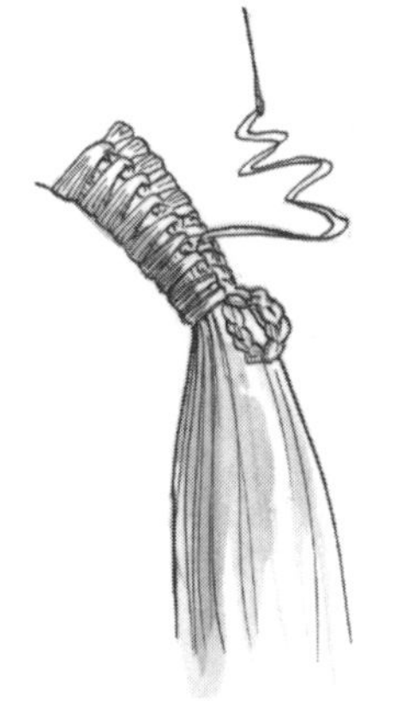

Fasten a small rubber band at the end and use a needle and heavy thread to pull the end of the braid under the "wrapped" section, bringing the needle out a few inches above the resulting loop. Tie off and cut the thread.

SMOOTHING UNRULY HAIRS

If your horse's tail head tends to get fuzzy and messy-looking, you can train the top hairs to lie smooth and stay neatly braided at a show. Resist the temptation to snip hairs that stick out of the braid, because then they will never grow. Instead, dampen the tail head with water or a tiny bit of conditioner, comb the hair smooth, and make a French braid for about 8 inches. Leave it this way for one or two days, if it is not braided too tightly, and then take it out and dampen and smooth the hairs again.

Tail braiding

Tail, care of

Just like human hair, horse tails get tangled and snarled if not regularly cared for. Left alone over a season of turnout, the tail can become pretty solidly matted with burrs, brush, and debris. Dealing with the mess will take hours. Start by finger combing the worst of the snarls and picking out all plant matter, with the aid of a commercial detangler. Trying to shampoo the tail at this stage would just produce a clean, wet, thoroughly tangled tail.

Once you can run your fingers through the tail hairs, you can actually wash the tail. After you get the tail completely wet, add some diluted shampoo directly to the tail head, which is mainly what needs to be cleaned. Scrub the shampoo into the dock with your fingertips, a procedure your horse will no doubt enjoy immensely because that's an itchy spot that is hard to reach. Before rinsing, part the hairs at the tail head to make sure you got all of the dirt and dandruff. Don't forget the skin under the dock (but be gentle). Work the lather down the skirt of the tail, taking care not to tangle the hairs.

Rinse the tail completely, leaving no residue to cause itching later on. Use a conditioner (some types can be left in, others need to be rinsed out) on the tail and comb it through with your fingers or just dunk the whole

BRAIDING A TAIL FOR THE WINTER

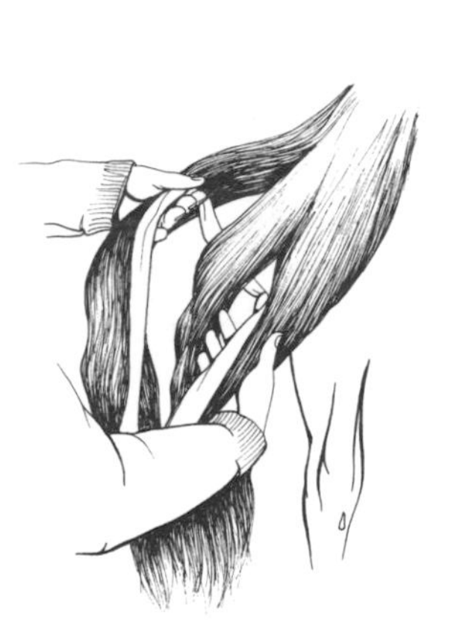

1. Start with a clean, dry, and well-brushed tail. Have ready a 10-foot length of gauze bandage (¼ to 1 inch wide), folded in the middle. At the *bottom* of the dock, divide the hair into three equal sections, with the midpoint of the gauze behind the middle section of hair.

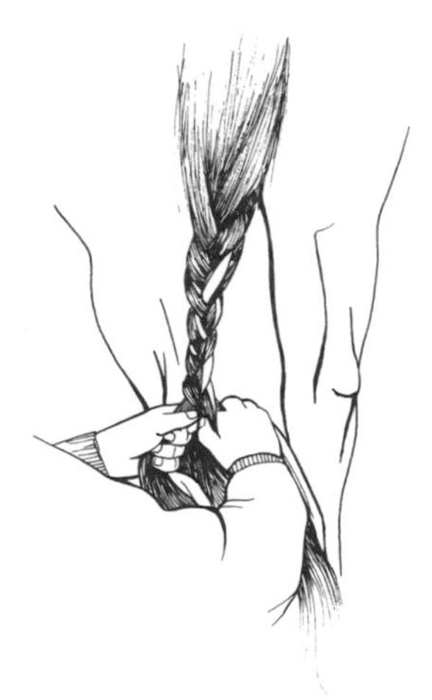

2. As you braid the tail, incorporate the bandage into the braid with the outer sections, stopping about a foot from the end of the tail and tying off the gauze with two overhand knots. There should be a foot or two of gauze left on each side.

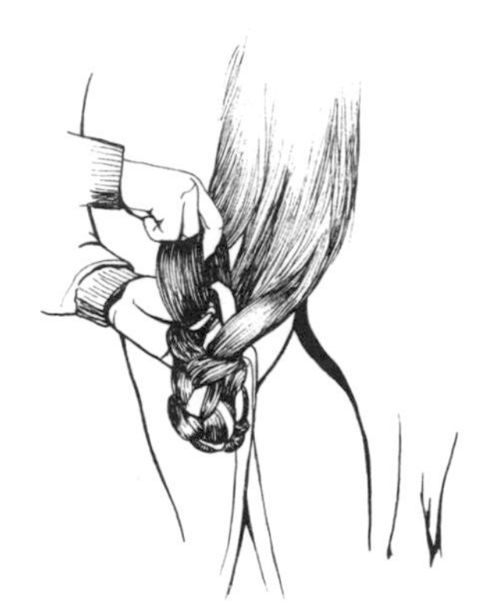

3. Now make a loop by poking the end of the braid between the hairs, just above where the braid starts, and pulling the unbraided end and the gauze strips through.

4. To finish the job, secure the loop in two places to close it up. You can leave the end tuft hanging loose or lay it over the braided part and tie it in two places.

tail into a bucket of conditioner solution. Strip most of the water out of the tail by hand and then use dry towels to remove as much of the remaining moisture as you can. Squeeze with the towels — don't rub! When the tail is almost dry (you may have to let it air dry for a while), you can work your fingers through the tangles, again using a spray-in conditioner if necessary. Then brush from the bottom up, using a human hairbrush with widely spaced bristles. Only use a comb (and only a wide-toothed one) at the very end or you'll pull out and break off hairs.

To prevent a truly tangled tail from happening in the first place, you can braid your horse's tail for long periods of turnout, as long as you take it out and rebraid it at least twice a month. This discourages and keeps ice balls and frozen mud from breaking off hairs.

Tailing

When going up steep hills, you can help your horse by dismounting and holding onto his tail as you climb (don't pull; just let him tow you). Naturally, this practice assumes that your horse is polite enough to stop at the summit and let you get back on.

Tail rope

A rope that is tied to a mare's tail and brought forward to a neck strap. A tail rope is sometimes used during breeding to keep the tail out of the stallion's way, but more often the mare's tail is just wrapped in a bandage.

Tail rubbing

Horses can't reach the top of their tails with their teeth, so if they're itchy, they will rub against a fence, a stall wall, a tree, or even another horse to get relief. An itchy tail head can be caused by soap residue, dirt, lice, ticks, or pinworms. Horses also rub their tails when their udders or sheaths are itchy, since that's the closest they can get to the problem. Shedding also makes horses feel itchy. Once a horse develops the habit of rubbing his tail, he may just keep doing it.

This is a problem because rubbing breaks off hairs and eventually creates bald patches. Keeping the tail and udder or sheath clean (and carefully rinsing after a bath) and regular deworming generally discourage tail rubbing.

Tail wrapping

Owners wrap their horse's tails for a variety of reasons. When a mare is being bred, her tail is wrapped to keep it clean and out of the way of the stallion. Sometimes a tail is wrapped to train the top hairs to lie flat. Most often, tails are wrapped to protect them during trailering.

You can use a self-closing type of tail wrap with a rubberized interior that grips the hairs, or you can wrap the tail with a stretchy bandage. In both cases, be careful to use light, even pressure when fastening the wrap. A tight bandage cuts off circulation and can damage the dock, causing hair to fall out. Tail wraps should only be left on for an hour or so at a time.

Tail Wraps

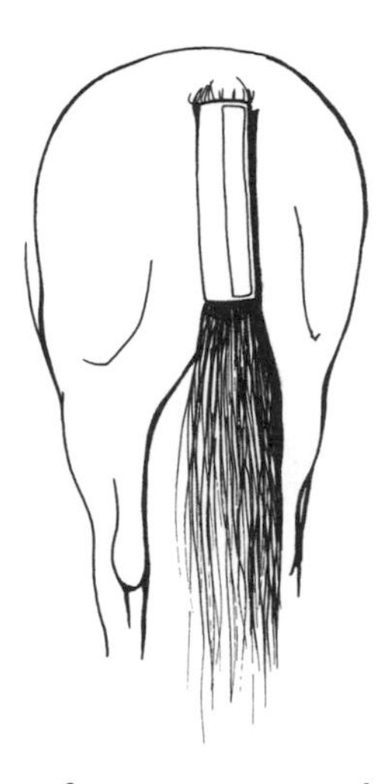

This type of wrap is convenient, but make sure that it reaches past the butt chain or it won't do the job. You can apply a bandage wrap to the lower part of the tail or braid it for further protection.

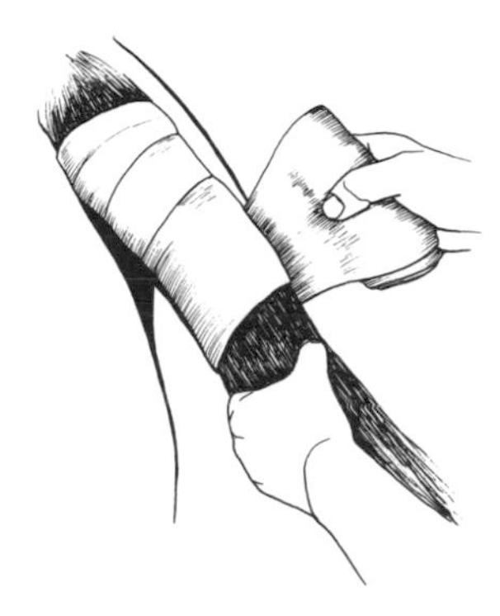

Starting at the very top of the tail (and being careful not to get kicked), wrap the bandage around a couple of times and then wind it down, using even tension. The bandage should cover the dock before being tied off. It must be tight enough to stay in place during the trip, but should slide right off with a gentle tug.

Talk, of horses

➢ *See Body language, of horses; Vocalizing, of horse, Whinnying*

Talking to horses

Keeping up a flow of conversation around a horse is a good idea. Your voice communicates your state of mind, lets the horse know where you are as you move around him in the stall or approach him in the paddock, and tells him if his behavior is acceptable ("What a great guy!") or not ("Cut that out!"). Your voice can reassure a timid or nervous horse that the terrifying tractor you're approaching will not, in fact, eat him for dinner. Conversely, if you're frightened yourself, maybe you'd better keep quiet!

Horses are capable of learning a number of words, as many an experienced school horse proves by responding to the instructor's verbal commands to the students rather than waiting for the aids. Having a vocabulary makes your horse much easier to handle, as he can respond to your voice when you might not be right next to him. For example, longeing a horse that understands the basic commands is a lot easier than working with one that hasn't made that connection. Adding "whoa," "trot," or "walk on" to your physical aids reinforces the concept and eventually allows you to use more subtle cues. The goal is for the rider's use of voice to fade out as mastery of the other natural aids improves.

Tall fescue

➢ *See Fescue grass; Poisonous plants*

Tamed iodine (Betadine)

A mild form of tincture of iodine that is effective in treating girth itch and other fungal infections.

➢ *See also Iodine, antiseptic; Skin problems*

Tapadero

On a Western saddle, a leather hood that attaches to the front of the stirrup and protects the foot against brush and thorns.

➢ *See also Stirrups; Western tack*

Tapeworms

Though not common in horses, tapeworms can inhabit the intestines and rob horses of nutrients. Tapeworms can cause colic if present in sufficient numbers to create a blockage in the cecum. Tapeworms are more prevalent in warmer climates, where their life cycle depends on orbatid mites, which take on the immature tapeworms that hatch in manure. Horses pick up the mites by grazing.

Tapeworms are resistant to many dewormers. Pyrantel pamoate given at high doses will kill them, as will the canine dewormer praziquantel (Droncit).

➢ *See also Deworming; Parasites*

THE TALLEST HORSE

According to *The Guiness Book of World Record*, the tallest horse on record was Samson, later renamed Mammoth, a Shire born in 1846 who stood 21.2 hands high. That's just over 7 feet at the withers!

THE TARPAN

The Tarpan was a wild primitive horse that roamed through Eastern Europe and the Ukraine until the late 1800s and influenced the bloodlines of many light horse breeds. In the 20th century breeders worked to bring back the Tarpan line, through its descendants, and now a semiwild herd lives in Poland.

Tartar buildup

Horses on pasture rarely develop tartar, as the roughage cleans the teeth and gums. Finely ground foods like pellets and mashes are more likely to create problems, most commonly on the canine teeth.

➢ *See also Dental care; Teeth*

Tarweed (*Amsinckia* spp.)

A plant commonly found in grain crops in the western and midwestern United States that can cause liver damage even in small amounts.

➢ *See also Poisonous plants*

TDN

➢ *See Total digestible nutrients (TDN)*

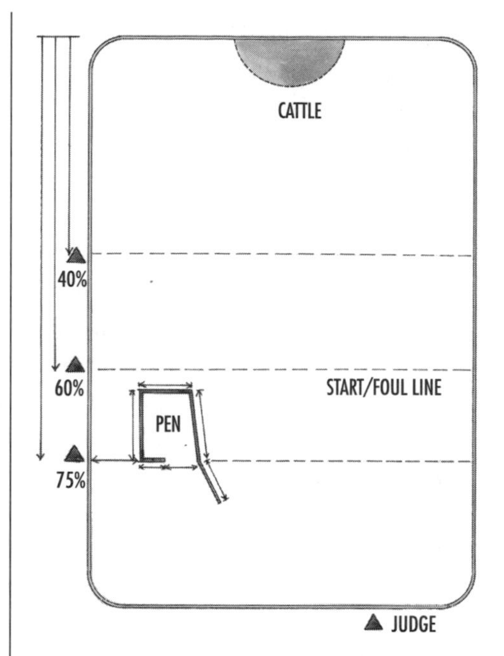

Team penning layout (U.S. Team Penning Association)

Team penning

This event pits a team of three riders against a herd of 30 often uncooperative cattle. The team must select out a designated trio of cows (marked with large numbers) and move them from one end of the arena to a pen at the other end. An experienced team on a lucky day can pen their cattle in about 30 seconds, but one balky steer can ruin everything.

The cattle are marked in threes with numbers from 0 to 9. Riders, called settlers, gather the herd after each round and keep it together while the next team gets ready to begin from behind the start line.

A penning horse needs to have good cow sense, quick reflexes, and speed. The riders must be alert to each others' whereabouts and have to communicate well. The team can call time if only one or two cows are in the pen. All undesignated cows must be in front of the start/foul line when time is called.

➢ *See also Western competition*

While breed associations allow a second throw, the U.S. Team Roping Championships do not.

Team roping

In this very popular Western competition, two riders (a header and a heeler), work together to rope a steer. The steer is released from a chute and gets a head start of between 10 and 25 feet. The riders start out simultaneously, but the header throws his rope first and pulls the steer around so that the heeler can rope the back legs. Once the heeler makes his catch, he "dallies" or wraps his rope around the saddle horn, and takes up the slack. In the meantime, the header turns his horse to face the steer, whereupon the timing flag is dropped.

Headers and heelers practice long hours on their technique to get it right every time on a moving target. A good horse can really help the rider by getting into the right position, rating the steer properly, and knowing when to back off or speed up to keep the rope taut.

➢ *See also Western competition*

Tearing eyes

Tearing eyes can indicate conjunctivitis or a blocked tear duct, or they can be a symptom of a more serious problem. Call your veterinarian.
➢ *See also Eyes*

Teasing

Teasing is a process used on breeding farms to aid in the detection of mares that are ready to breed (heat or estrus detection). A variety of methods are used to accomplish this, but most often mares are exposed to a stallion over a fencelike structure or teasing chute. This allows the handlers of both mare and stallion to remain safe, since striking, biting and kicking are common reactions of both stallion and mare during this process.

Some mares need to be teased at the time of breeding in order to make them more receptive to the process. This use of teasing is done in the same manner as teasing used for heat detection. Each mare responds differently to teasing and being bred, so it's important to be flexible in the approach used to both processes.
➢ *See also Breeding; Estrus*

Teasing wall

A solid wall about four feet high that separates a mare and stallion before mating. It allows the horses to indicate their level of interest without being able to injure one another.

Teeth

The expression "long in the tooth" comes from the age-old practice of checking a horse's teeth to tell his age. Equine teeth are made of three different materials that vary in hardness. The teeth wear away at varying rates over the years. To compensate for the constant grinding, the permanent teeth continue to erupt over the horse's lifetime. The amount and type of wear allows an experienced horseman to tell, within a year or so, what age a particular horse has reached.

While aging by the teeth is a complicated and inexact science, there are some basic facts and general guidelines that all horse owners should be familiar with. An adult horse has 36 teeth: 12 each of incisors, premolars, and molars. He may also have up to four wolf teeth, which generally erupt just in front of the premolars at about 5 or 6 months of age. The wolf teeth sometimes interfere with the bit, and many horse owners have them removed. Premolars are permanent teeth located in front of the molars. Between the premolars and the incisors is a section of gum called the interdental space. Male horses (and some mares) grow a set of four canine teeth in this space.

By nine months, a foal has all his deciduous, or baby, teeth (12 incisors and 12 premolars), and over the next three months will grow his first set of permanent molars. It takes four to five years for all the deciduous teeth to be replaced by the much larger permanent ones. By age 10, an

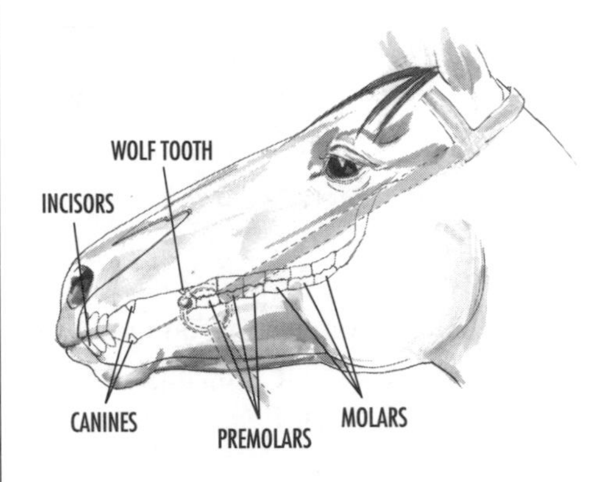

Teeth of a mature male. The wolf tooth can pinch the lip against the bit, making your horse resist your cues.

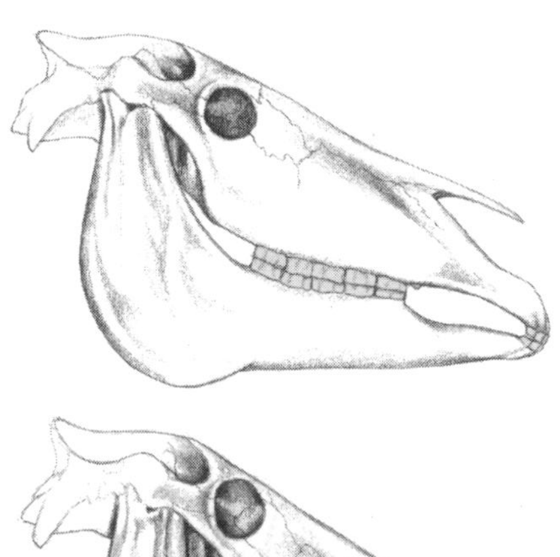

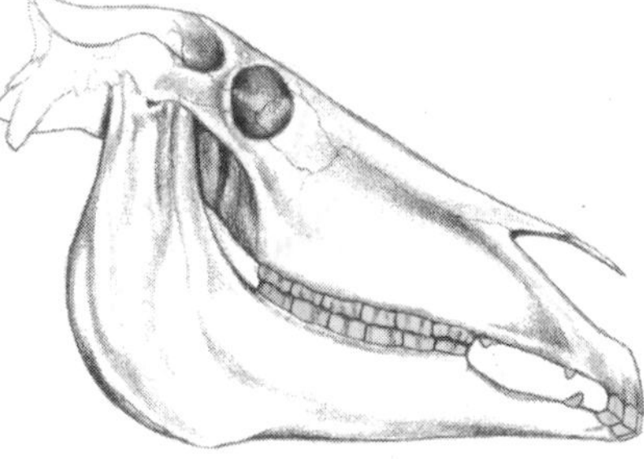

Deciduous vs. permanent teeth. Note the difference between the incisors of the 2-year-old filly *(top)* and the 20-year-old mare *(bottom)*. You can see Galvayne's groove extending over most of each tooth in the older mare.

indentation known as Galvayne's groove becomes noticeable in the incisors. As the horse ages this groove lengthens, and, eventually, as the teeth continue to erupt, reaches the lower edge of the tooth and is worn away. A very old horse may have only little stubs of teeth, which he may end up losing completely.

The upper jaw is approximately 30 percent wider than the lower jaw. Horses grind their food with a sideways action, so the teeth wear unevenly and eventually sharp ridges form on the outer edges of the upper molars and the inner edges of the lower molars. These ridges cut the inside of the mouth and make eating quite painful, so regular dental checks are important.

➢ *See also Age; Deciduous teeth; Dental care; Floating teeth; Galvayne's groove; Impaction; Overbite (parrot mouth); Shear mouth; Sow mouth; Wolf teeth*

Never look a gift horse in the mouth.

St. Jerome, *On the Epistle to the Ephesians*

Temperament, of horses

Horses are like people and most other animals in that different individuals have different personalities and characteristics. The temperament of an individual horse depends on his natural instincts, his particular genetic makeup, and his experience in the world.

Horses are genetically programmed to be herd animals, so they feel more comfortable in groups. A horse living with a herd of cattle is probably happier than one living alone in a pasture. This is why you often see a goat keeping a single horse company. Horses are also programmed to stay alert to potential danger and to run from threats. In general, they are curious, sociable, and very attuned to your emotional state.

Individual horses can be cheerful or crabby, bold or skittish, lively or dull, lazy or eager. Indifferent or abusive handling will exacerbate negative characteristics and subdue positive ones. Horses that are treated with respect, affection, and consistent firmness are more likely to be well-mannered and easy to handle.

➢ *See also Body language, of horses; Handling horses*

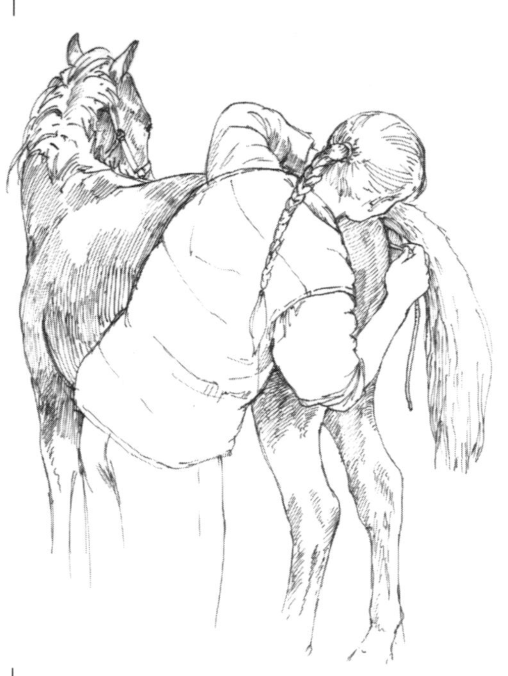

Stay close to your horse's hip when taking his temperature to prevent him from kicking. The thermometer should rest against the rectal wall; if it gets stuck in a fecal ball, remove it and try again.

Temperature, of horse

Along with pulse and respiration, temperature is a basic indicator of health, and you should be familiar with the process of taking your horse's temperature. A healthy horse has a normal temperature of 99° to 100.5°F, though it tends to be a little lower in the morning and a little higher in the afternoon or after exertion. Expect to see seasonal differences as well, with lower temperatures normal in the winter and higher ones in the summer.

Accustom your horse to having his temperature taken while he's in a good mood so that you don't have to introduce him to a rectal thermometer when he's already ill and might be feeling rather cranky. *Note:* Always tie a string to the thermometer (and a clamp to attach to the tail) before using it.

Begin with the horse tied or being held. Shake the thermometer down so it reads below 96°F. While standing close to his hip so he can't kick you, relax him by rubbing the top and sides of his tail — most horses love this because they can't reach that spot themselves. Gently move the tail aside and carefully insert the well-lubricated thermometer into the rectum, aiming it slightly upward. (Petroleum jelly and vaginal lubricants both work, but make sure any lubricant is at least at room temperature.) Twisting the thermometer as you insert it helps it go in more easily, but don't poke the sides of the rectum. Insert the thermometer to the last inch or so and leave it for three minutes for the most accurate reading.

➢ *See also Pulse and respiration; TPR*

Temperature, of leg tissues

Be familiar with your horse's legs when they are sound so that you can spot any signs of injury or lameness right away. Heat in the hoof and leg can be caused by infection, swelling, or overuse.

➢ *See also Leg, of horse; Lameness*

Tempi changes

Flying changes of lead after each stride or every few strides. The sequence and frequency of the flying lead changes is determined by the dressage test or by the choreographer to keep time with the music in a kur, or dressage Musical Freestyle.

➢ *See also Dressage; Flying lead change*

Tendons

Tendons attach muscles to bones, while ligaments connect bone to bone. Inflammation or injury to either causes pain and lameness. Tendons have a texture similar to rubber bands. Imagine a damaged rubber band if you want to picture what a damaged tendon is like. Just as a damaged rubber band will never regain its original elasticity, a damaged tendon will never be exactly the same.

➢ *See also Leg, of horse; Lameness*

Tennessee Walker, Tennessee Walking Horse

Like other gaited horses, the Tennessee Walker rarely trots. But in addition to a normal, flat walk and a smooth canter, this horse moves in an extended "running walk" that is famous for being comfortable and steady for the rider. The Tennessee Walker's stride is much longer than the American Saddlebred (whose origins he shares), with his front feet reaching way ahead of him and his hind feet landing as much as two feet ahead of where his front feet were with each stride. His head nods with each step and often his teeth clack as he goes along, covering 5 to 8 miles an hour.

Bred by plantation owners who spent many hours in the saddle, the Walker is known as an exceptionally gentle and pleasant horse, but not as flashy as his Saddlebred cousins. Walkers make wonderful trail horses. Medium to large in size, they come in all colors, including roan and true black.

➢ *See also Amble; Breed; Five-gaited horse; Gait; Gaited horse; Three-gaited horse*

TENNESSEE WALKER LINEAGE

The Tennessee Walking Horse, like the American Saddlebred, is descended from the Narragansett Pacer, a now-extinct American pacer developed in colonial Rhode Island. The foundation sires of the Tennessee Walker breed were Black Allan, a Standardbred stallion, and Roan Allan, his son.

TERMINO

A swinging motion of the horse's front leg from the shoulder that is characteristic of the Pervian Paso's gaits: flat walk, paso llano, and sobreandando.

Tetanus (lockjaw)

This fatal disease is particularly dangerous to humans and horses, and is much easier to prevent (by annual vaccination) than to treat. Tetanus is caused by a spore-forming bacterium called *Clostridium tetani,* which lives in the intestines of many animals and in their manure, as well as in soil rich in humus. Wounds, especially punctures, provide an entry for the toxin, and it spreads to the central nervous system after incubating for 1 to 3 weeks or even longer.

The term "lockjaw" comes from the paralysis that results, which makes it difficult for the horse to move his mouth and, eventually, to breathe. The symptoms begin with muscle stiffness and spasms; as the nerves are affected, the slightest stimulation can send the horse into a frenzied overreaction. Treatment is usually unsuccessful.

Caution: Everybody who works around horses should have a regular tetanus booster.

➢ *See also* Clostridium botulinum

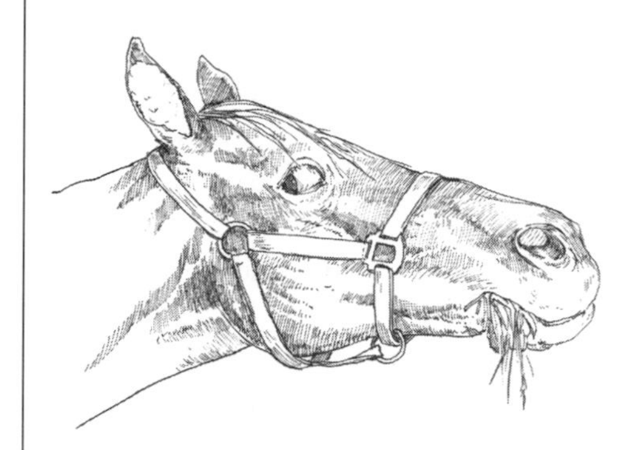

Signs of tetanus include rigid neck and head, staring eye with protruding third eyelid, flaring nostrils, and inability to chew.

Theiler's disease (serum hepatitis)

➢ *See Hepatitis, infectious hepatitis*

Therapeutic riding

Riding is a wonderful activity that can be enjoyed by anyone at any level, including people with disabilities. There are over 550 centers for therapeutic riding across the nation, according to the North American Riding for the Handicapped Association (NARHA). Riding offers children and adults with various limitations the chance to exercise, socialize, and gain strength and confidence.

Recognized as physical therapy by many insurance companies, hippotherapy (therapy from horse handling and riding) increases muscle tone and improves balance and coordination. Riding can benefit people with muscular dystrophy, multiple sclerosis, cerebral palsy, autism, epilepsy, blindness, and a host of other mental and physical disabilities. For more information about therapeutic riding centers, contact the NARHA.

Thermometer, use of

➢ *See Temperature, of horse*

Third eyelid

Many animals, including horses, have a nictitating inner membrane, which extends from the corner of the eye to cover the eyeball. It is usually visible only if there is infection or injury.

Thoroughbred

Today's Thoroughbred registry traces all of its members to three founding sires, imported from the Near East to England in the 17th century when the breed was established. These small, fiery stallions, named the Byerly Turk, the Darley Arabian, and the Godolphin Arabian, were crossed with heavier English horses to create an animal of great speed, stamina, and heart. The Thoroughbred is also admired for his versatility. In addition to thrilling thousands at the racetrack, Thoroughbreds are seen in showrings of every variety. They make fine hunters, polo ponies, and pleasure horses, although retired racehorses need special training to make the transition to "civilian" life.

Lean and athletic, Thoroughbreds stand anywhere from 15 to 17 hands high. Though quite variable in appearance, they generally have refined heads, prominent withers, and deep chests. They can be any solid color and often have white markings on the face and legs. Thoroughbreds can be high-strung and nervous, but much depends on how they are handled. A Thoroughbred cross often makes a wonderful mount for the less advanced rider.

➢ *See also Breed; Horse racing; Speed, of horse*

THOROUGHBRED PREPOTENCY

Thoroughbreds have improved the horse population worldwide, contributing size, stamina, speed, excellent conformation, and "heart" to hundreds of other breeds. This genetic strength and consistency, called prepotency, is thought to come from the 18th-century Arabian founding sires, although breeding of race horses in Great Britain dates from the time of Henry VIII.

Thoroughpin

A distension of the synovial membrane that protects the tarsal sheath. This swelling just above and in front of the point of the hock can be "pushed through" from the outside to the inside of the hock, and vice versa.

➢ *See also Lameness; Leg, of horse; Unsoundness*

Threadworms *(Strongyloides westeri)*

Horses develop immunity to threadworms several weeks after birth, but newborn foals are vulnerable. The dam should be dewormed right after giving birth, as the larvae lie dormant in the mammary tissues until the foal begins nursing, whereupon they migrate into the milk. Threadworms can affect a foal's lungs and liver and cause diarrhea.

In addition to deworming the dam regularly, keep her udder clean and pick up manure daily to prevent the foal from nibbling at it. Foals that develop heavy parasitic infestations may never reach their full health potential.

➢ *See also Deworming; Parasites*

Three-day eventing

➢ *See Combined training/eventing*

Three-gaited horse

Usually refers to Saddlebreds that perform in classes where only the walk, trot, and canter are called for.

➢ *See also Amble; American Saddlebred; Five-gaited horse; Gait; Gaited horse; Rack; Slow amble or slow gait*

Three-point contact

Used most commonly to differentiate a rider's position from two-point contact, in which a rider has lifted his seat out of the saddle, thus only having contact with the horse with two-points, his legs. Three-point contact uses the weight of the seat in the saddle.

➢ *See also Two-point seat (hunt seat, light seat)*

Three-track

➢ *See Shoulder-in*

Throatlatch

The strap on a bridle or halter that runs under the throat and prevents the tack from being pulled off over the head.

➢ *See also Bridle; Tacking up*

Throat ring

The metal ring on a halter that connects the noseband to the throatlatch.

➢ *See also Halter*

Thromboembolic colic

Colic caused by a blocked blood vessel in the abdomen or rear legs.

➢ *See also Colic*

Thrush

Hoof rot caused by either bacteria or fungi that thrive in wet, dirty conditions. Horses whose hooves are rarely cleaned and who stand in mud and manure a lot of the time are susceptible. The condition starts as black spots on the sole or a grimy layer on the frog, accompanied by a foul odor. If caught at this stage, the infection can be treated with iodine or bleach (every day until it clears up). But if it progresses, the frog can begin to decay. A nasty, black secretion indicates the necessity for a visit from the veterinarian, who will trim off the infected tissue and recommend a special disinfectant treatment. If untreated, thrush can cause lameness.

As with many ailments, it is easier to prevent thrush than to cure it. Clean your horse's hooves daily and make sure he is not standing on boggy or muddy ground for long periods of time. Winter and early spring are times to be particularly vigilant. Once thrush occurs, it can be hard to get rid of, especially if the underlying conditions persist.

A HOME REMEDY FOR THRUSH

Though there are many commercial thrush treatments available, this mixture is very effective. Called "sugardyne," it consists of ordinary sugar mixed with Betadine 10% Stock Solution to the consistency of molasses. Apply with a small brush. Check with your farrier or veterinarian if thrush is severe.

Ticks

Ticks are common in brushy pastures and tall grass and will attach themselves to passing horses. They carry a number of diseases that are dangerous to humans, notably Lyme disease (from tiny deer ticks) and Rocky Mountain spotted fever (from what are often called "dog ticks"), so be careful when removing and destroying them. Keeping horses away from tick-infested areas is the best cure, although heavy infestations can be sprayed with an insecticide that is safe for horses. Regular deworming with ivermectin can also help control tick problems.

➢ *See also Lyme disease; Parasites; Pest control; Piroplasmosis*

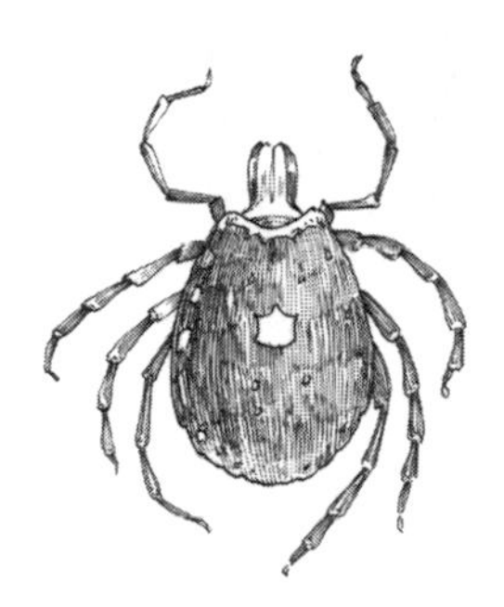

Rocky Mountain tick

Tied in knees or hocks

A horse that is tied in at the knee or hock has tendons and cannon bones that are too close together, giving the leg a round look rather than the flat appearance that is desirable. This common conformation fault usually leads to unsoundness from the friction between tendon and bone.

➢ *See also Leg, of horse*

Tie-down

A piece of equipment used in Western riding that serves some of the same functions as a standing martingale. A strap that connects from the noseband to the cinch, the tie-down keeps the horse from holding his head too high. It also helps him keep his head in position through tight turns and gives him something to brace against when holding a roped steer.

➢ *See also Martingale; Standing martingale*

Tie stall

➢ *See Stall; Straight stall (tie stall)*

Timed events

In many classes, competitors are racing the clock to complete the event at hand. Going over the allotted time can either disqualify you or detract from your score, depending on the event.

➢ *See also individual events*

Timothy

A kind of grass hay.

➢ *See also Hay*

Timothy

TOE-STIRRUPS

Indonesian ponies are still ridden with toe-stirrups, simply a knot in a rope, grasped by the rider's toe.

Tobiano

According to the American Paint Horse Association, a horse that meets the following specifications is considered tobiano:

- The horse is white, with white crossing the back or rump.
- The dark color usually covers one or both flanks.
- Generally, all four legs are white, at least below the hocks and knees.
- Generally, the spots are regular and distinct as ovals or round patterns that extend down over the neck and chest, giving the appearance of a shield.
- Head markings are like those of a solid-colored horse — solid or with a blaze, strip, star, or snip.
- The tail is often two colors.

➢ *See also Overo; Paint and Pinto; Tovero*

Toed-in or toed-out legs

When viewed from the front, a horse's hooves should point straight ahead, not toward each other (pigeon toes, also called toed-in) or away

from each other (splayed feet, also called toed-out). Having toed-in or toed-out feet/legs is considered a conformation fault.
➢ *See also Leg, of horse*

Tolt

The tolt is a gait peculiar to the Icelandic horse. Instead of trotting, these diminutive horses travel with a fast, running stride that is very comfortable to ride.
➢ *See also Icelandic horse; Gait; Gaited horse; Skeid*

Tom thumb

A mild Pelham bit with short shanks. It comes in a variety of mouthpiece styles.
➢ *See also Bit; Bridle; Pelham bit*

Tongue injuries

Although an infrequent occurrence, horses do bite their tongues — in a fall, for example. A severe bit improperly used can damage the tongue, and if the reins are dropped and the horse steps on them and jerks his head up, he can hurt himself quite badly. Most of these wounds, although they bleed profusely, will heal on their own. But if the horse is having trouble eating or accepting the bit, call your veterinarian.

Spare the lash, my boy, and hold the reins more firmly!

—Ovid, *Metamorphoses*

Topline, of horse

The topline runs from the withers to the dock. A well-proportioned horse has a shorter topline than underline. Each breed essentially has its own ideal topline. For example, the Arabian halter rules describe the ideal topline as "relatively level"; whereas the Thoroughbred ideal prefers a slightly sloped croup.
➢ *See also Conformation; Underline, of horse*

Torpedo grass *(Panicum repens)*

A toxic grass that causes anemia severe enough to kill a horse.
➢ *See also Poisonous plants*

Torsion, as cause of colic

Torsion of the intestine (twisted gut) is largely debated as a cause versus a complication of colic. Both scenarios are fairly equally supported. As a cause of colic, torsion essentially develops on its own through the intense muscle spasms that can occur within a nervous or high-strung horse. This twisted intestine leads to intense pain and constipation that are precursors to colic. As a complication of colic, the twisted intestine is actually seen as the result of a horse rolling around during colic, thus twisting the gut. Veterinarians do not easily agree upon the cause-effect relationship because horses that have had no history of such intense rolling have had torsion.
➢ *See also Colic; Digestive system*

Total digestible nutrients (TDN)

Refers to the amount of nutrients from a given feeding of hay and grain that the horse actually utilizes. A maintenance ration for a horse doing light work needs less TDN than one for a working horse or broodmare.
➢ *See also Feeding and nutrition*

Tovero

According to the American Paint Horse Association, a horse must meet the following specifications to be considered tovero:

- Dark pigmentation around the ears may expand to cover the forehead and/or eyes.
- One or both eyes are blue.
- Dark pigmentation around the mouth may extend up the sides of the face and form spots.
- Chest spot(s) vary in size and may extend up the neck.
- Flank spot(s) range in size and may be accompanied by smaller spots that extend forward across the barrel and up over the loin.
- Spots, varying in size, are located at the base of the tail.

➢ *See also Overo; Paint and Pinto; Tobiano*

Towing a trailer

➢ *See Trailering a horse*

Toxic substances

In addition to the many plants that horses shouldn't eat and the hazards of mold and fungi in feed, there a number of other substances that are harmful to horses. These include many obvious toxins, such as antifreeze, pesticides, fungicides, lead paint, and overdoses of medicine. Horses should also not eat any feed meant for cattle or other livestock unless it has been approved for use by a horse veterinarian.
➢ *See also Blister beetle poisoning; Fungi (mold) in feed; Monensin (Rumensin) poisoning; Poisonous plants; Silage for feed*

Toys, for horses

A variety of patented products, such as balls (some scented) and rolling grain-dispensing cylinders, are available to provide exercise and entertainment for a horse. Products such as these may keep a horse from knocking over and playing with his water bucket or from leaning out of his stall and grabbing whatever he can reach with his teeth.

TPR (temperature, pulse, respiration)

Become familiar with your horse's normal vital signs. Normal ranges for a horse that weighs 1,000 pounds are temperature between 99° and 100.5°, pulse of 36 to 40 beats per minute, and respiration of 8 to 15 breaths per minute. Some horses may deviate slightly from these norms but be perfectly healthy.
➢ *See also Pulse and respiration; Temperature, of horse*

Let us ride together, —
Blowing mane and hair,
Careless of the weather,
Miles ahead of care,
Ring of hoof and snaffle,
Swing of waist and hip,
Trotting down the
twisted road
With the world let slip.

—Anonymous, *Riding Song*

Trace

A heavy leather strap that connects the harness to the load or vehicle being hauled.

➢ *See also Harness*

Trace clip

A body clip that leaves hair on the horse's back, withers, and legs.

➢ *See also Blanket clip; Clipping; Hunt clip*

Trace minerals

Horses need a certain amount of minerals in their diet for proper body function and good health. These inorganic elements are usually found in sufficient quantities in the plants that horses eat, but in some cases supplementation is necessary to provide the required amount. Essential minerals are calcium, phosphorus, magnesium, sodium, chlorine, and potassium. Trace, or very small, amounts of the following are also necessary: iodine, cobalt, copper, iron, zinc, manganese, and selenium.

➢ *See also Calcium; Feeding and nutrition; Phosphorus; Salt (sodium chloride)*

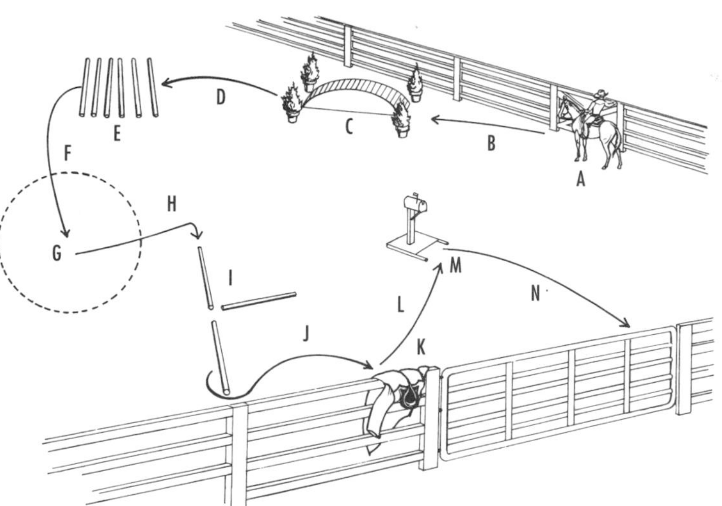

A typical trail class. (A) Open gate and enter. (B) Jog to bridge. (C) Cross bridge. (D) Lope to poles on left lead. (E) Walk over poles. (F) Lope in a circle. (G) Rider dismounts, leaves horse ground tied or hobbled, walks perimeter of circle, and remounts. (H) Jog to the poles shaped in a T. (I) Sidepass over all three poles, performing turns on the forehand and on haunches. (J) Walk to slicker on fence. (K) Rider dons slicker and removes it; horse stands still. (L) Jog to mailbox. (M) Open mailbox; remove, unfold, refold, replace newspaper. (N) Walk out of gate.

Track

A racing venue; the line of travel followed by a moving horse. For example, a horse is said to be "tracking left" if he is moving counterclockwise around the ring (that is, with his left side facing toward the center) or he is two-tracking during a half-pass.

Trail class

A Western competition that involves riding your horse over, around, and through a number of obstacles similar to ones you might find on a trail ride. Each competitor rides in a prescribed pattern, with the goal being to move smoothly through the course without stopping or hesitating. Tasks might include crossing a bridge, opening a gate, sidepassing and backing in complex patterns, stepping into a small square area and making a complete circle without touching the outlines of the space, and cantering over a series of poles.

➢ *See also Slicker-broke; Turn on the forehand; Turn on the haunches; Western competition*

Trail riding

For many riders, trail riding is the most fun you can have on horseback. It's a chance to use all the skills you've acquired in the ring, while meeting the challenge of going over uneven terrain with constantly changing surroundings. Before going out on the trail, you should be able to: negotiate small obstacles in your half-seat (practice over cavalletti); maintain a deep, balanced seat if your horse shies; and be able to handle a horse that might not want to go through a shallow stream or a scary-looking gap in the trees.

There are some basic guidelines for safe, enjoyable trail riding, the most basic of which is: Don't go out alone. If you take a bad fall, you could be pretty far from help, and your horse is unlikely to call 911. If you must go, tell someone at the barn where you're going and how long you expect to be gone. And carry a cell phone.

For group rides, the important thing to remember is to follow the leader, keeping a consistent pace and safe distance between horses (be aware of potential kickers in the group and stay well away from them). Don't rush past other riders because you might spook their horses. Don't let your horse snatch at branches and grass, either when you're riding or when you've halted for a breather. Scan the surroundings for potentially scary objects that could make your horse shy, but don't anticipate that he will.

➢ *See also Safety; Shying, spooking; Trail class*

TRAIL TIPS

- ❑ Respect NO TRESPASSING signs.
- ❑ Cede the right of way to bikers and hikers.
- ❑ Don't let branches snap behind you into the next rider's face.
- ❑ When going uphill, lean slightly forward (your horse may want to canter, as it's easier for him).
- ❑ When going downhill, lean slightly back and be prepared for a rush at the bottom as your horse seeks level ground.
- ❑ Return to the barn at a flat walk, making your horse travel in a zigzag line if necessary to distract him and keep him from picking up speed.
- ❑ Try to relax and stay alert at the same time.

Trailer

There are many different options when purchasing a trailer — more than there is room to discuss here. You will need to evaluate your own needs, number of horses, frequency of travel, and budget. There are trailers for every horse owner, from the casual competitor who hauls one or two horses to a few local shows every summer to the professional trainer who regularly brings a dozen horses and riders to high-level competitions in several states.

The two main types of trailers are tagalong (or straight-pull) trailers and gooseneck trailers. The former, usually built to carry two horses, are more common and less expensive. They attach to a trailer hitch welded to the towing vehicle. A gooseneck is pulled by a full-sized, open-bed truck. They are more expensive, but offer some advantages in terms of storage space, and they are often the only option for hauling four or more horses.

When purchasing any trailer, but especially a used one, carefully inspect it for quality of construction. It is especially important to check the floorboards of a used trailer for soundness. Make sure all the wiring works, that the hinges and latches are oiled and easy to manipulate, and that there is no rust or other obvious signs of decay.

Loading and unloading calmly is an essential skill for all horses.

Trailering a horse

It's very important to teach your horse to ride in a trailer and to load and unload calmly. Even if you never take him to shows, it's unlikely that he'll stay in one place for his whole life. A few hours spent familiarizing him with the process of trailering will save you endless hassles in the long run (and may save his life in an emergency).

Begin by leading your horse over and through a variety of obstacles, so that the trailer presents just one more challenge to overcome. Get him used to stepping up onto a platform and into narrow, covered spaces before asking him to try the trailer. When he's ready for the real thing, let him get used to the trailer by approaching it and sniffing it before asking him to get in. You will probably have to lead him in at first, but eventually, you should be able to send him in with a verbal command, staying outside to fasten the butt bar before going around to tie his head. (Loading a horse into a trailer is obviously easier with two people.)

Taking short trips with a seasoned companion helps a young or inexperienced horse learn that traveling by trailer isn't too scary. For longer trips, prepare an extensive checklist with everything you'll need for your trip. Make sure you're completely packed and ready to go (having checked over your list several times) before you load the horse(s), who should be well wrapped for protection against bumps and cuts that could happen along the way.

On long trips, you should stop periodically to check your rig and make sure the horses aren't too hot. Offer water every 2 or 3 hours.

➢ *See also Loading horses into trailers; Shipping boots and shipping wraps; Tail wrapping; Trailer*

ROAD TRIP

Whenever you travel, make sure you've got everything you need for your trip and for any emergencies that might arise. The list below doesn't include all the tack and equipment you need if you're going to a show!

- ❑ **For the trailer:** A spare tire and tools for changing it; jumper cables; a tow chain; extra bungee cords; a set of hand tools; flares; a flashlight with extra batteries; a fire extinguisher.
- ❑ **For the truck:** Human first-aid kit; all necessary documents, including health papers, maps, and important phone numbers; a cell phone or CB radio; enough money to handle unexpected problems (a towing fee, for example). If you'll be traveling a long distance or in a remote area, bring gear in case of a long breakdown in rough weather: a sleeping bag, extra clothes, a hat, etc.
- ❑ **For the horse:** Plenty of hay, feed, and water; a horse first-aid kit; emergency horseshoe supplies, including a hoof boot; some type of restraint, such as a halter chain, in case you have to unload after an accident or in a strange place; electrolytes; equipment to clean out the trailer; enough bedding for your trip, if it is long enough.

Trainer

A professional who works with you and your horse, usually with the specific goal of improving your performance in the showring.

Trakehner

A German warmblood developed in the early 18th century from Thoroughbred and Arabian stock. Originally used as cavalry mounts, these equines are now popular and elegant show and pleasure horses. A tall horse, standing 16 to 17 hands high, the Trakehner is gracefully built and seen usually in dark, solid colors.

➢ *See also Breed; Warmblood*

IMPORTANT WARNING

Your horse should never be tied unless the door or butt-bar/chain is closed behind him. So always close the door or butt-bar/chain behind him after loading *before* tying, and always untie him before opening the door to unload. (It is best to have a door behind the horse to prevent him from getting stuck under a bar or chain if he is frightened into moving backward suddenly after loading. Note that some multihorse trailers do not provide separate doors behind each horse.)

Tranquilizers, use of

It is sometimes necessary to use tranquilizers to calm a horse that isn't responding to other forms of restraint, such as a twitch or Stableizer. Tranquilizers can be useful for horses that become frantic or unmanageable during medical procedures (suturing a wound, floating teeth, and so forth), when being clipped, while traveling, or other stressful situations. Tranquilizers should only be used under the supervision of a veterinarian, because different horses will react differently to these medications.

➢ *See also Acetylpromazine maleate (Acepromazine); Stableizer; Twitch (restraint); Xylazine*

EQUINE TREADMILLS

While not a replacement for being ridden, equine treadmills can supplement a horse's exercise program if care is taken to keep the program short enough to prevent injury or sourness. Because a workout on a treadmill takes approximately half the time needed for other types of exercise, it can be useful on days when time is short.

Transition

A change from one gait to another. An upward transition is to a faster gait, while a downward transition is to a slower one. The aim is for transitions to be made smoothly, with subtle aids from the rider and prompt responses from the horse.

➢ *See also Aids; Gait*

Trappy

A course with sharp turns between jumps.

Trashy loper

The lope is a three-beat gait that is supposed to be slow but collected. A horse that is loping too slowly may slip into a four-beat gait and look as though he's loping in front while trotting behind. Some judges refer to a horse that does this as a "trashy loper." It's not a desirable characteristic.

Collected canter or lope

Traveling with a horse

➢ *See Trailering a horse*

Travers

The dressage equivalent of haunches in. This lateral movement requires that the horse move forward along the rail with his shoulder at the rail and his haunches inside the perimeter slightly so that the horse appears to be moving on two tracks. In dressage, the rider expects a real curve to the horse's body, rather than just having the front and back legs moving on different tracks.

➢ *See also Haunches in; Renvers; Shoulder-in; Two-track (half-pass)*

Treats, feeding to horses

Everyone likes giving their horses a reward for a good ride. In addition to hugs, pats, and praise, everyone knows that horses love carrots, apples, and other goodies. Horses show a distinct preference for sweets (many of them love the flavor of peppermint!), but since they can't brush their teeth, giving them sugar isn't the best choice.

Many barns discourage or outright forbid hand-feeding of lesson horses, for the simple reason that it can encourage nipping and biting. With his strong jaws and sharp teeth, a horse can break or cut a finger if one accidentally gets in his mouth. If he develops the habit of checking for possible treats, he might also begin nipping the hand that doesn't feed him. You can give treats in his feed bucket or in a bucket that you hold for him, and he'll still know whom to thank.

However, most owners feed their horses by hand anyway and allow others to do so as well. If you are going to do this, keep your fingers flat and together so the horse can take the treat without catching any part of you.

Warning: Never offer any horse a treat from between your teeth. You could be badly hurt, and it's not worth the risk for a "cute" trick.

Don't let your fingers get mistaken for a carrot — keep your hand flat and your fingers together. Better yet, feed the treats from a bucket.

Trees, toxic

➢ *See Poisonous plants*

Trip

In a show, one circuit of a course over fences. Announcers commonly say, "There are two trips remaining," when two competitors are still waiting to ride.

Triple bar

A type of jump with three bars at ascending heights. The standards can be placed farther away to give more spread.

➢ *See also Jumps, types of; Spread; Stadium jumping (show jumping)*

Triple Crown

This famous trio of races consists of the Belmont Stakes (established in 1867 in New York), the Preakness (first run in Baltimore in 1873), and the Kentucky Derby (1875). The first horse to win all three races was Sir Barton in 1919. There have only been 11 Triple Crown winners in more than 125 years.

WINNERS OF THE TRIPLE CROWN

1919 Sir Barton

1930 Gallant Fox

1935 Omaha (sired by Gallant Fox)

1937 War Admiral

1941 Whirlaway

1943 Count Fleet

1946 Assault

1948 Citation

1973 Secretariat

1977 Seattle Slew

1978 Affirmed

Trombidiform mites

➢ *See Mites*

Trophies

In addition to prize money, many shows offer trophies to the winners of certain classes. Depending on the show, the trophy might remain the property of the association, with the winner's name engraved on it and a replica taken home. In other cases, the trophy belongs to the winner. Sometimes a business or other sponsor will donate a challenge trophy. Trophies can be urns, bowls, or plates, in addition to prizes such as fancy saddles, blankets or coolers, halters, and other more practical items.

➢ *See also Prizes; Ribbons*

Trot

Faster than the walk but slower than the canter, the trot is a two-beat gait in which the horse's diagonal pairs of legs move together. The footfall pattern is left front and right hind touch the ground together, then the right front and left hind, or vice versa.

➢ *See also Gait; Posting trot; Sitting trot*

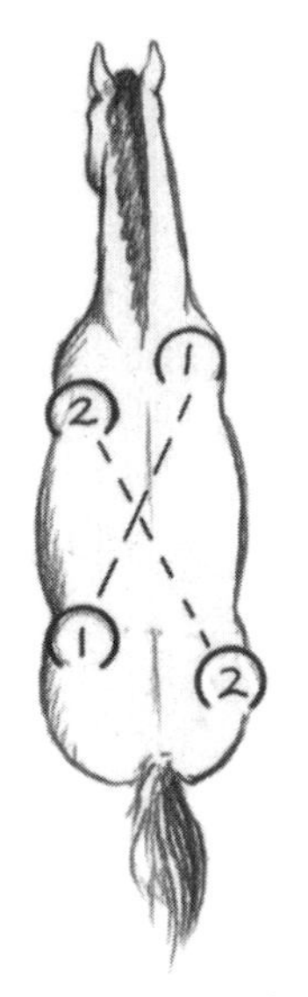

Horse moving at the trot

Trot, posting

➢ *See Posting trot*

Trot, sitting

➢ *See Sitting trot*

Turnback man

In a cutting class, the competitor is assisted by two herd holders and two turnback men (also on horseback). While the former keep the rest of the herd settled, the latter keep the chosen cow facing the competitor.

➢ *See also Cutting; Western competition*

Turn on the forehand

A maneuver that prepares the horse to learn lateral movements. In this turn, the horse pivots his hindquarters in 90-degree increments around his inside front foot. It is performed from a halt with the horse standing squarely on all four legs. The pivot leg does not get rooted to the ground, however. Instead, the horse moves in a walking sequence, just without the forward movement.

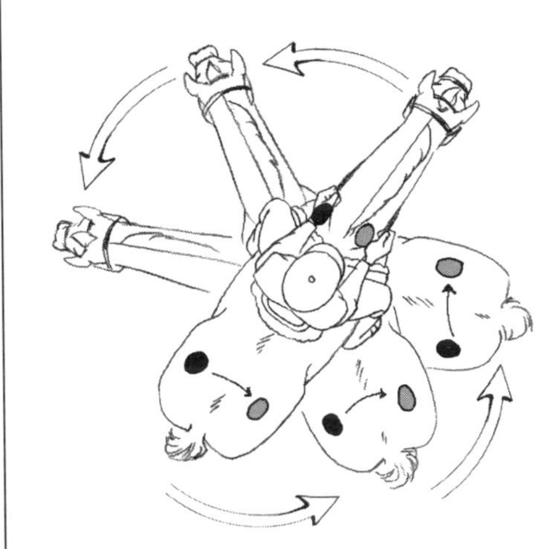

Turn on the forehand

Turn on the haunches

A maneuver that starts to build the foundation for working a horse's front end separately from his hind end. In this turn, the horse turns his forequarters in 90-degree increments around his inside hind foot. It is performed from a halt with the horse building some impulsion in order to achieve the proper crossing over of the outside foreleg in front of the inside foreleg. Both of the rider's legs are used to prevent the horse from stepping backward.

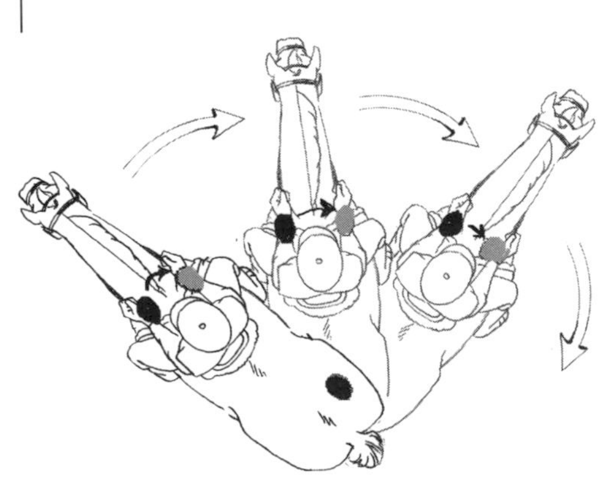

Turn on the haunches

Turnout

Turnout can refer to both the appearance of a rider and horse in a competition (correctness of attire and tack, cleanliness and neatness, appropriate braiding of mane and tail, and so on). It also means the time a horse spends out of his stall in a pasture or paddock, where he can roam at will.

Turnout blanket or rug

➢ *See Blankets and blanketing*

Turnout schedule

Horses should have some time every day (the more the better) in a safe, open area that includes shelter from bad weather. A turnout area can range from a small corral to acres of pasture, but the fresh air, exercise, and interaction with other horses are very important to maintaining a healthy horse.

Turnout schedules vary greatly, depending on the weather, the number of horses sharing the space, and the needs and whims of the animals involved. It could be a few hours on a sunny winter day (or even a snowy winter day; however, the only weather that really bothers most horses is wind-driven rain), an entire long summer night when the flies aren't as active, or even around the clock — as long as adequate shelter is provided.
➢ *See also Corral; Facilities; Fencing materials; Paddock; Pasture; Run (enclosure)*

Turnout sheet

➢ *See Sheet*

Twin foals

While mares often produce multiple eggs, they rarely sustain a twin pregnancy and twin foals are uncommon, though not unheard of. To develop normally, the fetus must have contact with the entire placenta, and twins do not have enough room for this contact. The mare will often absorb one of the embryos on her own. If she doesn't do so within the first several weeks, she is likely to abort one or both fetuses later on.
➢ *See also Foaling*

Twisted intestine

➢ *See Torsion, as cause of colic*

Twitch (restraint)

To twitch a horse means to grasp him firmly by the upper lip in order to make him stand still. This can be done by hand or with a special piece of equipment that is specially designed for this purpose.

Instead of relying on pain and fear to subdue a horse (that is probably already either fearful or in pain or he wouldn't need to be restrained), twitching releases endorphins that calm the horse and increase his ability to handle pain. Not all horses will react calmly to being twitched, however, and you should only try it with a mature horse that is used to being handled. If you use the twitch gently and massage his nose afterwards, he should accept the procedure.
➢ *See also Handling horses; Restraints; Stableizer*

The traditional twitch *(left)* consists of a loop of chain (or sometimes rope) attached to a wooden handle. A metal clamp or "humane twitch" *(right)* can be locked onto the horse's lip and fastened to the halter.

USING A TWITCH

To use the twitch, slide your hand through the loop of chain before grasping the horse's upper lip. Bring the loop over your hand onto the lip and tighten gently so that the horse cannot move away. You should always have two people around when twitching a horse; one to control the twitch and the other to do whatever it is the horse doesn't want done to him.

Alternative Method

A quick and easy way to twitch a horse is to grab a fold of skin on the shoulder and roll it over your fingers. This method also releases endorphins that help the horse relax.

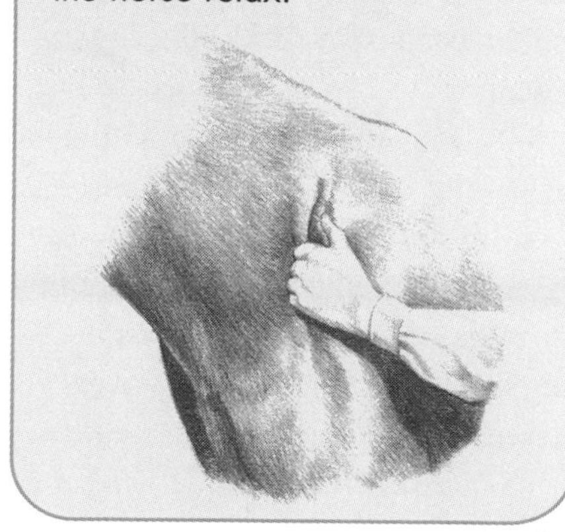

Two-handed riding

English riders almost always ride with one rein in each hand, but Western riders usually keep both reins in one hand. For some events, such as barrel racing, or for training purposes, a Western rider will ride two-handed; just as some English riders will ride with one hand on the reins to develop coordination of seat and body signals.

➢ *See also One-handed riding; Neck-reining*

The two-point seat keeps the rider balanced and secure while jumping and allows her to move with the horse as he goes over the fence.

Two-point seat (hunt seat, light seat)

In a two-point seat, the rider stands lightly in the stirrups with her weight balanced. Her third point of contact, her rear end, does not touch the saddle. This position allows her to absorb the motion of the trot or canter without bouncing on the horse's back. It also improves her balance.

➢ *See also Three-point contact*

Two-track (half-pass)

When two-tracking, or half-passing, the horse moves laterally with his forefeet and hind feet on two different tracks. This is the Western version of a half-pass.

➢ *See also Half-pass*

"Two wraps and a hooey"

In a calf roping class, the rider not only ropes the calf, but also she must immobilize it by tying three of its legs together. Using a short leather piggin string, she wraps it around the legs twice and fastens it with a hooey or half-hitch knot.

➢ *See also Calf roping; Piggin string*

Tying

Horses spend a lot of their time tied up, so it's important that they learn to do so patiently. It's also important that you follow several basic safety rules, because a tied horse can easily injure himself if he panics or tries to escape.

Always tie your horse with a halter (never by the bridle) and use strong rope that isn't frayed or rotten. Tie him in cross ties with panic snaps or to a solid post or ring using a quick-release knot. Never leave a tied horse alone — he could get into a lot of trouble in just a few minutes, including flipping himself over (rare, but it happens).

➢ *See also Cross ties; Moving a tied horse [box]; Panic snap; Quick-release knot; Safety*

TYING DOS AND DON'TS

Do use a halter and strong rope.
Don't tie your horse by the bridle.
Do use a quick-release knot or panic snaps.
Do tie your horse at about eye level (his) and at arm's length (yours).
Don't tie him too long (could get a leg over the rope) or too short (very uncomfortable for him).
Do train your horse to stand patiently for a reasonable period of time without fidgeting and fussing.
Don't expect him to stand tied for hours on end.

This horse is properly tied with a quick-release knot (note that the end of the rope runs through the loop so that he can't pull it free with his teeth). The rope is at a good height, with enough slack for him to be comfortable without hurting himself, and it is tied to a solid metal rail.

Tying up

➢ *See Azoturia; Rhabdomyolosis*

Tympanic colic

Colic caused by gas.

➢ *See also Colic*

Types of horses

Horses can be sorted into roughly three categories: heavy (or draft) horses, ponies, and light horses (which are the majority of riding horses). There is much variation within each category, particularly in light horses, because there are so many different breeds. Within the light horse group, horses can be further classified into six distinct types. Aside from Thoroughbred racehorses, each of the other five types encompasses a number of different breeds, and any given breed will have individuals fitting the description of one type or another. See the following box for more information.

➢ *See also Breed; English competition; Western competition; individual breeds*

TYPES OF LIGHT (RIDING) HORSES

Animated. Typified by the American Saddlebred, animated horses excel in the showring and are valued for their good looks, flashy action, and spirited demeanor. Park and show horses are included in this type.

Hunter. Hunters have strong backs and clean legs for smooth action, and they also have the ability to jump large obstacles at high speed. They are generally calm and willing. Many Thoroughbreds make fine hunters.

Pleasure. Nearly all breeds produce individuals that make good pleasure horses. The most important considerations are attractiveness and a comfortable ride at all three gaits. A pleasure horse must be well-balanced, trustworthy, and sensible.

Sport. Warmbloods, Thoroughbreds, and Arabians all fit into this category. Sport horses perform in a variety of disciplines, including dressage, jumping, combined training, driving, and endurance riding. With their long necks and short backs, high stepping action, and strong hindquarters, these horses love to compete and will give their best effort.

Stock. Also known as Western horses, the word "stock" refers to their original function as cow and sheep movers. However, stocky also describes these sturdy, well-muscled horses with their quick reflexes and levelheaded attitude. The Quarter Horse is the most popular of the stock types, but Appaloosas and Paints are also in this group.

u

Udder

The glandular organ of a mare (and all female mammals) that secretes milk. The physiology of a mare's udder is essentially the same as that of all other mammals. It is important for horse owners to become familiar with the normal feel of their nonpregnant mare's udder, so they are aware of changes that occur during pregnancy or as a result of injury or illness. Handling the udder helps prepare the mare to accept her foal's nursing and also the frequent washing of the udder that should be a routine part of grooming for mares that regularly work up a sweat.

Ride not a free horse to death.

—Miguel de Cervantes, *Don Quixote*

Udder problems

Most mares never experience problems with their udders. However, there are a few problems that owners should be aware of. Most common is irritation to the udder resulting from a buildup of smegma, a combination of dirt, sweat, and dead skin cells that accumulates between the teats. It is usually a problem only in mares who are worked to the point of sweating on a regular basis. This irritation can easily be avoided by washing the udder after workouts.

Any injury to the udder requires veterinary attention, as injury can result in serious damage to sensitive tissues. In addition, any major change in the udder as indicated by increased heat in the udder, or lumps or growths on or within the udder, should be reported to the veterinarian.

The most serious udder problem is **mastitis,** a relatively common bacterial infection of the udder in female mammals, from cats to dogs to cows to humans. The infection can be brought on by bacteria and lowered defenses, resulting from stress, exhaustion, or cracked nipples, and may be accompanied by fever, fatigue, and colic. Many mares will show a sudden dislike of having their foal nurse, often resorting to kicking or biting if the pain is severe enough.

Any of these symptoms should be reported to your veterinarian, who may recommend treatment with antibiotics. Warm, wet towels applied to the udder may help alleviate the mare's discomfort and enable her to accept the foal's nursing.

➢ *See also Mastitis*

Ulcer

A lesion in the digestive tract. Though many horses learn how to handle tension and anxiety, others will respond by developing gastric ulcers. Foals are even more vulnerable to long-term stress, which can be caused by confinement or illness. Symptoms of ulcers include grinding of the teeth, slobbering, and lying in odd positions to alleviate discomfort. These signs of gut pain will ease after feeding. Ulcers can be treated with cimetidine.

Umbilical cord, breaking

After a foal is born, the mare usually remains lying down for some time, allowing the blood from the placenta to pass to the newborn through the umbilical cord. The cord breaks naturally as the foal struggles to stand for the first time or the mare gets up to sniff at her baby. It's important that the cord not be cut prematurely, as up to 35 percent of the foal's blood supply comes from the placenta in the first few moments after he is born.

If a mare foals from a standing position, which sometimes happens, two people should be ready to catch the foal and hold him up to keep the umbilical cord from breaking too soon.

➢ *See also Foaling; Gestation; Joint ill*

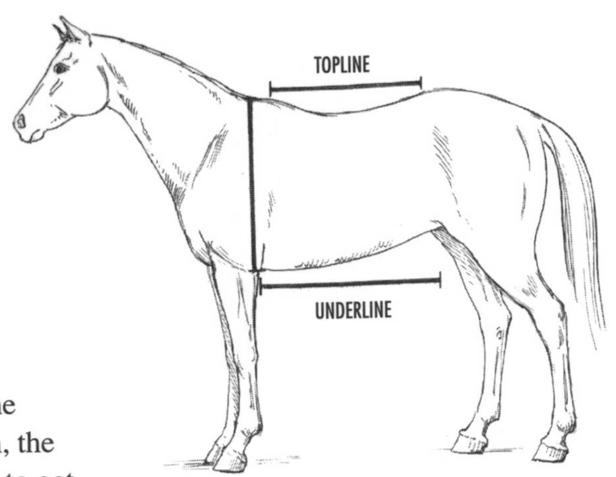

A well-proportioned horse, with a longer underline than topline.

Umbilical hernia

A soft swelling at the naval caused by an imperfect closure of the muscles at the naval. The opening allows a portion of the omentum, the protective tissue of the gut, or possibly even a part of the gut itself, to get caught between the abdominal wall and the skin. Most hernias resolve themselves or with minor assistance by pushing the swelling back into the abdominal cavity. However, if you sense any increase in heat or the size of the swelling call a veterinarian as this could indicate that the gut itself is involved and an impaction or infection might be developing.

Underline, of horse

The underline runs from the point of the elbow to the stifle joint. It should be longer than a line from the last cervical vertebra to the point of the hip. This allows the horse sufficient room for proper striding. Horses with underlines equal to, or shorter than, their toplines are prone to injuries from forging and overreaching.

➢ *See also Conformation; Topline, of horse*

UNDERWEAR

If you are not wearing comfortable underwear when you get to the barn, it's only going to get more uncomfortable once you're in the saddle. For men, briefs are probably less likely to chafe and bunch up than boxers, while many women find a fuller garment preferable to a bikini-type cut. If your seatbones get sore or you tend to bounce around a bit, you can actually buy padded underwear made especially for riding, which may help. A sports bra is probably a good idea as well, particularly for large-breasted women.

Underweight horse

You can tell a horse is too thin if you can see the shape of each rib. On a horse of the correct weight you will be able to see where the ribs end, but not their entire definition.

If your horse is underweight, increase his feed gradually or increase fat in the ration by adding a supplement like corn oil. Do not give him an excessive amount of food all at once, because his system will not be able to absorb it.

Uneven wear of teeth

➢ *See Dental care; Teeth*

Ungulate

A noun describing the group of hoofed, herbivorous, quadruped mammals that includes horses, camels, elephants, hippopotamuses, hyraxes, rhinoceroses, ruminants, swine, and tapirs. Also, an adjective meaning having hoofs.

A hoof, claw, or talon is called an **ungula** or an **unguis** (plural **ungues**).

Unhorse

To unseat a rider from a horse.

Unicorn

A magical, mythical, horselike creature with a long, straight horn growing from its forehead.

United States Cavalry

➢ *See Cavalry*

United States Combined Training Association (USCTA)

The organization that oversees all horse trials for U.S. Equestrian Team tryouts, as well as for competition at lower levels.

➢ *See appendix*

United States Department of Agriculture Horse Protection Act

Passed in 1970, the Horse Protection Act seeks to eliminate the practice of "soring" horses through mechanical or chemical means to enhance their performance in the showring.

➢ *See also Soring*

So we'll ride together,
Comrade, you and I,
Careless of the weather,
Letting care go by.

—Anonymous, *Riding Song*

United States Dressage Federation (USDF)

This national organization has dozens of chapters around the country and provides information on dressage competition. The USDF administers numerous educational and licensing programs for instructors, as well as competitors in the sport of dressage. It also provides incentive awards for riders to be recognized at various levels of competition; essentially year-end and career high-point awards.

➢ *See appendix*

United States Equestrian Team (USET)

Until 1949, the United States Cavalry provided the teams that represented the country in international competition. After the mechanization of the Cavalry, however, a group of concerned riders established the United States Equestrian Team in 1950. The organization now oversees the selection and training of teams in six disciplines: Combined Driving, Dressage, Endurance Riding, Eventing (Combined training), Reining (the most recent addition), and Show Jumping.

Riders, drivers, and horses for each discipline are chosen annually from numerous applicants who have qualified by accumulating enough points in competition or by winning in selected shows. Riders and drivers may be professionals, but do not need to own their own horse and in many cases will ride a horse owned by someone else. However, the best combinations tend to be a pair that have worked together for a long time in many different circumstances.

USET competitors have won many medals in international competition, including at the World Championships, the Pan-American Games, and the Olympic Games. A total of 30 Olympic medals have been earned.

Endurance riding has been particularly successful, with Becky Hart winning the World Championship in 1988, 1990, and 1992, and team members bringing home Olympic gold and silver for both individual and team achievements over the past decade. David O'Connor and Custom Made won the gold for Individual Three-Day Eventing competition in the 2000 Olympics in Sydney, Australia.

The success of USET at the Pairs World Championships in 1991 led to the hosting of the event in Gladstone, N.J., in 1993 and again in October, 2000. USET show jumpers have won seven Fédération Equestre Internationale (FEI) Show Jumping World Cup Finals, a very prestigious record.

➢ *See also Olympic equestrian events; individual event*

Dear to me is my bonny white steed;
Oft has he helped me at pinch of need.

—Sir Walter Scott, *The Lay of the Last Minstrel*

United States Pony Club

➢ *See Pony Club*

United States Team Penning Association

A national association for nonprofessional competitors.

➢ *See appendix*

Unloading horses

➢ *See Loading horses into trailers; Trailering a horse*

Unsound

Description of a horse that is unable to do the work expected of him. A horse may be unsound for a variety of reasons. He could be lame, have a sore back, or suffer from a respiratory ailment. Unsoundness can be either temporary or permanent, depending on the severity of the illness or injury.

➢ *See also Lameness; Respiratory system; Sound*

The horse on the right could be described as unthrifty, compared to the horse on the left, meaning that he is not receiving adequate nutrition from his food.

Unthriftiness

A horse is said to be unthrifty if he is not getting proper nutrition from his food — that is, if he is thin and his coat is harsh and dull. Parasites are a frequent cause of unthriftiness because they rob the horse of much-needed nutrients.

➢ *See also Deworming; Parasites*

Upward transition

➢ *See Transition*

Urinary stones

Mineral salts that crystallize in the urine. Horses, especially males, sometimes suffer from this condition. A urinary stone can block the urethra, causing painful urination and colic. If the stone is in the urethra, it can often be removed by catheter. Surgery may be required if the stone is in the bladder.

Urine

Normal horse urine is yellow, is cloudy, and has a strong odor.

Urine, brown

A sign that the muscles are breaking down and being excreted. Brown urine can occur during an episode of rhabdomyolosis (tying up) and also is a symptom of some forms of poisoning. In any case, it warrants an emergency veterinary call.

➢ *See also Poisonous plants; Rhabdomyolosis*

Urine removal

➢ *See Mucking out stalls*

Urticaria

An allergic reaction that causes small bumps or hives on the surface of the skin.

➢ *See also Skin problems*

Uterine prolapse

Very rarely, a mare's uterus will prolapse after she gives birth; that is, all or part of the uterus will emerge after the foal and hang down from the vulva. If uterine prolapse occurs, call your veterinarian immediately and, in the meantime, keep the mare quiet while you support the uterus (preferably as high toward the vulva as possible) in a clean blanket or towel soaked with sterile water or a mild disinfectant. (Disposable bed liners work well, if part of your first-aid kit.)

➢ *See also Foaling*

Uterine torsion

A twisting of the entire pregnant uterus, cutting off access to the cervix. A mild torsion may be corrected if the mare rolls during early labor, which is not uncommon, but a complete twist must be corrected by a veterinarian. Like uterine prolapse, this condition is quite rare.

➢ *See also Foaling*

Uterus (wethers)

The female reproductive organ that nurtures the fetus throughout the pregnancy.

➢ *See also Anatomy of the horse; Breeding; Foaling; Gestation; Reproductive tract*

Uveitis (moon blindness)

➢ *See Iridocyclitis; Periodic ophthalmia*

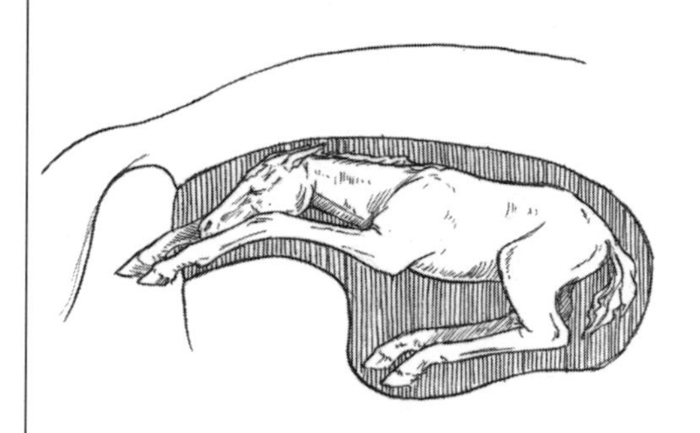

The normal position of a foal in the uterus just before birth.

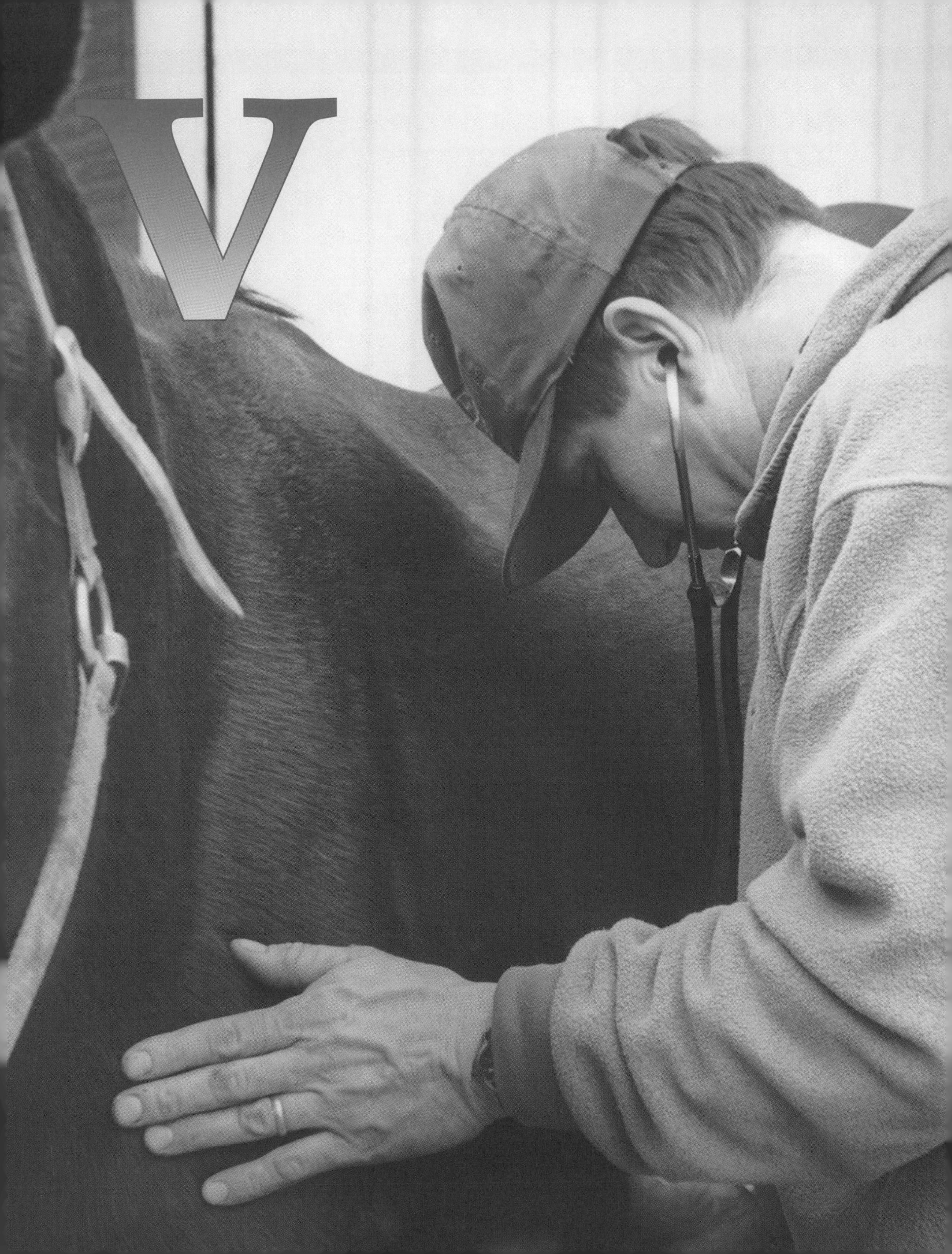
V

Vaccinations

Vaccinations work by stimulating antibodies in the blood to fight off the harmful effects of bacterial or viral infection. Once an animal is exposed to a mild form of a disease through an inoculation, he can better resist infection caused by the disease pathogens.

Horses are more capable of building immunities if they receive vaccinations at a time when they are already healthy and in good condition. An ill, stressed, or undernourished animal may not have the reserves to effectively produce antibodies and, therefore, may not be sufficiently protected if exposed to a contagious disease via vaccination.

Become familiar with the diseases that affect horses in your area. Learn everything you can about the risks of disease versus the risks of vaccination against the disease; sometimes the vaccine is not worth the slight risk that your horse would contract the disease. It is impossible to provide sound advice regarding the complex nature of diseases that affect horses in the various regions of the country. It is best to work with your veterinarian to develop an effective vaccination program for your horse.

Horse vans are often the most comfortable way for horses to travel.

Vagina

The passage through which the foal passes during birth.

➢ *See also Breeding; Foaling; Gestation; Reproductive tract*

Van, horse

A horse van is a single unit designed to transport horses, unlike a trailer, which disconnects from the towing vehicle. Horse vans can carry up to nine horses and are the most comfortable option for long-distance hauling. They have better suspension, more effective insulation, and usually give the horses a quieter and smoother ride. Naturally, vans are the most expensive option.

➢ *See also Trailer*

Vaquero

The original cowboys, Mexican *vaqueros* herded cattle throughout the Southwest during the 18th and 19th centuries. A corruption of this word inspired the slang term "buckaroo" to describe a cowboy.

➢ *See also Rodeo*

For riders who get tired of just sitting in the saddle, vaulting offers an exciting, challenging alternative.

Vaulting

An internationally recognized equine sport involving a variety of gymnastic maneuvers performed on and off the back of a trotting or cantering horse. Individual and team competitions are held at all levels, including the International World Championship. Contact the American Vaulting Association for more information.

➢ *See appendix*

VEE

➢ *See Venezuelan equine encephalomyelitis*

Venezuelan equine encephalomyelitis (VEE)

Like Eastern and Western equine encephalomyelitis, VEE spreads from birds to mosquitoes to horses. The Venezuelan strain came into Texas from Mexico in 1971, and thousands of horses died before an effective vaccine to contain the spread of the virus was developed.

➢ *See Equine encephalomyelitis; Vaccinations*

Ventilation in barns

Horses fare better in barns that have plenty of air and cross-ventilation. Stuffy, enclosed buildings encourage the spread of airborne pathogens and respiratory illness.

➢ *See also Barn; Shelter*

Vertical

A simple jump consisting of standards and poles.

➢ *See also Jumps, types of*

A **vertical** jump

Vesicular stomatitis (VS)

A highly contagious virus that is most commonly found in South and Central America and Mexico, but can spread into the United States. With symptoms similar to foot-and-mouth disease, VS isn't fatal but health officials are careful to limit its spread.

The virus is carried by insects and in the saliva of infected animals. It is also carried in the fluid of the blisters that are a symptom of the disease. The blisters form around the feet, teats, nostrils, mouth, tongue, and lips. As they swell and break, lesions form that make eating and drinking so painful that the horse may actually die of dehydration or starvation.

There is no cure for VS, but treatment includes using a mild antiseptic mouthwash or ointment on the blisters to relieve pain and to guard against secondary infection of the sores left by broken blisters. Thoroughly clean and disinfect any area used by an ill horse (including trailers) and wash and change your clothes and shoes before handling a healthy horse.

VESICULAR STOMATITIS WARNING

An outbreak of VS must be reported to local authorities, and all affected horses must be quarantined. A farm with a case of VS is not considered safe until 30 days after the lesions have healed.

Vests, protective

When jumping or riding an unpredictable or predictably unreliable horse, many riders choose to wear protective vests in case of a fall. Some stables require them for jumping students. Vests are mandatory in many cross-country competitions, and they are a good idea for rough rodeo riding. Some horsemen even wear them when working with unruly horses on the ground, to lessen risk of injury from being struck or trampled.

➢ *See also Helmet, safety*

Veterinarian

No matter how well you take care of your horse, you'll come to know your veterinarian well, so it makes sense to work with one you like and trust. The best way to find a good vet is to ask around for recommendations. Depending on the size of your barn and your geographic location, there may be several veterinarians from which to choose.

If you're lucky, you'll only see the vet a couple of times a year, when vaccinations and checkups are due. However, if you have an emergency or other consultation, make sure that you have the horse available when the veterinarian arrives. Make a note of any significant changes in behavior, as well as his current temperature, pulse, and respiration. Be prepared to hold and calm your horse during any procedures that need to be done.

When to Call the Vet

You know that you should call the veterinarian when your horse is lame, bleeding, or obviously colicking, but here are some other signs that a consultation is in order:

- Not eating or eating only hay or only grain
- Having trouble chewing; slobbering; dropping food
- Coughing
- Acting depressed or lethargic; acting anxious or uneasy
- Shivering or sweating inappropriately
- Lying down frequently; pawing excessively; rolling a lot
- Showing a change in stool: diarrhea; hard, dry manure; not passing manure at all
- Having difficulty urinating
- Standing in an odd or uncomfortable position
- Having a runny nose or discharge from eyes

You know your horse better than anyone else does, so trust your instincts. If you think your horse is not well, he probably isn't. You're better off calling the veterinarian early on than waiting to see what happens and risking further complications. Unless your horse is in a life-threatening situation, take note of the following before calling, so that your veterinarian has the best information possible:

- Note general health and vital signs (temperature, pulse, respiration).
- Check capillary refill time
- Check gut sounds if you suspect colic
- Assess the lameness, if this is the problem

The more information you can provide to your veterinarian via telephone, the better she can instruct you in how to manage the situation until she arrives. This information will also help her assess the critical level of your horse's needs. Many veterinarians receive multiple calls for help within the same timeframe. It is vital that she be able to care for the most critical first. (It also can save you money if the vet doesn't have to consider your visit an emergency. Many vets add a premium charge for emergency calls.)

➢ *See also Capillary refill time; Colic; First aid; Lameness; Parasites; Poisonous plants; Posture as sign of illness; Prepurchase exam; Pulse and respiration; Signs of illness; Temperature, of horse; Vital signs*

Vetting out

If a horse "vets out," he has passed his prepurchase exam.

➢ *See also Prepurchase exam*

FINDING A VET ON THE ROAD

If you need a vet while you're traveling with your horse, call 1-800-GET-A-DVM. This service, run by the Equine Connection, lists the names of local veterinarians who are members of the American Association of Equine Practitioners. On the World Wide Web, check www.getadvm.com.

WHEN IS IT AN EMERGENCY?

- ❑ When your horse's temperature is more than 3°F above normal.
- ❑ When your horse's resting pulse or respiration is more than 20 percent above normal.
- ❑ When respiration is greater than pulse (indicates serious shock).
- ❑ When capillary refill time is longer than three seconds.
- ❑ When there is a lack of normal gut sounds.
- ❑ When your horse can't bear weight or hops while trying to move.
- ❑ When an injury spurts blood (indicating an artery is involved).
- ❑ When an injury is near a joint and amber-colored fluid is seeping from wound (indicates joint capsule is injured).
- ❑ When an injury exposes bone, tendons, or ligaments.

See also Vital signs

HOW TO HANDLE VICES AND BAD HABITS

PROBLEM	SIGNS AND POTENTIAL RESULTS	PROBABLE CAUSE	PRESCRIPTION
Cribbing	Clamping jaws on a solid object, tensing the neck muscles, and gulping air. May cause colic or poor condition if horse gets too "addicted."	May start out of boredom, dietary deficiency, or parasite infestation. Thought to stimulate endorphin production, making it addictive.	Sometimes incurable, but cribbing straps can manage behavior. Muzzles or electrical shocking devices also effective in some cases. Drug treatments may be available in future; surgery is possible.
Pawing	Using hooves to dig holes or knock over feeders and water buckets. Can get foot caught in door or fence; wears down hooves and shoes; could injure someone if doing it while tied.	Usually horse is bored, confined, anxious; may be overfed and/or underexercised.	Rubber mats in stall help a lot, as does more exercise or toys. Keep feed and water off the ground; don't reward behavior by feeding. Teach the horse to stand patiently.
Self-mutilation	Happens primarily with stallions around age 2; biting at self with squeals, kicking out, and pawing.	Confinement, lack of exercise, sexual frustration, hormones. Can become an addictive behavior similar to cribbing.	Gelding helps if the stallion isn't to be bred, as does increasing exercise, finding larger turnout space, and providing companionship and/or toys. A neck cradle or muzzle may be effective; drug treatment may be available in future.
Stall kicking	Kicking at stall walls and doors. Loosens shoes; causes leg and foot injuries; damages facilities.	Boredom, confinement, dislike of neighbor, or may just like the noise. Usually gets attention, which rewards the behavior. Could have itchy heels or rodents in stall.	Curable if not too ingrained. More exercise helps, and walls and hooves can be padded. Kicking chains are available. Do not reward behavior by feeding.
Tail and mane rubbing	Routinely rubbing haunches or neck against rough surface. After a while, hair breaks off, raw patches appear, and infection can occur if the rubbing is not stopped.	Often indicates parasites or skin infection, dirty udder or sheath, soap residue. Could be itchiness from allergies or shedding.	Keep tail and mane clean; frequent shampooing is not necessary, but thorough rinsing is. Groom often and well and deworm regularly. Electric fencing will discourage the habit, as well.
Weaving/ pacing	Swaying head and neck side to side/walking back and forth in the same path.	Happens if a horse is high-strung or anxious; could be too much confinement or feed; not enough stimulation or exercise.	Lots of exercise, turnout time, and company of other horses usually alleviates this behavior. Stall toys, a doubled hay net, and special stall door will help, as well.
Wood chewing	Biting and gnawing wood fences, stall doors, and feeders. Can actually eat a fair amount of wood each day; causes a lot of damage to facilities. Other horses might imitate.	Lack of roughage or minerals in diet, boredom, stress, teething, or might be copying behavior from other horses.	Make sure horse is getting plenty of roughage and free access to minerals and salt. Apply antichew products or metal stripping on tempting targets. Increase exercise and turnout time.

Note: These stable vices are generally manageable with some changes to the physical environment. Unpleasant behaviors aimed at human handlers are also vices and should not be tolerated. Any horse that bites, kicks, strikes, bucks (more than the occasional hop after being confined for too long), crowds, or sits back when tied needs to be retrained, probably by a professional handler.

Vices (stereotypes)

If there were a vice squad in every stable, it would be on the lookout for a variety of equine behaviors that are dangerous, destructive, or just downright annoying. Many vices (such as cribbing and weaving) result from a horse seeking stimulation of some sort. Others may begin as dietary deficiencies (wood chewing) or parasite problems (tail rubbing), and they may continue even after the initial cause is removed.

➢ *See also individual vices*

Vinegar as fly repellent

➢ *See Pest control*

Viral disease

An infectious disease caused by a virus, a microorganism that grows and multiplies in living cells. Viral diseases that affect horses include rabies, equine viral encephalomyelitis (sleeping sickness), viral respiratory diseases (such as influenza, rhinopneumonitis, and equine viral arteritis), equine infectious anemia (swamp fever), and vesicular stomatitis.

When a horse becomes ill after exposure to a virus, he makes antibodies against that specific germ. When he recovers, the antibodies will protect him for a varying period of time, providing active immunity to that virus. Vaccination is another way to acquire active immunity, through exposure to a controlled amount of virus that stimulates antibody production but is not significant enough to cause illness. Horses that are rundown have more difficulty than healthy horses do in producing adequate antibodies to build immunity.

➢ *See also Vaccination; individual disease*

Depending on where the light strikes his eye, a horse focuses on objects at varying distances by moving his head up and down. With his head up, he can see things nearby; he holds it in a normal position to focus at a middle distance, and he lowers it to look farther away.

Vision, of horse

Like humans, horses are visual creatures, but they have much better eyesight than we do. They are able to see about 340 degrees around their heads because their eyes work independently. With their widely spaced eyes, horses have remarkable peripheral vision, a great advantage for a prey animal that spends much of its time grazing. The disadvantage is that each eye sees things separately, which explains why your horse might be fine going past a parked trailer from one direction but balk at it coming the other way. Until he processes the visual cues, the second view is of a completely different and potentially dangerous object.

Horses also need to move their heads to focus; they hold the head up to see close objects and lower it to see things that are far away. Equines do have blind spots; they can't see things that are directly in front of them (including jumps at the moment of take-off) or immediately behind them.

➢ *See also Eye; Pig-eyed; Shying, spooking*

Vital signs

If you own a horse it is essential to become familiar with his normal vital signs, so that you will recognize problems if they arise. Here are the normal vital signs for a resting horse that weighs 1,000 pounds. Some horses may deviate slightly from these norms, yet be perfectly healthy, which is another reason to get to know your own horse's particular physiology.

Temperature (rectal): 99–100.5°F
Pulse: 36–40 beats per minute
Respiration: 8–15 breaths per minute

Vitamins

A horse on good pasture or high-quality hay has no need of supplementary vitamins. If you determine that your horse requires additional vitamins because of poor feed or his current condition, consult with your veterinarian about appropriately supplementing the diet.

➢ *See also Feeding and nutrition; Supplements*

VITAMIN PROS AND CONS

VITAMIN	SOURCE	BENEFITS	CAUTIONS
A	Made by the horse's body from carotene in green pasture and green, leafy alfalfa hay	Keeps eyes healthy; important in reproduction; helps hoof growth. Lack of vitamin A can make horse prone to respiratory infections.	Fat-soluable and stored in body tissues. Too much can cause bone deterioration. Horses will store enough from summer pasture to last through the winter.
B	Several types; created by bacteria in the cecum	Keeps red blood cells healthy and helps energy metabolism.	The B vitamins include thiamine, riboflavin, niacin, pyridoxine, pantothenic acid, biotin, choline, folic acid, and B_{12}. No supplementation is necessary unless the horse is ill or anemic.
C	Synthesized by horse during digestion	An antioxidant beneficial for fats, proteins, and cell membranes; helps bones and teeth grow.	No supplementation necessary unless horse is ill.
D	Sunlight; sun-dried hay	Helps convert calcium and phosphorus into healthy bone. A deficiency leads to rickets or osteomalacia (soft bones).	Normally, horses receive all the vitamin D they need from being outside. Too much can cause calcification of bruises.
E	Natural pasture	Necessary for healthy reproduction as well as for muscle growth and function. Boosts immune-system and neurological function.	Usually sufficient in natural feeds, but may need to be supplemented during times of stress.
K	Produced in the intestines	Vital to blood clotting; activates certain proteins.	Fat-soluble; stored in tissues.

Vocalizing, of horse

In addition to their expressive body language, horses communicate with one another and with their human handlers through a variety of sounds. A low nicker is often heard at the approach of a familiar person (especially one with food) or when two equine friends greet each other.

Separating two horses that have bonded can produce endless, noisy whinnying as they pine to be together again.

Angry squeals often precede flying hooves and snapping teeth in paddock disputes. A truly furious horse will produce a sound often described as a scream, and fighting stallions make loud roaring noises as they struggle with each other. A horse may snort with alarm at a new sight or smell, but easily excited horses may snort often just because they want to.

➢ *See also Body language, of horses; Whinnying*

Voice, as an aid

Your voice is an important aid in handling and riding your horse. Your tone and inflection relay a great deal of information about the situation and what you expect of him.

➢ *See also Aids; Handling horses; Talking to horses*

Volte

A dressage term meaning a small circle (perhaps 20 feet in diameter) made by the horse.

Volvulus

A twist in the intestinal tract.

➢ *See also Torsion, as cause of colic*

Vomiting

Horses are unable to vomit, making them very susceptible to poisoning or digestive upsets. Many experienced horsepeople will tell you that a horse has inadequate reverse peristalsis — the vomiting mechanics of humans, dogs, cats — and therefore cannot vomit to cleanse his digestive tract of spoiled or toxic foods. In fact, food can return upward from the stomach, but structures in the back of a horse's throat keep it from exiting through the mouth and force it through the nostrils or, worse still, back down into the lungs, causing pneumonia. Vomiting is so difficult for horses, the stomach may actually rupture in the process. Feeding a horse properly and keeping an eye on his digestive function are essential.

➢ *See also Colic; Digestion; Digestive problems; Stomach*

VS

➢ *See Vesicular stomatitis (VS)*

Vulva

The outside opening of the female reproductive tract.

➢ *See also Reproductive tract*

When it comes to saddle hawses, there's a difference in steeds:
There is fancy-gaited critters that will suit some feller's needs;
There is nags high-bred an' tony, with a smooth an' shiny skin,
That will capture all the races that you want to run 'em in.
But fer one that never tires; one that's faithful, tried and true;
One that allus is a "stayer" when you want to slam him through -
There is but one breed o' critters that I ever come across
That will allus stand the racket: 't is the —
Ol' cow hawse!

—E. A. Brinninstool, *The Ol' Cow Hawse*

V

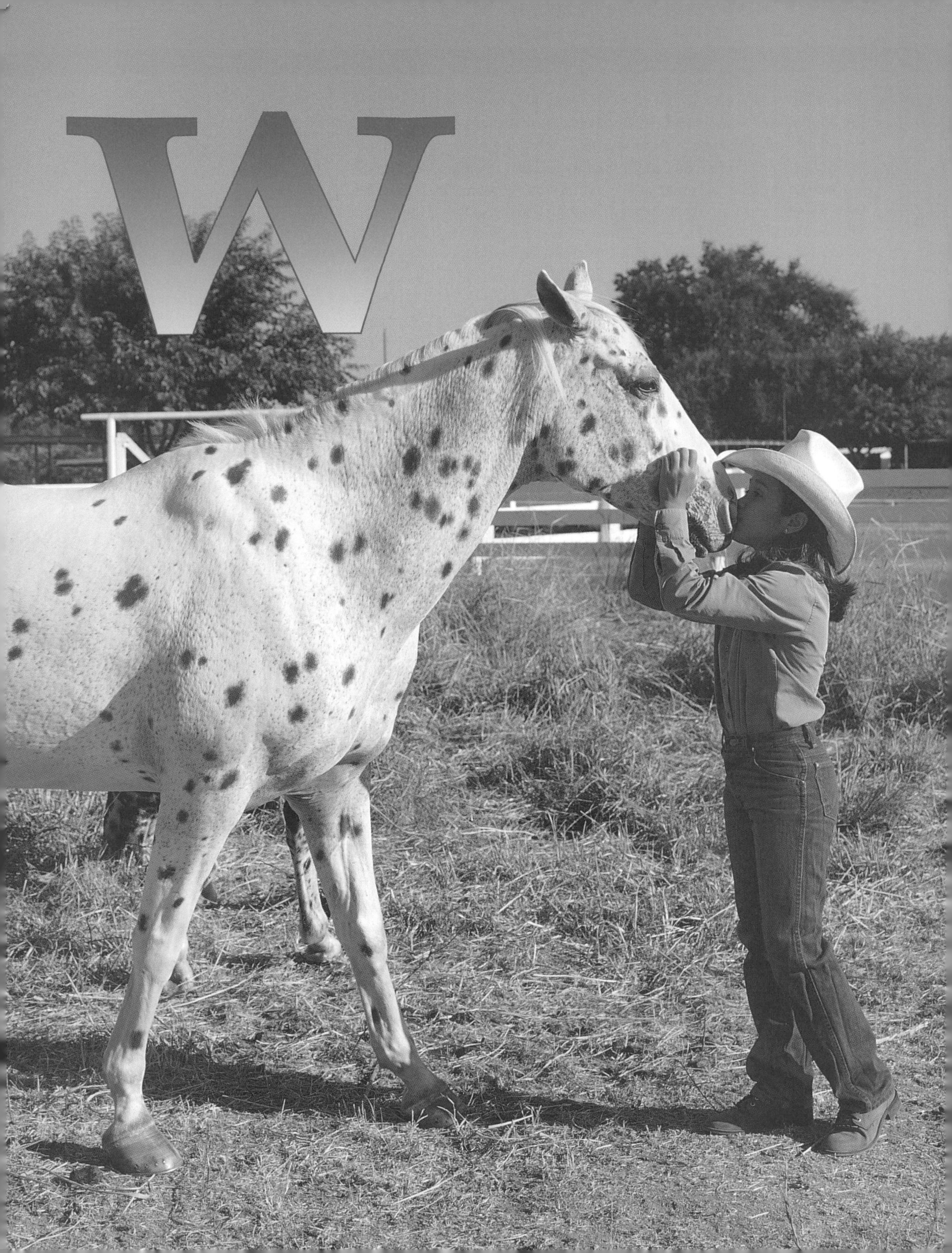
W

Waler

A light, mixed-breed saddle horse developed in New South Wales, Australia, exported to British military forces in India during the 19th century. The present-day Australian Stock Horse is the descendant of the Waler.

Walk

A slow four-beat gait in which each foot strikes the ground separately.
➢ See also Gait; Running walk

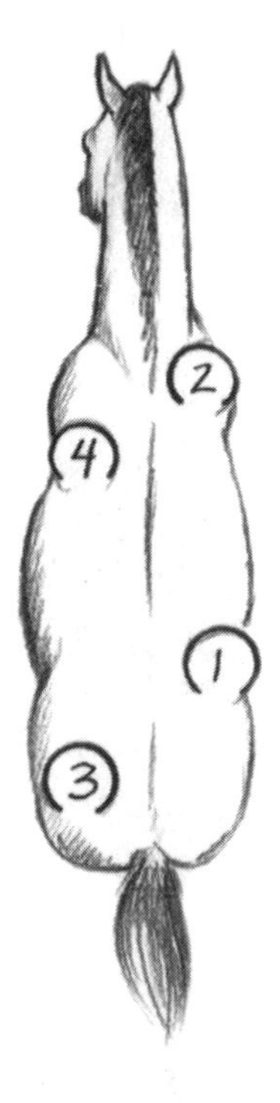

Moving at the walk

Walking a horse down

A phrase used interchangeably with "cooling out a horse." It is meant to describe the vital process of walking a horse after a vigorous workout. This helps his pulse, respiration and temperature return to normal. Although some horsemen actually measure these vital signs to determine when to stop walking their horse, most use a subjective measure of the horse's condition by feeling the chest or neck. If it seems relatively normal, the horse is considered "cooled out," and ready to be untacked and groomed.

A second meaning of the phrase refers to a method of catching a horse in a pasture. No one likes chasing a recalcitrant horse around a field trying to catch him. Since he's definitely got the advantage, this is one habit that needs correcting. Bribing with treats usually works but can be hazardous if there are other horses in the pasture that want a treat, too. Another method that takes a lot of time initially but is worth the effort is called "walking a horse down."

This term means that if you just keep following the horse, he will finally give up and let you approach him. Don't run after him and don't get mad and yell at him; just walk after him calmly. Keep your body aimed at his shoulder, not his head or hindquarters, and don't stare directly at him. Eventually he'll turn and let you come near. Reward the horse with a treat (kept hidden for this very moment) or just some pleasant attention. Then walk away from him (don't let him move away first). After a few sessions, he should learn that being caught isn't so bad after all.
➢ See also Handling horses; Leading

Walk-trot classes

Classes held at competitions of many breeds and disciplines, where the only gaits called for are the walk and trot. Most people view these classes primarily for beginners; however, many coaches encourage higher level riders to enter these classes whenever they need to "go back to basics." This might be due to a previous bad experience in competition, or the fact that the higher level rider has moved on to a less experienced horse. These classes are judged primarily on the manners of the horse and suitability of the horse to the rider, unless it is an equitation class in which the riders' skills are assessed.

The main exception to this definition is the Walk-Trot class held at Saddlebred shows. Animation and brilliance are considered more important than manners and suitability.
➢ *See also English competition; Western competition*

Wall of hoof

The hard outer lining that protects the frog and sole of the hoof. It should be relatively smooth and free from cracks and rings. Rings are deep ridges that run parallel to the coronary band; they may indicate previous founder. Grass rings are much shallower rings that might indicate stress in the animal's life, such as a change in diet or environment.
➢ *See also Foot care; Hoof*

Warbles

The larvae of the cattle grub, best known for burrowing into the backs of cattle and forming cysts. Sometimes they affect horses despite the best efforts of horsemen. Ivermectin is probably the best guard against this parasite. Although they do feed on the blood of the horse and can deplete certain nutrients, their primary source of trouble is that they rub against saddle pads. This can create terrible sores that become infected if not treated carefully. Cattle pastured near horses should also be treated.
➢ *See also Parasites*

War bridle

A type of restraint created by running a rope under the horse's upper lip and over his poll.

Warmblood

A type of horse developed in Europe from "cold-blooded" draft horses and "hot-blooded" Thoroughbreds and Arabs. There are a number of individual breeds of these tall horses, and all have smooth, elastic gaits and striking presence. They tend to be dark in color, with few white markings, but you may see Pintos and other colors. Popular in the showring, warmbloods are often seen in dressage classes, combined driving events, or show jumping.

Something to be aware of when purchasing a warmblood is that these large horses need more time to mature than lighter breeds do. In Europe, they often aren't started under saddle until they are five or six years old. If fed too much protein as foals and ridden too early, warmbloods can develop hock and other joint problems. Some warmbloods are hard to handle, as they can be stubborn and prone to fits of temper, especially when young. A few specific breeds are: Dutch Warmblood, Hanoverian, Holstein, Selle Francais, and Trakehner.
➢ *See also Breed; Coldblood; individual breeds*

Warming Up

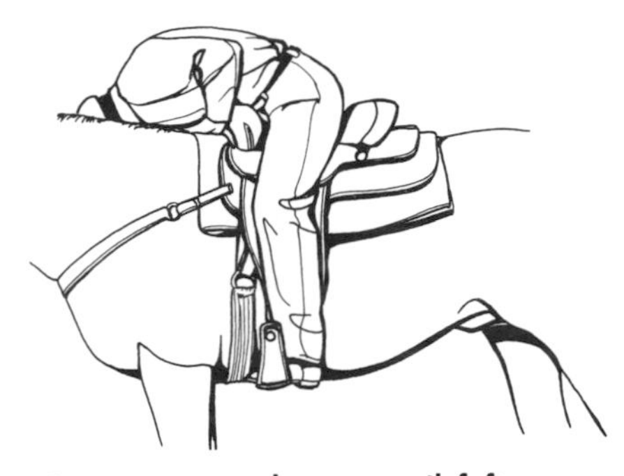

Lean over to touch your toes (left fingers to left toes; right fingers to right toes). This rider's legs have moved out of position.

Now lean over to the opposite side (left fingers to right toes; right fingers to left toes). This rider keeps her leg steady while stretching.

Warm-up exercises

At the beginning of every ride, both horse and rider need to warm up, stretch a little, and get their blood moving. Have your horse do a few circles at the walk with a relaxed rein before asking for a trot. When beginning to trot, post for a while before doing a sitting trot, which is harder on the horse's back.

There are several stretches and bends that can increase the rider's stability and help keep the legs in the correct position. These should be done on a calm horse, not one that spooks easily. While the horse is standing still or moving at a slow walk, move your arms in large circles (both directions). Stretch forward to touch your horse's ears, keeping your seat in the saddle and your feet where they belong. Slide each arm back to the tail without moving your legs. Reach down to touch your toes, first on the same side and then stretching each arm to the opposite side.

These are just a few possibilities. Your instructor may have other suggestions.

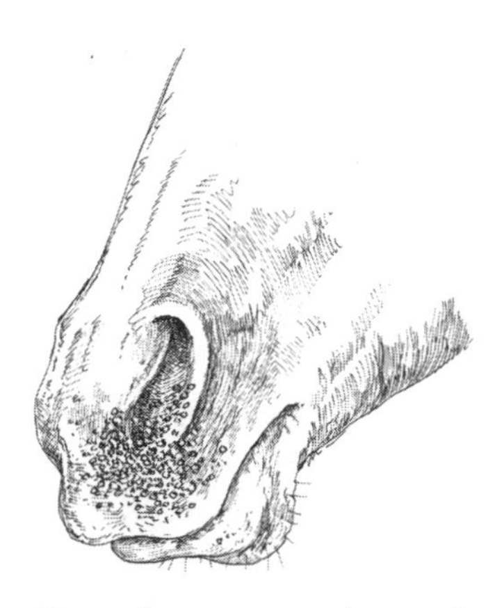

Warts **often appear on the muzzle.**

Warts

Caused by a virus, warts are not a serious problem unless they grow thickly enough to interfere with breathing. Warts are common in young horses that haven't built up immunity to the virus, and will generally disappear on their own after some time (several months to a year). They can spread from infected horses through small lesions and cuts in the skin where the virus can enter, as well as through shared grooming tools.

Warts generally appear around the muzzle and nostrils but somtimes are found in the ears or other parts of the body. Have your veterinarian examine larger warts and decide whether to deal with them surgically.

➢ *See also Skin problems*

Washing horses

➢ *See Bathing horses*

Water

Good-quality water is vital to your horse's health and should be available to him at all times. On average, a horse will drink between 8 and 12 gallons a day — more in the summer or after hard work, and less in the winter. However your horse gets his water (in a bucket or tub, from a stream or pond, or through an automatic waterer), you need to check the source daily. Containers need to be cleaned and refilled. Ponds can become stagnant, and even running brooks may freeze. Automatic waterers can malfunction, either flooding an area or leaving your horse without needed water.

After strenuous exercise, limit your horse's water intake to a few sips until he is completely cooled down; too much water too soon after exertion can lead to founder. A very warm horse can be sponged or hosed with tepid water to cool him off and wash the sweat from his coat. Cold water is not only uncomfortable for him, but also it can cause his back muscles to stiffen.

➢ *See also Facilities*

EASY WATERING

Automatic waterers for both fields and stalls are available, and they cut down on manual labor. But beware of several problems. If your horse is ill and is not drinking enough, you won't be able to notice as readily as you would with buckets. Automatic waterers can cease working, leaving your horse without water until you discover the problem. Or they can break, flooding the paddock or stall and leaving your horse ripe for problems such as scratches and intestinal disturbances. Finally, they can be dangerous if a hot horse has access to more water than is safe; for this reason automatic waterers should be outfitted with a shutoff valve.

Water bag rupture

Once a mare enters the second stage of labor, the water sac will break, releasing a gush of amber fluid. After this point, the foal should be born within 30 minutes (perhaps a little longer for a first-time mother). If it isn't, the mare is in trouble and a veterinarian should be called.

➢ *See also Foaling*

Water, crossing

Always cross a brook or stream at a shallow place where you can see the bottom. Deep, rushing water can be very dangerous. Footing in a brook can be treacherous, so go slowly and let your horse feel his way. It might make sense to dismount and lead your horse.

Let your horse stop for a drink if he wants, but be aware that some horses will lie down in water. Pawing might be a sign of imminent rolling, so get him moving! If your horse is nervous about crossing water, let a more experienced companion lead the way. A lot of horses are just like cats when it comes to getting their feet wet and may need plenty of encouragement before stepping into water. However, trust your horse's instincts if he's usually confident but becomes wary of crossing.

Be very careful when riding in a boggy area or swamp. A muddy path that would support your weight might give way under your horse and bog him down. Struggling to get out of a hole could cause your horse to panic and injure himself.

➢ *See also Trail riding*

A man may well bring a horse to the water,
But he cannot make him drink without he will.

—John Heywood,
Be Merry Friends

Waxing, as preparation for foaling

The mare's udder enlarges during the last month of gestation. A dry, waxy substance that is normally inside the various mammary ducts usually begins to push through the teats as milk replaces it. This process is known as "waxing," and is a classic sign that the mare will likely foal soon. However, some mares have so little wax in their ducts that it is pushed out without being noticed by humans. So it is not unusual for a mare to foal without ever showing signs of waxing. Other mares do not progress to labor until many days after waxing, perhaps even days after obvious milk flow has been noticed.

➢ *See also Foaling*

Way of going

The overall look of a horse as he moves, including his manners, natural impulsion, and presence.

Weaning foals

The proper time for weaning varies depending on the size and health of the foal, but it is generally done between 2 and 6 months of age. Keeping a foal with his dam longer than this time is difficult for the mare if she is pregnant again, and it can lead to a spoiled, badly behaved yearling. Traditionally, foals and mares were abruptly parted and put into separate pens where they could neither see nor hear each other. This is extremely stressful for the foal, in particular, who is still emotionally dependent on his mother. A foal that is removed abruptly may not eat for several days and could injure himself literally trying to climb the fence to find his dam.

A more humane way to accomplish weaning is to put mare and foal in adjacent pens with sturdy, high wire fencing between them (no gaps for the foal to get his head through). The foal doesn't need to nurse, but will be reassured by the sight and smell of his mother. It also helps for him to be with other weanlings or some other companion.

It will take a couple of weeks for the mare's milk to dry up. She should be on restricted food at this time, and unless it's very hot, you can limit her water, as well, to speed up the drying process. Hand-milking to relieve the pressure in her udder will only encourage further milk production.

➢ *See also Foal care*

This foal is taking the first step to adulthood by getting used to being separated from his mother. The fence must be high enough so that he won't try to scramble over it.

Weanling

A foal that is not yet a year old but is no longer nursing. Weanlings need lots of exercise, the company of other horses (preferably their own age), and a diet with adequate protein and energy for good growth. Natural foods are best: pasture and grass hay, with some alfalfa for protein. Grain requirements depend on breed and training expectations. A common mistake is feeding young horses too much protein, which causes rapid growth and can lead to a variety of skeletal problems.

Young horses need to socialize with others, and a group of weanlings get all the exercise they need if they have a large enough pasture. Feed them separately from mature horses to ensure that the youngsters are getting enough to eat. Fillies and colts should be separated fairly early on, as puberty can come early.

Contact with humans is important, too. It's never too early to start teaching ground manners. All foals should learn to lead, to stand tied, to pick up their feet, and to let their mouths be looked at. Grooming, sacking out, and introductions to new situations are all valuable training. Be aware, however, that work on the longe line is too stressful for young ligaments and should be avoided or limited until age two or older.

➢ *See also Feeding and nutrition; Foal care; Gelding; Protein; Sacking out; Weaning foals*

Weaving and stall walking

A horse that is confined too much may begin to sway back and forth continuously or walk back and forth, wearing a path in his bedding. These almost hypnotic behaviors are generally a sign of anxiety, restlessness, or boredom. Providing some toys for the stall and arranging for more exercise or turnout time will probably alleviate this problem.

➢ *See also Pacing, habit of; Vices, how to handle [chart]; Vices (sterotypes)*

Web halter

A halter made of strong nylon weave. Web halters are the most commonly used halters because they are inexpensive and come in a variety of colors. However, the wide webbing doesn't work for some horses that may become pushy. If the hardware on the halter is poorly made, the halter may break at one of those points. On the other hand, a nylon halter with heavy hardware may not break at all, which could be dangerous if the horse gets caught on something, so look for one with a leather crownpiece.

➢ *See also Halter*

Weed control

Weeds and tall grass should be cut around fence lines to prevent horses from getting their heads stuck while trying to see if the grass is indeed greener on the other side.

➢ *See also Facilities; Pasture; Poisonous plants*

Weighing feed

A horse's ration should be determined by weight, not volume. This is due to the fact that each crop of hay or grain has the potential of drastic differences in weight, due in part to variances in moisture and mineral content. So it is vital to your horse's well-being that you weigh your standard measure *each* time you receive a new load of feed. Weigh a bale of hay, then divide that weight by the number of flakes/slabs of hay in that bale to determine the number of flakes your horse needs at each feeding. Weigh a full grain scoop so you know how many scoops your horse needs of the new feed to equal the weight he received of your previous feed. You need to do this even if you don't change rations. A corn/oat mix can differ by as much as 25 percent depending on the moisture content and quality of the grains.

➢ *See also Feeding and nutrition; Grains; Hay; Pelleted feed; Scoop for grain; Total digestible nutrients (TDN)*

WEEDING OUT WEEDS

Weeds can gain a foothold in an overgrazed pasture, as the good grasses are eaten and bare ground appears. You can best maintain a good pasture by not overgrazing it. Check with your county extension agent to determine how many horses your pastures can support and in what way — full grazing, limited turnout, and so on. Ask, too, about weeds and how to eradicate them. Some weeds, such as dandelion and lamb's-quarter, are nutritious; still, they choke out the high-protein legumes.

Other plants are poisonous. Still others, such as foxtail, have sharp awns or seed pods that can lodge in a horse's mucous membranes and cause sores and abscesses. And certain plants, such as thistle, are neither poisonous nor dangerous, but they crowd out the beneficial plants.

WEIGHT AND NUTRITION INFORMATION FOR COMMON GRAINS

TYPE OF GRAIN	WEIGHT PER BUSHEL (LBS.)	TOTAL DIGESTIBLE NUTRIENTS	CRUDE FIBER	DIGESTIBLE PROTEIN	COMMENTS
Barley	48	78%	5.9%	10%	Some bulk added by hulls, but usually mixed with oats or wheat bran to aid digestion.
Corn	56	80%	6.7%	2%	High energy, low bulk and protein; too much makes a horse fat.
Oats	32–40	70%	11%	9.4%	Safest and most common grain fed; hulls give lots of bulk to keep digestive system working properly.
Sorghum (milo)	56	80%	2.3%	8.5%	Low bulk; needs to be mixed with other grains to prevent constipation.
Wheat	60	80%	2.6%	11%	Expensive where not grown; very high in protein and low in bulk, so it should make up less than a fifth of total ration.

Weight, as an aid

Along with your legs and hands, an important aid is your seat and the signals you give by shifting your weight. This is especially true in Western riding (where horses respond to neck-reining), but it is also a major component of nearly all high-level training and riding. Beginning with a deep, balanced seat, you can cue your horse to turn by shifting your weight slightly in the direction you want to go. For example, putting more weight on your outside stirrup can keep you horse from falling in to the center as he canters in a circle.

➢ *See also Aids; Seat*

Weight, of horse

Most riding horses weigh between 800 and 1,200 pounds, depending on height, body length, and overall fitness. Horses do get fat, even from just eating grass, and they sometimes need to be put on diets. There are several methods of calculating your horse's weight, which you'll need to know in order to estimate the correct dosage for deworming and other medications. The most accurate way is to use an actual livestock or truck scale. At a trucking weigh station, you can weigh your truck and trailer empty and then put in the horse and weigh it again.

A simpler, and more common, method is to use a specially calibrated weight tape, which measures the girth just behind the front legs. It should be wrapped snugly around the body. This method, however, gives only a rough estimate of weight.

A more accurate calculation is to measure the heart girth with a non-stretching measuring tape (a piece of nonstretching string will do, as well; it can then be measured on a carpenter's tape or yardstick, if that's what you've got). Then measure the horse's body length from shoulder to haunch. To calculate total weight, multiply the heart girth measurement by itself, then by the body length, and then divide that total by 330.

Weight tape

These specially calibrated measuring tapes can tell you roughly how much your horse weighs from the length of his heart girth.

Welsh pony

This popular pony is a wonderful first mount for a child, but, as with all ponies, temperaments vary among individuals. Welsh ponies are finely constructed, with clean lines and pretty heads. They are strong and athletic, however, with a reputation for soundness. Solid colors are the norm, with white markings permitted (no Pintos, though). Height can range from 11 to 14 hands high. Welsh ponies are seen in all kinds of showrings, including jumping and driving classes.

➢ *See also Breed; Pony*

CALCULATING HORSES' BODY WEIGHT

Heart girth x heart girth x body length ÷ 330 = total body weight

Example: Heart girth = 72"; body length = 61" inches

72 x 72 x 61 = 316,224
316,224 ÷ 330 = 958

AVERAGE WEIGHT OF HORSES

Type of Animal	Weight in pounds (lbs.)
Draft horse	1,500
Saddle horse	1,250
Colt	500
Pony	500

From *Small-Scale Livestock Farming* by Carol Ekarius

Welsh pony

W

Welsh Pony Society of America

➢ *See appendix*

Western banding

This is a show grooming process used by many of today's stock horse breeds in which the mane is banded into numerous small sections. Its purpose is similar to that for braiding a hunter's mane: to create a clean, smooth-appearing neck. Instead of being braided, the mane is divided into 30 or 40 tiny ponytails with small rubber bands. It is vital to practice banding until it is perfected. A sloppy banding job can distract from an otherwise excellent turnout.

At its highest level, banding can actually be used to create optical illusions to improve the appearance of conformation weaknesses. For example, a horse with a short, thick neck can be made to appear as if his neck is longer and slimmer by increasing the number of bands used. The more bands, the longer the neck appears. Likewise, a neck that is too slim can be made to appear thicker by reducing the number of bands used.

Western competition

Competitive Western classes are divided into three groups: in-hand classes (Showmanship and Conformation), in which the horse is shown wearing just a halter or bridle and led by a handler; equitation classes, which evaluate the rider's ability; and performance classes, which judge the horse's ability. Many of the specific classes measure the time-honored skills of the horses and riders of the American West such as cutting, reining, and so on.

➢ *See also Rodeo; Western riding*

WESTERN COMPETITIVE CLASSES

Conformation. Overall body structure of horse is judged, as well as action at the walk and trot.
Showmanship. Skill of handler and turnout of horse are judged, according to the rules of the specific show.
Western Horsemanship. Skill of rider is judged. Also called Stock Seat Equitation.
Western Performance. Horse's abilities are judged, according to guidelines of the specific show. Includes events such as Cutting, Reined Cow Horse, Reining, Western Pleasure, Trail, and Western Riding.

Western equine encephalomyelitis (WEE)

➢ *See Equine encephalomyelitis*

Western horse

The modern Western horse is the result of many years of careful selection for important traits. The tough little Spanish horses that first worked the American range have been mingled with the blood of the Arab and the Thoroughbred to produce a strong, smart horse with plenty of stamina.

The classic Western horse is the Quarter Horse, with his stocky build, well-muscled body, and sturdy legs. All breeds of this type share characteristics that make them able to work long days, carry heavy loads, react to unpredictable cattle, and respond with instant speed when asked. The Western horse tends to be intelligent and calm, with the ability to form a true partnership with his rider. Other popular breeds include the Spanish Barb and Spanish Mustang, as well as the color breeds, which are typified by the Appaloosa, Paint and Pinto, Palomino, and Buckskin.

➢ *See also Types of horse; individual breeds*

The modern Western horse combines the best of Arab, Spanish, and Thoroughbred bloodlines.

Western riding

Western riding, developed in the United States, Canada, and Australia from the work done on cattle ranches, was originally modeled after the riding styles of Mexican *vaqueros*. France, however, also has cowboys, working the wild black bulls in the south of France on indigenous semi-wild white Camargue horses.

Western horsemanship may ask the horse for similar skills as those required in English riding, except for jumping. The rider's body action for creating a particular movement may also be similar. However, the Western saddle offers substantially more security to the rider and the bits in the horse's mouth offer substantially more direction to the horse.

Western Tack

The prominent Western saddle horn is designed to anchor the rope for cowboys working cattle. Australian stock saddles are lighter than Western saddles but they are meant to offer many of the same benefits to the rider, including protection for the inside of the leg. Western stirrups are attached by a leg-width strip of leather, whereas English stirrups are attached by an inch-wide strip that can cut into the calf unless tall boots are worn. Western saddles also feature a deep seat, which provides stability for sitting the horse when a cow jerks against a rope, and provides relative comfort while galloping over long distances.

Western vs. English Riding

Western riding offers just as much in the way of specialization and competition as English riding, although the competition is based on Western ranch activities: roping, pole bending, and negotiating challenging terrain, rather than the more leisurely activities of fox hunts.

While some English riders like to think of their mounts as "hot" and demanding of courage and skill to ride, Western riders usually prefer to think of their mounts as willing, skilled, and pleasurable, even when faced with taxing sports such as pole-bending. In general, Western horses do not compete much until they are very skilled.

Because of their desire for a steady, even-tempered animal, Western trainers and riders spend a lot of time enhancing a horse's ability to do tasks in a relaxed way, with a long and low frame. English riders, on the other hand, want the horse "up" and "on the bit." Both disciplines demand that a horse go forward and stop on request.

A Western-trained horse carries his head lower to the ground. The proportions of rider and horse in Western riding are different from those in English. English riders tend to be about one-fifth or less of the size of their horses (but not too small, to avoid the rider's getting "lost" in the saddle). Western riding uses a ratio of about one-fourth, so the rider appears larger on an appropriate horse.

Western riding utilizes a balanced seat, so it is relatively easy for these riders to do dressage. Indeed, many Western riders bring a better, deeper seat to their initial dressage training than hunt- and forward-seat riders do.

COMPLEMENTARY COLORS

Unlike English competitions, riders in Western competitions can wear different colors. Here are some guidelines for choosing colors that will be most flattering to your horse's coat color.

- **Black, white, blue roan, silver gray, and bay horses.** Wear bright jewel tones such as royal blue, purple, red, emerald, teal, forest green, or turquoise.
- **Chestnut, liver chestnut, sorrel, and dun horses.** Wear earth tones, such as browns, golds, and tans, or cool jewel tones, such as emerald, teal, forest green, or turquoise.
- **Buckskin and palomino horses.** Any color looks good against these horses' warm neutral coats.
- Black looks good with any horse color and will project a dignified image.

W

Because the Western bits are so much less forgiving of rough hands than English bits, Western horsemen may also bring softer hands to the dressage ride.

Western Competitive Attire

Western competitive events, both for amateurs and professionals, borrow from ranch work. The events include cutting, calf roping, pole bending, team penning, obstacle courses (there may even be small "jumps," or obstacles the horse can either walk or spring over). Western pleasure classes assess the horse as a comfortable animal to take out for long trail rides or camping excursions. Slightly different attire is used for all of these events, but all apparel has a Western flavor. These days, some Western hats also have helmets at the crown, since riders have realized that they can get hurt during a fall. Western riders may also compete in endurance rides; often, in these cases, they switch to the lighter endurance saddle.

In Western riding, the horse may be as nattily attired as the rider. Western show saddles are usually tooled, and may be ornamented with silver and turquoise. The bridle and breastplate might also be decorated. Usually, the rider chooses colors for her shirt and chaps that enhance the horse's coat color.

The Western Horse

The original Western horse is the Quarter Horse. Usually, the type preferred is the "foundation" Quarter Horse, with massive hindquarters and a wide body on relatively short legs; this gives the horse good conformation for standing firm while, for example, a calf pulls at the end of a rope. Quarter Horses are also sprinters, good for fast spurts over short distances. They are flexible and can turn easily and quickly, and they can jump. Paint horses are much like Quarter Horses in conformation, as are many Appaloosas. Colorado Rangers, a minor breed that looks something like an Appaloosa, may be seen from time to time, and Palominos (which refers to their pale gold coloring and not their breed — usually Quarter Horse or Thoroughbred) are very popular.

There are, however, Western shows in some other breed organizations, notably in the Arabian horse world. In an open show, horses of any breed that can do the work compete equally, whereas in a breed show only registered members of the breed may compete. For example, if you own a Quarter Horse and show him in open Western (or English) shows, you don't need papers to prove his genetic heritage. But in a Quarter Horse breed show, you would need the papers to compete.

➢ *See also Appaloosa; Attire; Bending; Bit; Breastplate [box]; Bridle; Calf roping; Chaps; Cutting; Paint and Pinto; Palomino; Pole bending; Quarter Horse; Rodeo; Stock saddle; Team penning*

ATTIRE GUIDELINES

The rider will have various levels of attire for different events in which he or she participates. Male riders generally wear a string tie with a shirt; they may wear a vest, but usually jackets of any sort are not worn to show. Women may wear bow ties, string ties, or rosettes at the closed collar of their show shirts. Some events call for jeans and Western boots; others call for chaps over jeans with shorter boots.

Gloves are not necessary, except in roping events where hand protection is desired. Boots, of course, are Western style, and they are chosen to complement the rest of the outfit. For hacking and schooling, riders may wear jeans, chaps, and short boots, or they may wear jeans and Western boots, depending on their preference and what they are working on. Check rules of the competition.

Western tack

The basic differences between Western and English saddles are quite obvious. The Western saddle was developed for sturdiness, comfort, and utility. It needed to handle the strain and stress of roping, be able to pack overnight gear, and keep the rider comfortable for many hours. Today there are a number of variations on the standard Western saddle. Which one you choose depends on the event you're most interested in. In addition to the ranch or working saddle, specialized saddles are available for cutting, reining, roping, team penning, barrel racing, pleasure riding, horsemanship, or equitation.

This lovely Palomino is wearing a typical Western saddle with decorative tooling around the edges and a colorful wool blanket. Note the long saddle strings at the shoulder and behind the cantle for attaching gear. The bridle is shown with no noseband and split reins, a common Western style.

Roping and team penning saddles have high horns and deep seats to absorb the shock of the cow hitting the end of the rope. Reining saddles have lower horns to stay out of the way of the reins. Barrel racing saddles are lightweight with deep seats and wide swells to keep the rider in place. Pleasure saddles tend to be richly decorated with tooling, fancy stitching, or silver medallions.

For fast work that requires many changes of direction and sudden stops, saddles generally have both a cinch and a flank cinch for added security. The cinch for a Western saddle is usually made of wool or mohair webbing, not leather, although the flank cinch consists of a leather strap.

Western bridles, also called headstalls, come in several types, but they generally have no nosebands. The hackamore does have a noseband; however, it doesn't have a bit. Some bridles have brow bands that pass between the horse's ears instead of across the forehead. Reins can be of three styles: open or split, closed, and roping. Most Western riders use open reins, which are long, separate straps that are not attached at the ends. Closed reins are connected with a three-foot-long romal or quirt at the end. Roping reins are a continuous loop running from one side of the bit to the other.

➢ *See also Bit; Hackamore; Saddle; Tack; Tacking up*

West Nile virus

West Nile encephalitis is a disease that has been present in parts of Africa, Asia, and Europe for more than half a century. In recent years it has appeared in the northeastern United States. Mosquitoes carry the virus from birds to other animals, including horses and humans. Only a small percentage of those infected actually show symptoms of illness, but the virus can be deadly. Affected horses become feverish and unable to control their muscles.

Wethers

Another term for uterus.

➢ *See also Anatomy of the horse; Breeding; Foaling; Gestation; Reporductive tract*

W

Wheals

➢ *See Hives (wheals); Skin problems*

Wheat and wheat bran

Wheat is not commonly fed to horses because of its cost, but in areas where it is grown, it can make up part of the grain ration (no more than 20 percent). Wheat is high in protein and energy but it isn't very bulky and should be mixed with other grains. It must be rolled or crushed to make it easier to chew and digest.

Wheat bran, the rough outer casing of the kernel, adds bulk to concentrated foods and has a laxative effect. A bran mash can be a nice pick-me-up for a tired or sick horse in the winter and it can help stop constipation. Wheat bran should not be fed dry unless it's given in very small amounts and is well-mixed with other grains.

➢ *See also Feeding and nutrition; Grains; Weighing feed*

I'd horsewhip you if I had a horse.

—Groucho Marx, in "Horsefeathers," 1932; written by Bert Kalmar, Harry Ruby, and S. J. Perelman

Whinnying

One of the many vocalizations that horses make as they communicate with each other and with humans. Horse pals will whinny to one another when they are separated, as will a mare and foal, and you will often hear horses neigh to each other as riders return from a trail ride. A horse's whinny or neigh is as individual as the human voice.

➢ *See also Body language, of horses; Vocalizing, of horse*

This handler has her crop in the right position to give the horse a tap on the hindquarters in case he moves out of line.

Whippletree (wiffletree)

➢ *See Harness; Swingletree*

Whips, use of

A whip can be an effective addition to the natural aids, which should always be used first. If a horse does not respond to the touch of the lower leg or to the voice, a tap with a whip can reinforce the desired response. In longeing, the long whip keeps the horse from moving in towards the handler. A long whip can also help in teaching ground manners.

Whips and crops should never be used to punish or frighten a horse. If improperly applied, these aids create fear and resentment, and horses don't forget such treatment. Overuse leads to insensitivity and decreasing responsiveness to aids.

➢ *See also Aids; Crop; Longeing*

Whiskers

Like a cat, a horse needs those whiskers to feel his way in enclosed spaces (like a feed bucket). However, grooming requirements for most shows are for a clean-shaven muzzle, so most show riders trim the whiskers off their horses' muzzles and chins. If you're not showing your horse, trimming his whiskers is unnecessary. If you want to tidy him up, go ahead, but don't trim them too short!

White line

White border found between the wall of the hoof and the sole of the foot.

➢ *See also Anatomy of the horse; Foot care; Hoof*

White snakeroot *(Eupatorium rugosum)*

Found in sandy areas in the Midwest, this poisonous plant often stays green into the early fall when other plants are drying up. Symptoms of poisoning include stiffness, lethargy, clumsiness, and weakness, and can show up after less than 10 pounds are consumed.

➢ *See also Poisonous plants*

"Whoa" command

Whoa means "stop." Some people say "ho" instead. Your horse should learn to associate the word with the action so that eventually he will halt on a vocal cue alone. It may save your life.

➢ *See also Aids; Leading; Voice, as an aid*

Wiffletree (whippletree)

➢ *See Harness; Swingletree*

Wild cherry *(Prunus avium* and *Prunus serotina)*

A common plant found all over North America in both wet and dry areas. In arid regions it grows like a shrub, but with plenty of water it will grow into a tree. Sometimes mistakenly called chokecherry *(Prunus virginiana)*, it is very poisonous to horses, especially when the leaves are wilted.

➢ *See also Poisonous plants*

Windbreaks, natural

Horses can stand nearly any kind of weather if they have sufficient shelter from the worst of wind and rain. After all, wild horses don't have stables and run-in sheds to protect them. They find shelter in stands of trees, in canyons and gullies, and behind tall rock formations. A pasture with a grove of trees or patch of brush offers sufficient shelter for most horses with heavy winter coats.

➢ *See also Shelter*

Wind-sucking

➢ *See Cribbing; Vices, how to handle [chart]; Vices (stereotypes)*

Winging

A movement fault associated with feet that turn in. The horse swings his hooves toward each other when moving, and can injure himself.

➢ *See also Leg, of horse*

POISONOUS PLANT

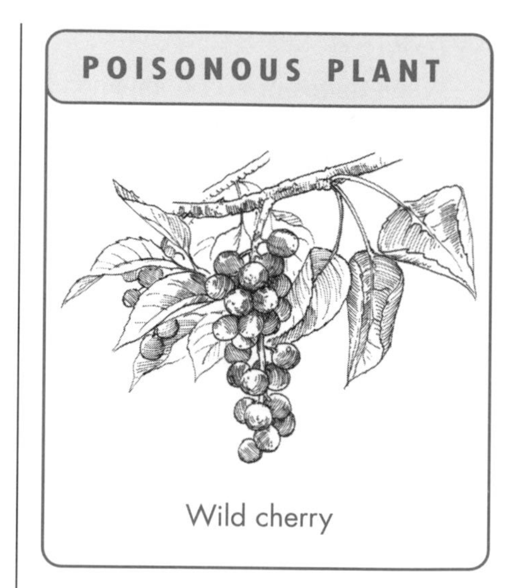

Wild cherry

Winking

When mares in heat open and close the lips of the vulva to expose the clitoris and indicate readiness to mate.

➢ *See also Breeding; Estrus*

Winter care

➢ *See Seasonal care*

Wisps, making and use of

A simple handful of hay can be a useful grooming tool, or you can weave a sturdy pad that can be used for a stimulating massage. Using a wisp can be a nice addition to your grooming routine. For a vigorous stropping, smack the wisp flat against the larger muscles (neck, shoulders, hindquarters) and whisk it along the direction of the hair growth. This gentle banging causes the muscles to contract and then relax, increasing circulation and carrying away waste products.

➢ *See also Grooming*

WEAVING A WISP

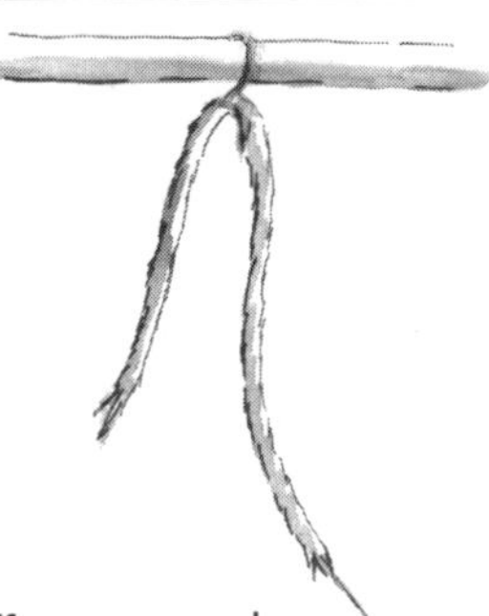

1. Fluff up a couple of flakes of soft, long-stemmed hay or straw and dampen it well. Take a two-foot length of twine and tie it to a wall, leaving about six inches hanging on one side. Starting with the longer end, twist a handful of hay onto the twine; keep adding handfuls as you go. You'll have to use both hands to keep the twist tight. You'll run out of twine, but continue to twist hay until you've got about nine feet of "rope." It should be about ¾ inch thick. Twist another length of twine into the end of the rope for about a foot.

2. You'll need a second person to help with the rest of the process. Detach the end of the rope from the wall and maintain the tension on the twisted hay. Lay the twine into a rough pretzel as shown. The short end of the pretzel should have about six inches of twine left free.

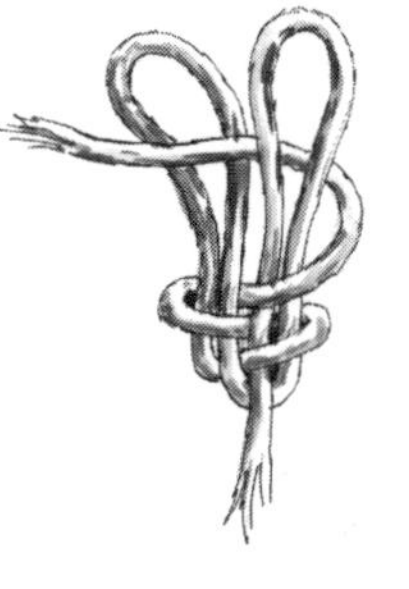

3. Take the longer end of the rope and begin to weave it behind one loop and in front of the other.

4. When you reach the top of the loops, run the end through both loops twice.

5. You should have a woven pad of hay with two ends of twine hanging out. Tie the twine into a handle (custom fit to your hand).

Withers

The bone structure at the top of the horse's shoulder, between his neck and his back. A horse needs well-defined withers to keep a saddle securely in place. Poorly defined withers are known as **mutton withers,** because they appear very similar to the withers on sheep. Mutton withers can be a serious fault for riding horses, especially those used for working Western events. The saddle often will slide all the way to the side of the barrel on horses with mutton withers, unless you use breastplates to prevent it.

➢ *See also Anatomy of the horse; Conformation; Mutton-withered*

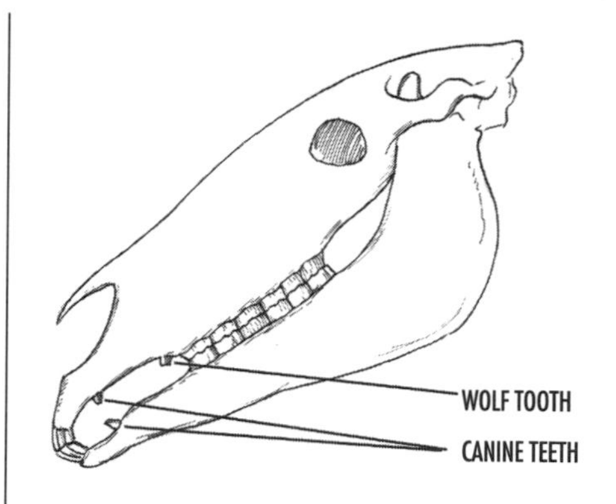

Wolf teeth emerge just in front of the molars, usually just the upper molars. Canine teeth are located behind the incisors on the top and bottom jaw.

Wolf teeth

Often confused with canine teeth, wolf teeth are harder to see, as they sit farther back (in front of the upper molars). Usually small with shallow roots, wolf teeth can interfere with the action of the bit, and many owners have them removed.

➢ *See also Dental care; Teeth*

Wood fencing

Wood fences are the classic look for a horse farm, but they have advantages and disadvantages. When properly maintained, they can last for years, especially in a dry climate (wet climates require periodic application of preservatives). Maintenance is the key — check regularly for nails, splinters, cracks, loose boards, and other hazards. An electric wire along the top and just inside the fence line keeps cribbers and chewers away.

➢ *See also Fencing materials*

WORKING RANCHES

There are working ranches that still perform a lot of work on horseback, and amateur riders can often arrange to help for a week, as a vacation. Riders are expected to have at least minimal skill and be capable of spending eight hours in the saddle doing real ranch work.

Working cow horse

The working cow horse needs innate "cow sense" in addition to quick reflexes and rapport with the rider. Together, rider and horse complete a reining pattern and then have two minutes to move a single cow through a predetermined sequence. They must "box" the cow, or keep it at one end of the arena, drive it along the fence line, and turn it in circles.

➢ *See also Cutting; Team penning; Team roping; Western competition*

Working gait

In dressage, a gait that is free and easy, steady and elastic, but without the tightly controlled impulsion of a collected gait.

➢ *See also Dressage*

Working horse

In Western competition, working horse events include cutting, reining, and working cow horse. These classes showcase the skills you and your horse would use when actually working with stock on a ranch.

➢ *See also Western competition; individual classes*

Working hunter

A type of horse judged on his ability to negotiate a jump course.

➢ *See also English competition; Hunter*

xyz

X

Xenophon

A 4th century Greek soldier, Xenophon was an avid horseman whose book *On Horsemanship* contains many principles that are still standard practice today. Horseback riding, he claimed, "makes the body healthy, improves the sight and hearing, and keeps men from growing old."

Xylazine

A popular tranquilizer for horses. It is a controlled substance, so it should be administered only by a licensed veterinarian. It can cause severe damage to muscle tissue if injected improperly. Beware of working around the hindquarters of horses on xylazine since they are known to kick aggressively, even if they've never been prone to kicking before.

Y

Yearling

A foal that has passed its first birthday. All registered horses share the same birthday: January 1st.

➢ *See also Foal care*

Yellow burr weed (*Amsinckia* spp.)

A common weed in wheat and grain crops that causes liver damage in horses.

➢ *See also Poisonous plants*

Yellow star thistle *(Centaurea solstitialis)*

A plant common to the Pacific Coast and other areas of the United States that causes brain damage if eaten in sufficient quantities.

➢ *See also Poisonous plants*

Yew (*Taxus* spp.)

A dark evergreen tree or shrub with red berries. This ubiquitous decorative shrub is very poisonous to both horses and humans. A mouthful can kill a horse within minutes.

➢ *See also Poisonous plants*

Yielding to pressure

➢ *See Leg yield*

Young horse care

➢ *See Foal care*

Youth

In most competitions, the youth division is for riders under age 18. However, be aware that each ruling organization has the right to organize and name its divisions however it chooses. So one horse show may have

POISONOUS PLANT

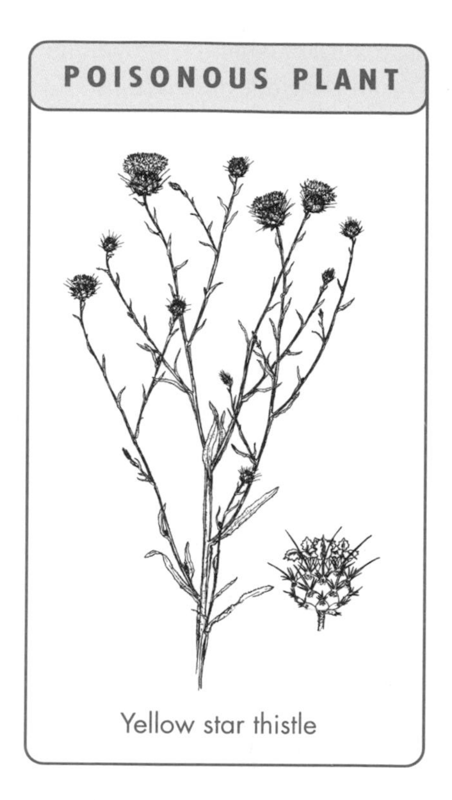

Yellow star thistle

POISONOUS PLANT

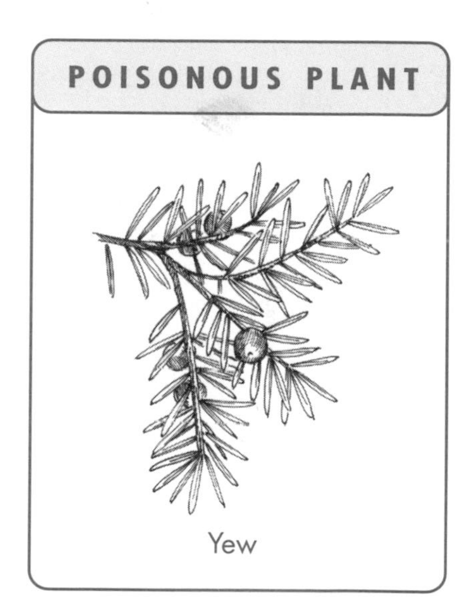

Yew

a Junior Division for the youngest exhibitors (under 10 years), plus a Youth Division for exhibitors from 11 to 17 years. Another show may have various Junior Divisions, without any mention of a "Youth" division. Check the rules of your organization before entering a show.

A zebra's life in the herd is similar to that of the horse.

Z

Zebra

This wild, boldly striped *Equus* comes from the same genus and family (Equidae) as the horse. There are several branches of the zebra line, all native to Africa, including *Equus zebra* (mountain zebra), *Equus burchelli* (common or Burchell's zebra), and *Equus greyvi* (Grey's zebra). Zebras can have stripes of black or dark brown and white or buff. All equids can interbreed to produce hybrids, which are normally sterile.

The zebra's stripes are among the few features that truly distinguish it from the horse, since they share many other traits. Their similarities include:

- They live in herds, headed by a single stallion.
- Younger stallions will remain with their family herd until they are old enough to establish their own breeding herd. Aggression between stallions increases during breeding season.
- Females establish a pecking order with a lead female just below the stallion.
- The lead mare guides the herd when it is fleeing predators. The stallion is the rear guard.
- Gestation is a little less than one year.
- The female guards the foal when it is first born.
- Foals remain bonded to their mother until the next foal is born.
- Their natural diet is made up of available grasses, nonselectively grazed.

illustration credits

Cathy Baker: 109, 119, 427

Sarah Brill: 16 (bottom), 230, 346

Jeffrey Domm: 253

Beverly Duncan: 353

Jim Dyekman: 15, 19 (bottom), 21 (bottom), 22, 24, 29 (bottom), 43, 55 (top), 67, 69 (bottom), 72 (top), 83, 87, 104, 108, 121, 132, 137 (top), 156, 166, 181, 196, 226, 231, 236 (bottom), 238 (top), 245, 249, 282 (top), 283 (top), 286 (bottom), 293 (top), 321 (top row), 340 (top), 341 (bottom), 343 (bottom), 344 (top box: dressage, jumping, and Western saddles), 355 (top), 360 (bottom), 366, 378 (box: right), 380 (bottom), 398, back cover (center left)

Brigita Fuhrmann: 99

Chuck Galey: 8 (bottom), 19 (top), 95 (top), 348 (bottom)

Peggy Judy: 410

Carl Kirkpatrick: 36 (bottom), 168

Richard Klimesh: 124

Alison Kolesar: 37, 70 (top), 112, 137 (bottom), 194, 215 (bottom), 235, 359 (bottom), 391 (bottom)

JoAnna Rissanen-Welch: 4–6, 7 (top), 8, 10 (top), 13, 14, 21 (top and middle), 28, 34 (bottom), 36 (top), 44, 45, 48 (bottom), 50, 54 (bottom), 63, 64, 65 (bottom), 70 (bottom), 72 (bottom), 75, 86, 89, 95 (bottom), 100 (bottom), 128 (bottom), 148, 165 (bottom), 173, 175, 178,179, 200, 214, 215, 250, 262, 264, 265, 268, 279, 280, 285 (top), 287, 292, 294, 297, 298 (top), 300, 308, 309, 315, 317, 321 (second row and bottom), 323, 325, 328 (top), 333–337, 340 (bottom), 341 (top), 344 (top box: racing and endurance saddle, bottom), 345, 348 (top), 349, 355 (bottom), 356–358, 359 (top), 362–365, 370, 372, 375, 376, 378 (box: left and middle), 379, 380 (top), 383, 390, 391 (sidebar), 392 (top and middle), 393 (right), 394, 395, 397 (bottom), 398 (bottom), 411, 412 (middle and bottom), 413 (sidebar), 414, 423, 424 (top), 432 (bottom), 435, 437, 441, 442, 444 (steps 1 and 5), 448

Elayne Sears: 2, 7 (bottom), 10 (bottom), 17 (top), 18, 20, 25, 29 (top), 34 (top), 42, 46 (top), 48 (top), 49, 51, 52, 54 (top), 56, 58, 65 (top), 71, 76, 78, 80, 81, 85, 94 (bottom), 100 (top), 101, 103, 116, 122, 123, 134, 135, 144 (bottom), 147, 149–153, 155, 158 (top), 174, 184, 185, 189, 192, 199, 202, 203, 205, 206, 210, 215 (top), 220, 229, 233, 234, 237, 238 (bottom), 239, 241, 242, 252 (top and bottom), 255, 256, 272, 274, 275, 281, 288, 290 (top), 293 (bottom), 307 (bottom), 310, 316, 319, 328 (bottom), 360 (top), 361, 369, 373, 377, 392 (bottom), 393 (left), 399, 401 (top), 403, 412 (top), 413 (bottom), 415, 418, 420, 421, 431 (top), 433, 443, 444 (steps 2–4), 445, 447 (bottom), back cover (top left, bottom right)

N. J. Wiley: 68, 73, 94 (top), 105, 113, 146, 165 (top 3), 291, 371, 407, back cover (top right, bottom left)

resources

If your organization is not listed here, or if the listing needs to be updated, please send your contact information to: Equestrian Editor, Storey Books, 87 Marshall Street, North Adams, MA 01247, or deb.burns@storey.com.

General Organizations

American Association of Horsemanship Safety
Drawer 39
Fentress, TX 78622
512-488-2220

American Horse Council
1700 K Street NW, Suite 300
Washington, DC 20006
202-296-4031
202-296-1970 (fax)
www.horsecouncil.org

American Horse Shows Assoc.
220 East 42nd Street
New York, NY 10017-5876
212-972-2472

American Medical Equestrian Assoc.
103 Surrey Road
Waynesville, NC 28786
704-456-3392

American Riding Instructors Assoc.
28801 Trenton Court
Bonita Springs, FL 34134
941-948-3232
941-948-5053 (fax)

British Horse Society
Stoneleigh Deer Park
Kenilworth
Warwickshire CV8 2XZ
08701-202-244
01926-707-800 (fax)
www.bhs.org.uk

Bureau of Land Management National Wild Horse and Burro Program
P.O. Box 12000
Reno, NV 89520-0006
775-861-6583 or 866-4MUSTANGS
www.wildhorseandburro.blm.gov/

Fédération Equestre Internationale
Avenue Mon Repos 24
P.O. Box 157
1000 Lausanne 5
Switzerland
41-21-310-47-47
41-21-310-47-60 (fax)
www.horsesport.org

Horsemanship Safety Assoc.
517 Bear Road
Lake Placid, FL 33852
800-798-8106

National 4-H Council
7100 Connecticut Avenue
Chevy Chase, MD 20815-4999
301-961-2937
www.4-H.org
International 4H Youth Exchange
www.ifyeusa.org
For your local chapter of 4-H, look in your telephone directory under Cooperative Extension Office, and through it contact your County 4-H Extension Agent. Individual state listings can be found at www.4-H.org/fourhweb/statelist

National FFA Organization
P.O. Box 68960
6060 FFA Drive
Indianapolis, IN 46268
317-802-6060
317-802-6061
www.ffa.org

North American Riding for the Handicapped Assoc.
P.O. Box 33150
Denver, CO 80233
800-369-RIDE (7433)
303-252-5610 (fax)
www.narha.org

Royal Canadian Mounted Police
www.rcmp-grc.gc.ca

U.S. Pony Clubs, Inc.
The Kentucky Horse Park
4041 Iron Works Parkway
Lexington, KY 40511-8462
859-254-7669
www.ponyclub.org

Breed Associations

American Connemara Pony Society
2360 Hunting Ridge Road
Winchester, VA 22603
540-662-5953
www.acps.com

American Hackney Horse Society
4059 Iron Works Parkway, Suite 3
Lexington, KY 40511
859-255-8694
www.hackneysociety.com

American Hanoverian Society
4067 Iron Works Parkway, Suite 1
Lexington, KY 40511
859-255-4141
www.hanoverian.org

American Holsteiner Horse Assoc.
222 East Main Street, Suite 1
Georgetown, KY 40324
502-863-4239
www.holsteiner.com

American Miniature Horse Assoc.
5601 South Interstate 35 West
Alvarado, TX 76009
817-783-5600
www.amha.com

American Morgan Horse Assoc.
P.O. Box 960
Shelburne, VT 05482
802-985-4944
www.morganhorse.com

American Paint Horse Assoc.
P.O. Box 961023
Fort Worth, TX 76161-0023
817-834-APHA
www.apha.com

American Quarter Horse Assoc.
P.O. Box 200
Amarillo, TX 79168
806-376-4811
www.aqha.com

American Saddlebred Horse Assoc.
4093 Iron Works Parkway
Lexington, KY 40511
859-259-2742
www.saddlebred.com

American Shetland Pony Club
81 B Queenwood Road
Morton, IL 61550
309-263-4044
www.shetlandminiature.com

American Trakehner Assoc.
1520 West Church Street
Newark, OH 43055
740-344-1111
www.americantrakehner.com

Appaloosa Horse Club
2720 West Pullman Road
Moscow, ID 83843
208-882-5578
www.appaloosa.com

Cleveland Bay Horse Society of North America
P.O. Box 221
South Windham, CT 06266
860-423-9457

Friesian Horse Assoc. of North America
P.O. Box 1809
Sisters, OR 97759
541-549-4272
www.fhana.com

Gaited Horse International Assoc.
507 North Sullivan Road, Ste. A-3
Veradale, WA 99037-8531
www.gaitedhorse.com

Haflinger Assoc. of America
14570 Gratiot Road
Hemlock, MI 48626

Haflinger Breeders Organization
14640 State Route 83
Coshocton, OH 43812-8911
740-829-2790
www.stallionstation.com

Haflinger Registry of North America
14640 State Road #83
Coshocton, OH 43812

International Arabian Horse Assoc.
10805 East Bellamy Drive
Aurora, CO 80014-2605
303-696-4500
www.iaha.com

International Quarter Pony Assoc.
P.O. Box 125
Sheridan, CA 95681
916-645-9313
www.netpets.org/~iqpa

The Jockey Club (Thoroughbreds)
821 Corporate Drive
Lexington, KY 40503
www.equineline.com

Masters of Foxhounds Assoc.
P.O. Box 2420
Leesburg, VA 20177
703-771-7442
www.mfha.com

Missouri Fox Trotting Horse Breed Assoc.
P.O. Box 1027
Ava, MO 65608
417-683-2468
www.mfthba.com

National Mustang Assoc.
P.O. Box 1367
Cedar City, UT 84721-1367

NA/WPN Dutch Warmblood
P.O. Box 0
Sutherlin, OR 97479
541-459-3232
www.nawpn.org

North American Trakehner Assoc.
P.O. Box 12172
Lexington, KY 40581
502-867-0375
www.trakehner.org

Northeast Fjord Horse Assoc.
P.O. Box 59
Nottingham, NH 03290

Oldenburg Horse Breeders Society
150 Hammocks Drive
West Palm Beach, FL 33413
561-969-0709
www.Oldenburghorse.com

Palomino Horse Breeders Assoc. of America, Inc.
15253 East Skelly Drive
Tulsa, OK 74116-2637
918-438-1234
www.palominohba.com

Paso Fino Horse Assoc., Inc.
101 North Collins Street
Plant City, FL 33566-3311
813-719-7777
www.pfha.org

Performance Horse Registry
c/o American Horse Shows Assoc.
4047 Ironworks Parkway
Lexington, KY 40511
859-258-2472
www.ahsa.org

Peruvian Paso Horse Registry of North America
3077 Wiljan Court, Suite A
Santa Rosa, CA 95407
707-579-4394
www.pphrna.org

Pinto Horse Assoc. of America
1900 Samuels Avenue
Fort Worth, TX 76102
817-336-7842
www.pinto.org

Pony of the Americas Club, Inc.
5240 Elmwood Avenue
Indianapolis, IN 46203
317-788-0107
www.poac.org

The Racking Horse Breeders Assoc. of America
67 Horse Center Road
Decatur, AL 35603
www.rackinghorse.com

Rocky Mountain Horse Assoc.
P.O. Box 286
Paris, KY 40362-0286
859-987-9285
www.rmhorse.com

Tennessee Walking Horse Breeders' and Exhibitors' Assoc.
P.O. Box 286
Lewisburg, TN 37091
800-359-1574
www.twhbea.com

United States Icelandic Horse Congress
38 Park Street
Montclair, NJ 07042
973-783-3429
www.icelandics.org

United States Trotting Assoc. (Standardbreds)
750 Michigan Avenue
Columbus, OH 43215
614-224-2291
www.ustrotting.com

Welsh Pony & Cob Society of America
P.O. Box 2977
Winchester, VA 22604
540-667-6195
www.scendtek.com/wpcsa

Associations for Specific Types of Riding

American Cutting Horse Assoc.
P.O. Box 2568
Wimberley, TX 78676
512-847-6168

American Endurance Ride Conference
701 High Street, Suite 203
Auburn, CA 95603
916-823-2260
www.aerc.org

American Remount Assoc.
P.O. Box 28
Mercersburg, PA 17236
301-971-1238
e-mail: amremount@aol.com

American Team Penning Championships
1776 Montano Road NW, Bldg. 3
Albuquerque, NM 87107
505-344-1776

Intercollegiate Horse Show Assoc.
2109 Stillwell Beckett Road
Hamilton, OH 45013

International Halter-Pleasure Quarter Horse Assoc.
11182 U.S. Highway 69 No.
Tyler, TX 75706

Mustang and Burro Organized Assistance
P.O. Box 601
Barrington, NH 03825
603-664-2787

National Barrel Horse Assoc.
P.O. Box 1988
Augusta, GA 30903
706-722-7223
www.nbha.com

National Cutting Horse Assoc.
4704 Hwy 377 South
Fort Worth, TX 76116-8805
817-244-6188
817-244-2015 (fax)
www.nchacutting.com

National High School Rodeo Assoc.
11178 North Huron, Suite 7
Denver, CO 80234
303-452-0820
www.nhsra.org

National Intercollegiate Rodeo Assoc.
1815 Portland Avenue, #3
Walla Walla, WA 99362
509-529-4402
www.collegerodeo.com

National Little Britches Rodeo Assoc.
1045 West Rio Grande
Colorado Springs, CO 80906
719-389-0333
www.rodeoattitude.com

National Reined Cow Horse Assoc.
1318 Jepsen Avenue
Corcoran, CA 93212
209-992-9396
www.nrcha.com

National Reining Horse Assoc.
3000 NW 10th Street
Oklahoma City, OK 73107-5302
405-946-7400
www.nrha.com

National Show Horse Registry
10368 Bluegrass Parkway
Louisville, KY 40299
502-266-5100
www.nshregistry.org

North American Trail Ride Conference
P.O. Box 2136
Ranchos de Taos, NM 87557
505-751-4198
www.natrc.org

Professional Rodeo Cowboys Assoc.
101 ProRodeo Drive
Colorado Springs, CO 80919
719-593-8840
www.prorodeo.com

Ride and Tie Assoc.
1865 Indian Valley Road
Novato, CA 94947
414-897-1829
www.rideandtie.org

United States Combined Training Association
525 Old Waterford Road, NW
Leesburg, VA 20176
703-779-0440
www.eventingusa.com

United States Dressage Federation
P.O. Box 6669
Lincoln, NE 68506-0669
402-434-8550
www.usdf.org

United States Equestrian Team
Pottersville Road
Gladstone, NJ 07934
908-234-1251
908-234-9417 (fax)
www.uset.org

United States Team Penning Assoc.
P.O. Box 161848
Fort Worth, TX 76161-01848
800-848-3882
www.tpaa.com

United States Team Roping Championships
P.O. Box 7651
Albuquerque, NM 87194
505-899-1870

Academic Programs

California State Polytechnic University, Pomona
College of Agriculture
3801 W. Temple Avenue
Pomona, CA 91768
909-869-2200
909-869-4454 (fax)

Colorado State University at Fort Collins
Department of Equine Sciences
Fort Collins, CO 80523
970-491-8373
www.colostate.edu

Meredith Manor International Equestrian Centre
Route 1, Box 66
Waverly, WV 26184
800-679-2603
www.meredithmanor.com

Stephens College
1200 E. Broadway
Columbia, MO 65215
800-876-7207
573-876-7237 (fax)
www.stephens.edu

University of Kentucky
Lexington, KY 40506
859-257-9000
www.uky.edu

University of Wisconsin — River Falls
College of Agriculture, Food, and Environmental Sciences
410 South Third Street
River Falls, WI 54022-5001
715-425-3784
www.uwrf.edu

William Woods University
One University Avenue
Fulton, MO 65251
800-995-3159
www.wmwoods.edu

Equestrian Expositions

Devon Horse Show
Box 865
Devon, PA 19333
610-964-0550
Fax 610-964-1608
www.thedevonhorseshow.com
Held in Devon, Pennsylvania, at the end of May and beginning of June

Equine Affaire
136 East High Street
London, OH 43140
740-845-0085
www.equineaffaire.com
Held in Pomona, California, in February; Columbus, Ohio, in April; and West Springfield, Massachusetts, in November

Equitana Asia Pacific
APN Miller Freeman Exhibitions Pty Ltd
Suite 4, 214 Bay Street
Brighton 3186
Victoria
Australia
+ 61 (0) 3 9596 8744
Fax: + 61 (0) 3 9596 0936
www.equitana.com.au
Held in Melbourne, Australia, in November

Equitana Germany
Miller Freeman
Blenheim International (Deutschland) GmbH
Projekt EQUITANA
Volklinger Strasse 4
D-40219 Dusseldorf, Germany
www.equitana.de
Held in Essen, Germany, in March of odd-numbered years

Equitana USA
Miller Freeman, Inc.
Equitana USA
P.O. Box 612288
Dallas, TX 85261
888-HORSES-1 or 800-487-1212
Fax: 888-811-7391 or 972-906-6890
www.equitanausa.com
Held in Louisville, Kentucky, in June

EqWest
P.O. Box 612288
Dallas, TX 75261-2288
888-HORSES-1
www.equitanausa.com/eqwest
Held in Del Mar, California, in November

Grand National Rodeo, Horse and Stock Show
Grand National Tickets
Cow Palace
P.O. Box 34206
San Francisco, CA 94134
www.grandnationalrodeo.com
Held in San Francisco, California, in November and December

National Finals Rodeo
Professional Rodeo Cowboys Assoc.
101 ProRodeo Drive
Colorado Springs, CO 80919
719-593-8840
www.prorodeo.com
Held in Las Vegas, Nevada, in December

National Horse Show
888-MS-GARDEN (888-674-2733)
516-484-1982 (fax)
www.nhs.org
Held in New York, New York, in November

Quarter Horse Congress
101 Tawa Road
P.O. Box 209
Richwood, OH 43344-0209
740-943-2346
E-mail: qtrhorse@oqha.com
Held in Columbus, Ohio, in October

index

Index of Quotations

General Index

Numbers in *italics* indicate illustrations; numbers in **bold** indicate charts.

a

d

g

m

n

o

X

Y

Z

Other Storey Titles You Will Enjoy

101 Arena Exercises by Cherry Hill. A ringside exercise book for riders who want to improve their own and their horses' skills. Classic exercises and original patterns and drills presented in a unique "read and ride" format. Comb-bound to allow hanging in the barn or lying flat on a barrel for easy reference. 224 pages. Paperback. ISBN 0-88266-316-X.

101 Horsemanship & Equitation Patterns by Cherry Hill. A sequel to *101 Arena Exercises,* this book presents patterns geared toward the most popular classes in English and Western competition. Comb-bound. 224 pages. Paperback. ISBN 1-58017-159-1.

The Basics of Western Riding by Charlene Strickland. A complete guide to the exciting world of Western riding, including choosing a horse, selecting and fitting tack, training, problem-solving, and competing. 144 pages. Paperback. ISBN 1-58017-030-7.

Becoming an Effective Rider by Cherry Hill. Teaches riders how to evaluate their own skills, plan a work session, get maximum benefit from lesson time, set goals and achieve them, and protect themselves from injury. 192 pages. Paperback. ISBN 0-88266-688-6.

Fences for Pasture & Garden by Gail Damerow. The complete guide to choosing, planning, and building today's best fences: wire, rail, electric, high-tension, temporary, woven, and snow. 160 pages. Paperback. ISBN 0-88266-753-X.

From the Center of the Ring by Cherry Hill. The inside secrets of equestrian competition, including Western and English events, protocol, tack and clothing, grooming, and etiquette. 192 pages. Paperback. ISBN 0-88266-494-8.

Getting Your First Horse by Judith Dutson. Everything a new or prospective horse owner needs to know about selecting a horse, the needs of a horse, terminology, horse-handling techniques, and more. 176 pages. Paperback. ISBN 1-58017-078-1.

Getting the Most from Riding Lessons by Mike Smith. An introduction to horseback riding with information on safety, horse behavior, conditioning for the rider, and preparation for the first show. 160 pages. Paperback. ISBN 1-58017-082-X.

Horse Handling and Grooming by Cherry Hill. A user-friendly guide, complete with 350 how-to photographs, presenting correct techniques for leading, haltering, tying, grooming, clippng, bathing, braiding, hoof handling, and more. 160 pages. Paperback. ISBN 0-88266-956-7.

Horse Health Care by Cherry Hill. Practical advice, complete with 350 how-to photographs, on dozens of essential skills, including daily examination, restraint, leg wrapping, hoof care, administering shots, dental care, wound care, and more. 160 pages. Paperback. ISBN 0-88266-955-9.

Horsekeeping on a Small Acreage: Facilities Design and Management by Cherry Hill. How to design safe and functional facilities for your horse. 192 pages. Paperback. ISBN 0-88266-596-0.

Horse Sense by John J. Mettler, Jr., D.V.M. Advice from a country veterinarian on preventive health care, breeding, foaling, feeding, and more. 160 pages. Paperback. ISBN 0-88266-545-6.

Safe Horse, Safe Rider by Jessie Haas. A young rider's guide to safe horsekeeping. 160 pages. Paperback. ISBN 0-88266-700-9.

Stablekeeping by Cherry Hill. A photographic guide to providing a safe, healthy, and efficient environment for a horse. Includes information on stalls, tack rooms, work and storage areas, sanitation and pest control, feeding practices, safety and emergencies, and more. 160 pages. Paperback. ISBN 1-58017-175-3.

Starting & Running Your Own Horse Business by Mary Ashby McDonald. How to run a successful business and make the most of your investments in horses, facilities, and equipment. 160 pages. Paperback. ISBN 0-88266-960-5.

Storey's Guide to Raising Horses by Heather Smith Thomas. The complete reference for the horse owner with detailed coverage of feeding and nutrition, foot care, disease prevention, dental care, breeding, foaling, and caring for the young horse. 512 pages. Paperback. ISBN 1-58017-127-3.

Trailering Your Horse by Cherry Hill. A photographic guide to low-stress traveling, including selecting a trailer, training, loading, packing, safe driving, and care en route. 160 pages. Paperback. ISBN 1-58017-176-1.